Practical Metallurgy and Materials of Industry

Practical Metallurgy and Materials of Industry

John Neely

Machine Technology
Lane Community College

John Wiley and Sons

New York/Chichester/Brisbane/Toronto

Library of Congress Cataloging in Publication Data

Neely, John, 1920-
Practical metallurgy and materials of industry.

Includes index.
1. Metallurgy. 2. Materials. I. Title.

TN665.N3 669 78-19166
ISBN 0-471-02962-9

Printed in the United States of America

10 9 8 7 6 5 4 3 2 1

Preface

This book is a metallurgy and materials textbook for metals shop students and industrial workers. Most metallurgical texts are intended for engineering or metallurgy students preparing for a career. The primary purpose of this book is to relate some of the basic principles of metallurgy to shopwork instead of teaching only metallurgy or materials. Machinists, welders, and most metal workers need to gain a basic understanding of the behavior of metals and materials without having to take a traditional course in these specialized fields.

It is often the responsibility of welding technicians, millwrights, and journeymen or master machinists to make decisions on the selection of materials and their processing. With a basic understanding of the behavior and characteristics of metals and other materials, these decisions will not be made blindly. Of course, in many places material selection is the responsibility of the engineering staff, but even there, the workers who handle the materials should understand their behavior to avoid difficulties.

This book is intended to be easy to read. I have used language that the average reader can understand and shop terms whenever possible. The terms used by the metallurgist can be baffling to the student. However, terms such as pearlite, cementite, austenitizing, and eutectoid are necessary in referring to a concept because they avoid extensive explanation whenever the concept is referred to. Only those terms that are necessary to convey the thought or idea are used, and all unnecessary metallurgical jargon is omitted. An effort is made to explain each new word in the text; a complete glossary is included.

The highly visual approach in this book uses graphs, drawings, and photographs of actual shop applications of metallurgy. They will help the reader to comprehend the many abstractions that must be used. The first few chapters deal with a basic approach to the study of metals and to their identification. The next chapters deal with the application of metallurgy to actual shop work, welding of various metals, and service problems such as metal fatigue and corrosion. This is followed by chapters on nonmetal materials, other common materials used in shop work.

To the casual reader, some chapters in the book may seem to be out of sequence or to repeat information already covered in earlier chapters.

This is because the reinforcement approach to instruction is used, an arrangement that works extremely well for vocational students. The rationale for using this approach is that one application of a particular concept is insufficient to enable students to retain the knowledge they have gained, especially if many ideas are involved. An initial basic approach to an idea, followed by an interim of study in other areas, and a return to the subject at a higher level or with different applications of the prior presentation gives the kind of reinforcement that induces retention of knowledge. An example of this is in Chapter 9 in which a student is given basic instruction on the hardening and tempering of a punch or chisel. The deeper concepts of hardenability and the effects of alloying elements on hardening and tempering are reserved for Chapters 12 and 13. These principles are again used and reinforced in later chapters on welding metallurgy, but with a different application. Although a chapter on the metallurgy of machining is included, machining characteristics on materials are discussed throughout the book, where they are directly applicable.

This book may be used for a supplementary or related course for machine shop and welding students and for apprenticeship classes in schools and industrial situations. Community education for welders, millwrights, and other metal workers is an ideal use for this textbook since much of its emphasis is directed toward problem solving. This is one of the major reasons that industrial workers want to attend classes in shop metallurgy.

The format of the book is adaptable to either conventional lecture-lab or to an individualized approach to instruction. Objectives, information, self-tests, and worksheets are intended to help both the student and instructor.

An instructor's manual is also available for this book and it contains post-tests for most chapters and answer keys. Instructions are given on how to set up a simple metallurgical laboratory and on the various methods of using the book for setting up teaching programs. A list of available metallurgy films is also included.

Some of the text content and illustrations used in this book were adapted from the section on materials in *Machine Tools and Machining Practices* that I wrote.

It is my hope that this book will help many vocational students and other skilled metal workers to be better machinists, welders, or metal workers by enabling them to gain a better understanding of the behavior of the materials they work with.

JOHN E. NEELY

Acknowledgments

I wish to express my sincere thanks and appreciation to Doris A. Neely for her untiring devotion in assembling, organizing, handling the correspondence, and typing the manuscript. Without her help, this book could not have been produced.

I especially thank Thomas J. Nordby, a former student, who took an interest in my work on the book and helped me in preparing lab specimens for photomicrography. I am grateful to all individuals who served as models for photographs. I would also like to express my appreciation to the reviewers of the manuscript.

The following industries have been most helpful in supplying illustrations and technical information.

American Chain & Cable Company, Inc., Wilson Instrument Division, Bridgeport, Connecticut
Aluminum Company of America, Pittsburgh, Pennsylvania
American Iron & Steel Institute, Washington, D.C.
American Society for Metals, Metals Park, Ohio
Armco Steel Corporation, Middletown, Ohio
Bethlehem Steel Corporation, Bethlehem, Pennsylvania
Bohemia Incorporated, Eugene, Oregon
Buehler, Ltd., Evanston, Illinois
Chilton Company, Radnor, Pennsylvania
W. C. Dillon & Company, Inc., Van Nuys, California
DoAll Company, Des Plaines, Illinois
Dover Publications, Inc., New York, New York
Hitchiner Manufacturing Company, Inc., Milford, New Hampshire
HPM Corporation, Mount Gilead, Ohio
John J. Hren, Department of Materials Science and Engineering, University of Florida, Gainesville, Florida
The International Nickel Company, Inc., New York, New York
Investment Casting Corporation, Springfield, New Jersey
John Wiley & Sons, Inc., New York, New York
Kennecott Copper Corporation, Salt Lake City, Utah
Lane Community College, Eugene, Oregon
Lindberg Sola Basic Industries, Chicago, Illinois
The Lincoln Electric Company, Cleveland, Ohio
Linn-Benton Community College, Albany, Oregon

Magnaflux Corporation, Chicago, Illinois
Megadiamond Industries, New York, New York
Mt. Hood Community College, Gresham, Oregon
Pacific Machinery & Tool Steel Company, Portland, Oregon
R. A. Parham of the Institute of Paper Chemistry, Appleton, Wisconsin
Republic Steel Corporation, Cleveland, Ohio
The Shore Instrument & Manufacturing Company, Inc., Jamaica, New York
Tinius Olsen Testing Machine Company, Inc., Willow Grove, Pennsylvania
United States Steel Corporation, Pittsburgh, Pennsylvania
Western Wood Products Association, Portland, Oregon
Weyerhaeuser Company, Springfield, Oregon

J.E.N.

Contents

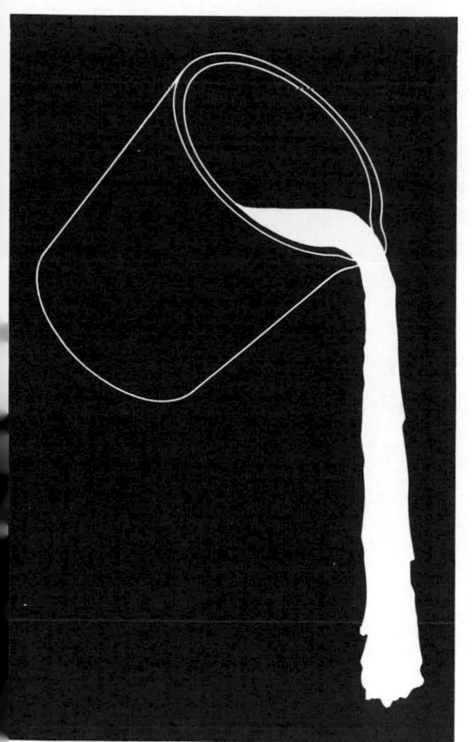

Introduction

The ability of human beings to make and use tools is probably the single most important reason for the tremendous progress leading to our present technological age. The first humans used tools made of wood, bone, and stone (Figure 1). Without metals, however, these tools were primitive and crude, thus hindering further progress. After the discovery of useful metals such as copper and tin (both are needed to make bronze) and iron, a completely new world of techological development was made possible.

Nearly everything we need for our present civilization depends on metals. Vast amounts of iron and steel are used for automobiles, ships, bridges, buildings, machines, and many other products (Figure 2). Almost all uses of electricity depend on copper. All around us we see the utilization of aluminum, copper, and many other metals. Some metals that were impossible to smelt or extract from ores a few years ago are now used in large quantities. These are usually called "space age" metals. There are also hundreds of combinations of metals called alloys, and many tool steels.

We have come a long way from the first iron that was smelted, as some people believe, by the Hittites about 3500 years ago (Figure 3). Their new iron tools, however, were not much better

Figure 1. *(a)* Stone saws and *(b)* stone hatchets (Courtesy of DoAll Company).

Figure 2. Large scale production of metal products is achieved in modern steel mills (Bethlehem Steel Corporation).

Figure 3. One of the first methods of making iron was by heating the ore over a fire. This produced a small white hot lump of iron known as sponge, which was hammered into shape on an anvil (*Machine Tools and Machining Practices*).

than those of the softer metals, copper and bronze, that were already in use at the time. This was because tools made from wrought iron would bend and would not hold an edge. The steel making process, which uses iron to make a strong and hard material by heat treatment, was still a long way from being discovered.

Metallurgy may be defined as (1) the technology of extracting metals from their ores and refining them for use and (2) the science or study of the behavior, structure, properties, and composition of metals. *Extractive* or *process metallurgy* deals with how ores are mined and processed in furnaces and rolling mills into useful metals. *Physical metallurgy* is the study of the properties and compositions of metals. These studies often lead to the development of new alloys that are suited for a particular application.

In this book you will learn how metallic ores are smelted into metals, and how these metals are then formed into the many different products needed in our society. Many metallic ores exist in nature as oxides, that is, in compounds in which metals are combined chemically with oxygen. Most iron is removed from the ore by a process called oxidation reduction. Metallic ores are also found in other combinations such as carbonates and silicates. The smelting and processing of many metals other than iron and steel are also considered.

Modern metallurgy stems from the ancient desire to understand fully the behavior of metals. Long ago, the art of the metal worker was enshrouded in mystery and folklore. Crude methods of making and heat treating small amounts of steel were discovered by trial and error. Unfortunately these methods were forgotten and had to be discovered again (Figure 4). Our progress has led us from those early open forges that produced 20 or 30 pounds of soft wrought iron a day to our modern production marvels that produce more than 100 million tons of steel yearly in the United States.

The modern story of iron and steel begins with the raw materials, iron ore, coal, and limestone. From these ingredients pig iron is extracted. Pig iron is the source of almost all our ferrous (iron containing) metals. The steel mill then refines it in furnaces to produce steel, after which it is cast into ingot molds to solidify. By various techniques the ingot is then used to produce the many steel products that are so familiar to us.

When the only available steel for heat treating was a medium or high carbon steel, only a limited understanding of the properties of metals was needed by the heat treater. The metal simply had to be heated to a cherry red color and dunked or quenched in water or brine, followed by tempering (that is, further heat treatment, usually by the oxide color method) to give it the proper toughness for its intended purpose. Present day requirements, however, demand a variety of tool steels including deep hardening steels for making products such as die blocks (Figure 5) that must be hardened completely even through its thick sections. Plain carbon or water hardening steels will harden only thin sections about $\frac{1}{8}$ in. deep on the surface of thicker pieces and so are not suitable for hardening thick sections clear through. Many tool

Figure 4. Just prior to the industrial revolution, iron working had become a highly skilled craft (Dover Publications, Inc.).

steels are used for such purposes as wear resistance, shock (Figure 6), or heat resistance.

Heat treating is no longer a simple process, but actually consists of very complex procedures and processes that must be followed carefully to avoid difficulties and failures, such as quench cracking (Figure 7). A good heat treater should not only know and understand the behavior of metals, but he or she should also be able to test their mechanical properties such as hardness, ductility, and toughness.

When you have studied these concepts, you will be shown how to heat treat a specific type of tool steel. You will be able to predict the outcome of its hardness and depth of hardening by the use of tables and graphs and by understanding the type of steel that you are using.

Metal workers are often confronted with the sometimes mysterious behavior of metals in such areas as welding, forming, machining, and heat treating. Sometimes during service operations, mechanisms will fail by fatigue or, as some mechanics often erroneously call it, by crystallizing. In many of these cases, the corrections are roughly made by the cut and try method or by attempting to change to a heavier part or a tougher material although the cause is not really understood. Often these processes create a greater problem.

Service problems often occur in metals such as a crack down the center of a weld in soft steel, or a clean break in a steel shaft only recently put into service under a minimum load (Figure 8). You need to know how to solve these and other shop problems, and to understand the reasons behind these service failures.

Some basic instruction in process and physical metallurgy is necessary before any application to shop problems can be made. The smelting of nonferrous metals, the making of pig iron, and the manufacture of steel form a necessary base for understanding the

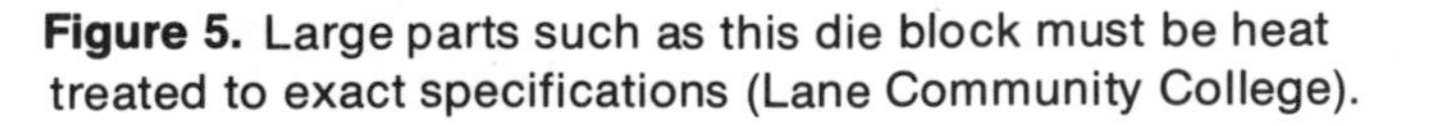

Figure 5. Large parts such as this die block must be heat treated to exact specifications (Lane Community College).

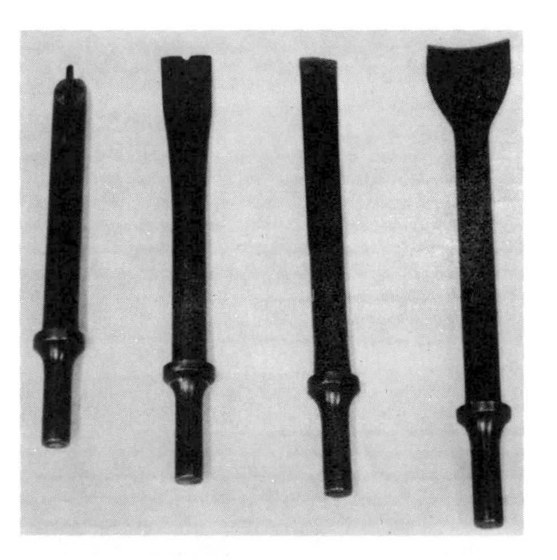

Figure 6. Air hammer tools that must resist shock require special steels (*Machine Tools and Machining Practices*).

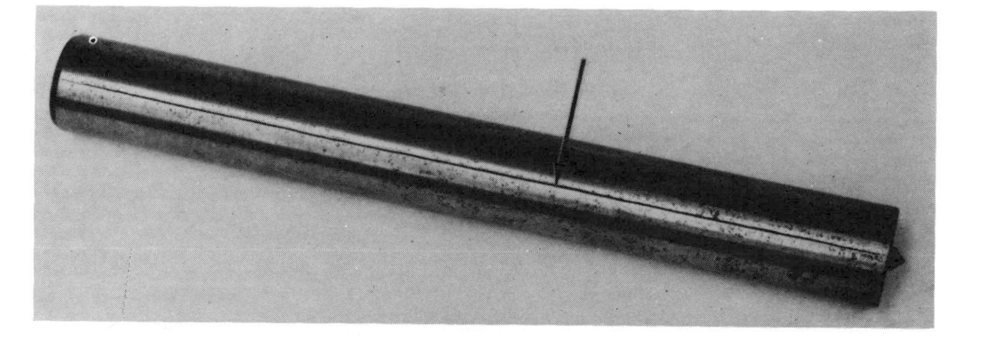

Figure 7. Quench cracking occurred in this part because the proper heat treating procedures were not followed (Lane Community College).

nature of metals. Some new terms must be learned, though these are kept to a minimum. Once this foundation is laid, practical shop solutions to problems in heat treatment, machinability, welding, and metal corrosion can be made.

The metal worker must be able to select and identify a great variety of metals. There are thousands of alloy steels and nonferrous alloys to select from. Shop methods, such as spark testing, are often used to separate one type of metal from another (Figure 9).

All metals are classified for industrial use by their specific working qualities; you will learn to select and identify materials using the tables in your handbook. Some systems use numbers to classify metals, whereas others use color codes, which consist of a brand painted on the end of the piece of material.

Testing metals is not only detailed because of the use of testing machines and other metallurgical equipment, but also be-

Figure 8. Fatigue failure that started at the end of keyways (Lane Community College).

cause of simple shop tests that must be made where this equipment is not available. The worker may thus determine the hardness or weldability of a piece of steel in the field or laboratory, depending on the level of precision needed for the job. It is necessary to understand the microscopic structure of metals in order to understand their behavior in terms of hardening, fatigue, and other properties.

As you progress, you will investigate the many materials you will use in the shop. For instance, cast iron is brittle while soft iron is easily bent and the difference results from how they are made. You will learn the characteristics of many metals and become familiar with hardening and tempering processes that strengthen and harden steel by heating and rapid cooling. For tool making, some steels can be made as hard as needed by heat treatment. You will learn several hardness tests that enable you to measure the durability and resistance of a metal (Figure 10), and such tests as tensile tests or notch toughness tests that are used to determine other properties (Figures 11*a* and 11*b*). You will also look into the basic structure of metals from the atomic to the crystalline structures, seeing how they are arranged to make up the particular substances that we know as metals (Figure 12).

Nondestructive testing is used extensively in industry to check for flaws in machine parts, especially in sensitive areas such as aircraft parts. There are several methods of nondestructive testing discussed in this book. These include the Magnaflux, Zyglo, and Spotcheck penetrant methods as well as ultrasonic and X ray testing

Figure 9. Spark testing is often done on a pedestal grinder in the shop (Lane Community College).

Where appropriate, we give safety precautions for the safe handling of materials and metallurgical equipment such as abrasive cutoff saws, acids, and the other reagents that are used for etching specimens. We refer also to the toxicity of certain metals and materials.

Several chapters are devoted to the metallurgy of welds in ferrous and nonferrous metals. This is because of the wide use of welding processes and the important part metallurgy plays in welding. The study of welding metallurgy is not only needed by the welding student but also by the machine shop student since the strength and machinability of welded parts are greatly affected by the welding process and heat treatments (Figures 13*a* and 13*b*).

Nonmetallic materials are discussed according to their machinability, tool shape, handling, and uses. Most plastic parts, for example, are produced by the process of injection molding, but machining is commonly practiced on many specific forms of plas-

Figure 10. Rockwell hardness testing machine (Courtesy of Wilson Instrument Division of Acco. *Note:* Rockwell is a registered trademark of American Chain & Cable Company, Inc., for hardness testers and test blocks.)

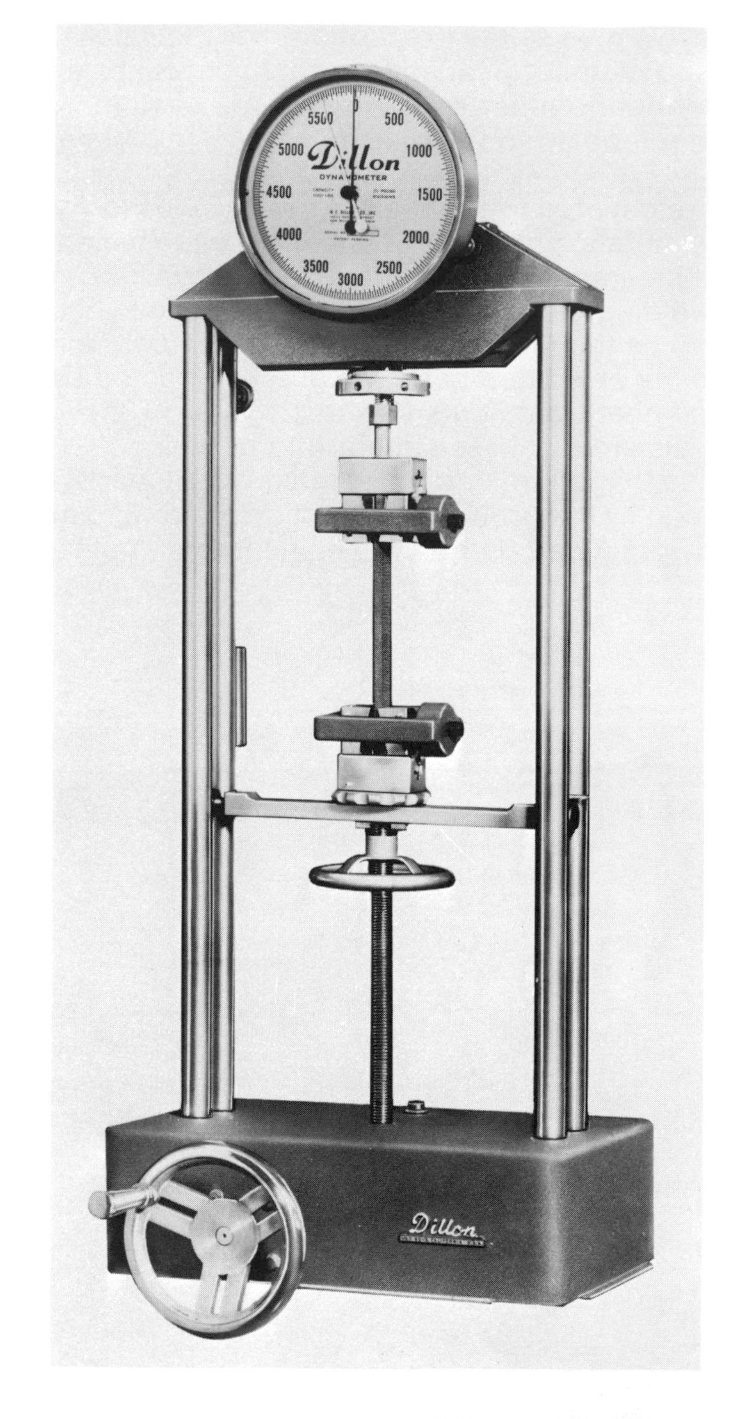

Figure 11*a*. Tensile tests are made on this machine (Photo courtesy of W. C. Dillon & Company, Inc.).

Figure 11*b*. Notch toughness is measured on this Charpy-Izod testing machine (The Tinius Olsen Testing Machine Company, Inc.).

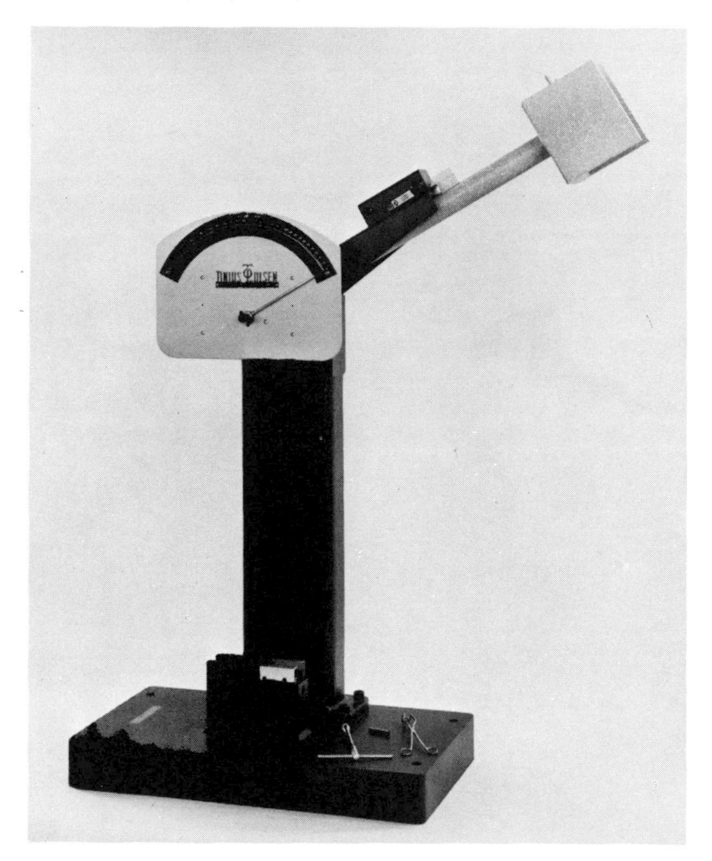

Figure 12. Photomicrograph of cartridge brass, 70 percent (75 ×) (By permission, from *Metals Handbook*, Volume 7, Copyright American Society for Metals, 1972).

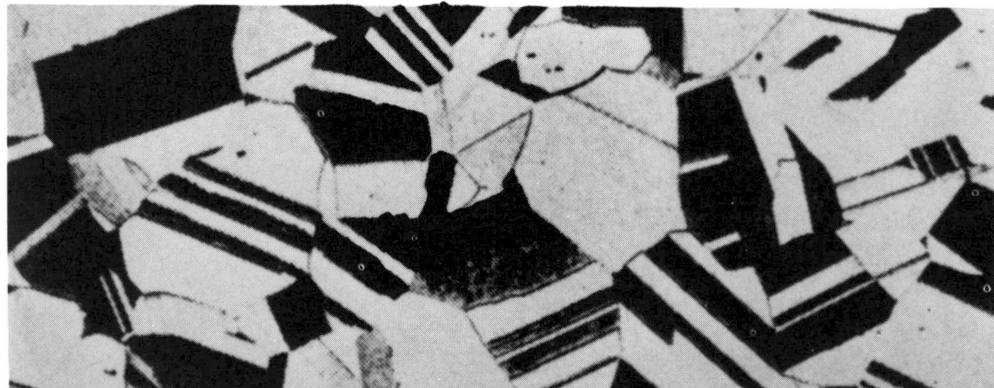

Figure 13*a*. Welder making welds on hard-to-weld material, which is here plow and abrasive resistant steel being welded to mild steel. He is using the gas metal arc process with wire drive.

Figure 13*b*. A machinist at work (Lane Community College).

tic. Wood, cement, and other materials of industry are discussed to give you a broader perspective of industrial processes (Figure 14).

This textbook is intended to give you a greater understanding of the materials of your trade and an awareness of their behavior. This knowledge will help you become far more capable and efficient in your chosen trade.

Figure 14. Many materials are used in this housing project, such as foundations and the building structure.

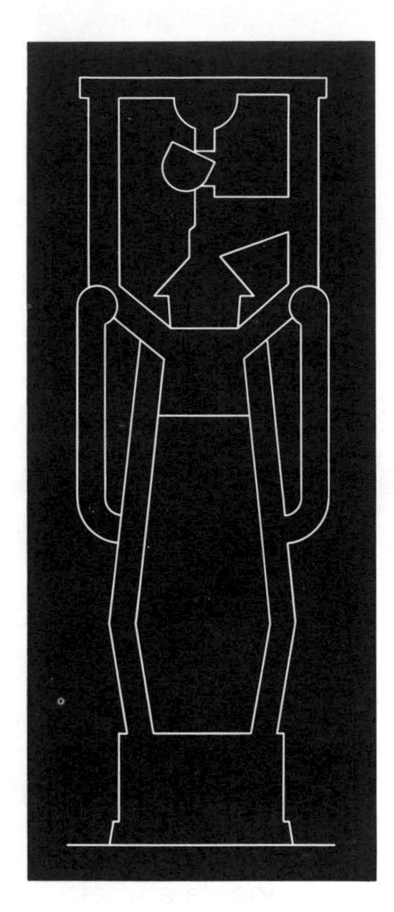

1 Extracting Metals from Ores

Iron is the fourth most plentiful element in the earth's crust. It is almost never found in its native or metallic state, but as part of various mineral compounds called ores. The separation of metals from their ores is known as extractive metallurgy. This chapter will describe how iron ores are processed and converted into pig iron in blast furnaces. It will also describe various methods of converting pig iron into steel. A brief description of the extraction of copper and aluminum is also included.

OBJECTIVES

After completing this chapter, you will be able to:

1. List the various steps, basic materials, and principles involved in the making of pig iron.
2. Identify various steel making processes.

INFORMATION

Mining Natural Ores

Iron ores are found all over the world, but in the past only certain deposits were considered rich enough in iron to be mined. Only two decades ago, most iron and steel makers in the United States would not have considered mining an ore whose iron content was less than 30 percent, particularly if the mineral were difficult to process. Today, however, a mineral called taconite is one of the primary sources for making pig iron in blast furnaces in this country. From Table 1, you can see that the iron content of taconite ranges from 25 to 35 percent. In the United States, for many years the Lake Superior Mesabi Range has produced hematite ore, which contained a high iron content up to 68 percent, but this source has since been depleted. Table 1 lists some minerals of iron and their chemical formulas.

These ores are removed by open-pit mining (Figures 1*a* and 1*b*) as contrasted with the more

Table 1 Some Natural Ores of Iron

Name	Formula	Iron Content (Percent)
Magnetite	Fe_3O_4	72.4
Hematite	Fe_2O_3	70.0
Limonite	$2Fe_2O_3, 3H_2O$	59.8
Goethite	Fe_2O_3, H_2O	62.9
Siderite	$FeCO_3$	48.2
Taconite		25-35

Figure 1*a*. Open pit mining showing a 12¼ inch rotary drill in the foreground (Kennecott Copper Corporation photograph by Don Green).

costly and dangerous underground mining method. When the ore has been removed from the mine, it is cleaned and separated from the gangue, or worthless rock, by a process called ore dressing. This process could be carried out by any of several methods such as floatation, agglomeration, and magnetic separation. By these processes, low grade ores such as taconite are upgraded and pelletized before being shipped to the steel mill.

Production of Pig Iron

The iron ore is converted into pig iron in a blast furnace. Figure 2 shows the operation of a blast furnace. The three raw materials, **iron ore, coke,** and **limestone,** are put into the furnace alternately at intervals, thus making the process continuous. About two tons of ore, one ton of coke, and a half ton of limestone are required to produce one ton of iron.

One of the three major ingredients in the production of pig iron is coke, a residue left after certain soft coals have been heated in the absence of air. When coal is heated in coke ovens and the resulting gases are driven off (Figure 3), coke is the result. Coke is a hard, brittle, and porous material containing from 85 to 90 percent carbon, together with some ash, sulfur, and phosphorus. An older type of coke oven, called a beehive oven because of its shape (Figure 4), is now obsolete because it

Figure 1*b*. Twenty-five-yard shovel loading a 150 ton truck in an open pit mine (Kennecott Copper Corporation photograph by Don Green).

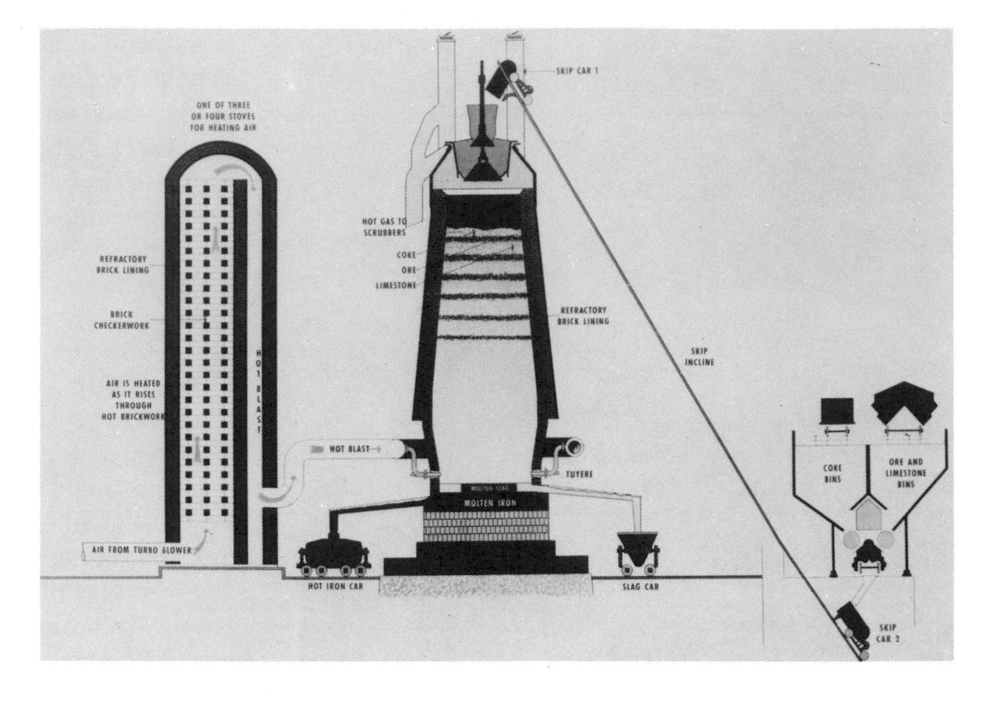

Figure 2. Blast furnace. Iron ore is converted into pig iron by means of a series of chemical reactions that take place in the blast furnace (Bethlehem Steel Corporation).

wasted gases produced during the process. Many useful products are made from the coke oven gas that is driven off: fuel gas, ammonia, sulfur, oils, and coal tars. From the coal tars come many important products such as dyes, plastics, synthetic rubbers, perfumes, sulfa drugs, and aspirin.

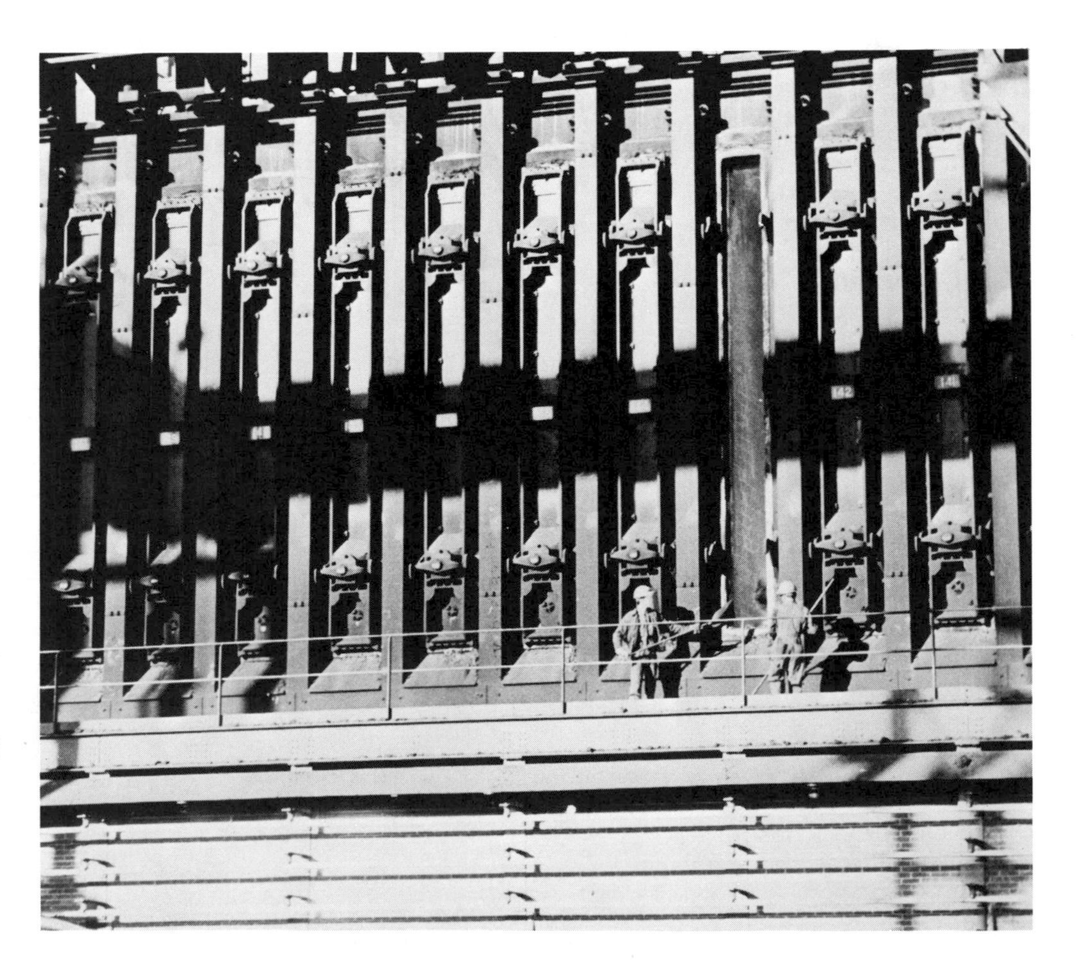

Figure 3. Between each pair of vertical dividers in this coke oven battery is an individual oven in which coal is heated and converted to coke (Courtesy of American Iron & Steel Institute).

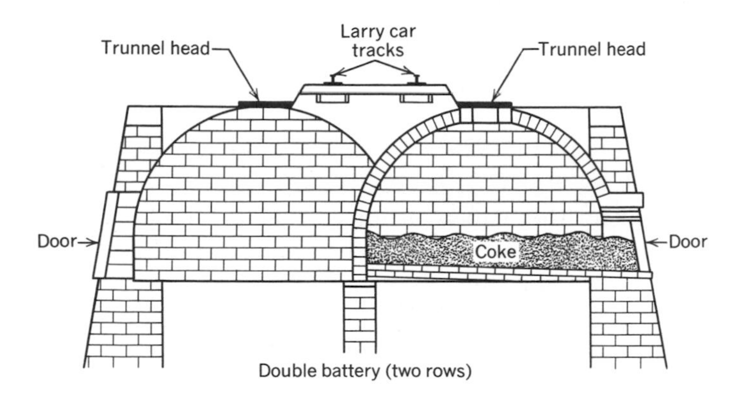

Figure 4. Beehive coke oven. The coal is charged through the trunnel head from which the gases escape. This method is obsolete, since it was wasteful. The gases are not collected and utilized, but are lost through the upper vent (*Machine Tools and Machining Practices*).

Reduction is a process by which oxygen (O_2) is removed from a compound, such as an iron ore, and combined with the carbon (C). Hence when iron ore and coke are put into the blast furnace, the metallic iron is released from its oxide state (formulas in Table 1) by reduction. The solid materials, coke, limestone, and ore, enter the blast furnace through the hopper at the top, and heated air is blasted in at the bottom. This air is heated in stoves composed of refractory brick before it enters the furnace. Refractory materials have a high melting point to enable them to retain their shape and strength at high temperatures.

In the furnace, the fuel burns near the bottom, and the heat rises to meet the descending charge of coke and ore. At very high temperatures, the coke unites with the oxygen in the iron ore and is converted to carbon monoxide gas. As the process continues, iron is released to the bottom of the furnace, where it remains a molten mass. Here the limestone is used to separate the impurities (mostly SiO_2) from the iron by combining to form a lower temperature melting compound. Since this waste slag is lighter in weight than the iron, it floats on top and is drawn off periodically to be hauled away in slag cars. The slag is sometimes ground into an aggregate that is used for asphaltic concrete roads and in concrete building blocks.

The molten iron at the bottom of the furnace contains from 3 to 4.5 percent carbon and other unwanted impurities such as phosphorus and sulfur. These are later removed to some extent by other refining processes. The reduction of iron ore into pig iron is actually an intermediate step in the manufacture of steel. The iron is tapped (drawn off) at intervals and collected in a transfer car, which is insulated so that the molten metal will stay hot (Figure 5). It is then moved to the steel furnaces and added to the charge of scrap steel, ore, and limestone. Sometimes the iron is not made into steel, but is instead poured directly into molds. Before pig casting machines were developed, the iron was poured in open sand molds consisting of a groove or trough with many small molds on each side, reminding one of a sow and pigs, hence, the name "pig iron" (Figure 6). Iron pigs are remelted in cast iron foundries and in steel mills.

Wrought Iron

In ancient times iron was produced by the process of direct reduction; that is, the ore was heated in a forge to a white heat to remove the impurities and to produce iron from the ore. The charcoal fire in the forge did not get hot enough to melt the iron, but resulted in a pasty mass of "sponge" iron that was then hammered to remove the molten "gangue" or slag. The result was soft wrought iron containing little or no carbon. This method of

Figure 5. Tapping the blast furnace. Blast furnaces operate continuously. The molten iron is tapped every five to six hours (Bethlehem Steel Corporation).

Figure 6. Some blast furnace iron is poured to solidify in molds in the automatic pig casting machine. Iron foundries are major users (Courtesy of American Iron & Steel Institute).

smelting iron ore has not been used commercially since the Middle Ages.

Before the processes of modern steel making were known (over a hundred years ago), wrought iron was used extensively for bars, plates, rails, and structural shapes for bridges, boilers, and many other uses. Wrought iron was then made by a puddling process in which pig iron was melted in an open hearth type of furnace. Iron oxide was then added to form a slag. The iron was slowly cooled to a pasty mass and puddled or stirred with "rabble arms" by hand. The carbon and other impurities were removed by the iron oxide slag. The mass of iron was removed and squeezed to remove the slag; however, much of it remained in the iron. The result was a very low carbon, fibrous (from the trapped slag), soft iron. This process is obsolete, but wrought iron is now produced by the Aston process in which molten pig iron and steel are poured into an open hearth furnace in a prepared slag; this cools the steel to a pasty mass that is later squeezed in a hydraulic press. Even though direct reduction of iron ore to produce wrought iron as it was practiced in ancient times had become obsolete, recently the direct reduction process has been greatly improved, and currently several processes are being used.

Sponge Iron and Powder

Iron powder has been produced directly from the ore since the 1920s and is used to produce small parts by forming them under high pressure and sintering them in a furnace, that is, heating them without melting. Many other systems in operation today convert the ore directly into the pellet form (Figure 7). The HyL, Midrex, and SL/RN systems are some of the many processes in which small pellets of sponge containing about 0.95 percent iron are produced. The pellets can be remelted in steel making furnaces such as the basic oxygen furnace, but most of this iron is shipped to electric furnaces. Only a few million tons of sponge iron are produced yearly at present as compared to over 125 million tons of pig iron produced in blast furnaces in the United States. There is an increasing demand for sponge and a rapid increase in its production can be expected in the near future. This is a result of the fact that iron can be mined in remote areas in the world and be shipped economically as sponge pellets to industrial nations.

Steel Making Processes

Since pig iron contains too many impurities to be useful, it must be refined to produce steel or cast iron of various types. Steel is simply an alloy of iron with most of the impurities removed; that is, it contains 0.05 to 2 percent carbon plus any other

Figure 7. Iron pellets produced by direct reduction.

alloying elements. Over 90 percent of all steel produced is classified as plain carbon steel, since its carbon content is controlled, usually under 1 percent. Small amounts of manganese are added to control the sulfur and other impurities that still remain. Different alloy steels comprise the remaining 10 percent of all the steels produced.

The major steel making processes are **basic oxygen, open hearth, Bessemer,** and **electric.** The Bessemer converter (Figure 8) uses air to burn out excess carbon and some impurities. Although the converter is fast (20 minutes to a blow, that is, an air blast forced through the molten iron), very little steel (none in the United States) is made by this process today, as it produces a low grade product. The Bessemer process is limited to using pig iron for steel making. Small measured amounts of carbon are also added to the molten metal after the blow, thus producing a carbon steel.

A modern development that resembles the old Bessemer converter is the basic oxygen furnace (BOF) (Figures 9*a* to 9*c*). It is designed to make steel of high quality in a very short period of time compared to the open hearth process (see below). The basic oxygen process uses a lance that blows

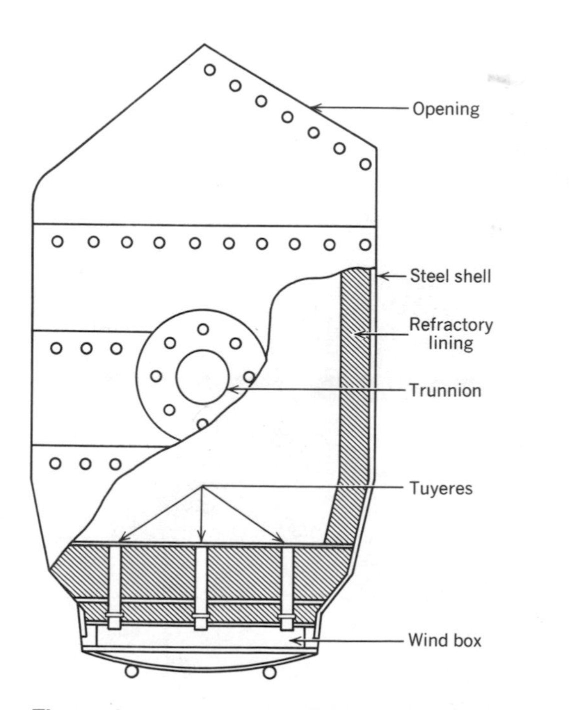

Figure 8. Cutaway view of Bessemer converter (*Machine Tools and Machining Practices*).

oxygen down from the top of the furnace to burn out the impurities (Figure 10). It has the added advantage of being able to use fairly large amounts of scrap, unlike the old Bessemer process. About 60 percent of the total steel production in the United States is made using oxygen furnaces.

The basic oxygen furnace can turn out steel at the rate of about one 200 ton heat per hour. A heat is a melting period of time in which alloying ingredients are added under closely controlled conditions. The furnace is charged with molten pig iron, iron ore, steel scrap, and fluxing materials such as limestone that react with the impurities and form a slag on top of the molten metal. The water cooled lance is lowered to within a few feet of the charge and then it blows a stream of oxygen at more than 100 psi on the surface of the bath. The oxidation of carbon and impurities causes a violent agitation of the molten bath, bringing all the metal into contact with the oxygen stream. The ladle is first tipped to remove the slag and then rotated to pour out the molten steel into a ladle. Carbon and other elements are added to produce the desired quality of steel.

Today, steel is still produced in the United States by the open hearth process (Figure 11), which produces a steel of high quality. Open hearth furnaces produce 100 to 375 tons per heat (Figure 12) with each heat taking from 8 to 10 hours. During this time molten pig iron, scrap, iron ore, and other elements such as manganese are added to control the condition of the melt or heat. Ferrosilicon is added if the steel is to be killed. A **killed steel** is one that has been sufficiently deoxidized in the ladle or ingot to prevent gas evolution in the ingot mold. This makes a uniform steel. If the heat is not killed, gas is formed in the ingot and the resultant holes remain in the ingot when the steel hardens. This is called **rimmed steel** because the rim or outside surface of the ingot is free from defects while the flaws remain in the center. These flaws, however, are mostly removed by rolling processes, discussed in the next chapter.

When the contents in the melt have been controlled and the mixture and temperature are right, the furnace is tapped and the molten metal poured into a ladle (Figure 13). Adjustments in carbon content are made in the ladle, usually by adding pellets of anthracite coal. The slag on top of the

Figure 9*a*. Molten iron from the blast furnace is charged into one of two basic oxygen furnaces at Bethlehem Steel Corporation's Bethlehem, Pa. plant. After the charge has been completed, the vessel will return to its upright position for the oxygen "blow" during which the blast furnace iron, mixed with scrap and selected additives, will be refined into steel (Bethlehem Steel Corporation).

Figure 9*b*. One of two basic oxygen furnaces is charged with scrap. The charging machine empties scrap into the vessel before the hot metal is added. In addition to positioning and charging the loaded boxes of scrap, the charging machine also positions the heat shield that is utilized for testing during furnace turndowns (Bethlehem Steel Corporation).

Figure 9c. Heat shield protects members of the basic oxygen furnace crew who are taking steel samples and a temperature measurement from the bath of molten metal. The steel sample is sent by pneumatic tube to the basic oxygen furnace (BOF) shop spectrometer laboratory for analysis (Bethlehem Steel Corporation).

Oxygen lance
Water cooled hood
Steel Shell
Tap Hole
Refractory lining

Figure 10. The basic oxygen furnace resembles the old Bessemer converter in appearance, but it is an entirely new development, designed expressly to get the best results by the use of oxygen in steel making. Steel of excellent quality can be made at a high rate of speed—about one heat an hour (Bethlehem Steel Corporation).

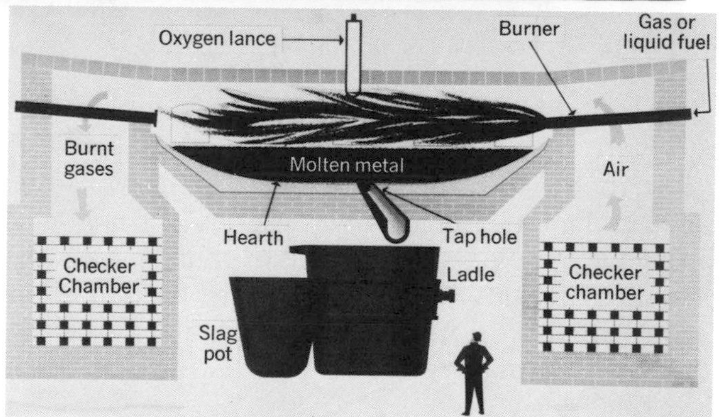

Figure 11. Simplified cutaway diagram of a typical open hearth furnace. Oxygen may be injected through one or more lances. In some cases it is introduced through the burners to improve combustion (Bethlehem Steel Corporation).

Figure 12. Open hearth charging floor. Molten cast iron being poured from ladle into the hearth furnace (Bethlehem Steel Corporation).

Figure 13. Tapping an open hearth furnace. Molten steel fills the ladle; the slag spills over into the slag pot (Bethlehem Steel Corporation).

Figure 14. Teeming ingots. The steel is poured into molds where it solidifies (Bethlehem Steel Corporation).

molten steel overflows into the slag spout and into a smaller slag pot.

About 100 tons of molten steel are picked up by a crane and brought over a series of heavy cast iron ingot molds. It is teemed (poured) into the molds by means of an opening in the bottom of the ladle (Figure 14). Aluminum is sometimes added in the mold to make the steel finer grained and less porous.

Electric furnace steel makes up about 20 percent of the total production in the United States. As with the open hearth process, electric furnaces (Figures 15 and 16) use blast furnace iron, selected scrap, and other control elements. Where very little coal and iron ore are found, and where considerable steel scrap and cheap electricity are available, the electric furnace is a competitive producer of high quality steel. Most of the special alloys such as stainless steels and tool steels are produced by this method. Steel foundries, where steel is melted down and remolded, use electric furnace steel to produce steel castings.

Electric furnaces produce up to 100 tons per heat. The time is determined by the amount of cold scrap and the amount of the furnace current. Scrap is charged through a door in the side or through the furnace top that swings to one side. The entire furnace tilts in order to pour the molten steel into the ladle.

The direct arc electric furnaces are most popu-

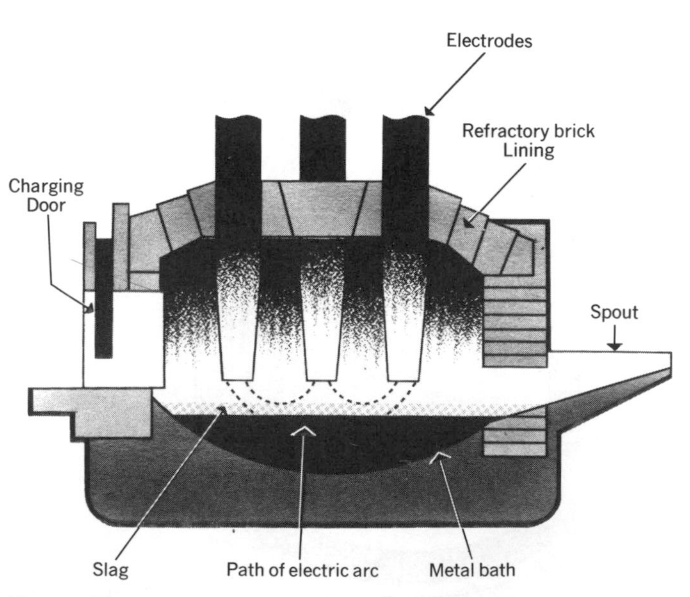

Figure 15. Electric furnace. Because both the temperature and atmosphere can be closely controlled in an electric furnace, it is ideal for producing steel to exacting specifications. The entire process takes from 4 to 12 hours, depending on the type of steel to be produced (Bethlehem Steel Corporation).

Figure 16. Tapping an electric furnace, which is tilted to pour its steel into the huge ladle below (Bethlehem Steel Corporation).

lar for making alloys of steel. In this type of furnace the current passes from one electrode through an arc to the metal charge and then through an arc to another electrode. Induction heating furnaces are often used for remelting metals such as cast iron in an iron foundry. In this type a coil surrounds a crucible and a high frequency electric current is passed through the coil, causing the charge to heat. Combinations of both direct arc and induction furnaces are also used.

The production of iron and steel from the raw materials to the final finished products may be followed in the flow chart (Figure 17). The following chapter deals with steel making from the ingot stage to the formation of the many types of steel products.

Another type of electric furnace in use today is the consumable electrode unit (Figure 18). Two basic designs of consumable electrode furnaces are the vacuum arc and vacuum induction models. High purity steels, titanium, zirconium, and various other metals are melted by these units. The metal used to charge the furnace is fashioned into long rods by methods such as briquetting sponge metal, as is done with titanium, and then welding these briquettes into an electrode. The vacuum in the furnace is constantly maintained in order to remove any gases before they can be absorbed by the metal. The metal electrode is fed into the melt at a controlled rate by which the arc length is automatically adjusted. By this process of melting in a water cooled mold at the base of the furnace an ingot is gradually formed. The finished ingot is removed from the mold and processed.

Copper

The use of copper dates from prehistory. Neolithic communities of Eskimos and ancient dwellers in Turkey, Egypt, and North America hammered native copper into tools and ornaments. Copper is one of the comparatively few metals found in the "native" or metallic state. Most copper used today is extracted from various ores by smelting. Sulfide ores are ground (Figures 19*a* to 19*c*) and then require a roasting process in which the sulfur is removed, leaving copper oxides. The fume given off in the process is destructive to the environment surrounding the smelter, so that collection of the fume is required. The copper oxide also contains gold, silver, and other impurities, which are later removed.

The copper ore is then smelted in a blast furnace to produce an impure alloy called matte. Air is then blown through the molten matte to produce a refined copper known as "blister" copper (Figures 20*a* and 20*b*). Further refinement is necessary to produce copper for electrical and other uses; this is done by electrolysis. It is at this stage that impurities such as gold and silver are collected at the bottom of the tanks and removed from the sludge by a separate process. Copper 99.9 percent pure is produced by the electrolytic process (Figures 21*a* and 21*b*).

Lead and Zinc

Lead and zinc are refined by a process similar to that used in copper refinement. Lead ores mostly occur as galena (lead sulfide) and zinc ores either as oxides or sulfides. Roasting furnaces, blast furnaces, hearth furnaces, and the electrolytic process are used to smelt and refine these metals.

Aluminum

Of all the structural metals used, such as iron, aluminum, and copper, aluminum is by far the

Figure 17. Flow chart of the steel finishing process, showing the route from ore to iron to finished steel (Bethlehem Steel Corporation).

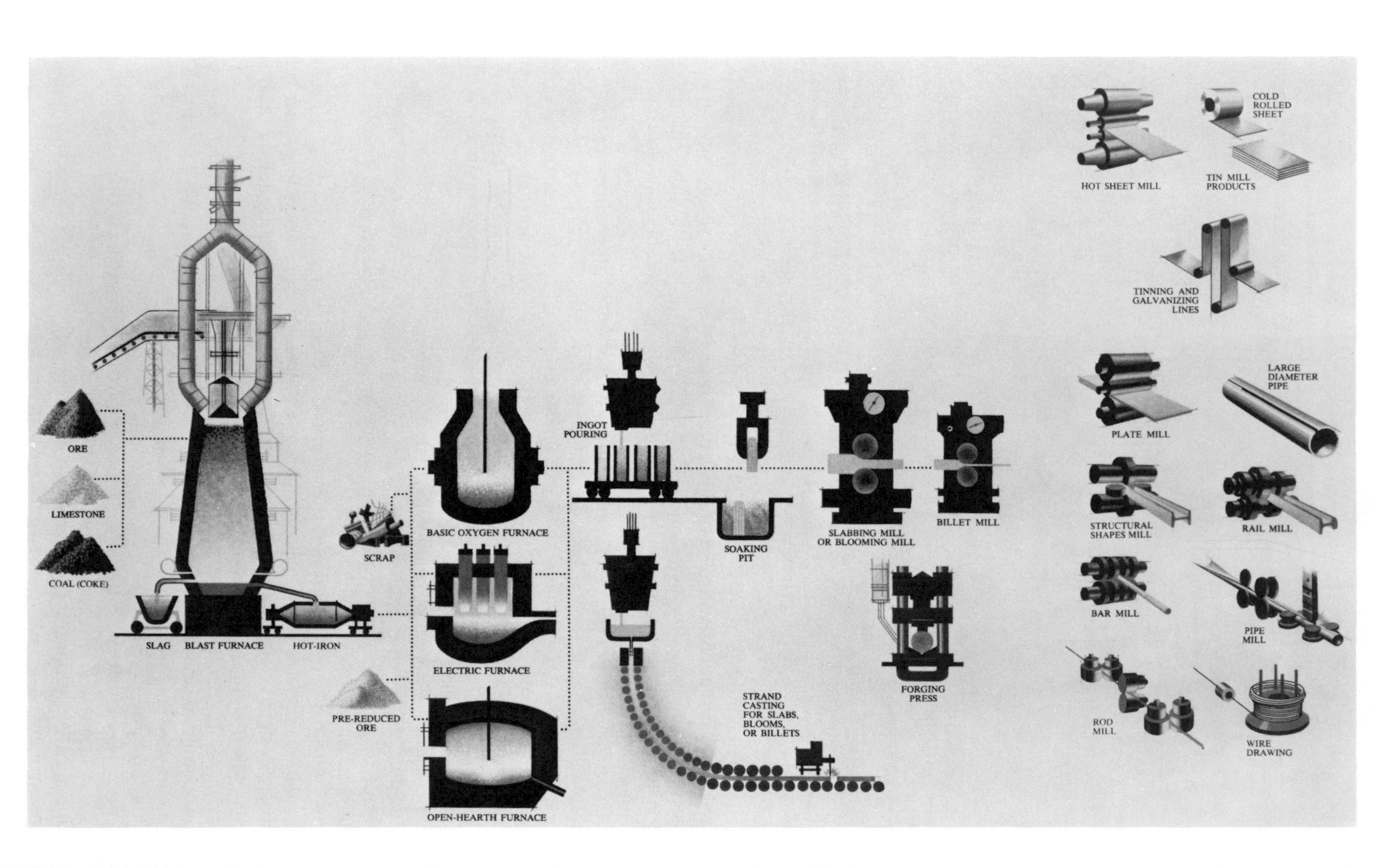

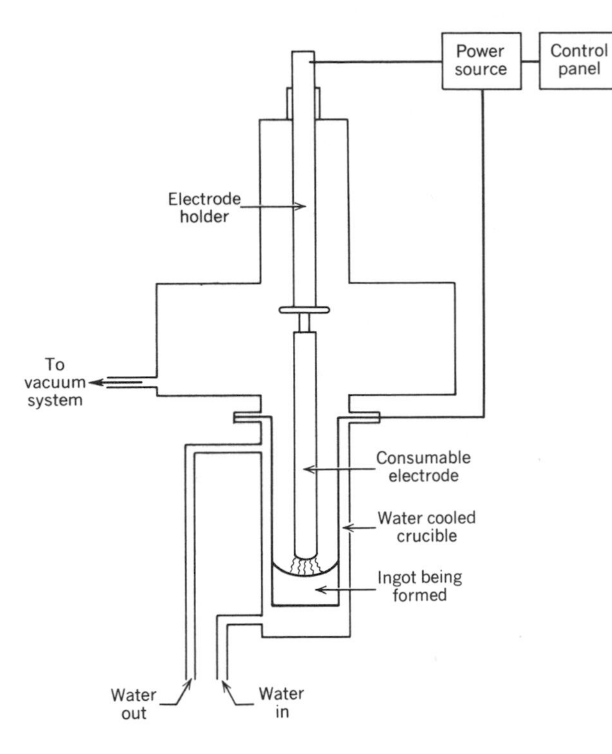

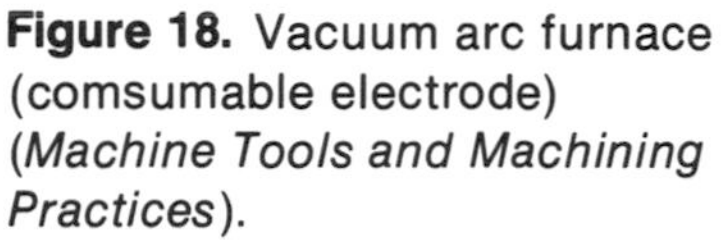
Figure 18. Vacuum arc furnace (comsumable electrode) (*Machine Tools and Machining Practices*).

Figure 19*a*. Copper ores are first ground in the primary crusher (Kennecott Copper Corporation photograph by Don Green).

Figure 19*b*. The copper ore is transferred by conveyer from the primary crusher to the next step in the grinding of the ores (Kennecott Copper Corporation photograph by Don Green).

Figure 19*c*. The ore is finally ground in these ball mills (Kennecott Copper Corporation photograph by Don Green).

Figure 20*a*. Copper converter (Kennecott Copper Corporation photograph by Don Green).

Figure 20*b*. Copper matte being charged into converter from a 50 ton ladle. The converter produces "blister" copper (Kennecott Copper Corporation photograph by Don Green).

most abundant in the earth's crust. Approximately 7.5 percent of the earth's crust is made up of aluminum. Many of the earth's ores contain aluminum, but the ore most used in the production of aluminum is bauxite, and is found in relatively few places.

The bauxite ore basically goes through two refining processes. First, a refining process breaks the ore down into its components: silicon, iron, titanium oxide, water, and aluminum oxide. This is done by crushing the bauxite to a powder and mixing it with caustic soda (sodium hydroxide, NaOH). The aluminum oxide is dissolved by the caustic soda leaving the impurities. The aluminum hydroxide then settles to the bottom of large tanks to which it is pumped. It is then washed to remove the caustic soda. Heat is applied to drive off the water, leaving a white powder: aluminum oxide (Figure 22).

In the second stage, the aluminum oxide (Al_2O_3), called alumina, is refined into metallic aluminum by the electrolytic reduction process. This is done in an electrolytic cell (Figure 23). Cryolite, a material mined in Greenland, is melted by an electric current in the cell and the alumina is placed in the cryolite bath.

The reaction that takes place in the electrolytic cell is quite simple. The electrolyte (molten

Figure 21*a.* These 600 pound copper anodes are produced in the casting machine at the rear. These will later be placed in electrolytic cells (Kennecott Copper Corporation photograph by Don Green).

Figure 21*b.* Electrolytic refining cells. The operator is using an electrical short detector (Kennecott Copper Corporation photograph by Don Green).

cryolite) dissolves the alumina. An electrical current is forced to flow from the cathode through the alumina to the anode. By the process of electrolysis, the alumina breaks down to form metallic aluminum and oxygen. The oxygen goes off as a gas, leaving metallic aluminum. The aluminum, because of its density, sinks to the bottom of the bath, where it is siphoned off at intervals and cast into ingots.

Many other complex processes are used to extract and process metals such as nickel, titanium, zirconium, tungsten, columbium, cerium, and many rare metals. Many publications including those of the Bureau of Mines that explain these processes in detail are available.

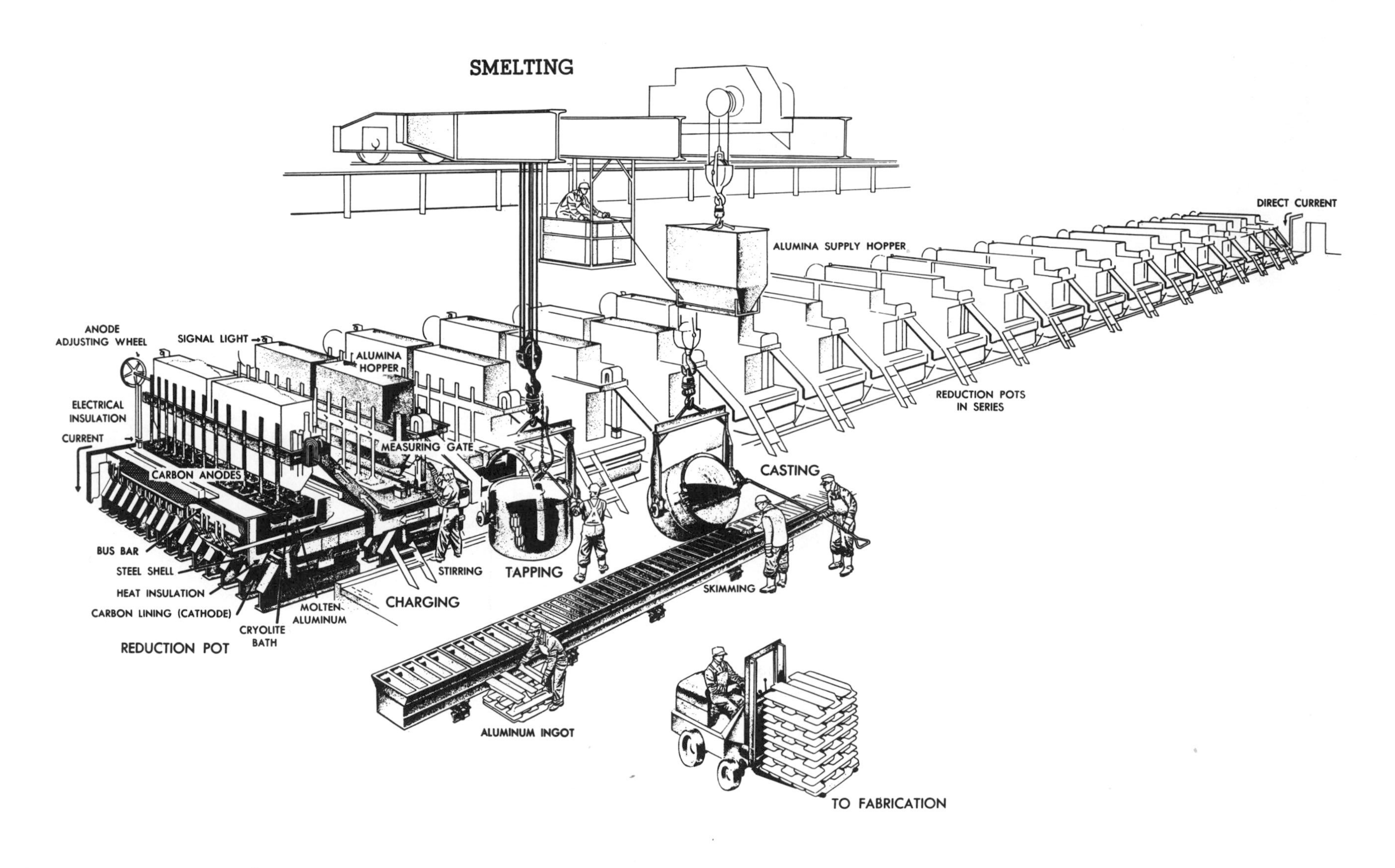

Figure 22. Process of making aluminum (Courtesy of the Aluminum Company of America).

Figure 23. An electrolytic cell showing a cutaway view of the reduction pot and the carbon anodes (Courtesy of the Aluminum Company of America).

Self-Evaluation

1 What needs to be removed from the iron ore to produce pig iron? How is this done?

2 Is the pig iron made into useful articles? How is it used?

3 What are the necessary ingredients to make pig iron?

4 Is the carbon content of pig iron the same as steel? Explain.

5 What is the purpose of ore dressing?

6 In what way does the steel making furnace affect pig iron in order to make steel? Are any other changes effected?

7 List some of the advantages of the electric furnace.

8 What is "killed" steel?

9 Explain the important differences between the Bessemer and the basic oxygen processes of making steel.

10 Most stainless steels and tool steels are made in the ______ furnace. Why is this done?

11 Name the heating process by which sulfide ores of copper, lead, and zinc must first go through prior to oxidation reduction in a blast furnace. What is the product called?

12 By what process is copper finally refined to obtain 99.9 percent pure copper?

13 What is the name of the ore that is commonly used to produce aluminum?

14 The refined aluminum oxide (alumina) is placed in an electrolytic cell in a bath of molten ______ .

15 By the process of electrolysis, oxygen is removed from the aluminum oxide. What happens to the free oxygen and aluminum?

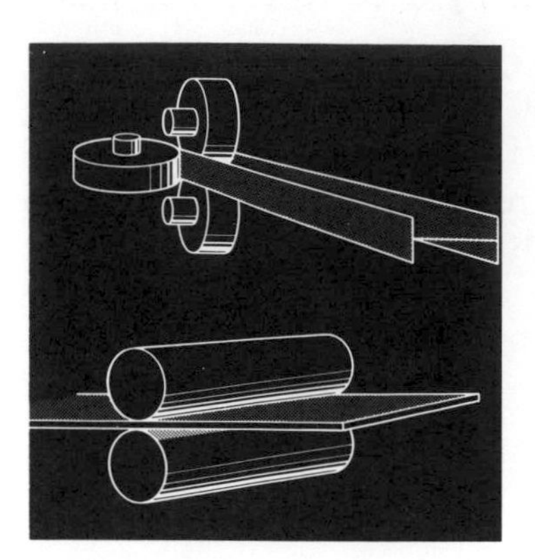

2 The Manufacture of Steel Products

What happens to the massive blocks of steel produced in steel making processes, and how are they formed into the many steel products you are familiar with? This chapter introduces you to the major steel finishing processes that take these massive forms of metal and shape them into workable materials for manufacture.

OBJECTIVE

After completing this chapter, you will be able to:

1. Describe how steel is formed into various shapes and products.
2. List the advantages of some processes over others for a given product.

INFORMATION

Almost all steel that is produced in steel making furnaces is teemed, or poured, into cast iron ingot molds, where it solidifies as it cools. From these steel ingots finished steel products such as structural shapes, plate, and pipe are made. The ingot mold is removed by "stripping" when the steel has solidified (Figure 1). A crane lifts the mold off the ingot.

Before rolling, the ingot must be reheated uniformly throughout as it has become too cold to work. It is placed in a furnace called a soaking pit (Figure 2), where it is heated to about 2200°F (1204°C) for four to eight hours. The grain structure of the ingot is coarse and columnar, making it weak (Figure 3*a*). Hot rolling breaks down the coarse grains and reforms them to make a stronger, finer grained steel (Figure 3*b*).

Ingots are formed into one of three different shapes, depending on the final products to be made from the steel. These basic forms are **slabs, blooms,** and **billets,** as shown in Figure 4. Any piece of steel to be rolled into a plate is first formed into a slab. Slabs are made in a primary mill called a slabbing mill. Blooms are made in a blooming mill, which gives the ingot a cross section of about 6 to 12 inches on a side. They are mainly used in the production of rails and other structural shapes. Billets are usually made directly from the bloom and are round or square; their cross section ranges from 2 to 5 inches on a side. Round billets are used in making seamless tubing, while bar stock of various sizes and shapes also are rolled from square billets.

Continuous Casting

A relatively new process of steel production that bypasses the ingot stage is called continuous or strand casting. The molten steel is converted directly into slabs, blooms, or billets in this process.

Figure 1. After steel has solidified in the ingot mold, the mold is pulled off the ingot by a stripper (Courtesy of American Iron & Steel Institute).

Figure 2. Powerful tongs lift an ingot from the soaking pit where it was thoroughly heated to the rolling temperature (Bethlehem Steel Corporation).

Figure 3*a*. Cross section of ingot showing columnar grains at the rim growing toward the center.

Slabs can be produced in 45 minutes compared to 12 hours for the conventional process. The steel is cooled as it passes through a water cooled mold, and the continuous slab is rolled to size in a series of mill stands. Although some difficulties have been experienced in the use of strand casting for steel, it is becoming more commonly used. Continuous casting of nonferrous metals has been very successful.

Rolling Mills

Structural rolling mills (Figure 5) produce shapes such as I-beams, channels, angles, and wide flange beams, as well as some special sections such as zees, tees, H-piles, and sheet piling. All these shapes are produced from blooms in a series of mill stands, each mill stand contributing to the final shape. Ordinary structural shapes are made of low

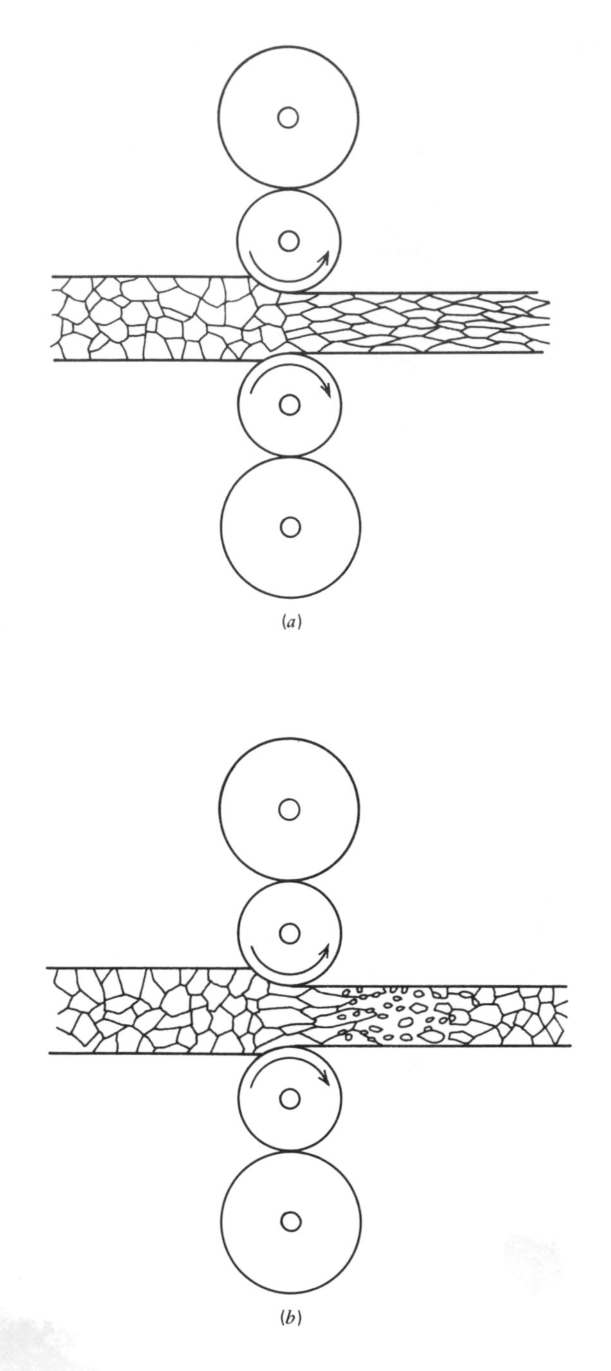

Figure 3b. Contrast of grain structure in (a) cold rolling and (b) hot rolling steel.

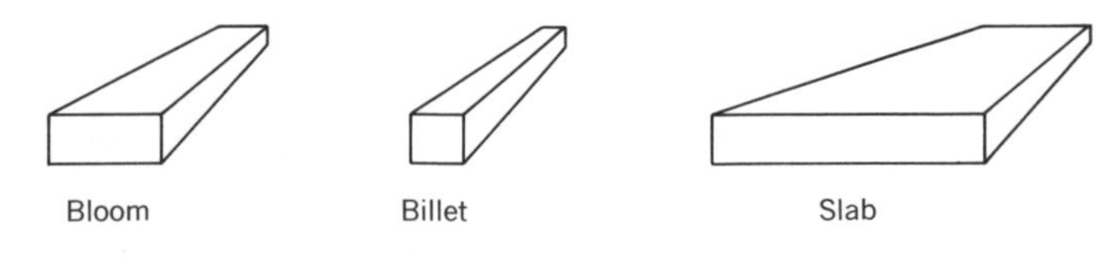

Figure 4. Ingot formed into blooms, billets, and slabs (*Machine Tools and Machining Practices*).

Figure 5. Structural steel and rails are rolled from blooms. Standard shapes are produced on mills equipped with grooved rolls. Wide-flange sections are rolled on mills, which have ungrooved horizontal and vertical rolls (Bethlehem Steel Corporation).

carbon steel containing from 0.10 to 0.25 percent carbon. This steel, which has been hot rolled, has a low strength compared to alloy steels. **Hot rolled** products have a gray or black mill scale on the surface, which must be removed before further finishing of the steel is done. Hot rolled plate, unlike structural steel, can be formed from a slab of low carbon steel, an alloy steel, or a high carbon steel (Figure 6). Hot rolled plate is manufactured as either sheared or universal milled (UM). The sheared plate is cut to the desired size, but the UM plate is formed on the edges by a series of rolls. Plate is designated as shapes from $\frac{1}{4}$ inch thick and under 48 inches wide or $\frac{3}{16}$ inch thick and over 48 inches wide. Thinner products are called sheets.

Cold rolled sheet makes up a large part of steel production (Figure 7). Hot rolled sheet is cleaned with an acid dip called pickling, followed by a dip in lime water. The sheet is then cold rolled under very heavy pressure, after which it is wound into coils. Some of this sheet steel is used to produce household appliances such as ranges, washers, and

Figure 6. Hot rolled plate emerging from a mill stand (Bethlehem Steel Corporation).

Figure 7. Coils of cold rolled sheet stock. These coils are rolled from descaled hot rolled sheet on high speed cold reduction mills (Bethlehem Steel Corporation).

dryers, while a vast tonnage of it is used for auto bodies. Very narrow sheets, or strip steel, are usually wound on a roll and used for such manufacturing purposes as press work.

Flat sheet stock is rolled without reheating, a process that permanently deforms and elongates the grain structure of the steel. This process toughens and strengthens the metal and gives it a smooth bright metallic finish, but reduces its ductility, that is, its ability to be deformed or stretched without breaking. The grains are elongated in the direction of the rolling, making the metal more ductile in one axis than the other. This characteristic of cold rolled metals makes them more liable to crack when they are bent in a small radius along the direction of rolling than across the direction of rolling (Figure 8).

Annealing, a heat treating operation, can restore the ductility at the expense of strength. In one such method, the steel is heated to approximately 1600°F (871°C) and very slowly cooled to room temperature. The steel can then be further cold worked without danger of cracking or splitting.

Galvanizing

When iron is in the presence of oxygen and moisture, it begins to rust. If it is not protected in some way, the iron will eventually revert back to its original state of iron oxide. Steel products are dipped in hot zinc to give them a coating resistant to the corrosive effects of the everyday atmosphere. Sheet steel is galvanized to make garbage cans, parts of autos, furnace and ventilating systems. Many products such as water pails are galvanized after manufacture.

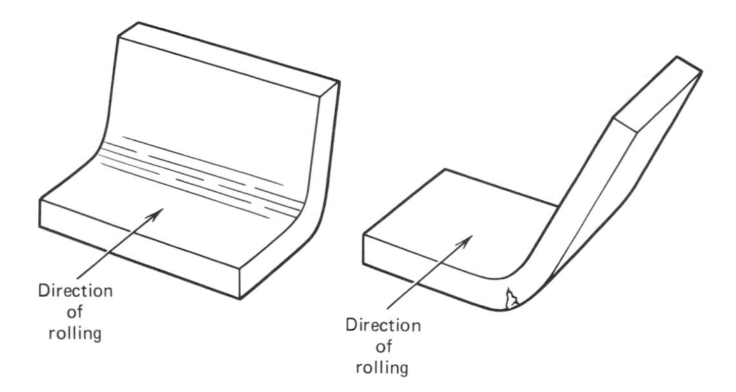

Figure 8. This shows the fibrous quality of rolled steel called anistropy. The metal is stronger in one axis than the other.

Tin Plate

Much of the food we eat is preserved in cans made from tin plate. These "tin cans" are actually 99 percent steel and only 1 percent tin. The thin plating of tin protects both the inside and outside of the can from corrosion, while giving it a bright attractive surface. The tin coating also is applied to the steel by a hot dip process.

Bars

Hot rolled bars (Figure 9) may be round, square, flat, and hexagonal in shape. Bars, like plate, can be low carbon of 0.10 to 0.20 percent, often called mild steel, or they can be alloy or tool steel. Tremendous numbers of steel bars are used to manufacture such things as axles, gears, and connecting rods in automobiles. Tools, as wrenches and pliers, are also made from steel bars. Tons of steel bars are used every year as reinforcers in concrete construction.

Cold finished bars are made from hot rolled bar stock and are designated as cold rolled (CR) or cold finished (CF). The hot rolled bar is given an acid bath to remove the black scale and is then washed to remove all traces of the acid.

Round bar stock is usually drawn through a die that accurately sizes it and gives a good finish.

Figure 9. Steel bars are shaped by rolling between pairs of grooved rails in this high speed hot mill (Courtesy of American Iron & Steel Institute).

Cold rolled or cold drawn steel posseses superior machining qualities compared to hot rolled steel. Cold drawn or rolled steel cuts clean producing a good finish when machined, but hot rolled steel is softer and somewhat gummy. This makes it difficult to get a good finish on these steels.

Forging

Forging is a process by which steel is heated and then pressed or hammered into specific shapes. Forging is used to produce tools or crank shafts. This is done by heating previously sized pieces of metal and forcing them into a die. By contrast, very large parts, such as huge generator shafts, are hot forged before they are machined to the finished size (Figure 10). Forging enables a metal to retain the grain flow in such a way that it makes a stronger part than one of equivalent shape that has been cut or machined from bar stock (Figure 11). Metals are stronger in the direction in which they have been forged; but, if a piece is cut across its grain by machining, it loses some of its strength.

Forged products are full of residual stresses that are often released as the material is being machined. This causes the material to warp during each heavy cut. For this reason forgings and other hot rolled products should be completely roughed out oversize before any finishing cut is taken. Cold upset forming is similar to hot forging in that a blank is forced into a desired shape in a die (Figure 12).

Tubular Products

A round billet is pierced and rolled over a tube mandrel (Figure 13) to form seamless tube. The process produces a rough tube that is finished in a tube mill by passing it through rolls over a ball.

Butt welded pipe used for water lines in homes is not made from a solid billet, but is rolled from flat strip called skelp. The skelp is heated and formed in a series of rolls (Figure 14). The hot metal edges are continuously pressed together through a set of pressure rolls and welded securely, thus forming the pipe. Pipes are then cut to length by a flying saw (a metal cutting circular saw that travels with the moving pipe while cutting it off).

Pipe and tubing are also made by two electric welding processes. Small size tubing is made by

Figure 10. Between the jaws of a 7500 ton hydraulic forging press, what was once a large glowing ingot is now taking on the desired shape and strength (Courtesy of American Iron & Steel Institute).

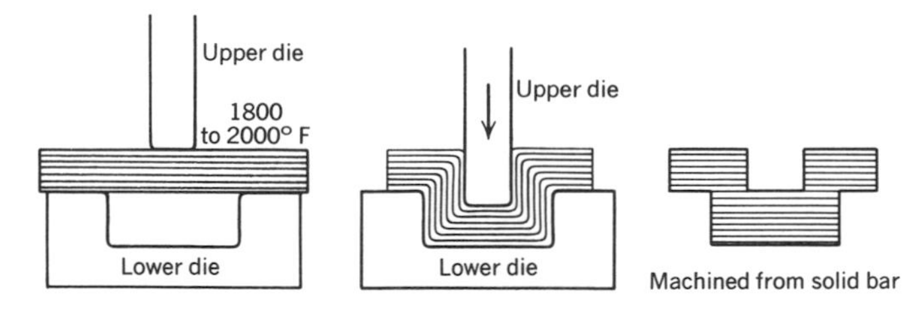

Figure 11. Grain flow in a solid bar as it is being forged compared to a machined solid bar (*Machine Tools and Machining Practices*).

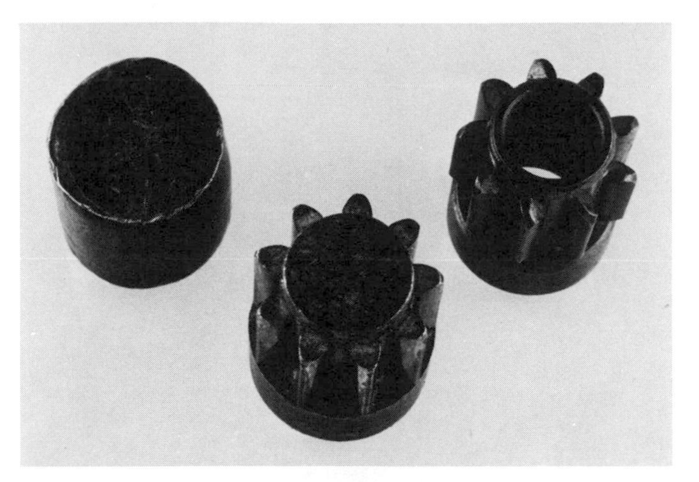

Figure 12. This gear was cold upset formed in two operations. The blank is at the left, the partially formed gear is at the center, and the finished gear is on the right. An anneal is sometimes needed between operations.

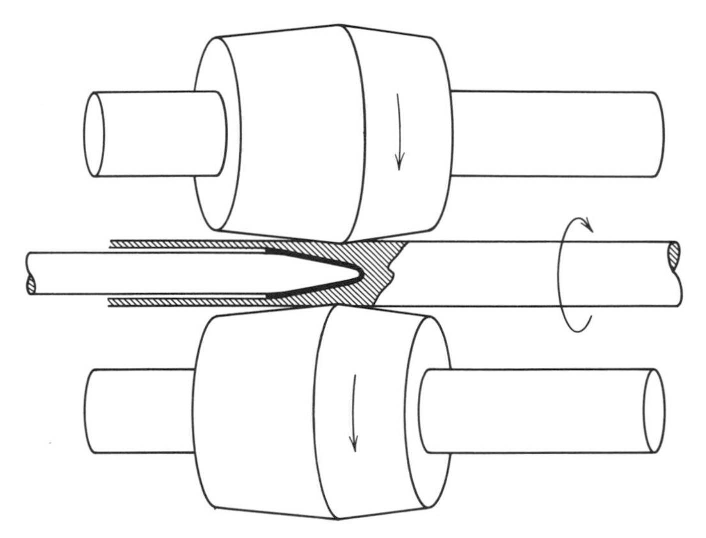

Figure 13. Piercing of solid billet to make seamless tubing (*Machine Tools and Machining Practices*).

Figure 14. Small diameter pipe formed by continuous butt welding. Skelp is welded into pipe as it passes through sets of rolls. A flying saw cuts the pipe into lengths (Bethlehem Steel Corporation).

Figure 15. In the electric-resistance welding process, skelp from the forming rolls (left) is welded automatically (Bethlehem Steel Corporation).

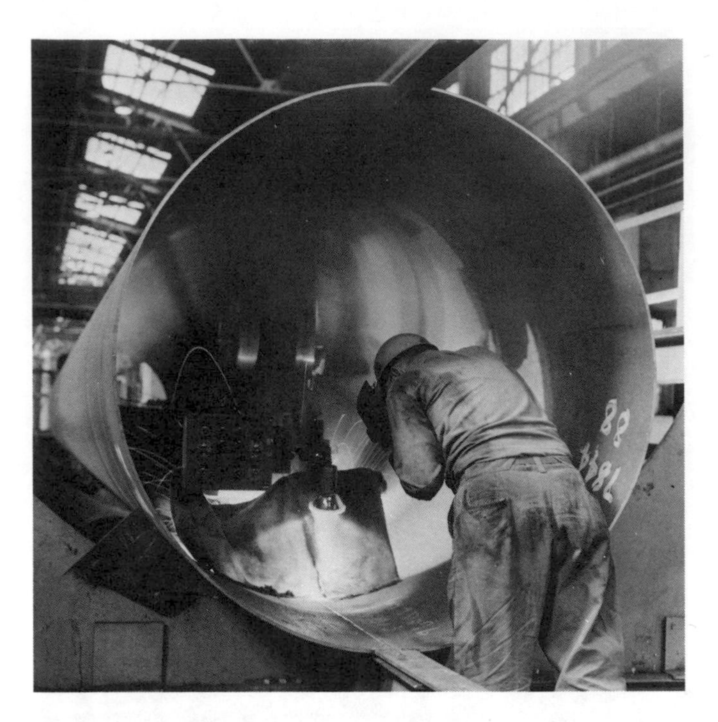

Figure 17. The final step in making large diameter steel pipe is welding the curved plates (Bethlehem Steel Corporation).

Figure 16. Large diameter pipe is made from plate that has been formed in successive shaping processes in the O-ing press (Courtesy of American Iron & Steel Institute).

the electric resistance weld (Figure 15). This process uses the resistance of the material when an electric current is applied to generate the heat required for welding. The forming of the tube from hot rolled or cold rolled strip is similar to that of butt welded pipe except that the material is not preheated.

Electric arc welding is used for the manufacture of large diameter pipe (Figure 16). The edges are usually joined by the submerged arc method (Figure 17). This automatic method of welding uses a spool of wire for an electrode and granulated flux, which is distributed over the weld. This is often a continuous process as the pipe moves along a conveyer. Pipe is usually tested hydrostatically, that is, by capping the ends and pumping in water under pressure. For some uses, pipe is galvanized, or zinc coated, to protect it against corrosion.

Wire Products

Wire is made by drawing a steel rod through a succession of dies (Figure 18), each die being slightly smaller in diameter than the previous one.

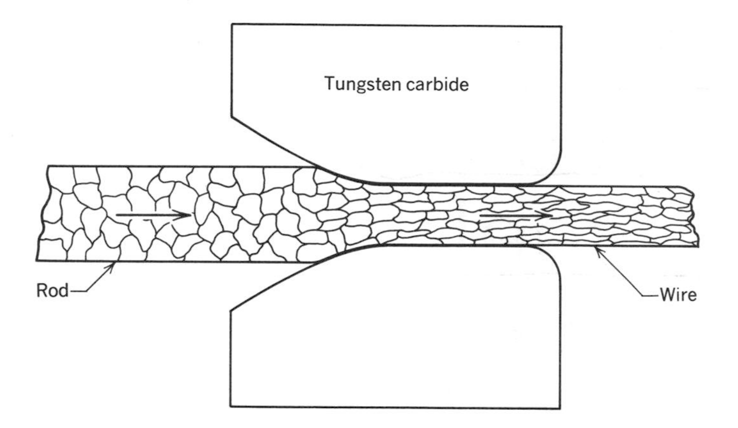

Figure 18. Enlarged cross section of wire drawing through a die (*Machine Tools and Machining Practices*).

The wire is pulled by a rotating capstan or drum that is between each die. The wire is wrapped one or more times around the capstan so that it will not slip. Coolant, which also acts as a lubricant, is used, since this is a cold forming operation. The finished wire is wound on a reel (Figure 19).

Figure 19. Battery of modern wire drawing machines. Rod enters at the left, is reduced in size as it passes through successive dies, and is coiled at the right (Bethlehem Steel Corporation).

Self-Evaluation

1 Why is an ingot heated in a soaking pit before it is rolled into a semifinished steel?

2 How does hot rolling improve the steel in the ingot?

3 What is the approximate carbon content of a mild steel bar?

4 Would a cold finished bar have more or less strength than a hot rolled bar of equal size, shape, and carbon content? Explain.

5 What is the difference in surface texture between hot rolled steel and cold rolled steel?

6 What advantage does forging have over other steel forming processes?

7 What is the major difference between seamless and butt welded pipe?

8 What are two methods used to electric weld pipe and tubing?

9 When machining forgings and other hot rolled products, the roughing cuts should all be made before any finish cuts. Why is this?

10 How can a piece of mild steel rod be turned into a small diameter wire?

3 Identification and Selection of Iron and Steel

When the village smithy plied his trade, there was only wrought iron and plain carbon steel for making tools, implements, and horsehoes; so this made the task of separating metals relatively simple. As industry began to need more alloy steels and special metals, they were gradually developed; today many hundreds of these metals are in use. Without some means of reference or identification, work in the welding and machine shop would be confusing. Therefore, this chapter will introduce you to several systems used for identifying steels and to some ways of choosing between them.

OBJECTIVES

After completing this chapter, you will be able to:

1. Identify different types of ferrous metals by various means of shop testing.
2. Select several commercial shafting alloys with various surface finishes.

INFORMATION

Classification of irons and steels makes it easier for the producer and user to identify the hundreds of different compositions available today. A few of the more common classification systems are given for plain carbon and low alloy steels, stainless steels, and cast irons.

Plain Carbon and Low Alloy Steels

The most common systems in the United States used to classify steels by chemical composition were developed by the Society of Automotive Engineers (SAE) and American Iron and Steel Institute (AISI). The SAE and AISI systems use a four or five digit number (Table 1). The first number indicates the type of steel. Carbon, for instance, is denoted by the number 1, 2 is a nickel steel, 3 is a nickel-chromium steel, and so on. The second digit indicates the approximate percentage of the predominant alloying element. The third and fourth digits, represented by x, always denote the percentage of carbon in hundredths. For plain carbon steel it could by anywhere from 0.08 to 1.70 percent. For example, SAE 1040 denotes plain carbon steel with 0.40 percent carbon; SAE 4140 denotes a chromium-molybdenum steel containing 0.40 percent carbon and about 1.0 percent of the major alloy (molybdenum).

The AISI numerical system is basically the same as the SAE system with certain capital letter prefixes. These prefixes designate the process used to make the steel.

AISI prefixes:

B. — Acid Bessemer, carbon steel
C — Basic open hearth carbon steel
CB — Either acid Bessemer or basic open hearth carbon steel at the option of the manufacturer
D — Acid open hearth carbon steel
E — Electric furnace alloy steel

all

Stainless Steel

It is the element chromium (Cr) that makes stainless steels stainless. Steel must contain a minimum of about 11 percent chromium in order to gain resistance to atmospheric corrosion. Higher percentages of chromium make steel even more resistant to corrosion at high temperatures. Nickel is added to improve ductility, corrosion resistance, and other properties.

Table 1 SAE-AISI Numerical Designation of Alloy Steels (x Represents Percent of Carbon in hundredths)

Carbon Steels	
Plain carbon	10xx
Free-cutting, resulfurized	11xx
Manganese Steels	13xx
Nickel Steels	
0.50% nickel	20xx
1.50% nickel	21xx
3.50% nickel	23xx
5.00% nickel	25xx
Nickel-Chromium Steels	
1.25% nickel, 0.65% chromium	31xx
1.75% nickel, 1.00% chromium	32xx
3.50% nickel, 1.57% chromium	33xx
3.00% nickel, 0.80% chromium	34xx
Corrosion and Heat-Resisting Steels	303xx
Molybdenum Steels	
Chromium	41xx
Chromium-nickel	43xx
Nickel	46xx and 48xx
Chromium Steels	
Low-chromium	50xx
Medium-chromium	51xx
High-chromium	52xx
Chromium-Vanadium Steels	6xxx
Tungsten Steels	7xxx
Triple Alloy Steels	8xxx
Silicon-Manganese Steels	9xxx
Leaded Steels	11Lxx (example)

There are three basic types of stainless steel: the martensitic and ferritic types of the 400 series, and the austenitic types of the 300 series. Among these are the precipitation hardening types that harden over a period of time after solution heat treatment.

Since the martensitic, hardenable type, has a carbon content up to 1 percent, it can be hardened by heating to a high temperature and then quenched (cooled) in oil or air. The grades of stainless steel for cutlery are to be found in this group. The ferritic type contains little or no carbon. It is essentially soft iron that has 12 percent or more chromium content. It is the least expensive of the stainless steels and is used for building trim, pots, and pans, for example. Both ferritic and martensitic types are magnetic.

Austenitic stainless steel contains chromium and nickel, little or no carbon, and cannot be hardened by quenching, but it readily work hardened. while retaining much of its ductility. For this reason it can be work hardened until it is almost as hard as hardened martensitic steel. Austenitic stainless steel is somewhat magnetic in its work hardened condition, but is nonmagnetic when annealed or soft.

Table 2 illustrates the method of classifying the stainless steels. Only a very few of the basic types are given here. Consult a manufacturer's catalog for further information. See Chapter 15 for the machining characteristics of stainless steels.

Tool Steels

Special carbon and alloy steel, called tool steels, have their own classification. There are seven major tool steels for which one or more letter symbols have been assigned.

1 Water hardening tool steels
W — high carbon steels

2 Shock resisting tool steels
S — Medium carbon, low alloy

3 Cold work tool steels
O — Oil hardening types
A — Medium alloy air hardening types
D — High carbon, high chromium types

Table 2 Classification of Stainless Steels

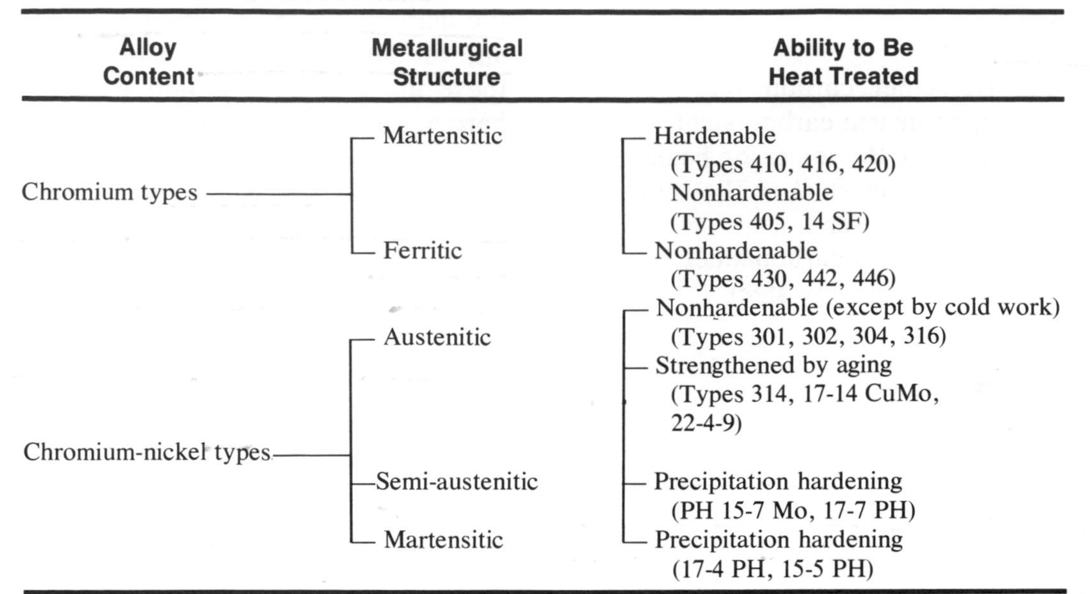

Alloy Content	Metallurgical Structure	Ability to Be Heat Treated
Chromium types	Martensitic	Hardenable (Types 410, 416, 420) Nonhardenable (Types 405, 14 SF)
	Ferritic	Nonhardenable (Types 430, 442, 446)
Chromium-nickel types	Austenitic	Nonhardenable (except by cold work) (Types 301, 302, 304, 316) Strengthened by aging (Types 314, 17-14 CuMo, 22-4-9)
	Semi-austenitic	Precipitation hardening (PH 15-7 Mo, 17-7 PH)
	Martensitic	Precipitation hardening (17-4 PH, 15-5 PH)

Source. Armco Steel Corporation, Middletown, Ohio, *Armco Stainless Steels,* 1966. The following are registered trademarks of Armco Steel Corporation: 17-4 PH, 15-5 PH, 17-7 PH, and PH 15-7 Mo.

4 Hot work tool steels
 H — H10 to H19, inclusive, chromium base types
 H20 to H39, inclusive, tungsten base types
 H40 to H59, inclusive, molybdenum base types

5 High speed tool steels
 T — Tungsten base types
 M — Molybdenum base types

6 Special purpose tool steels
 L — Low alloy types
 F — Carbon tungsten types

7 Mold steels
 P — P1 to P19, inclusive, low carbon types
 P20 to P39, inclusive, other types

Several metals can be classified under each group, so an individual type of tool steel will also have a suffix number that follows the letter symbol of its alloy group. The carbon content is given only in those cases where it is considered an identifying element of that steel.

Type of Steel	Examples
Water hardening: straight carbon tool steel	W1, W2, W4
Manganese, chromium, tungsten: oil hardening tool steel	O1, O2, O6
Chromium (5.0%): air hardening die steel	A2, A5, A10
Silicon, manganese, molybdenum: punch steel	S1, S5
High speed tool steel	M2, M3, M30 T1, T5, T15

Unified Numbering System

A new numbering system, called the Unified Numbering System for Metals and Alloys (UNS), provides a designation system for all present and future metals and alloys. The system was published by the SAE in 1975. Both SAE and ASTM (American Society for Testing Materials) will use these numbers from now on. The system will also be proposed to the ISO (International Standards Organization).

The Unified Numbering System (Table 3) establishes 15 series of numbers for metals and alloys. Each UNS number consists of a single letter prefix followed by five digits. In most cases, the letter is suggestive of the family of metals identified; for example, A for aluminum and P for precious metals.

Whenever feasible, identification numbers from existing systems are incorporated into the UNS numbers. For example, carbon steel, presently identified by AISI 1020, is covered by UNS G11020; free cutting brass, now identified by CDA 360, is covered by UNS C36000.

Cast Iron

There are several forms of cast iron: white, gray, malleable, and nodular. Cast irons contain more carbon (2 to 4.5 percent) than do steels, and are easily cast into molds. Cast irons have a melting temperature of 2100°F (1149°C) in contrast to the melting temperature of steel, which is 2500 to 2800°F (1371 to 1538°C). Cast iron may be identified by its brittle fracture and by spark testing (Figure 1).

Gray cast irons may be ferritic or pearlitic and may contain alloying elements, such as nickel or chromium and varying amounts of silicon. Gray cast irons are classified by the American Society for Testing Materials (ASTM) according to their tensile strength. Table 4 compares the ASTM number to its tensile strength in pounds per square inch (psi). Gray cast iron is very brittle, and its tensile strength is much lower than its compressive strength (Figure 2). See Chapter 18, "Metallurgy of Welds: Cast Iron."

Table 3 Unified Numbering System for Metals and Alloys

Ferrous Metals	
000001–D99999	Specified mechanical properties steels
F00001—F99999	Cast irons: Gray, malleable, pearlitic malleable, ductile (nodular)
	Carbon steel castings, low alloy steel castings
G00001–G99999	AISI–SAE carbon and alloy steels
H00001–H99999	AISI–"H" steels
K00001–K99999	Miscellaneous steels and ferrous alloys
S00001–S99999	Stainless steels
T00001–T99999	Tool steels
Nonferrous Metals	
A00001–A99999	Aluminum and aluminum alloys
C00001–C99999	Copper and copper alloys
E00001–E99999	Rare earth and rare earthlike metals
L00001–L99999	Low melting metals
M00001–M99999	Miscellaneous metals
N00001–N99999	Nickel and nickel alloys
P00001–P99999	Precious metals
R00001–R99999	Reactive and refractory metals

Source. Reprinted by John E. Neely from *Iron Age*, December 16, 1974, Chilton Company, 1976.

Shop Tests for Identifying Steels

Steel is usually identified by placing a color code at the end of a shaft. One of the disadvantages of this method is that the marking is often lost. If the marking is obliterated or cut off and the piece is separated from the rack where it is stored, it is very difficult to ascertain its carbon content and alloy group. This points up the necessity of returning stock material to its proper rack. It is also good practice always to leave the identifying mark on one end of the stock material and always to cut off the other end.

Unfortunately, in the shop there are always some short ends and otherwise useful pieces of steel that have lost their identifying marks. In addition, when repairing or replacing parts of old or

Table 4 Classes of Gray Iron

ASTM Number	Minimum Tensile Strength psi
20	20,000
25	25,000
30	30,000
35	35,000
40	40,000
45	45,000
50	50,000
60	60,000

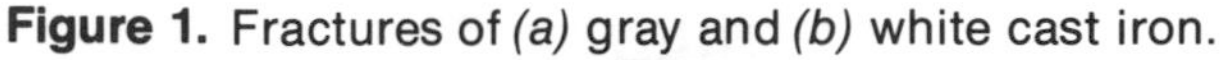

Figure 1. Fractures of *(a)* gray and *(b)* white cast iron.

Figure 2. This cast iron pulley has been ruined as a result of being hammered on the thin edges of the pulley. The hub also shows signs of being hammered. Wheel pullers placed on the thin edges can also break cast iron pulleys. The strain should have been applied to the hub when it was removed.

nonstandard machinery, there is usually no record available for material selection. There are many shop methods a machinist may use to identify the basic type of steel in an unknown sample. The machinist can compare the unknown sample with each of the several steels used in the shop. The following are several methods of shop testing that you can use.

Observation Some metals can be identified by visual observation of their finishes. Heat scale or **black mill scale** is found on all hot rolled (HR) steels (Figure 3). These can be either low carbon (0.05 to 0.30 percent), medium carbon (0.30 to 0.60 percent), high carbon (0.60 to 1.75 percent), or alloy steels. Other surface coatings that might be detected are the **sherardized, plated, case hardened,** or **nitrided surfaces.** Sherardizing is a process in which zinc vapor is inoculated into the surface of iron or steel. Figure 4 reveals the appearance of some of these textures.

Cold finish (CF) steel usually has a metallic luster. Ground and polished (G and P) steel has a bright, shiny finish, with closer dimensional tolerances than CF. Also cold drawn ebonized, or black, finishes are sometimes found on alloy and resulfurized shafting.

Chromium nickel stainless steel, which is austenitic and nonmagnetic, usually has a white appearance. Straight 12 to 27 percent chromium is called ferritic and is magnetic with a bluish-white color. Manganese steel is blue when polished but

Figure 3. Rolled bars: (top) hot rolled, (center) cold rolled, and (bottom) ground and polished.

Figure 4. Round bars having various surface finishes. Left to right: aluminum, ground and polished, cold finished steel, hot rolled steel, sherardized surface, and zinc dip surface.

copper colored when oxidized. White cast iron fractures will appear silvery or white. Gray cast iron fractures appear dark gray and will smear a finger with a gray graphite smudge when touched.

Magnet Test Most ferrous metals such as iron and steel are magnetic (that is, they are attracted to a magnet), but most nonferrous metals are nonmagnetic. An exception is nickel, which is nonferrous but magnetic. Since United States "nickel" coins contain about 25 percent nickel and 75 percent copper, they do not respond to the magnet test, but Canadian "nickel" coins are attracted to a magnet since they contain more nickel. Ferritic and martensitic (400 series) stainless steels are also attracted to a magnet and thus cannot be separated from other steels by this method. (See *Chemical Tests below*.) Austenitic (300 series) stainless steel is not magnetic unless it is work hardened.

Hardness Test Wrought iron is very soft since it contains almost no carbon or any other alloying element. Generally speaking, the *more* carbon (up to 2 percent) and other elements that steel contains, the *harder*, stronger, and less ductile it becomes, even if the steel has been annealed (softened). Thus, the hardness of a sample can help us to separate low carbon steel from an alloy steel or a high carbon steel. Of course, the best way to check for hardness is with a hardness tester. The Rockwell, Brinell, and other types of hardness testing will be studied in Chapter 5. Not all machine shops have hardness testers available, in which case the following shop methods can prove useful.

Scratch Test Geologists and "rock hounds" scratch rocks against items of known hardness for identification purposes. The same method can be used to check metals for relative hardness. Simply scratch one sample with another and the softer sample will be marked (Figures 5*a* and 5*b*). Be sure all scale or other surface impurities have been removed before scratch testing. A variation of this method is to strike two similar edges of the two samples together. The one sample receiving the deepest identation is the softer. (Figures 6*a* and 6*b*).

File Tests Files can be used to establish the relative hardness between two samples, as in the scratch test, or the test can be used to determine the approximate hardness of a piece of steel (Figure 7). Table 5 gives the Rockwell and Brinell hardness numbers for this file test when using new files. This method, however, can only be as accurate as the skill that the user has acquired through practice. Care must be taken not to damage the file, since filing on hard materials may ruin the file. Testing should be done on the tip end or on the edge of the file.

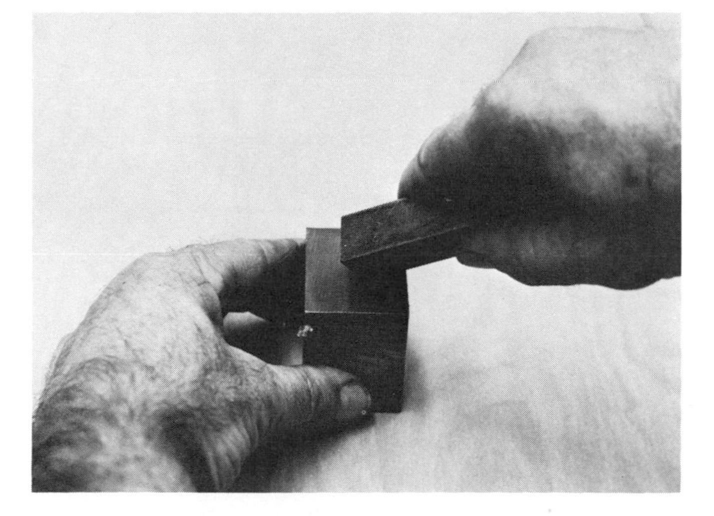

Figure 5*a*. A piece of keystock (mild steel) is scratched across an unknown metal sample. Since the sample is not scratched, it is harder than the keystock and probably is an alloy or tool steel.

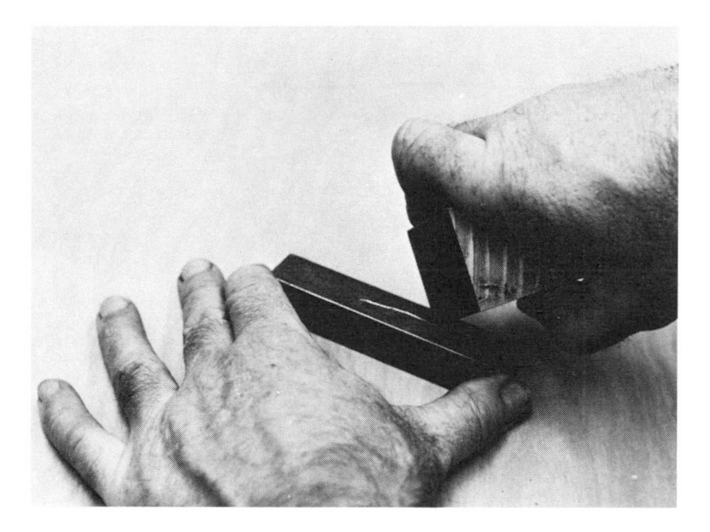

Figure 5*b*. The sample is now scratched against the keystock as a further test and it does scratch the keystock.

Chemical Tests Many shop chemical tests are very simple, most of them using chemicals normally found in the shop. More complex chemical testing should be done by a chemist.

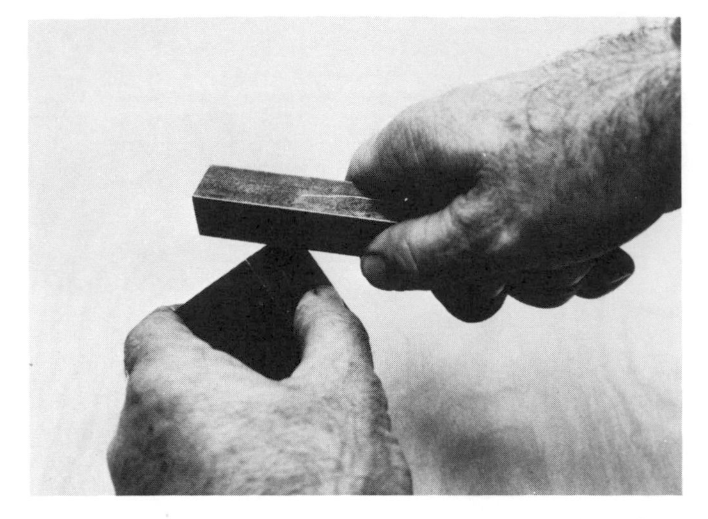

Figure 6*a*. An alternate hardness test is to strike the edges of the keystock and unknown sample together.

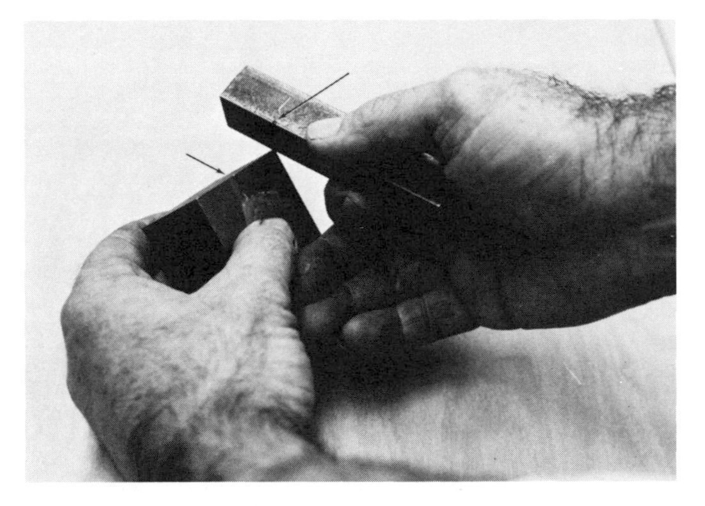

Figure 6*b*. It can be seen that only the keystock received an indentation where they were struck together, signifying that the keystock was the softer of the two samples.

Figure 7. File testing for hardness (Lane Community College).

Table 5 File Test and Hardness Table

Type Steel	Rockwell *B*	Rockwell *C*	Brinell	File Reaction
Mild steel	65		100	File bites easily into metal. (Machines well but makes built up edge on tool.)
Medium carbon steel		16	212	File bites into metal with pressure. (Easily machined with high speed tools.)
High alloy steel High carbon steel		31	294	File does not bite into metal except with difficulty. (Readily machinable with carbide tools.)
Tool steel		42	390	Metal can only be filed with extreme pressure. (Difficult to machine even with carbide tools.)
Hardened tool steel		50	481	File will mark metal but metal is nearly as hard as the file, and machining is impractical; should be ground.
Case hardened parts		64	739	Metal is as hard as the file; should be ground.

The surface to be tested must be very clean and free from all scale or oil. Apply only one drop of the acid with an eye dropper. A solution of 6 percent nitric acid in methanol (wood alcohol) will etch, or darken, carbon steel, but will not discolor stainless steels (Figure 8). A 10 percent nitric acid solution will etch mild steel almost immediately, while the 6 percent solution takes about one minute. A drop of concentrated copper sulfate solution will leave a copper-colored spot on clean iron or steel, but not on austenitic stainless steel.

Some stainless steels react to sulfuric and hydrochloric acids. Types 302 and 304 are strongly attacked by sulfuric acid, which leaves a dark surface with green crystals. Types 316 and 317 are attacked more slowly, leaving a tan surface.

Hydrochloric acid (HCl) reacts with types 304, 321, and 347 very rapidly, releasing gas. Type 302 leaves a pale blue-green solution on the surface, while types 303, 414, and 430F have a spoiled egg odor and leave a heavy black smudge. Steels containing selenium emit a garlic odor when attacked by HCl.

Commercial spot testers are available in kits. Stainless steel and many ferrous and nonferrous alloys can be readily identified using these test kits.

Spark Testing Spark testing is a useful way to test for carbon content in many steels. The metal tested, when held against a grinding wheel, will display a particular spark pattern depending on its content. Spark testing provides a convenient means of distinguishing between tool steel (of medium or high carbon) and low carbon steel. High carbon steel (Figure 9) shows many more bursts than low carbon steel (Figure 10).

Almost all tool steel contains some alloying elements besides the carbon that affect the carbon burst. Chromium, molybdenum, silicon, aluminum, and tungsten suppress the carbon burst. For this reason, spark testing is not always dependable for determining the carbon content of an unknown sample of steel unless it is plain carbon steel. It is useful, however, as a comparison test. Comparing the spark of a known sample to that of an unknown sample can be an effective method of identification for the trained observer. Cast iron may be distinguished from steel by the characteristic spark stream (Figure 11). High speed steel can also be readily identified by spark testing (Figure 12).

The various types of stainless steel may be identified with spark testing. Types 302, 303, 304,

Figure 8. Identifying stainless steels with an acid test. A 6 percent solution of nitric acid in methanol (nital) has been applied to the stainless steel on the left and is being applied to the mild steel on the right. The mild steel is discolored, but the stainless steel is not.

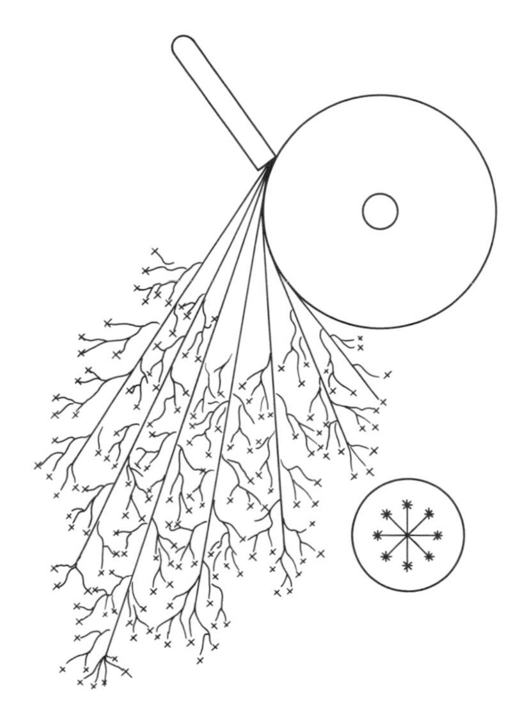

Figure 9. High carbon Steel. Short, very white or light yellow carrier lines with considerable forking, having many starlike bursts. Many of the sparks follow around the wheel. Inset shows spark structure (*Machine Tools and Machining Practices*).

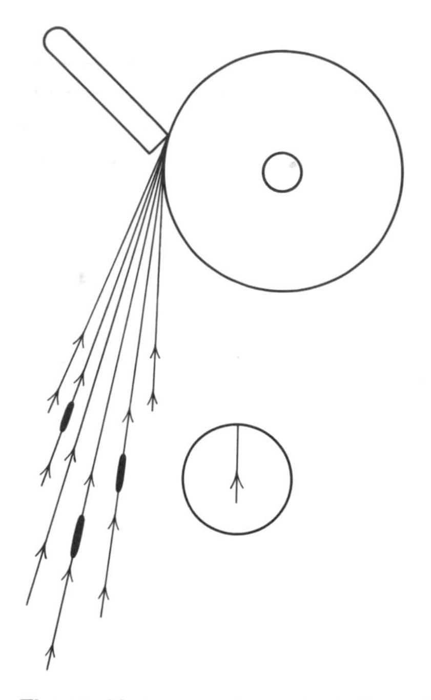

Figure 10. Low carbon steel. Straight carrier lines having a yellowish color with very small amount of branching and very few carbon bursts (*Machine Tools and Machining Practices*).

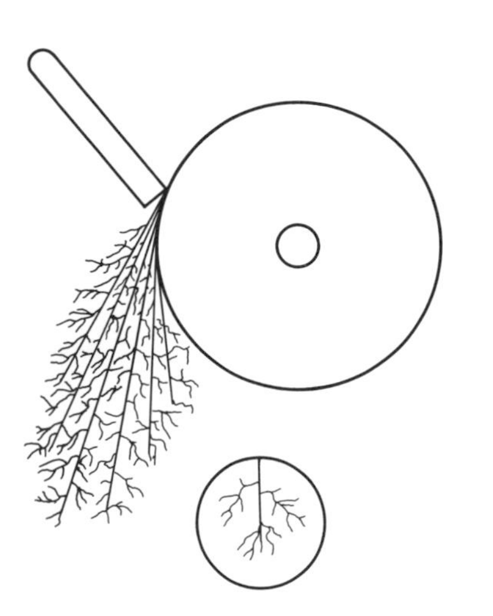

Figure 11. Cast iron. Short carrier lines with many bursts, which are red near the grinder and orange-yellow further out. Considerable pressure is required on cast iron to produce sparks (*Machine Tools and Machining Practices*).

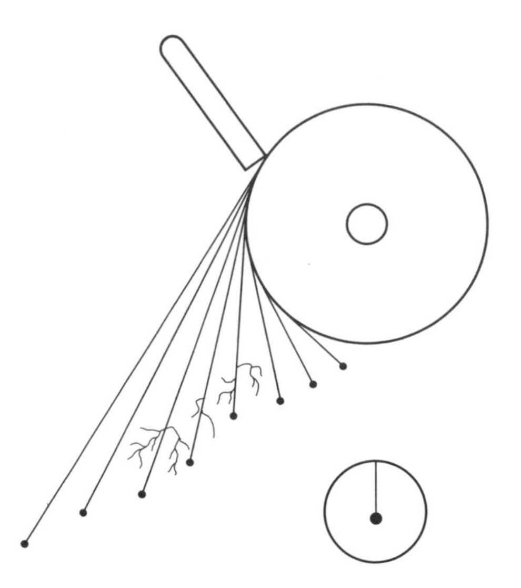

Figure 12. High speed steel. Carrier lines are orange, ending in pear-shaped globules with very little branching or carbon sparks. High speed steel requires moderate pressure to produce sparks (*Machine Tools and Machining Practices*).

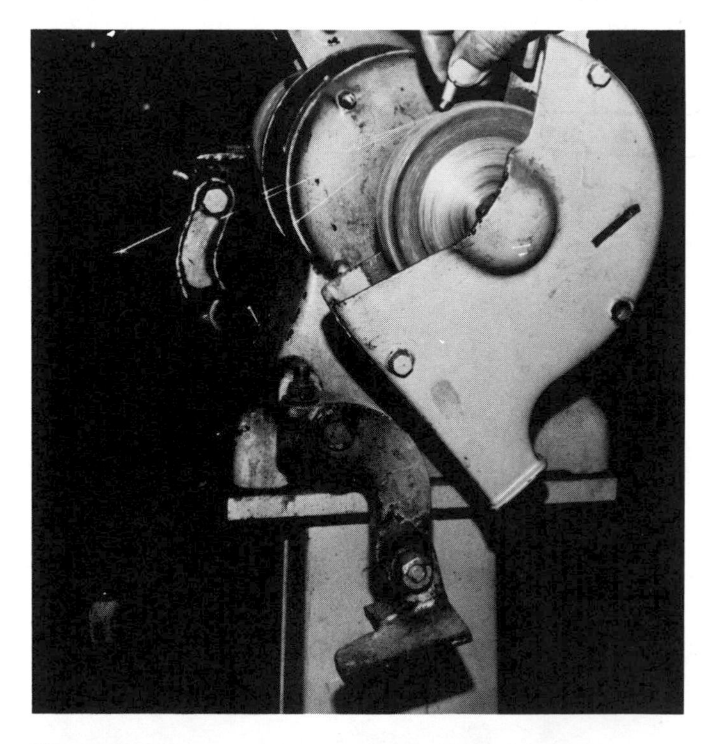

Figure 13. The proper way to make a spark test on a pedestal grinder (Lane Community College).

and 316 show short reddish carrier lines with few forks. Types 308, 309, and 310 have full red carrier lines with very few forks. Of the martensitic types of stainless steel, types 410, 414, 416, and 431 have long white carriers with few forks. Types 420, 440A, 440B, and 440C have long white-red carriers with bursts. Ferritic types range from white to red carriers with some forks.

When spark testing always wear safety glasses or a face shield. Adjust the wheel guard so that the spark will fly outward and downward, and away from you (Figure 13). A coarse grit wheel that has been freshly dressed to remove contaminants should be used.

Machinability Test As a simple comparison test, machinability can be useful to help determine a specific type of steel. For example, two unknown samples identical in appearance and size can be test cut in a machine tool using the same speed and feed for both of them. The ease of cutting should be compared and chips observed for heating, color, and curl (Figure 14). Hardness is generally related to ease of machining; harder materials are generally more difficult to machine. See Table 2 in Chapter 15.

Selection for Uses

There are a great variety of carbon alloy, and tool steels from which to choose when planning a project. Several types of cast irons are also available to the welder, machinist, or machine designer. Some of these are listed according to their carbon content and use in Table 6. In later chapters in this book, we will present many more principles and concepts regarding the selection of metals for a particular job. For example, many more alloy steels than plain carbon steels are used in the manufacture of tools as the alloy steel gives them special properties.

Figure 14. Machinability test. When one of two samples having the same feed and speed shows a darker color (blue) on the chip, it can be assumed that it is a harder, stronger metal (Lane Community College).

Several properties should be considered when selecting a piece of steel for a job: **strength, machinability, hardenability, weldability, fatigue resistance,** and **corrosion resistance.**

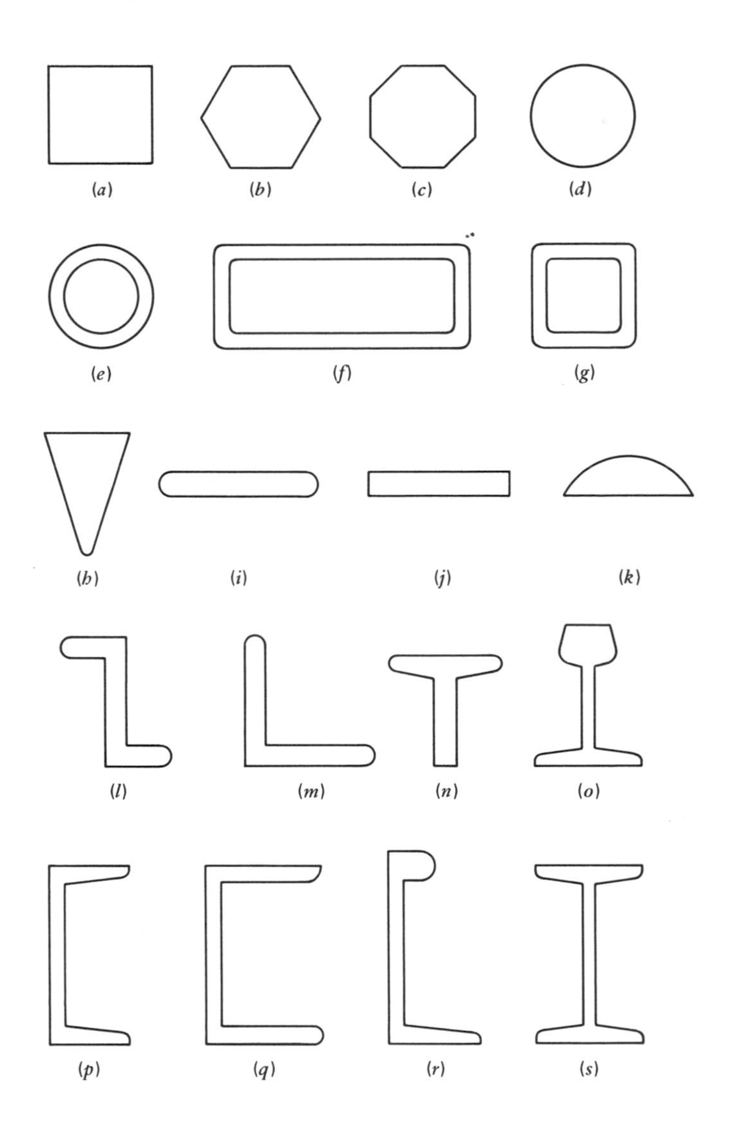

Figure 15. Steel shapes used in manufacturing.

(*a*) Square HR or CR
(*b*) Hexagonal
(*c*) Octagon
(*d*) Round
(*e*) Tubing and pipe (round)
(*f*) HREW (hot rolled electric welded) rectangular steel tubing
(*g*) HREW square steel tubing
(*h*) Wedge
(*i*) HR flat bar (round edge spring steel flats)
(*j*) Flat bar (CR and HR)
(*k*) Half round
(*l*) Zee
(*m*) Angle
(*n*) Tee
(*o*) Rail
(*p*) Channel
(*q*) Car and ship channel
(*r*) Bulb angle
(*s*) Beams—I, H, and wide flange

Table 6 Uses of Ferrous Metals by Carbon Content

Type	Carbon Range Percent	SAE Number	Typical Uses
Carbon Steels			
Low	0.05 to 0.30	1006	For cold formability
		1008	Wire, nails, rivets, screws
		1010	Sheet stock for drawing
		1015	Fenders, pots, pans, welding rods
		1020	Bars, plates, structural shapes, shafting
		1030	Forgings, carburized parts, keystock
		1111	Free machining steel
		1113	Free machining steel
Medium	0.30 to 0.60	1040	Heat treated parts that require moderate strength and high toughness such as bolts, shafting, axles, spline shafts
		1060	Higher strength, heat treated parts with moderate toughness such as lock washers, springs, band saw blades, ring gears, valve springs, snap rings
High	0.60 to 2.0	1070	Chisels, center punches
		1080	Music wire, mower blades, leaf springs
		1095	Hay rake tines, leaf springs, knives woodworking tools, files, reamers
		5210	Ball bearings, punches, dies
Cast Irons			
Gray	2.0 to 4.5		Machinable castings such as engine blocks, pipe, gears, lathe beds
White	2.0 to 3.5		Nonmachinable castings such as cast parts for wear resistance
Malleable castings	2.0 to 3.5		Produced from white cast iron Machinable castings such as axle and differential housings, crankshafts, camshafts
Nodular iron (ductile iron)	2.0 to 4.5		Machinable castings such as pistons, cylinder blocks and heads, wrenches, forming dies

Manufacturer's catalogs and hand reference books are available for the selection of standard structural shapes, bars, and other steel products (Figure 15). Others are available for the stainless steels, tool steels, and finished carbon steel and alloy shafting. Many of these steels are known by a trade name.

A machinist is often called on to select a shaft material from which to machine finish a part. Shafting is manufactured with two kinds of surface finish: cold finished (CF) found on low carbon steel, and ground and polished (G and P) found mostly on alloy shafts. Tolerances are kept much closer on G and P shafts. The following are some common alloy and carbon steels:

1 SAE 4140 is a chromium-molybdenum alloy with 0.40 percent carbon. It lends itself readily to heat treating, forging, and welding. It provides a high resistance to torsional and reversing stresses such as in drive shafts.

2 SAE 1140 is a resulfurized, drawn, free machining bar stock. This material has good resistance to bending stresses because of its fibrous qualities and it has a high tensile strength. It is best used on shafts where the RPM is high and the torque moderately low. SAE 1140 is also useful where stiffness is a requirement. It should not be welded.

3 Leaded steels have all the free machining qualities and finishes of resulfurized steels. Leaded alloy steels such as SAE 41L40 have the superior strength of 4140, but are much easier to machine.

4 SAE 1040 is a medium carbon steel with a normalized tensile strength of about 85,000 psi. It can be heat treated, but large cross sections can be hardened only on the surface while the core will be in a normalized condition. Its main advantage is that it is a less expensive way to obtain a higher strength part.

5 SAE 1020 is a low carbon steel that has good machining characteristics. It normally comes as CF shafting. It is very commonly used for shafting in industrial applications. It has a lower tensile strength than the alloy steels or higher carbon steels.

Costs of Steel

Since steel prices, like the prices of most other products, change constantly, costs can be shown only in example. Steel is usually priced by its weight. A cubic foot of mild steel weighs 489.60 pounds; therefore a square foot one inch thick weighs 40.80 pounds. From this, you can easily compute the weights for flat materials such as plate. For hexagonal and rounds, it would be much easier to consult a table in a catalog or handbook. Given a price per pound, you should then be able to figure the cost of a desired steel product.

EXAMPLE

A 1 by 6 inch mild steel bar is 48 inches long. If current steel prices are \$0.30/lb, how much does the bar cost?

$$\frac{6 \text{ in.} \times 48 \text{ in.}}{144 \text{ in.}^2/\text{ft.}^2} = 2 \text{ square ft, 1 in. thick}$$

$2 \times 40.80 = 81.6$ lb

81.6 lb × \$0.30/lb = \$24.48 (cost of steel)

also Bob write questions and then go to page 378 for answers JC

Self-Evaluation

1 By what universal coding system are carbon and alloy steels designated?

2 What are three basic types of stainless steels and what is the number series assigned to them? What are their basic differences?

3 If your shop stocked the following steel shafting, how would you determine the content of an unmarked piece of each, using shop tests as given in this chapter?
a. AISI C1020 (CF)
b. AISI B1140 (G and P)
c. AISI C4140 (G and P)
d. AISI 8620 (HR)
e. AISI B1140 (Ebony)
f. AISI C1040

4 A small part has obviously been made by a casting process. How can you determine whether it is a ferrous or a nonferrous metal, stainless steel, or white or gray cast iron?

5 What is the meaning of the symbols O1 and W1 when applied to tool steels?

6 A $2\frac{7}{16}$ inch diameter steel shaft weighs 1.322 pounds per linear inch, as taken from a table of weights of steel bars. A 40 inch length is needed for a job. At \$0.30 per pound, what would the shaft cost?

7 When checking the hardness of a piece of steel with the file test, the file slides over the surface without cutting.
a. Is the steel piece readily machinable?
b. What type of steel is it most likely to be?

8 Steel that is nonmagnetic is called ______________________.

9 What nonferrous metal is magnetic?

10 List at least four properties of steel that should be kept in mind when you select the material for a job.

11 What alloying element does all stainless steel contain in large amounts that makes it corrosion resistant? What other element does stainless steel sometimes contain in fairly large amounts?

12 State a general rule to use when machining stainless steels.

13 Some stainless steels are difficult to tap. Name one solution to this problem.

14 Which type of cast iron is the hardest to machine?

15 A machinist must make a cut through a weld on gray cast iron. What problem could arise? What is the cause?

16 When the work is extremely hard, as would be the case with white cast iron, and the job calls for machining, what procedure should be followed?

Worksheet

Objective
Correctly identify specimens by comparison by using various tests described in this chapter.

Materials
A box of numbered specimens and a similar set of known and marked specimens.

Conclusion
Record your results.

Item Number	Test Used	Kind of Metal
1.		
2.		
3.		
4.		
5.		
6.		
7.		
8.		
9.		
10.		

4 Identification of Nonferrous Metals

Nonferrous metals such as gold, silver, copper, and tin were in use hundreds of years before the smelting of iron, yet some nonferrous metals have appeared relatively recently in common industrial use. For example, aluminum was first commercially extracted from ore in 1886 by the Hall-Heroult process, and titanium is a space age metal that has been produced in commercial quantities only after World War II.

In general, nonferrous metals are more costly than ferrous metals. It is not always easy to distinguish a nonferrous metal from a ferrous metal, nor to separate one from another. This chapter should help you to identify, select, and properly use many of these metals.

OBJECTIVES

After completing this chapter, you will be able to:

1. Identify and classify nonferrous metals by a numerical system.
2. List the general appearance and use of the various nonferrous metals.

INFORMATION

Metals are designated as either ferrous or nonferrous. Iron and steel are ferrous metals, and any metal other than iron or steel is called nonferrous. The testing methods given in Chapter 3 for identifying ferrous metals will be used in this chapter; in addition, differences in specific gravity may be used as there is considerable difference in density (pounds per cubic inch ($lb/in.^3$) at 68°F or grams per cubic centimeter (g/cm^3) at 20°C, or relative weight in these metals. If the specific gravity of a material is known, the weight per cubic foot (ft^3) can be found by multiplying the specific gravity by 62.355. To find the weight of a cubic inch, multiply the specific gravity by 0.0361. Note that with a few exceptions, such as nickel and cobalt, nonferrous metals are not attracted to a magnet.

Aluminum

Aluminum is white or white-gray in color and can have any surface finish from dull to shiny and polished. Aluminum weighs 168.5 lb/ft^3 as compared to 487 lb/ft^3 for steel, and pure aluminum has a melting point of 1220°F (660°C). Aluminum alloys have a specific gravity of 2.55 to 2.77. Aluminum and its alloys are readily machinable and can be manufactured into almost any shape or form (Figure 1).

Classification of Aluminum

There are several numerical systems used to identify aluminum, such as federal specifications, mili-

Figure 1. Structural aluminum shapes used for building trim provides a pleasing appearance.

tary specifications, the American Society for Testing & Materials (ASTM), and the Society of Automotive Engineers (SAE) specifications. The system most used by manufacturers, however, is one that was adopted by the Aluminum Association in 1954.

From Table 1, you can see that the first digit of a number in the aluminum alloy series indicates the alloy type. The second digit, represented by an x in the table, indicates any modifications that were made to the original alloy. The last two digits identify either the specific alloy or aluminum impurity.

Table 1 Aluminum and Aluminum Alloys

Code Number	Major Alloying Element
1xxx	None
2xxx	Copper
3xxx	Manganese
4xxx	Silicon
5xxx	Magnesium
6xxx	Magnesium and silicon
7xxx	Zinc
8xxx	Other elements
9xxx	Unused (not yet assigned)

EXAMPLES

An aluminum alloy numbered 5056 is an aluminum-magnesium alloy, where the first 5 represents the alloy magnesium, the second digit represents modifications to the alloy, and 56 are numbers of a similar aluminum alloy of an older marking system. An aluminum numbered 1120 contains no major alloy, and has 0.20 percent pure aluminum above 99 percent.

Aluminum and its alloys are produced as **castings** or as **wrought** (cold worked) shapes such as sheets, bars, and tubing. They can be either cold worked by rolling, drawing, or extruding, or hot worked by forging. Cast aluminum is not generally as strong as wrought aluminum. Aluminum alloys are harder than pure aluminum and will scratch the softer (1100 series) aluminum. A 10 percent solution of sodium hydroxide (caustic soda) will not stain pure aluminum but will leave a dark stain on the aluminum alloy. Machinability of aluminum alloys is discussed in Chapter 15 and its weldability in Chapter 19.

Pure aluminum and some of its alloys cannot be hardened by heat treating, but they can be annealed to soften them by heat treatment. This means that hardening them can only be done by cold working (strain hardening). These temper (hardness) designations are made by a letter that follows the four digit alloy series number:

—F As fabricated. No special control over strain hardening or temper designation is noted.
—O Annealed, recrystallized wrought products only. Softest temper.
—H Strain hardened, wrought products only. Strength is increased by work hardening.

This letter—H is always followed by two or more digits. The first digit, 1, 2, or 3, denotes the final degree of strain hardening.

—H1 Strain hardened only
—H2 Strain hardened and partially annealed
—H3 Strain hardened and stabilized

The second digit denotes higher strength tempers obtained by heat treatment.

2 ¼ hard
4 ½ hard
6 ¾ hard
8 Full hard
9 Extra hard

Example

5056-H18 is an aluminum-magnesium alloy, strain hardened to a full hard temper.

Some aluminum alloys can be hardened to a great extent by a process called **solution heat treatment and precipitation hardening or aging.** This process involves heating the aluminum alloy to a temperature where the alloying element is dissolved into a solid solution. The aluminum alloy is then quenched in water and allowed to age or is artificially aged by heating slightly. The aging produces an internal strain that hardens and strengthens the aluminum. Some other nonferrous metals are also hardened by this process. See Chapter 15 "Heat Treating of Nonferrous Metals." For these aluminum alloys the letter —T follows the four digit series number. Numbers 2 to 10 follow this letter to indicate the sequence of treatment.

—T2 Annealed (cast products only)
—T3 Solution heat treated and cold worked
—T4 Solution heat treated, but naturally aged
—T6 Solution heat treated and artifically aged
—T8 Solution heat treated, cold worked, and artificially aged
—T9 Solution heat treated, artificially aged, and cold worked.
—T10 Artificially aged and then cold worked

Example

2024-T6 Aluminum-copper alloy, solution heat treated and artificially aged.

Cast aluminum alloys generally have lower tensile strength than wrought alloys. Sand castings, permanent mold, and die casting alloys are of this group. They owe their mechanical properties to solution heat treatment and precipitation or to the addition of alloys. A classification system similar to that of wrought aluminum alloys is used (Table 2).

Table 2. Cast Aluminum Alloy Designations

Code Number	Major Alloy Element
1xx.x	None, 99 percent aluminum
2xx.x	Copper
3xx.x	Silicon with Cu and/or Mg
4xx.x	Silicon
5xx.x	Magnesium
6xx.x	Zinc
7xx.x	Tin
8xx.x	Unused series
9xx.x	Other major alloys

The cast alloy 108F, for example, has an ultimate tensile strength of 24,000 psi in the as-fabricated condition and contains no alloy. The 220.T4 copper aluminum alloy has a tensile strength of 48,000 psi.

Cadmium

Cadmium has a blue-white color and is commonly used as a protective plating on parts such as screws, bolts, and washers. It is also used as an alloying element to make metal alloys that melt at low temperature, such as bearing metals, solder, type casting metals, and storage batteries. Cadmium compounds such as cadmium oxide are toxic and can cause illness when breathed. These toxic fumes can be produced by welding, cutting, or machining on cadmium plated parts. Breathing the fumes should be avoided by using adequate ventilation systems. The melting point of cadmium is 610°F (321°C). Its specific gravity is 8.648 and its density is 539.6 lb/ft^3.

Copper and Copper Alloys

Copper is a soft, heavy metal that has a reddish color. It has high electrical and thermal conductivity when pure, but loses these properties to a certain extent when alloyed. It must be strain hardened when used for electric wire. Copper is very ductile and can be easily drawn into wire or tubular products. It is so soft that it is difficult to machine and it has a tendency to adhere to tools. Copper can be work hardened or hardened by solution heat treatment when alloyed with beryllium. The melting point of copper is 1981°F (1083°C). The specific gravity is 8.89 and its density is 554.7 lb/ft^3.

Beryllium Copper Beryllium copper is an alloy of copper and beryllium that can be hardened by heat treating for making nonsparking tools and other products (Figure 2). Machining of this metal should be done after solution heat treatment and aging, and not when it is in the annealed state. Machining or welding beryllium copper can be very hazardous, if safety precautions are not followed. Machining dust or welding fumes should be

Figure 2. Beryllium copper chisel being used to cut a chip in mild steel.

removed by a heavy coolant flow or by a vacuum exhaust system, and a respirator type of face mask should be worn. The melting point of beryllium is 2345°F (1285°C), its specific gravity 1.847, and its density is 115 lb/ft³.

Brass Brass is an alloy of zinc and copper. Brass colors usually range from white to yellow, and in some alloys red to yellow. Brasses range from gilding metal used for jewelry (95 percent copper, 5 percent zinc) to Muntz metal (60 percent copper, 40 percent zinc) used for bronzing rod and sheet stock. Brasses are easily machined. The melting point of brass ranges from about 1700 to 1900°F (927 to 1038°C). Brass is usually tougher than bronze and produces a stringy chip when machined. The melting point of brasses ranges from 1616 to 1820°F (880 to 993°C), and their densities range from 512 to 536 lb/ft³.

Bronze Bronze is found in many combinations of copper and other metals, but copper and tin are the original elements combined to make bronze. Bronze colors usually range from red to yellow. Phosphor bronze contains 92 percent copper, 0.05 percent phosphorus, and 8 percent zinc. Aluminum bronze is often used in the shop for making bushings or bearings that support heavy loads

Figure 3. Flanged bronze bushing.

(Figure 3). (Brass is not normally used for making antifriction bushings.) The melting point of bronze is about 1841°F (1005°C) and its density is about 548 lb/ft³. Bronzes are usually harder than brasses, but are easily machined with sharp tools. The chip produced is often granular. Some bronze alloys are used as brazing rods.

Die Cast Metals

Finished castings are produced with various metal alloys by the process of die casting. Die casting is a method of casting molten metal by forcing it into a mold. After the metal has solidified, the mold opens and the casting is ejected. Carburetors, door handles, and many small precision parts are manufactured using this process (Figure 4). Die cast alloys, often called "pot metals," are classified in six groups:

1 Tin base alloys
2 Lead base alloys
3 Zinc base alloys
4 Aluminum base alloys
5 Copper, bronze, or brass alloys
6 Magnesium base alloys

The specific content of the alloying elements in each of the many die cast alloys may be found in handbooks or other references on die casting. See Chapter 24 "Casting Processes."

Lead and Lead Alloys

Lead is a heavy metal that is silvery when newly

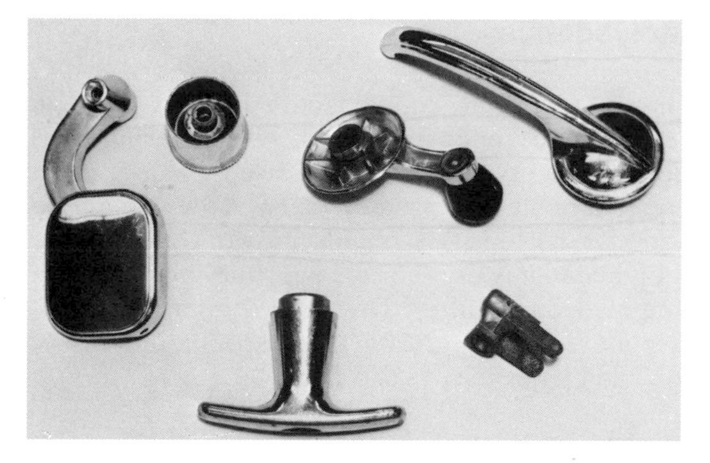

Figure 4. Die cast parts (*Machine Tools and Machining Practices*).

cut and gray when oxidized. It has a high density, low tensile strength, low ductility (cannot be easily drawn into wire), and high malleability (can be easily compressed into a thin sheet).

Lead has a high corrosion resistance and is alloyed with antimony and tin for various uses. It is used as a shielding material for nuclear and X ray radiation, for cable sheathing, and battery plates. Lead is added to steels, brasses, and bronzes to improve machinability. Lead compounds are very toxic and they are also cumulative in the body. Small amounts ingested over a period of time can be fatal. The melting point of lead is 621°F (327°C); its specific gravity is 11.342, and its density is 707.7 lb/ft^3.

A **babbitt metal** is a soft, antifriction alloy metal often used for bearings and is usually tin or lead based (Figure 5). Tin babbitts usually contain

Figure 5. Babbitted pillow block bearings (*Machine Tools and Machining Practices*).

from 65 to 90 percent tin with antimony, lead, and a small percentage of copper added. These are the higher grade and the generally more expensive of the two types. Lead babbitts contain up to 75 percent lead with antimony, tin, and some arsenic making up the difference.

Cadmium base babbitts resist higher temperatures than other tin and lead base types. These alloys contain from 1 to 15 percent nickel or a small percentage of copper and up to 2 percent silver. The melting point of babbitt is about 480°F (249°C).

Magnesium

When pure, magnesium is a soft, silver-white metal that closely resembles aluminum, but is lighter in density. In contrast to aluminum, magnesium will readily burn with a brilliant white light. Cast and wrought magnesium alloys are designated by SAE and ASTM numbers, which may be found in metals reference handbooks such as the *Machinery's Handbook*.

Magnesium is also a much lighter metal than aluminum. The density of magnesium is 108.6 lb/ft^3. Many aircraft parts are made from magnesium alloys. In order to distinguish between magnesium and aluminum, it is sometimes necessary to make a chemical test. Nitric acid will turn magnesium gray; aluminum will remain unchanged. A zinc chloride solution in water, copper sulfate, or 50 percent hydrochloric acid (muriatic acid such as is used in soldering fluxes) will blacken magnesium immediately, but will not change aluminum (Figure 6*a* and 6*b*).

Magnesium, although similar to aluminum in density and appearance, presents some quite different machining problems. Although magnesium chips can burn in air, applying water will only cause the chips to burn more fiercely. Sand or special compounds should be used to extinguish these fires. Thus, when working with magnesium, a water based coolant should never be used. Magnesium can be machined dry when light cuts are taken and the heat is dissipated. Compressed air is sometimes used as a coolant. Anhydrous (containing no water) oils having a high flash point and low viscosity are used in most production work. Magnesium is machined with very high surface speeds and with tool angles similar to those used for aluminum. The melting point of magnesium is 1204°F (651°C), and its specific gravity is 1.741.

Figure 6*a*. Testing to distinguish aluminum from magnesium is done here by applying a weak nitric acid solution. The material at the bottom is blackened when the acid was applied, indicating that it is magnesium. Acid that is applied to the upper block did not blacken, indicating that it is aluminum.

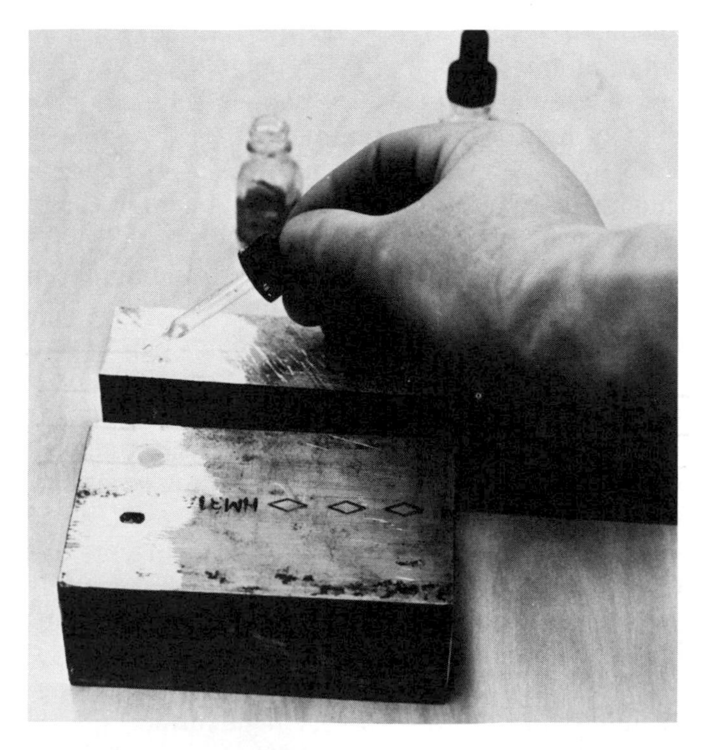

Figure 6*b*. The same test made with copper sulfate. A much darker color may be observed on the magnesium, while the aluminum is not affected.

Molybdenum

As a pure metal, molybdenum is used for high temperature applications and, when machined, it chips like gray cast iron. It is used as an alloying element in steel to promote deep hardening and to increase its tensile strength and toughness. Pure molybdenum is used for filament supports in lamps and in electron tubes. The melting point of molybdenum is 4748°F (2620°C). Its specific gravity is 10.2 and its density is 636.5 lb/ft^3.

Nickel

Nickel is noted for its resistance to corrosion and oxidation. It is a whitish metal used for electroplating and as an alloying element in steel and other metals to increase ductility and corrosion resistance. It resembles pure iron in some ways but has a greater corrosion resistance. Electroplating is the coating or covering of another material with a thin layer of metal, using electricity to deposit the layer.

When spark testing nickel, it throws short orange carrier lines with no sparks or sprigs (Figure 7). Nickel is attracted to a magnet, but becomes nonmagnetic near 680°F (360°C). The melting point of nickel is 2646°F (1452°C), its specific gravity is 8.8, and the density is 549.1 lb/ft^3.

Nickel Base Alloys

Monel is an alloy of 67 percent nickel and 28 percent copper, plus impurities such as iron, cobalt, and manganese. It is a tough, but machinable, duc-

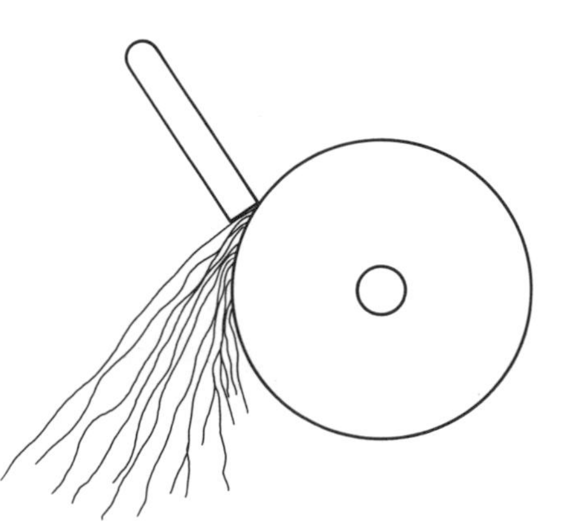

Figure 7. Nickel. Short carrier lines with no forks or sprigs. Average stream length is 10 inches, having an orange color (*Machine Tools and Machining Practices*).

tile, and corrosion resistant alloy. Its tensile strength (resistance of a metal to a force tending to tear it apart) is 70,000 to 85,000 lb/in.^{2} Monel metal is used to make marine equipment such as pumps, steam valves, and turbine blades. On a spark test, monel shoots orange colored, straight sparks about 10 inches long, similar to those of nickel. K-monel contains 3 to 5 percent aluminum and can be hardened by heat treatment.

Monel can be distinguished from nickel with the acid test for copper content. Apply one or two drops of nitric acid to the surface of the sample. Next, rub a clean steel rod, such as a nail, in the solution and observe. If the metal is monel, the steel will become copper colored where it was in the solution; if the steel did not become copper colored, the metal may be nickel. This same test can also be used to determine the copper content of ferrous metals. They will show a brown color instead of a copper color if no copper is present.

Chromel and nichrome are two nickel-chromium-iron alloys used as resistance elements in electric heaters and toasters. Nickel-silver contains nickel and copper in similar proportions to monel, but also contains 17 percent zinc. Other nickel alloys such as inconel are used for parts that are exposed to high temperatures for extended periods.

Inconel, a high temperature and corrosion resistant metal (Figure 8) consisting of nickel, iron, and chromium, can be distinguished from monel by the acid test. Apply one drop of dilute nitric acid to test samples of inconel and monel. The acid will turn blue-green in about one minute on monel, but will show no reaction on inconel. The nickel alloys melting point range is 2425 to 2950°F (1329 to 1621°C).

Precious Metals

Gold has a limited industrial value and is used in dentistry, electronic and chemical industries, and jewelry. In the past, gold has been used mostly for coinage. Gold coinage is usually hardened by alloying with about 10 percent copper. Silver is alloyed with 8 to 10 percent copper for coinage and jewelry. Sterling silver is 92.5 percent silver in English coinage and has been 90 percent silver for American coinage. Silver has many commercial uses, as an alloying element for mirrors, photographic compounds, and electrical equipment. It has a very high electrical conductivity. Silver is used in silver solders that are stronger and have a higher melting point than lead-tin solders.

Platinum, palladium, and iridium, as well as other rare metals are even more expensive than gold. These metals are used commercially because of their special properties such as extremely high resistance to corrosion, high melting points, and high hardness. The melting points of some precious metals are: gold 1945°F (1063°C), iridium 4430°F (2443°C), platinum 3224°F (1773°C), and silver 1761°F (961°C). Gold has a specific gravity of 19.3 and a density of 1204.3 lb/ft^3. Specific gravity

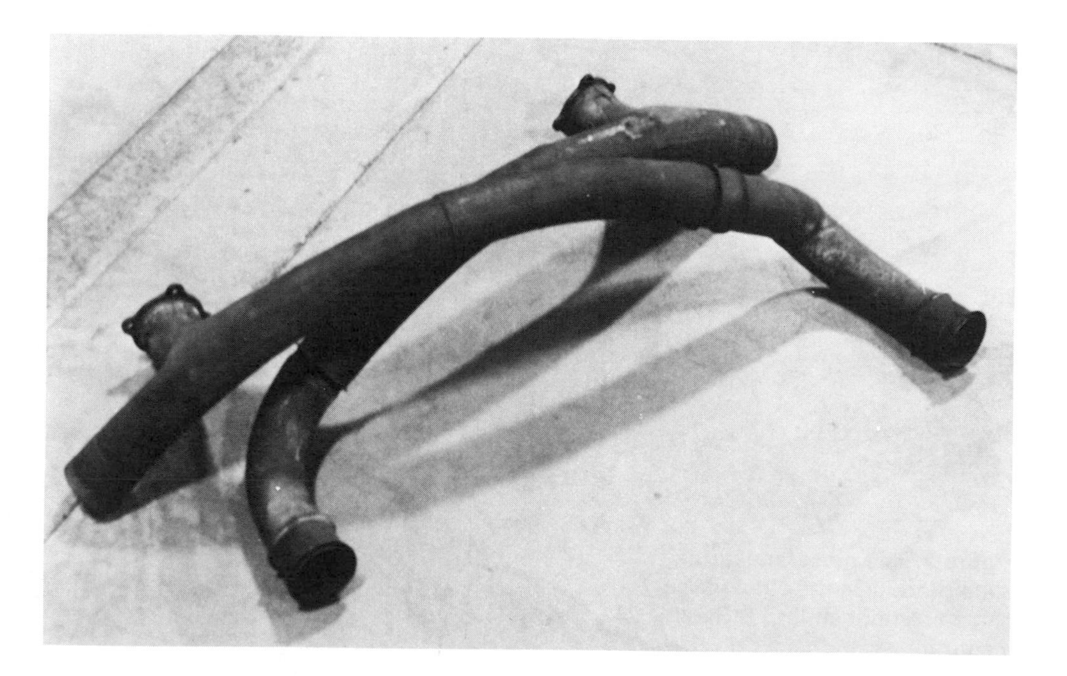

Figure 8. Inconel exhaust manifold for aircraft engines.

for silver is 10.50; its density is about 654 lb/ft³. Platinum is one of the heaviest of metals with a specific gravity of 21.37 and a density of 1333.5 lb/ft³. Iridium is also a heavy metal with a specific gravity of 22.42 weighing 1397 lb/ft³.

Tantalum

Tantalum is a bluish-gray metal that is difficult to machine because it is quite soft and ductile and the chip clings to the tool. It is immune to attack from all corrosive acids except hydrofluoric and fuming sulfuric acids. It is used for high temperature operations above 2000°F (1093°C). It is also used for surgical implants and in electronics. Tantalum carbides are combined with tungsten carbides for cutting tools that have extreme wear resistance. The melting point of tantalum is 5162°F (2850°C). Its specific gravity is 16.6 and it weighs 1035.8 lb/ft³.

Tin

Tin has a white color with a slightly bluish tinge. It is whiter than silver or zinc. Since tin has a good corrosion resistance, it is used to plate steel, especially for the food processing industry (Figure 9). Tin is used as an alloying element for solder, babbitt, and pewter. A popular solder is an alloy of 50 percent tin and 50 percent lead. Tin is alloyed with copper to make bronze. The melting point of tin is 449°F (232°C). The specific gravity of tin is 7.29 and its density is 454.9 lb/ft³.

Titanium

The strength and light weight of this silver-gray metal make it very useful in the aerospace industries for jet engine components, heat shrouds, and rocket parts. Pure titanium has a tensile strength of 60,000 to 110,000 psi, similar to that of steel; by alloying titanium, its tensile strength can be increased considerably. Titanium weighs about half as much as steel and, like stainless steel, is a relatively difficult metal to machine. Machining can be accomplished with rigid setups, sharp tools, slower surface speed, and the use of proper coolants. When spark tested titanium throws a brilliant white spark with a single burst on the end of each carrier. The melting point of titanium is 3272°F (1800°C). Its specific gravity is 4.5 and its density is 280.1 lb/ft³.

Tungsten

Typically, tungsten has been used for incandescent light filaments. It has the highest known melting point (6098°F or 3370°C) of any metal, but is not resistant to oxidation at high temperatures.

Figure 9. The most familiar tin plate product is the steel based tin can (American Iron & Steel Institute).

Tungsten is used for rocket engine nozzles and welding electrodes and as an alloying element with other metals. Machining pure tungsten is very difficult with single point tools, and grinding is preferred for finishing operations. Tungsten carbide compounds are used to make extremely hard and heat resistant lathe tools and milling cutters by compressing the tungsten carbide powder into a briquette and sintering it in a furnace. The specific gravity of tungsten is about 18.8 and it weighs about 1180 lb/ft^3.

Zinc

The familiar galvanized steel is actually steel plated with zinc and is used mainly for its high corrosion resistance. Zinc alloys are widely used as die casting metals. To distinguish zinc from aluminum or magnesium (all three are whitish colored metals), the following spot test is used. Nitric acid does not discolor aluminum, but does react with zinc, changing it to a gray color. Hydrochloric acid reacts violently with both zinc and magnesium, leaving a black deposit on both, and it reacts slightly with aluminum, leaving a clean surface.

Zinc and zinc based die cast metals conduct heat much more slowly than aluminum. The rate of heat transfer on similar shapes of aluminum and zinc is a means of distinguishing between them. The melting point of zinc is 787°F (419°C). Its specific gravity is about 7.10 and it weighs about 440 lb/ft^3.

Zirconium

Zirconium is similar to titanium in both appearance and physical properties. It was once used as an explosive primer and as a flashlight powder for photography. Machining zirconium, like titanium, requires rigid setups and slow surface speeds. Zirconium has an extremely high resistance to corrosion from acids and sea water. Zirconium alloys are used in nuclear reactors, flash bulbs, and surgical implants such as screws, pegs, and skull plates. When spark tested, it produces a spark that is similar to that of titanium. The melting point of zirconium is 3182°F (1750°C). Its specific gravity is 6.4 and its density is 399 lb/ft.^{3}.

Self-Evaluation

1. What advantage do aluminum and its alloys have over steel alloys? What disadvantages?
2. Describe the meaning of the letter "H" when it follows the four digit number that designates an aluminum alloy. The meaning of the letter "T."
3. Name two ways in which magnesium differs from aluminum.
4. What is the major use of copper? How can copper be hardened?
5. What is the basic difference between brass and bronze?
6. Name two uses for nickel.
7. Lead, tin, and zinc all have one useful property in common. What is it?
8. Molybdenum and tungsten are both used in __________ steels.
9. Babbitt metals, used for bearings, are made in what major basic types?
10. What type of metal can be injected under pressure into a permanent mold?
11. Which is stronger, cast or wrought (worked) aluminum?
12. What can be done to avoid building up an edge on the tool bit when machining aluminum?
13. Should a water based coolant be used when machining magnesium? Explain.
14. Should the rake angles on tools for brasses and bronzes be zero, positive, or negative? Explain.
15. How is tungsten used for cutting tools?

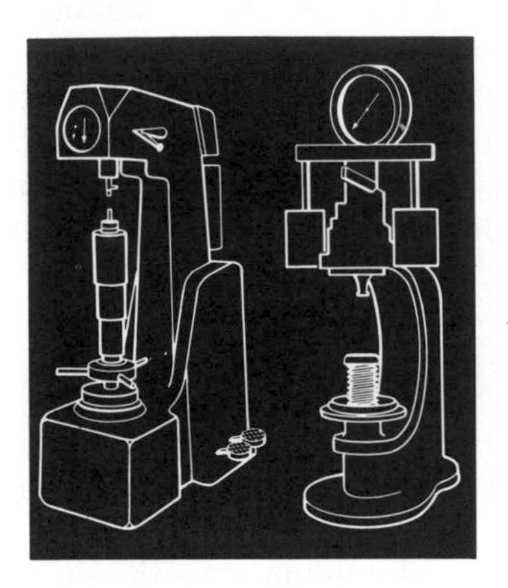

5 Using Rockwell and Brinell Hardness Testers

Rockwell and Brinell hardness testers are the most commonly used types of hardness testers for industrial and metallurgical purposes. Heat treaters, inspectors, and many others in industry often use these machines. This chapter will direct you to a proper understanding and use of both Rockwell and Brinell hardness testers.

OBJECTIVES

After completing this chapter, you will be able to:

1. Make a Rockwell test on three specimens using the correct penetrator, major and minor loads, and scale.
2. Make a Rockwell superficial test on two specimens using the correct penetrator, major and minor loads, and scale.
3. Make a Brinell test on three specimens, read the impression with a Brinell microscope, and determine the hardness number from a table.

INFORMATION

The **hardness** of a metal is its ability to resist being **permanently deformed.** There are three ways that hardness is measured: **resistance to penetration, elastic hardness,** and **resistance to abrasion.** In this chapter you will study the hardness of metals by their resistance to penetration.

Hardness varies considerably from material to material. This variation can be illustrated by making an indentation in a soft metal such as aluminum and then in a hard metal such as alloy tool steel. The indentation could be made with an ordinary center punch and a hammer, giving a light blow of equal force on each of the two specimens (Figure 1). In this case just by visual observation you can tell which specimen is hardest. Of course, this is not a reliable method of hardness testing, but it does show one of the principles of both the Rockwell and Brinell hardness testers: measuring penetration of the specimen by an indenter or penetrator, such as a steel ball or diamond point.

Using the Rockwell Hardness Tester

The Rockwell hardness test is made by applying two loads to a specimen and measuring the difference in depth of penetration in the specimen between the **minor load** and the **major load.** The minor load is used on the standard Rockwell tester to eliminate errors that could be caused by specimen surface irregularities. The minor load is 10 kgf (kilograms of force) when used with 60, 100, or 150 kgf major load and 3 kgf on the superficial tests where the major loads are 15, 30, and 45 kgf.

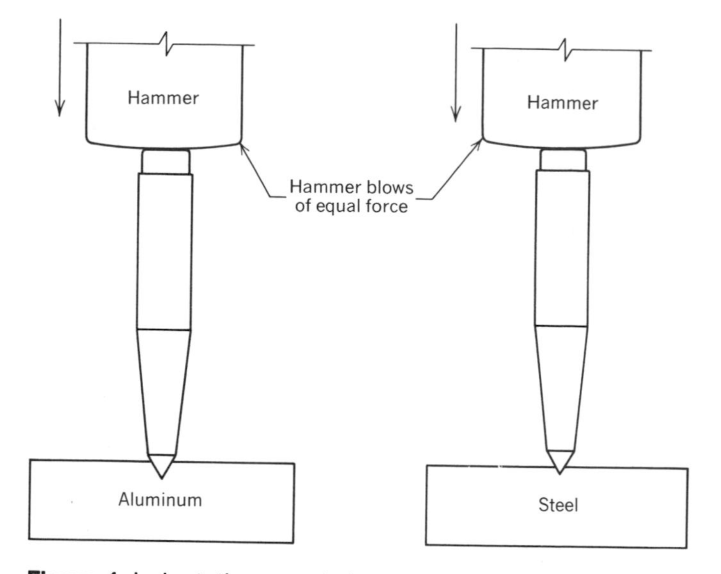

Figure 1. Indentations made by a punch in aluminum and alloy steel *(Machine Tools and Machining Practices).*

The major load is applied after the minor load has forced the indenter to a given depth. The Rockwell Hardness reading is based on the additional depth to which the penetrator is forced by the major load (Figure 2). The depth of penetration is indicated on the dial when the major load is removed. The amount of penetration decreases as the hardness of the specimen increases. Generally, the harder the material, the greater its tensile strength, that is, its ability to resist deformation and rupture when a load is applied. Table 1 compares the hardness by Brinell and Rockwell testers to tensile strength. Also see Table 4 in Appendix 1.

The Rockwell C scale is used for hard metals such as heat treated steels. A file, for example, may measure RC 65, a steel spring about RC 45, and a good knife between RC 52 and 58. Since the C scale below RC 15 is somewhat unreliable, the B scale should be used for these softer metals. Soft iron may read from 0 to 10 on the C scale but would read from 80 to 90 on the B scale.

There are two basic types of penetrators used on the Rockwell tester (Figure 3). One is a sphero-conical diamond called a **brale** that is used only for hard materials; that is, for materials over B-100 such as hardened steel, nitrided steel, and hard cast irons. When the C brale diamond penetrator is used, the recorded readings should be prefixed by the letter "C." The major load used is 150 kgf. The C scale is not used to test extremely hard materials such as cemented carbides or shallow case hardened steels and thin steel. An A brale penetrator is used in these cases and the A scale is used with 60 kgf major load.

The second type of penetrator is a $\frac{1}{16}$ inch diameter **steel ball** that is used for testing material in the range of B-100 to B-0, including such relatively soft materials as brass, bronze, and soft steel. If the ball penetrator is used on materials harder than B-100, there is a danger of flattening the ball. Ball penetrators for use on very soft bearing metals are available in sizes of $\frac{1}{2}$, $\frac{1}{4}$, and $\frac{1}{8}$ inches (Table 2).

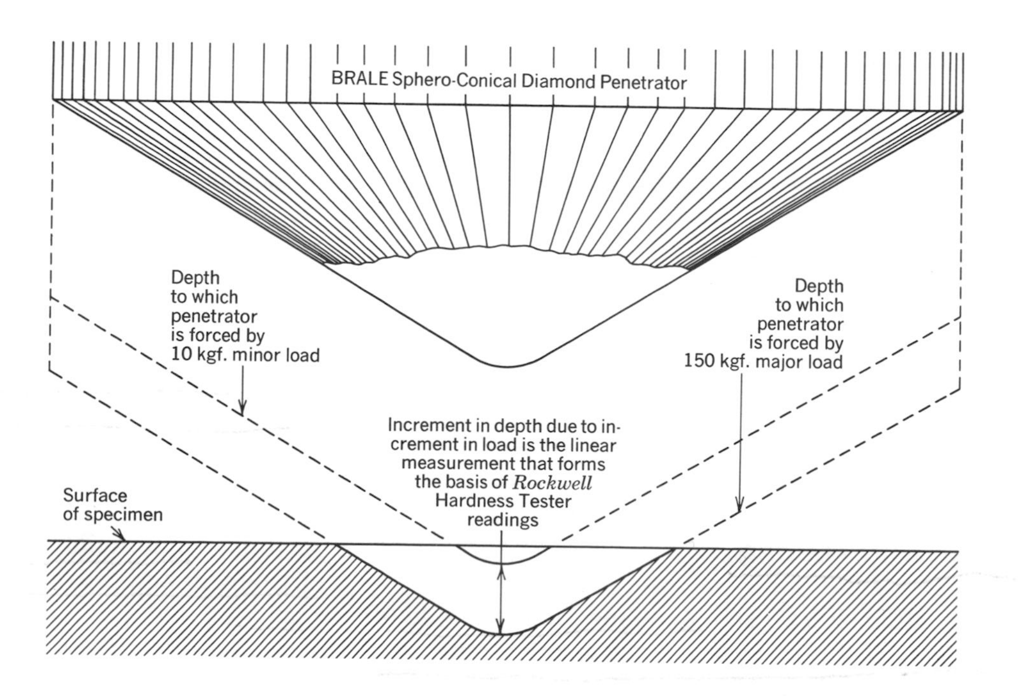

Figure 2. Schematic showing minor and major loads being applied (Wilson Instrument Division of Acco).

Table 1 Hardness and Tensile Strength Comparison Table

Hardness Conversion Table

Brinell		Rockwell			Brinell		Rockwell		
Indentation Diameter, mm	No.*	B	C	Tensile Strength, 1000 psi Approximately	Indentation Diameter, mm	No.*	B	C	Tensile Strength, 1000 psi Approximately
2.25	745		65.3		3.75	262	(103.0)	26.6	127
2.30	712		—		3.80	255	(102.0)	25.4	123
2.35	682		61.7		3.85	248	(101.0)	24.2	120
2.40	653		60.0		3.90	241	100.0	22.8	116
2.45	627		58.7		3.95	235	99.0	21.7	114
2.50	601		57.3		4.00	229	98.2	20.5	111
2.55	578		56.0		4.05	223	97.3	(18.8)	—
2.60	555		54.7	298	4.10	217	96.4	(17.5)	105
2.65	534		53.5	288	4.15	212	95.5	(16.0)	102
2.70	514		52.1	274	4.20	207	94.6	(15.2)	100
2.75	495		51.6	269	4.25	201	93.8	(13.8)	98
2.80	477		50.3	258	4.30	197	92.8	(12.7)	95
2.85	461		48.8	244	4.35	192	91.9	(11.5)	93
2.90	444		47.2	231	4.40	187	90.7	(10.0)	90
2.95	429		45.7	219	4.45	183	90.0	(9.0)	89
3.00	415		44.5	212	4.50	179	89.0	(8.0)	87
3.05	401		43.1	202	4.55	174	87.8	(6.4)	85
3.10	388		41.8	193	4.60	170	86.8	(5.4)	83
3.15	375		40.4	184	4.65	167	86.0	(4.4)	81
3.20	363		39.1	177	4.70	163	85.0	(3.3)	79
3.25	352	(110.0)	37.9	171	4.80	156	82.9	(0.9)	76
3.30	341	(109.0)	36.6	164	4.90	149	80.8		73
3.35	331	(108.5)	35.5	159	5.00	143	78.7		71
3.40	321	(108.0)	34.3	154	5.10	137	76.4		67
3.45	311	(107.5)	33.1	149	5.20	131	74.0		65
3.50	302	(107.0)	32.1	146	5.30	126	72.0		63
3.55	293	(106.0)	30.9	141	5.40	121	69.8		60
3.60	285	(105.5)	29.9	138	5.50	116	67.6		58
3.65	277	(104.5)	28.8	134	5.60	111	65.7		56
3.70	269	(104.0)	27.6	130					

Note 1. This is a condensation of Table 2, Report J417b, SAE 1971 Handbook. Values in ( ) are byond normal range, and are presented for information only.

*Values above 500 are for tungsten carbide ball; below 500 for standard ball.

Note 2. The following is a formula to approximate tensile strength when the Brinell hardness is known:

$$\text{Tensile strength} = BHN \times 500$$

Source. Bethlehem Steel Corporation, *Modern Steels and Their Properties,* Seventh Edition, Handbook 2757, 1972.

Figure 4 points out the parts used in the testing operation on the Rockwell hardness tester. You should learn the names of these parts before continuing with this chapter.

Superficial Testing

After testing sheet metal, examine the underside of the sheet. If the impression of the penetrator can be seen, then the reading is in error and the superficial test should be used. If the impression can still be seen after the superficial test, then a lighter load should be used. A minor load of 3 kgf and a major load of 30 kgf is recommended for most superficial testing. Superficial testing is also used for case hardened and nitrided steel having a very thin case.

A brale marked N is needed for superficial

Figure 3. Brale and ball. These two penetrators are the basic types used on the Rockwell Hardness Tester. (*Note. Brale* is a registered trademark of American Chain & Cable Company, Inc., for sphero-conical diamond penetrators.)

testing as A and C brales are not suitable. Recorded readings should be prefixed by the major load and the letter "N" when using the brale for superficial testing; for example, 30N78. When using the $\frac{1}{16}$ inch ball penetrator, the same as that used for the B, F, and G hardness scales, the readings should always be prefixed by the major load and the letter "T," for example, 30T85. The $\frac{1}{16}$ inch ball penetrator, however, should not be used on material harder than 30T82. Other superficial scales, such as W, X, and Y should also be prefixed with the major load when recording hardness. See Table 3 for superficial test penetrator selection.

The basic **anvils** used for Rockwell testing are shown in Figure 5. Flat anvils are used for specimens with flat surfaces and V-type anvils for round specimens. A spot anvil is used when the tester is being checked on a Rockwell test block. The spot anvil should not be used for checking cylindrical surfaces. The diamond spot anvil (Figure 6) is similar to the spot anvil, but it has a diamond set into the spot. The diamond is ground and polished to a flat surface. This anvil is used **only** with the superficial tester, and then **only** in conjunction with the steel ball penetrator for testing soft metal.

Table 2 Penetrator and Load Selection

Scale Symbol	Penetrator	Major Load, kgf	Dial Figures	Typical Applications of Scales
B	$\frac{1}{16}$ in. ball	100	Red	Copper alloys, soft steels, aluminum alloys, malleable iron, etc.
C	Brale	150	Black	Steel, hard cast irons, pearlitic malleable iron, titanium, deep case hardened steel, and other materials harder than B-100
A	Brale	60	Black	Cemented carbides, thin steel, and shallow case hardened steel
D	Brale	100	Black	Thin steel, medium case hardened steel, and pearlitic malleable iron
E	$\frac{1}{8}$ in. ball	100	Red	Cast iron, aluminum and magnesium alloys, and bearing metals
F	$\frac{1}{16}$ in. ball	60	Red	Annealed copper alloys, thin soft sheet metals
G	$\frac{1}{16}$ in. ball	150	Red	Phosphor bronze, beryllium copper, malleable irons. Upper limit G-92 to avoid possible flattening of ball
H	$\frac{1}{8}$ in. ball	60	Red	Aluminum, zinc, lead
K	$\frac{1}{8}$ in. ball	150	Red	
L	$\frac{1}{4}$ in. ball	60	Red	Bearing metals and other very soft or thin materials. Use the smallest ball and heaviest load that does not give an anvil effect.
M	$\frac{1}{4}$ in. ball	100	Red	
P	$\frac{1}{4}$ in. ball	150	Red	
R	$\frac{1}{2}$ in. ball	60	Red	
S	$\frac{1}{2}$ in. ball	100	Red	
V	$\frac{1}{2}$ in. ball	150	Red	

Source. Wilson Instruction Manual, "Rockwell Hardness Tester Models OUR-a and OUS-a," American Chain & Cable Company, Inc., 1973.

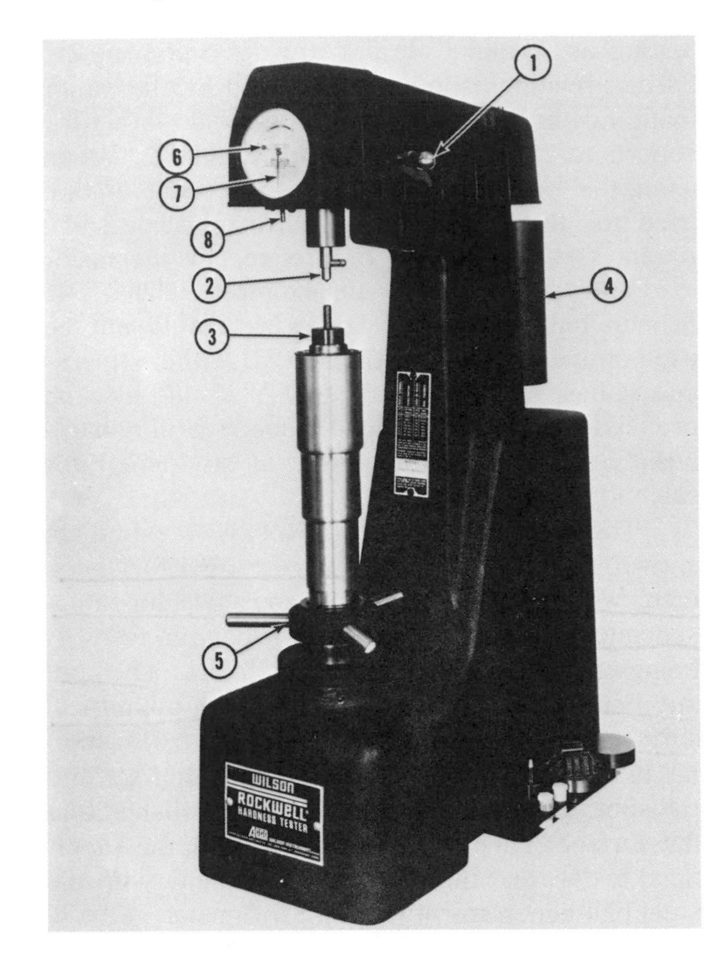

Figure 4. Rockwell Hardness Tester listing the names of parts used in the testing operations (Wilson Instrument Division of Acco).

1. Crank handle
2. Penetrator
3. Anvil
4. Weights
5. Capstan handwheel
6. Small pointer
7. Larger pointer
8. Lever for setting the bezel

(*Note. Rockwell* is a registered trademark of American Chain & Cable Company, Inc., for hardness testers and test blocks.)

(a)

(b) *(c)*

(d)

Figure 5. Basic anvils used with Rockwell Hardness Testers. *(a)* Plane, *(b)* Shallow V, *(c)* Spot, *(d)* Cylindron Jr. (Wilson Instrument Division of Acco).

Surface Preparation and Proper Use

When testing hardness, a surface condition is important for accuracy. A rough or ridged surface caused by coarse grinding will not produce results that are as reliable as a smoother surface. Any rough scale caused by hardening must be removed before testing. Likewise, if the workpiece has been decarburized by heat treatment, the remaining softer "skin" should be ground off the test area.

Figure 6. Diamond spot anvil (Wilson Instrument Division of Acco).

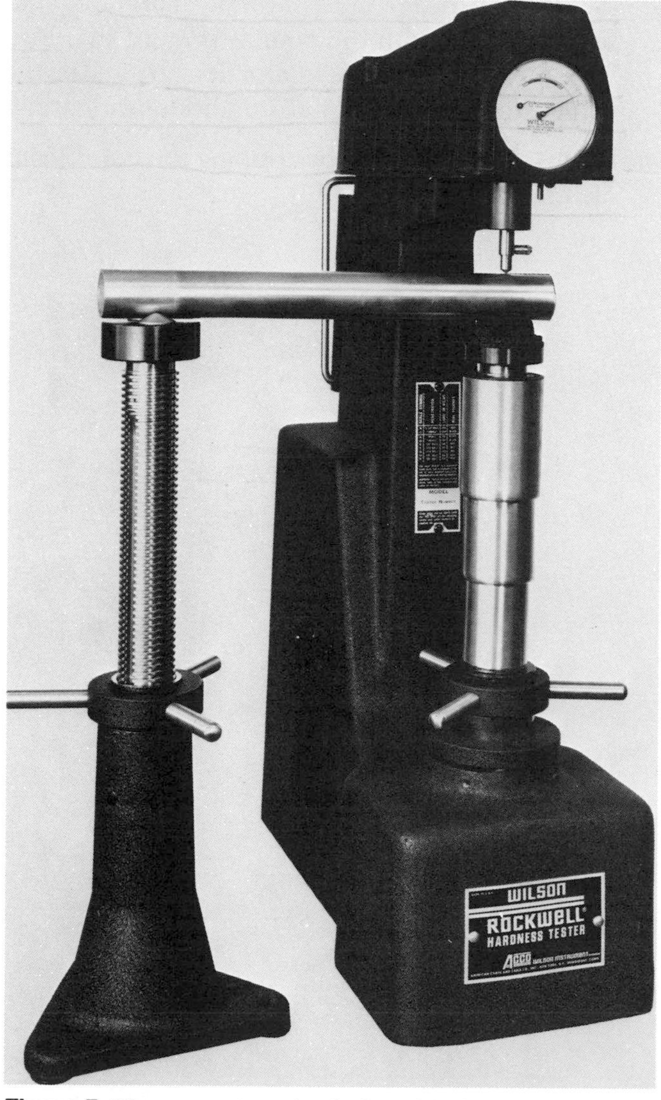

Figure 7. The correct method of testing long, heavy work requires the use of a jack rest (Wilson Instrument Division of Acco).

Error can also result from testing curved surfaces. This effect may be eliminated by grinding a small flat spot on the specimen. Cylindrical workpieces must always be supported in a V-type centering anvil, and the surface to be tested should not deviate from the horizontal by more than five degrees. Tubing is often so thin that it will deform when tested. It should be supported on the inside by a mandrel or gooseneck anvil to avoid this problem.

Several devices are made available for the Rockwell hardness tester to support overhanging or large specimens. One type, called a jack rest (Figure 7), is used for supporting long, heavy parts such as shafts. It consists of a separate elevating screw and anvil support similar to that on the tester. Without adequate support, overhanging work can damage the penetrator rod and cause inaccurate readings.

No test should be made near an edge of a specimen. Keep the penetrator at least $\frac{1}{8}$ inch away from the edge. The test block, as shown in Figure 8, should be used every day to check the calibration of the tester if it is in constant use.

Setting Up the Rockwell Hardness Tester and Making the Test

1. Using Table 2, select the proper weight (Figure 9) and penetrator. Make sure the crank handle is pulled completely forward.

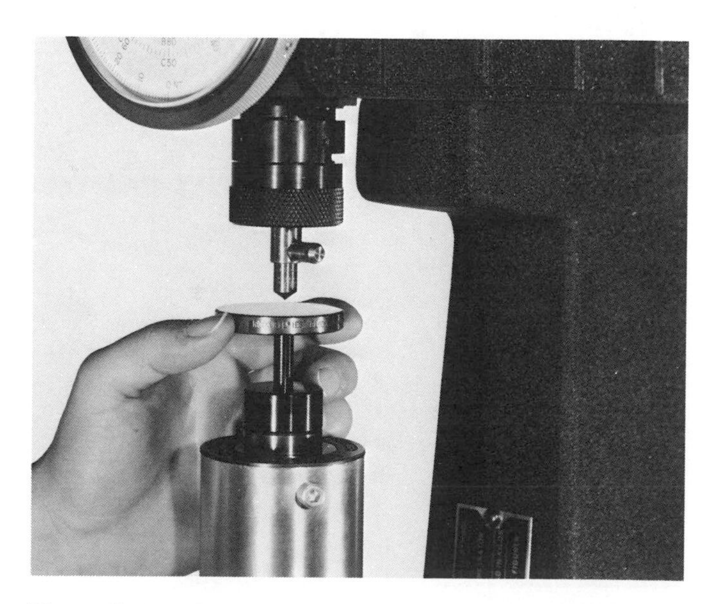

Figure 8. Placing the test block in the machine (Wilson Instrument Division of Acco).

Table 3 Superficial Tester Load and Penetrator Selection

Scale Symbol	Penetrator	Load in Kilograms
15N	Brale	15 kgf
30N	Brale	30 kgf
45N	Brale	45 kgf
15T	$\frac{1}{16}$ in. ball	15 kgf
30T	$\frac{1}{16}$ in. ball	30 kgf
45T	$\frac{1}{16}$ in. ball	45 kgf
15W	$\frac{1}{8}$ in. ball	15 kgf
30W	$\frac{1}{8}$ in. ball	30 kgf
45W	$\frac{1}{8}$ in. ball	45 kgf
15X	$\frac{1}{4}$ in. ball	15 kgf
30X	$\frac{1}{4}$ in. ball	30 kgf
45X	$\frac{1}{4}$ in. ball	45 kgf
15Y	$\frac{1}{2}$ in. ball	15 kgf
30Y	$\frac{1}{2}$ in. ball	30 kgf
45Y	$\frac{1}{2}$ in. ball	45 kgf

Source. Wilson Instruction Manual, "Rockwell Hardness Tester Models OUR-a and OUS-a," American Chain & Cable Company, Inc., 1973.

2 Place the proper anvil on the elevating screw, taking care not to bump the penetrator with the anvil. Make sure that the specimen to be tested is free from dirt, scale, or heavy oil on the underside.

3 Place the specimen to be tested on the anvil in the same way the test block is being placed in Figure 8. Then by turning the handwheel, gently raise the specimen until it comes in contact with the penetrator (Figure 10). Continue turning the handwheel slowly until the small pointer on the dial gage is nearly vertical (near the dot). Now watch the long pointer on the gage and continue raising the work until it is approximately vertical. It should not vary from the vertical position by more than 5 divisions on the dial. Set the dial to zero on the pointer by moving the bezel until the line marked "zero set" is in line with the pointer (Figure 11). You have now applied the minor load. This is the actual starting point for all conditions of testing.

4 Apply the major load by tripping the crank handle clockwise (Figure 12).

Figure 9. Selecting and installing the correct weight (Lane Community College).

Figure 10. Specimen being brought into contact with the penetrator. This establishes the minor load (Wilson Instrument Division of Acco).

Figure 11. Setting the bezel (Wilson Instrument Division of Acco).

5. Wait two seconds after the pointer has stopped moving, then remove the major load by pulling the crank handle forward or counterclockwise.
6. Read the hardness number in Rockwell units on the dial (Figure 13). The black numbers are for the A and C scales and the red numbers are for the B scale. The specimen should be tested in several places and an average of the test results taken, since many materials vary in hardness even on the same surface.

Figure 12. Applying the major load by tripping the crank handle clockwise (Wilson Instrument Division of Acco).

Figure 13. Dial face with reading in Rockwell units after completion of the test. The reading on the dial is RC 55 (Lane Community College).

Using the Brinell Hardness Tester

The Brinell hardness test is made by forcing a steel ball, usually 10 millimeters in diameter, into the test specimen by using a known load weight and measuring the diameter of the resulting impression. The Brinell hardness value is the load divided by the area of the impression, expressed as follows:

$$\text{BHN} = \frac{P}{\frac{\pi D}{2}\left(D - \sqrt{D^2 - d^2}\right)}$$

BHN = Brinell hardness number in kilograms per square millimeter
D = diameter of the steel ball in millimeters
P = applied load in kilograms
d = diameter of the impression in millimeters

A small microscope is used to measure the diameter of the impressions (Figure 14). Various loads are used for testing different materials: 500

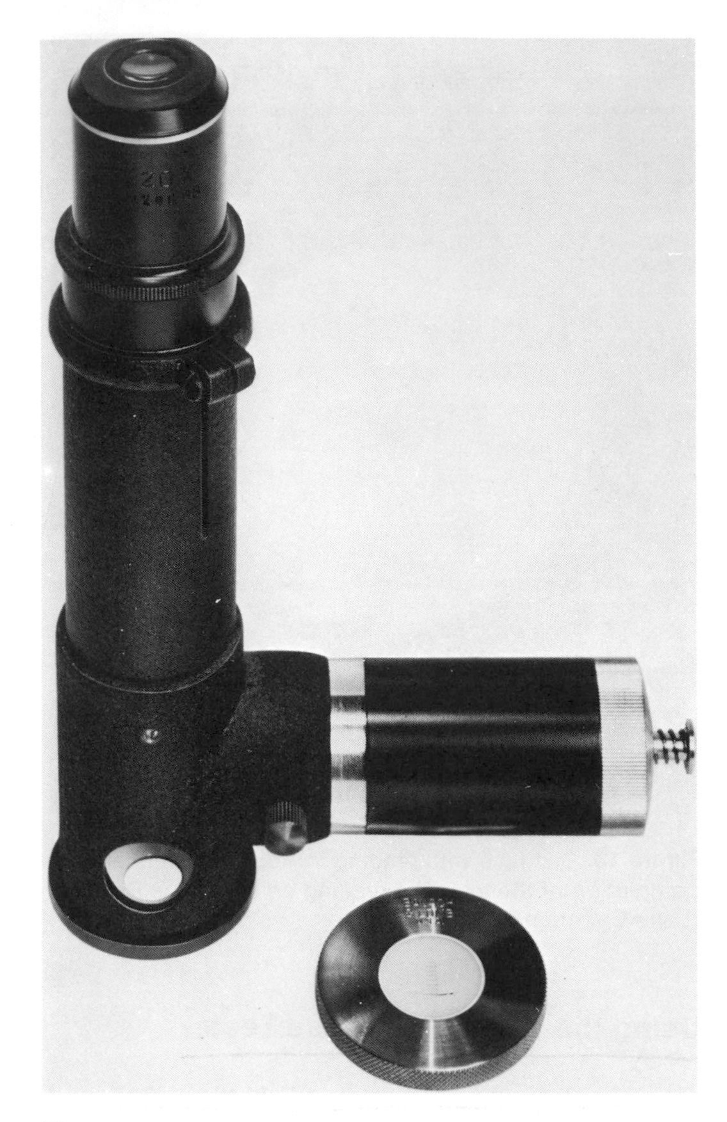

Figure 14. The Olsen Brinell microscope provides a fast, accurate means for measuring the diameter of the impression for determining the Brinell hardness number (Tinius Olsen Testing Machine Company, Inc.).

kilograms for soft materials such as copper and aluminum, and 3000 kilograms for steels and cast irons. For convenience, Table 1 gives the Brinell hardness number and corresponding diameters of impression for a 10 millimeter ball and a load of 3000 kilograms. The Brinell hardness numbers are obtained by reading the indentation diameter in millimeters on the tested specimen and reading the Brinell number in the table under the appropriate load. Conversion to Rockwell hardness numbers is easily done by reading the appropriate Rockwell numbers across from the Brinell numbers. Tensile strengths are also shown. Just as for the Rockwell tests, the impression of the steel ball must not show on the underside of the specimen and the same surface preparation is made before testing. Tests should not be made too near the edge of a specimen. Figure 15 shows an air operated Brinell hardness tester.

The Testing Sequence of the Brinell Hardness Tester

1. The desired load in kilograms is selected on the dial by adjusting the air regulator (Figure 16).
2. The specimen is placed on the anvil. Make sure the specimen is clean and free from burrs. It should be smooth enough so that an accurate measurement can be taken of the impression.
3. The specimen is raised to within $\frac{5}{8}$ inch of the Brinell ball by turning the handwheel.

Figure 15. Air-O-Brinell air-operated metal hardness tester (Tinius Olsen Testing Machine Company, Inc.).

Figure 16. Select load. Operator adjusts the air regulator as shown until the desired Brinell load in kilograms is indicated (Tinius Olsen Machine Company, Inc.).

Figure 18. Release load. As soon as the plunger is depressed, the Brinell ball retracts in readiness for the next test (Tinius Olsen Testing Maching Company, Inc.).

Figure 17. Apply load. Operator pulls out plunger type of control to apply load smoothly to specimen (Tinius Olsen Testing Machine Company, Inc.).

4 The load is then applied by pulling out the plunger control (Figure 17). Maintain the load for 30 seconds for nonferrous metals and 15 seconds for steel. Release the load (Figure 18).

5 Remove the specimen from the tester and measure the diameter of the impression.

6 Determine the Brinell hardness number (BHN) by calculation or by using the table. Soft copper should have a BHN of about 40, soft steel from 150 to 200, and hardened tools from 500 to 600. Fully hardened high carbon steel would have a BHN of 750. A Brinell test ball of tungsten carbide should be used for materials above 600 BHN.

Brinell hardness testers work best for testing softer metals and medium hard tool steels.

Self-Evaluation

1 What one specific category of the property of hardness do the Rockwell and Brinell hardness testers use and measure? How is it measured?

2 State the relationship that exists between hardness and tensile strength.

3 Explain which scale, major load, and penetrator should be used to test a block of tungsten carbide on the Rockwell tester.

4 What is the reason that the steel ball cannot be used on the Rockwell tester to test the harder steels?

5 When testing with the Rockwell superficial tester, is the brale used the same one that is used on the A, C, and D scales? Explain.

6 The $\frac{1}{16}$ inch ball penetrator used for the Rockwell superficial tester is a different one than that used for the B, F, and G scales.
True ______________ False ______________

7 What is the diamond spot anvil used for?

8 How does roughness on the specimen to be tested affect the test results?

9 How does decarburization affect the test results?

10 What does a curved surface do to the test results?

11 On the Brinell tester, what load should be used for testing steel?

12 What size ball penetrator is generally used on a Brinell tester?

Worksheet 1

Objective Make three hardness tests on the standard Rockwell tester and two tests on the superficial tester.

Material Nonferrous metal, hardened steel, tungsten carbide insert, 18 gage sheet steel, and a small part with nitrided steel case.

Procedure

1 Before starting, see that the crank handle is forward.

2 Select the proper penetrator and insert it in the plunger rod.

3 Place the proper anvil on the elevating screw.

4 Select the proper weights.

5 Place the specimen on the anvil.

6 Raise the specimen into contact with the penetrator by turning the capstan handwheel clockwise. Continue motion until the small pointer is near the dot. Continue until the larger pointer is in a vertical position. The minor load is now applied.

7 Turn the bezel of the dial gage until the "SET" line is directly behind the large pointer.

8 Release the weights (major load) by tripping the crank handle clockwise. Do not force the crank. Allow it to come to rest.

9 When the large pointer has come to rest, return the crank handle to the starting position. This removes the major load; the minor load is still applied.

10 Read the scale letter and Rockwell hardness number from the dial gage.

11 Remove the minor load by turning the capstan handwheel counterclockwise to lower the elevating screw and specimen.

12 Remove the specimen. Repeat the test procedure in one or more locations on the test piece.

Conclusions Record your results and show them to your instructor.

Material	Surface Condition	Rockwell Scale	Penetrator	Weight	Hardness Reading 1	2	3	Average
1. Nonferrous metal								
2. Hardened steel								
3. Tungsten carbide insert								
4. 18 gage sheet steel								
5. Nitrided steel part								

Worksheet 2

Objective Make three hardness tests on the Brinell hardness tester.

Materials Nonferrous material (aluminum or copper), heat treated steel, and soft steel.

Procedure

1 Place the specimen on the anvil.

2 Raise the specimen until it touches the ball.

3 Apply the load for 30 seconds. Release.

4 Remove the specimen from the tester and measure the diameter of the impression.

5 Determine the Brinell hardness number.

6 Check for hardness in two or three locations on each specimen.

Conclusions Record your results and show them to your instructor.

Material	Surface	Weight	Reading
1. Nonferrous metal			
2. Hardened steel			
3. Soft steel			

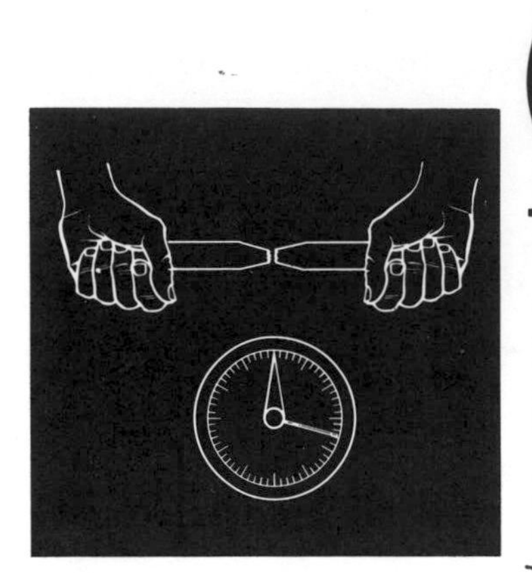

6 The Mechanical and Physical Properties of Metals

The mechanical properties of a material determine its usefulness for a particular job. An understanding of the nature of the mechanical properties and how they are measured will help in the selection of materials in the shop. This chapter introduces the concepts of material strength, elasticity, brittleness, plasticity, creep, corrosion, and others. An understanding of these characteristics will help you be more aware of problems in some severe environments.

All metals conduct both heat and electricity. They also expand when heated and contract when cooled. Conductivity and expansion of metals and other physical properties will be compared and their effect on shop work will be noted as you study this chapter.

OBJECTIVES

After completing this chapter, you will be able to:

1. Correctly define and describe the mechanical and physical properties of metals.
2. Describe the various testing machines and their uses, including the formulas and calculations needed.
3. Prepare specimens for the tensile tester and make tests and evaluations.
4. Prepare specimens for the Izod-Charpy tester and make tests and evaluations.
5. Conduct an experiment to demonstrate differences in thermal conductivity between two metals.
6. Perform an experiment that demonstrates the scaling characteristics of mild steel and stainless steel.

INFORMATION

Before reading this chapter, you should look up the following definitions of the mechanical properties of metals in the glossary: brittleness, ductility, elasticity, hardness, malleability, plasticity, strength, toughness, fatigue strength, oxidation corrosion of metals at high temperature, creep, thermal conductivity, and expansion of metals. Each mechanical property may be tested by mechanical means and evaluated to determine the usefulness of the metal or the correct heat treatment for a particular application. Physical properties such as electrical and thermal conductivity or thermal expansion may also be measured by various means.

Hardness

The property of hardness as tested by Brinell and Rockwell instruments is seen as the resistance to penetration. Microhardness testers that also measure resistance to penetration are used in metallurgical laboratories. An example is the Tukon instrument (Figure 1) that uses the knoop indenter and scale. Some of the advantages are:

1 Built-in microscopic inspection and measurement of the indentation.

2 Greater sensitivity of the tester. Smaller, thinner sections may be tested (Figure 2).

3 A Polaroid* camera may be attached directly onto the microscope (Figure 3) for making micrographs of the test area.

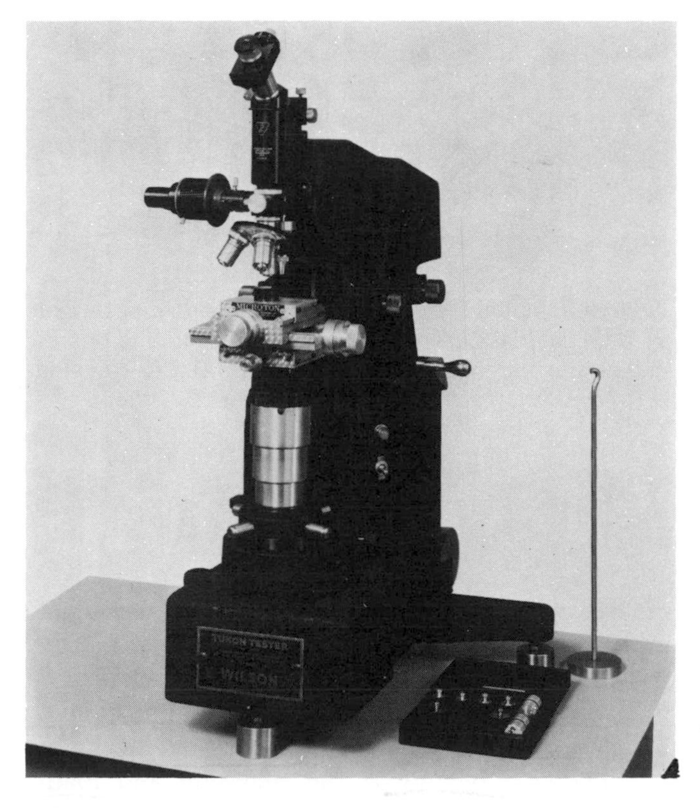

Figure 1. Tukon hardness tester (Wilson Instrument Division of Acco).

Figure 2. Indentations in teeth of hacksaw blade. Note that indentations get smaller at the cold worked edge. In mounting, the two sets of saw teeth were interlaced together to prevent edge from rounding when the sample was polished (Wilson Instrument Division of Acco).

Elastic hardness is measured by an instrument called a Shore Scleroscope (Figures 4*a* and 4*b*), which measures the height of rebound of a small diamond-tipped hammer after it falls by its own weight from a measured height. Hardness as related to resistance to cutting and abrasion is measured in the shop with the file test (Figure 5). This test is used mainly as a means of accepting or rejecting the part for a machining operation.

Strength

The **strength** of a metal is its ability to resist changing its shape or size when external forces are applied. There are three basic types of stresses: **tensile, compressive,** and **shear** (Figure 6). When we consider strength, the type of stress to which the material will be subjected must be known. Steel has equal compressive and tensile strength, but cast iron has low tensile strength and high compressive strength. Shear strength is less than tensile strength in virtually all metals. See Table 1.

The strength of materials is expressed in terms of pounds per square inch (psi). This is called **unit stress** (Figure 7). The unit stress equals the load divided by the total area:

*Polaroid is a trademark of the Polaroid Corporation.

Figure 3. Camera is mounted on the microscope in order to take photomicrographs (Wilson Instrument Division of Acco).

Figure 4*a*. Model D scleroscope. Small parts may be tested with this model. Scleroscopes are very simple to operate and do not mar finished surfaces (Shore U.S.A. Trademark # 757760, Scleroscope U.S.A. Trademark # 723850).

$$\text{Unit stress} = \frac{\text{Load}}{\text{Area}}$$

When **stress** is applied to a metal, it changes shape. For example, a metal in compressive stress will shorten and metal in tension is stretched longer. This change in shape is called **strain** and is expressed as inches of deformation per inch of material length. As stress increases, strain increases by direct proportion within the elastic range.

Metals are "pulled" on a machine called a tensile tester (Figure 8). A specimen of known dimension is placed in the machine and loaded until it breaks (Figure 9). Instruments are sometimes used to make a continuous record of the load

TABLE 1 Material Strength

Material	Modulus of Elasticity (psi)	Allowable Working Unit Stress: Tension	Compression	Shear	Extreme Fiber in Bending
Cast iron	15,000,000	3,000	15,000	3,000	
Wrought iron	25,000,000	12,000	12,000	9,000	12,000
Steel, structural	29,000,000	20,000	20,000	13,000	20,000
Tungsten carbide	50,000,000				

TABLE 1 (Continued)

Material	Elastic Limit (psi): Tension	Compression	Ultimate Strength (psi): Tension	Compression	Shear
Cast iron	6,000	20,000	20,000	80,000	20,000
Wrought iron	25,000	25,000	50,000	50,000	40,000
Steel, structural	36,000	36,000	65,000	65,000	50,000
Tungsten carbide	80,000	120,000	100,000	400,000	70,000

and the amount of strain (Figure 10). This information is put on a graph called a stress-strain diagram (Figure 11).

Elasticity

The ability of a metal to strain under load and then return to its original size and shape when unloaded is called elasticity. The **elastic limit** (proportional limit) is the greatest load a material can withstand and still spring back into its original shape when the load is removed. The elastic limit is easy to identify on any stress-strain diagram. It is the end of the straight line portion of the stress-strain curve (Figure 12).

The **yield point** (yield strength) is a point slightly higher than the elastic limit and, for most cases, they can by considered the same. The allowable (safe) load for a metal in service should be well below the elastic limit or yield strength. Mechanical properties charts in metals handbooks contain data such as yield point, ultimate strength, and hardness.

Plasticity

A perfectly plastic substance such as modeling clay will not return to its original dimensions when the load is removed, regardless of how small the load. Metals undergo plastic flow when stressed

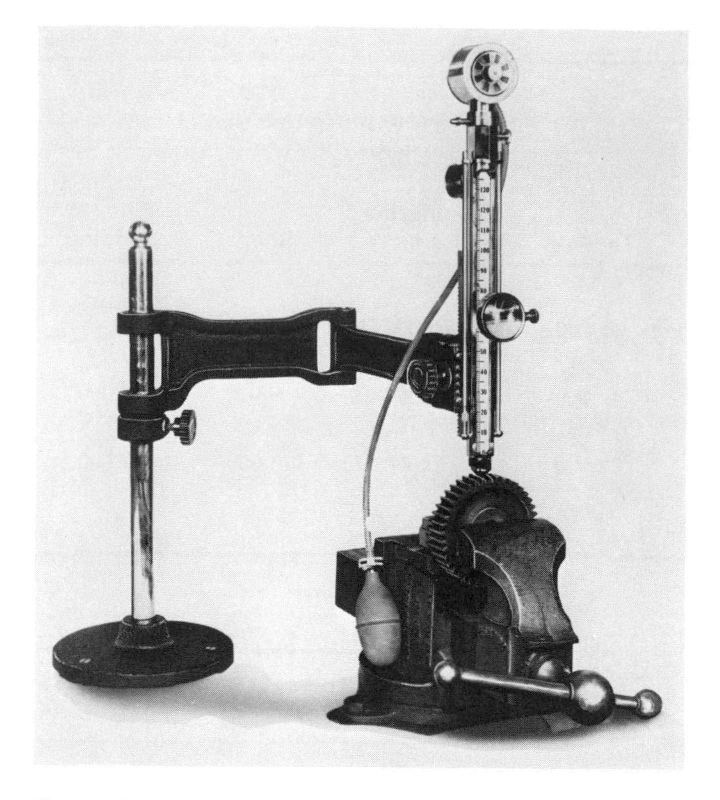

Figure 4*b*. The clamping stand is used with the Model C-2 scleroscope to make a hardness test on a gear (Shore U.S.A. Trademark # 757760, Scleroscope U.S.A. Trademark # 723850).

Figure 5. Before a cut with a machine is taken a file test should be made to determine the machinability of a piece (Lane Community College).

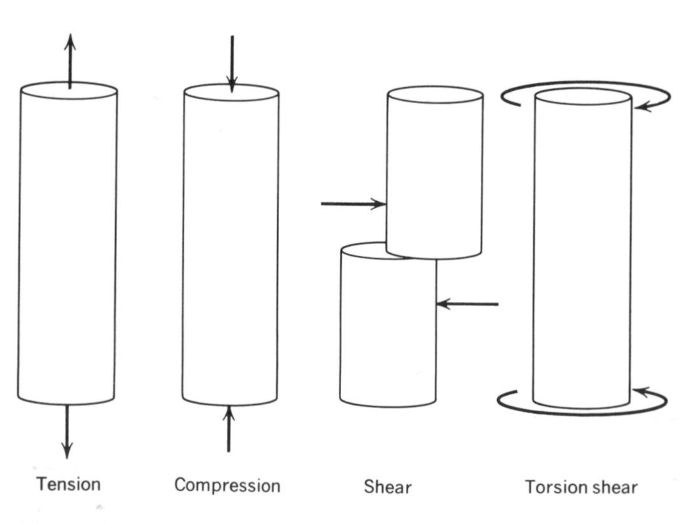

Figure 6. The three types of stresses *(Machine Tools and Machining Practices).*

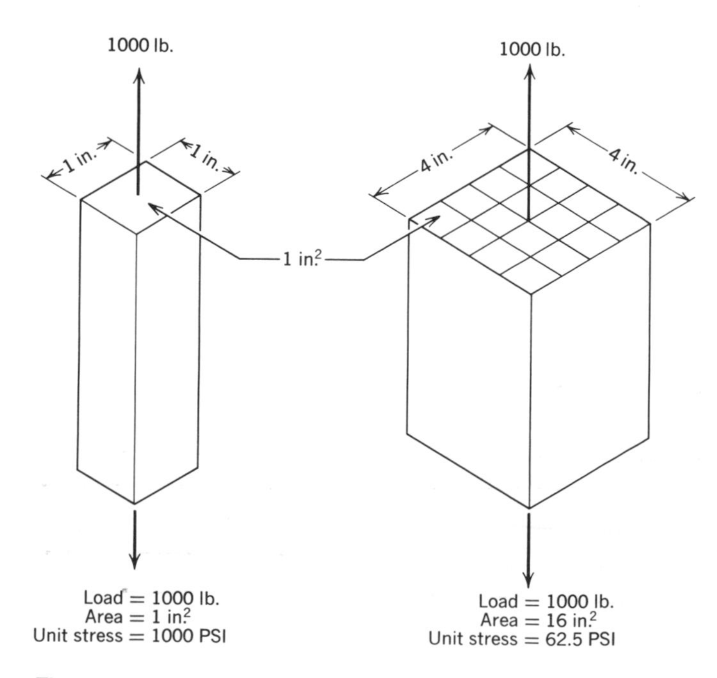

Figure 7. Unit stress *(Machine Tools and Machining Practices).*

Figure 8. Universal tensile tester (Photo courtesy of W. C. Dillon & Company, Inc.).

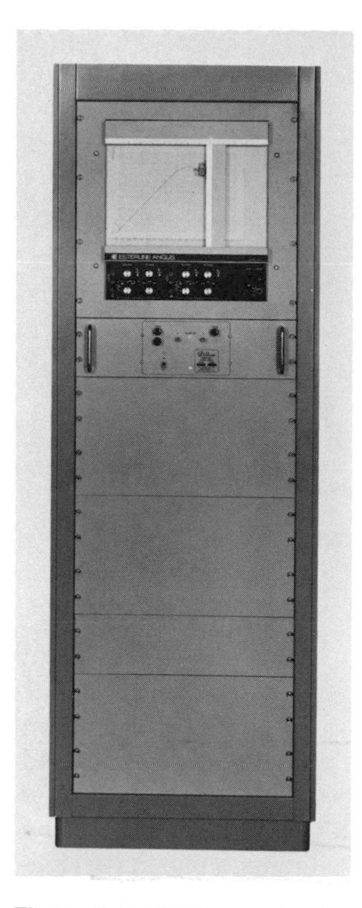

Figure 10a. X-Y Recorder. Stress-strain curves are recorded on machines such as this (Photo courtesy of W. C. Dillon & Company, Inc.).

beyond their elastic limits. For this reason, the area of the stress-strain curve beyond the elastic limit in Figure 11 is called the **plastic range.** It is this property that makes metals so useful. When enough force is applied by rolling, pressing, or hammer blows, metals can be formed when hot or cold into useful shapes. Many metals tend to work harden when cold formed, which in most cases increases their usefulness. They must be annealed for further cold work when a certain limit has been reached.

Brittleness

A material that will not deform plastically under load is said to be **brittle.** Excessive cold working

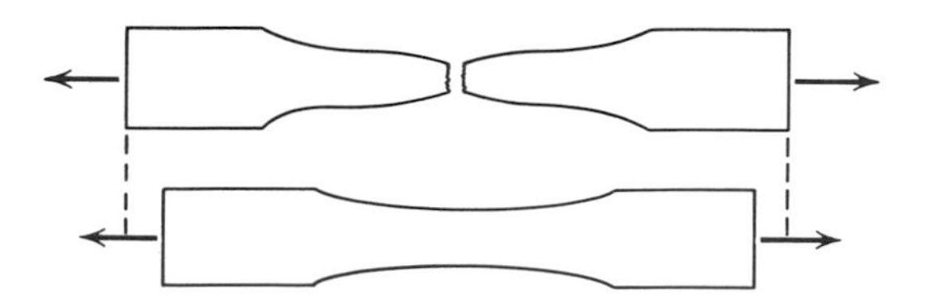

Figure 9. A tensile specimen of a ductile material before pull and after pull *(Machine Tools and Machining Practices).*

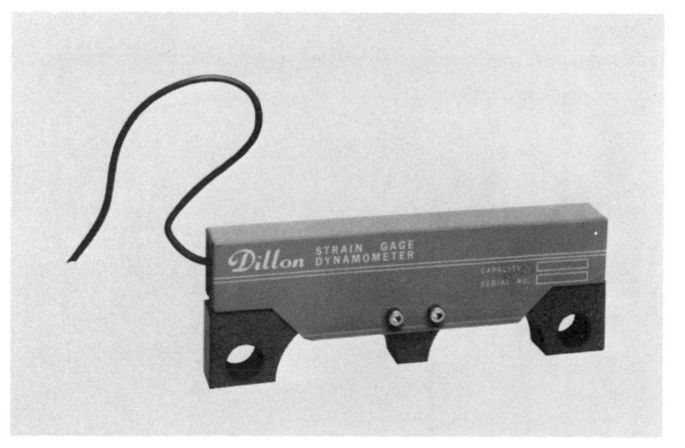

Figure 10b. The strain gage used in conjunction with the X-Y recorder (Photo courtesy of W. C. Dillon & Company, Inc.).

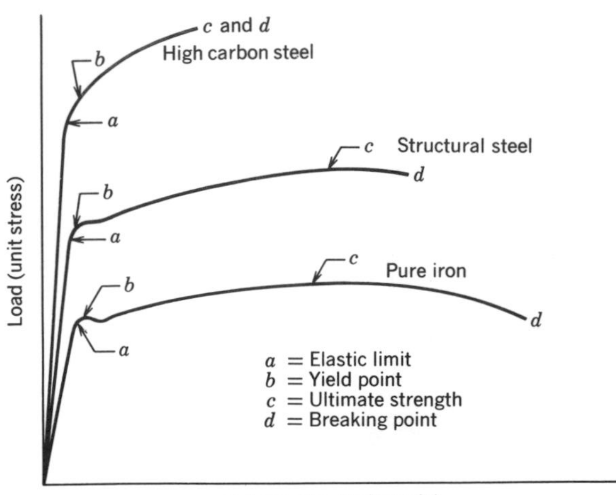

Figure 11. Stress-strain diagram. Several stress-strain curves are shown on this diagram *(Machine Tools and Machining Practices).*

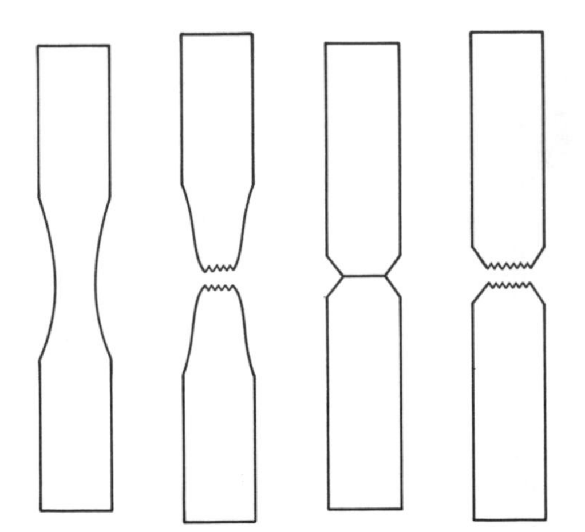

Figure 13. Notching and its effect on plasticity in an otherwise ductile metal can behave in a brittle manner when a stress raiser is present *(Machine Tools and Machining Practices).*

causes brittleness and loss of ductility. Cast iron does not deform plastically under a breaking load and is therefore brittle.

A very sharp "notch" that concentrates the load in a small area can also reduce plasticity (Figure 13). Notches are common causes of premature failure in parts. Weld undercut, sharp shoulders on machined shafts, and sharp angles on forgings and castings are examples of unwanted notches (stress raisers).

Stiffness (Modulus of Elasticity)

Stiffness is expressed by the **modulus of elasticity**, also called Young's Modulus. Within the elastic range, if the stress is divided by the corresponding strain at any given point, the result will be the modulus of elasticity for that material. Therefore, the modulus of elasticity is represented by the slope of the stress-strain curve below the elastic limit.

$$\text{Modulus of elasticity in psi} = \frac{\text{Stress}}{\text{Strain}}$$

The modulus of elasticity for some common materials is given in Table 1.

For an example of stiffness, two rods of equal dimensions are suspended horizontally on one end

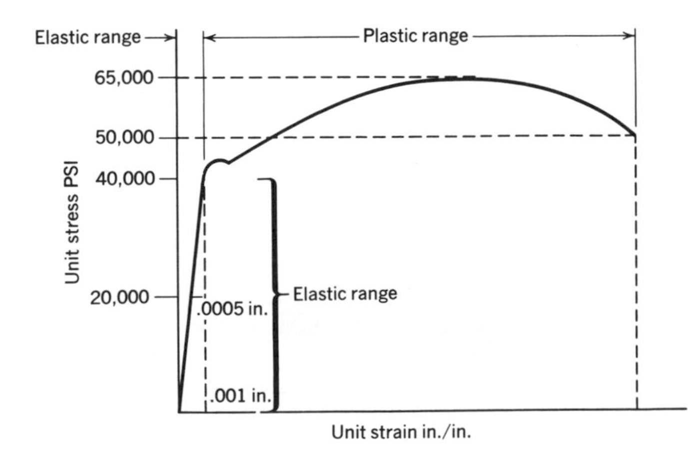

Figure 12. Stress-strain diagram for a ductile steel *(Machine Tools and Machining Practices).*

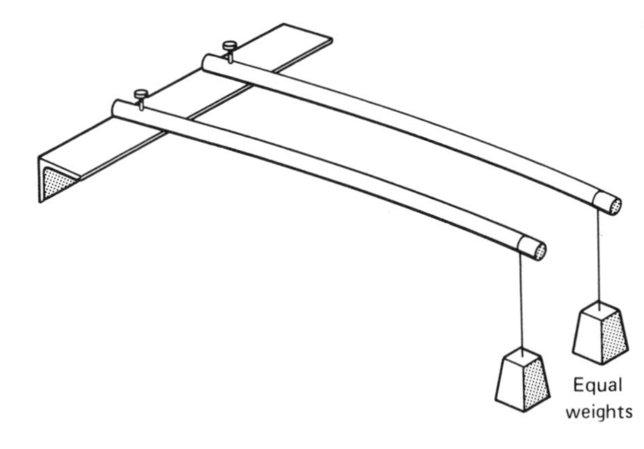

Figure 14. Deflection of two steel bars having the same modulus of elasticity. One is made of hardened alloy steel and the other is of low carbon steel. Both bars deflect the same amount.

with equal weights hanging on the other end. Of course, both rods will deflect the same amount if they are made of the same steel. Even if one rod were made of mild steel and the other of hardened tool steel, both would still deflect the same amount within the elastic range (Figure 14). The reason is that all steels have about the same modulus of elasticity. If one rod were made of tungsten carbide, the results would be quite different because the carbide rod would deflect much less than the steel since its modulus of elasticity is considerably higher than that of steel.

Ductility

The property that allows a metal to deform permanently when loaded in tension is called **ductility.** Any metal that can be drawn into a wire is ductile. Steel, aluminum, gold, silver, and nickel are examples of ductile metals.

The tensile test is used to measure ductility. Tensile specimens are measured for area and length between gage marks before and after they are pulled. The **percent of elongation** (increase in length) and the **percent of reduction** in area (decrease of area at the narrowest point) are the values for ductility. A high percent elongation (about 70 percent) and reduction in area indicate a high ductility. The method for calculating these values is shown in Figure 15. A metal showing less than 20 percent elongation would have low ductility.

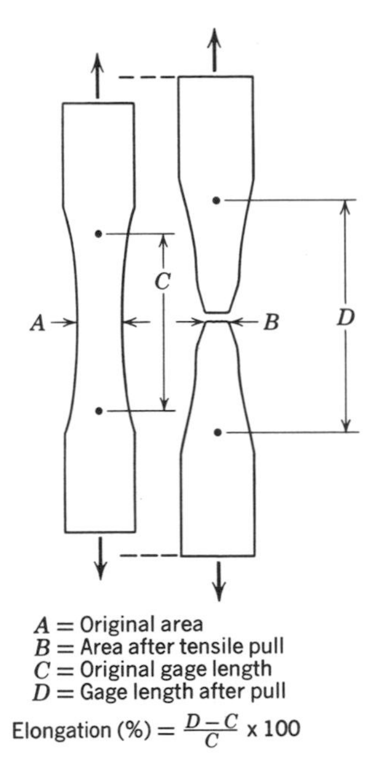

A = Original area
B = Area after tensile pull
C = Original gage length
D = Gage length after pull

Elongation (%) $= \frac{D-C}{C} \times 100$

Reduction in area (%) $= \frac{A-B}{A} \times 100$

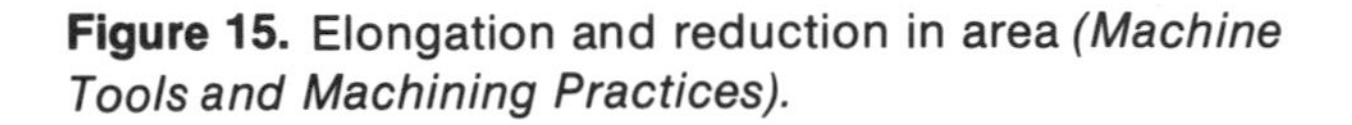

Figure 15. Elongation and reduction in area *(Machine Tools and Machining Practices).*

Malleability

The ability of a metal to deform permanently when loaded in compression is called **malleability.** Metals that can be hammered or rolled into sheets are malleable. Most ductile metals are also malleable, but some very malleable metals such as lead are not very ductile, and could not be drawn into wire easily. Metals with low ductility, such as lead, can be extruded or pushed out of a die to form wire and other shapes. Some very malleable metals are lead, tin, gold, silver, iron, and copper.

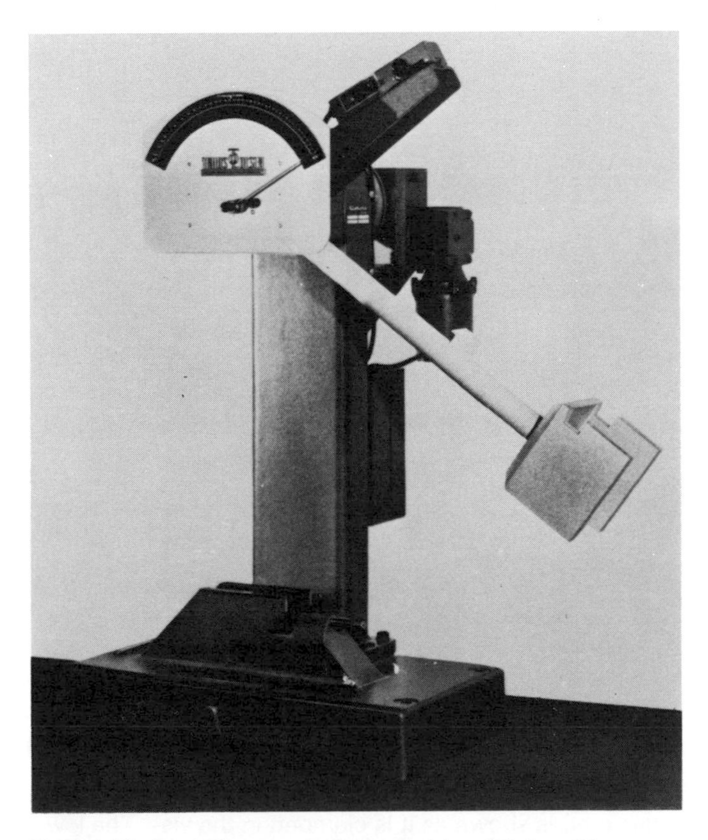

Figure 16. Izod-Charpy testing machine (Tinius Olsen Testing Machine Company, Inc.).

Notch Toughness

Notch toughness is the ability of a metal to resist rupture from impact loading when there is a notch

Figure 17*a*. The vertical mounting position for the Izod test specimen is shown as it is clamped in the vise. The lower photo shows the underside of the striking bit (Tinius Olsen Testing Machine Company, Inc.).

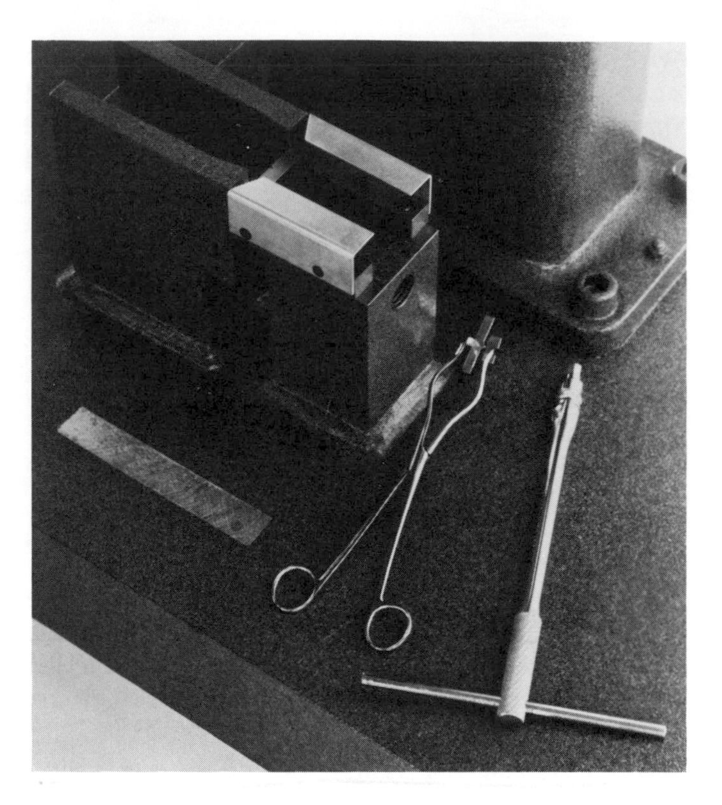

Figure 17*b*. The horizontal mounting position for the Charpy test specimens is shown as it is clamped in the vise. The lower photo shows the underside of the striking bit (Tinius Olsen Testing Machine Company, Inc.).

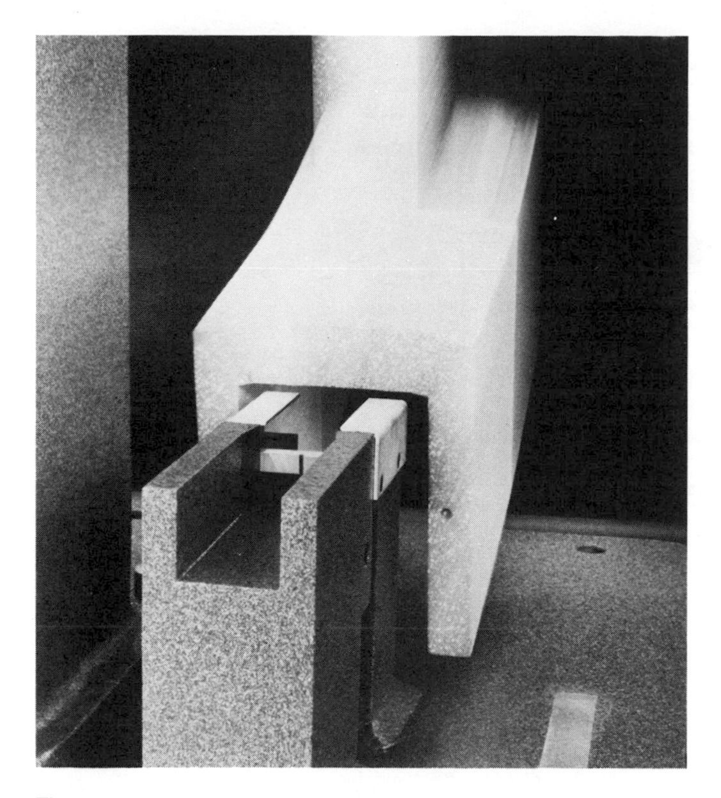

Figure 18. Base showing leveling pads. Hammer is dropping and about to strike the specimen (Tinius Olsen Testing Machine Company, Inc.).

or stress raiser present. The device used to measure toughness is called the Izod-Charpy testing machines (Figure 16). The method of supporting and differences in specimens distinguishes between the Izod and Charpy methods (Figures 17*a* and 17*b*), but the results are practically the same. The base (Figure 18) has two leveling pads for adjusting the machine. The hammer straddles the anvil support and the striking bit is in the hammer. Standard test specimens (Figure 19) are used for either the Izod or Charpy test. The testing machine consists of a vise where the test specimen is clamped. A weight on a swinging arm is allowed to drop (Figure 20). Note that the specimens are of standard geometry. The pendulum drops, strikes the specimen, and continues to swing forward. But it will not swing up as high as the starting position. The difference between the pendulum's beginning height and ending height indicates how much energy was absorbed in breaking the specimen. This energy is measured in foot-pounds. Tough metals absorb more ft-lb of energy than brittle metals and, therefore, the pendulum will not swing so far for tough metals.

Fatigue

When metal parts are subjected to repeated loading and unloading, they may fail at stresses far below their yield strength with no sign of plastic deformation. This is called a **fatigue failure.** When designing machine parts that are subject to vibration or cyclic loads, fatigue strength may be more important than ultimate tensile or yield strength.

Fatigue-testing machines put specimens through many cycles of loading at a given stress. The results of repeated tests at different stresses can be plotted on a graph called a stress cycle diagram (Figure 21). The **fatigue limit** is the maximum load in pounds per square inch that can be applied an infinite number of times without causing failure. But 10 million loading cycles are usually considered enough to establish fatigue limits.

Fatigue life can be enhanced by smooth design. Avoiding undercut in welds, sharp corners and shoulders, and deep tool marks in machined parts will help eliminate stress raisers (notches) and thereby increase fatigue life.

Creep Strength

Creep is a continuing slow plastic flow at a stress below the yield strength of a metal. Creep is usually associated with high temperatures, but it does occur to some extent at normal temperature. Low temperature creep can take months or years to alter machine parts that are habitually left in a stressed condition (Figures 22*a* and 22*b*). As the temperature increases, creep becomes more of a problem. Creep strength for a metal is given in terms of an allowable amount of plastic flow (creep) per 1000 hour period. Table 2 gives creep strength for some alloys. Note that strengths are given for 1 percent creep per 10,000 hours at 800°F (427°C) and for 1 percent creep per 100,000 hours at 1200°F (649°C). Stress to failure is also given.

Scaling

When metals are subjected to high temperatures, they often form heavy coatings of oxide. This coating is called **scale.** If the metal has a low resistance to scaling and if it is allowed enough time, it will eventually be entirely converted to scale. Resistance to scaling is usually achieved in steels by the

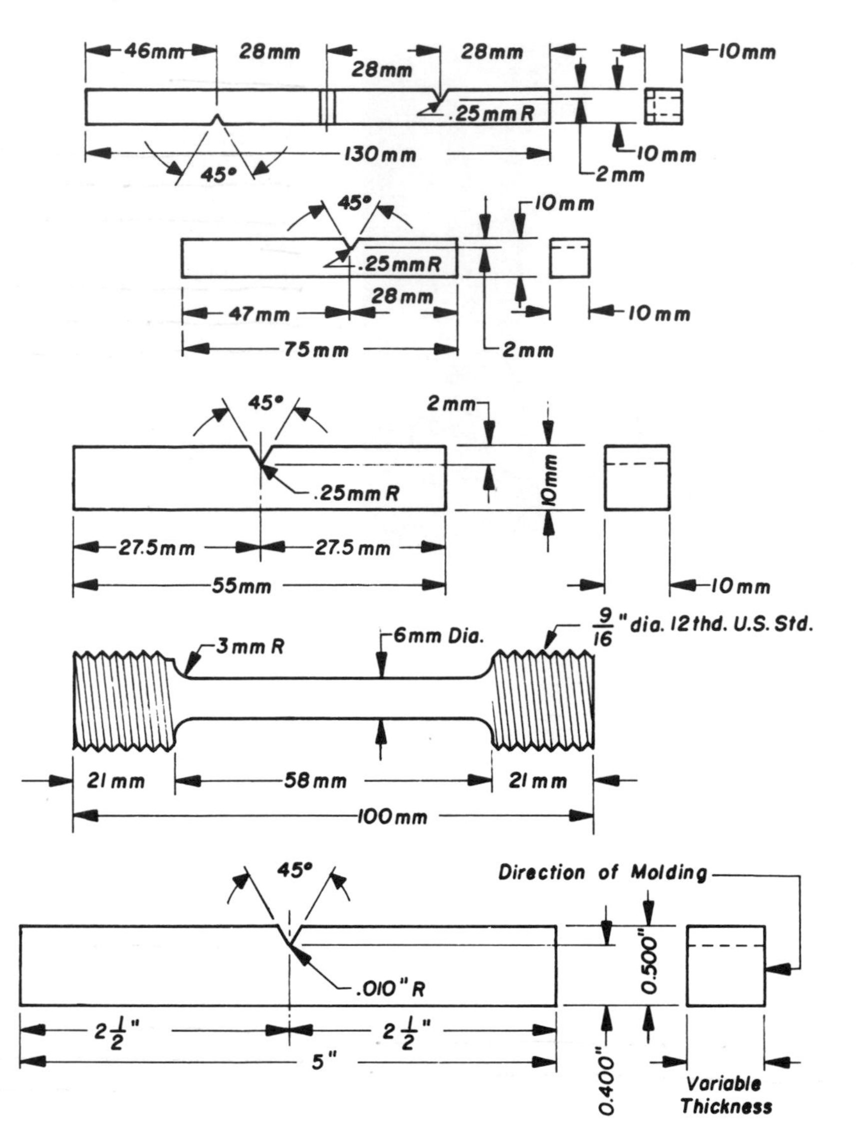

Figure 19. Test specimen specifications for Charpy and Izod test (Tinius Olsen Testing Machine Company, Inc.).

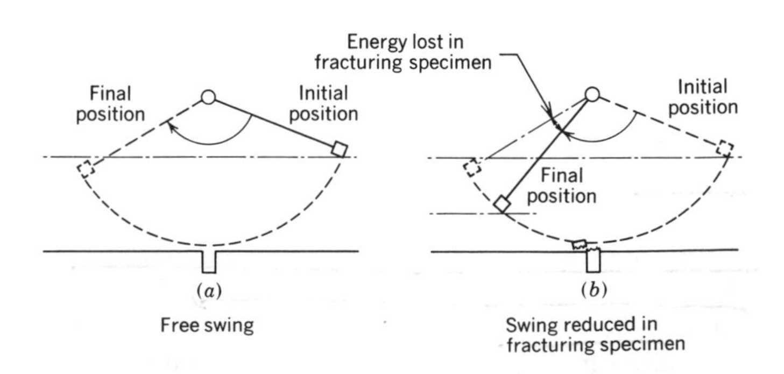

Figure 20. The method by which (Izod) impact values are determined *(Machine Tools and Machining Practices)*.

addition of chromium or nickel or both. These elements tend to form an oxide skin with a high melting temperature. This skin protects the metal and retards further scaling. For example, some stainless steels resist scaling at high temperatures. As temperatures increase in metals, ductility, malleability, and plasticity increase while the modulus of elasticity, hardness, and strength decrease.

Corrosion Resistance

The ability of a metal to resist attack by chemical action is called corrosion resistance. Some metals corrode easily. Iron, for instance, needs only water to corrode (rust). Other metals like gold show a strong resistance to almost any chemical environment. Some more common metals that are noted for corrosion resistance are nickel, zirconium,

Table 2 Creep Strengths for Several Alloys

Alloy	70°F Tensile Strength (psi)	CREEP STRENGTH (PSI) 800°F—Stress for 1 percent Elongation per 10,000 h	1200°F—Stress for 1 percent Elongation per 100,000 h	1500°F—Stress to Failure
0.20 percent carbon steel	62,000	35,100	200	1500
0.50 percent molybdenum 0.08 percent to 20 percent carbon steel	64,000	39,000	500	2600
1.00 percent chromium 0.60 percent molybdenum 0.20 percent C steel	75,000	40,000	1500	3500
304 stainless steel 19 percent chromium 9 percent Nickel	85,000	28,000	7000	15,000

chromium, cobalt, and manganese. Some alloys such as stainless steels have an excellent ability to resist attack by the environment. This is due to the formation of a thin film of protective oxide on the metal surface. More information on corrosion is given in Chapter 23, "Corrosion of Metals."

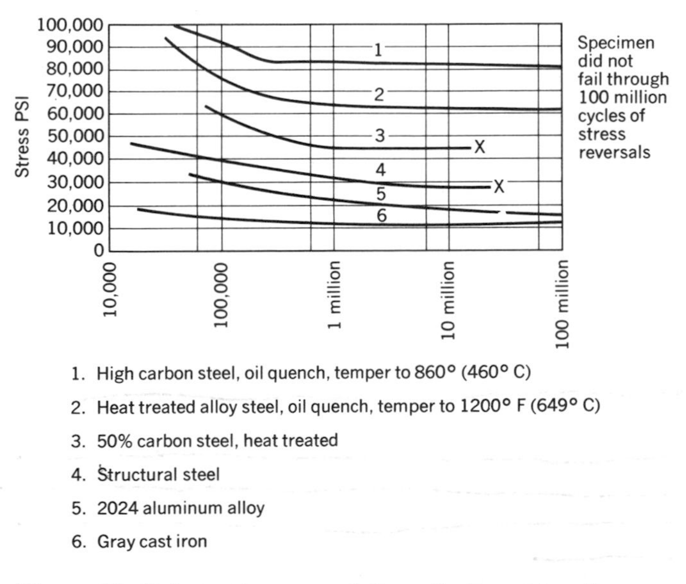

Figure 21. Relation between fatigue limit and tensile strength. The fatigue limit of steel is approximately 45 to 50 percent of its tensile strength up to about 200,000 pounds. Repeated stresses in excess of this fatigue limit causes ultimate failure *(Machine Tools and Machining Practices).*

Metals at Low Temperatures

As the temperature decreases, the strength, hardness, and modulus of elasticity increase for almost all metals. The effect of temperature drop on ductility separates metals into two groups: those that become brittle at low temperatures and those that remain ductile. Figure 23 shows examples of these two groups.

Metals of the group that remain ductile show a slow steady decrease in ductility with a drop in temperature. Metals in the group that become brittle at low temperatures show a temperature range where ductility and, most important, toughness drop rapidly. This range is called the transition zone. The Izod-Charpy impact test is the most common method used to determine the transition zone of metals. When the notch-bar specimens show half brittle failures and half ductile failures, the transition temperature has been reached. When parts for low temperature service are designed, the operating temperature should be well above the transition temperature. Nickel is one of the most effective alloying elements for lowering the transition temperature of steels. The following are some examples of operating temperatures for a few alloys and metals.

Figure 22*a*. This milling machine was left with the table positioned to one side with a heavy milling machine vise weighing it down. Leaving a machine in this position can cause the table to deflect through creep, causing binding of the ways (Lane Community College).

Figure 22*b*. The machine should always be left in this position when not in use (Lane Community College).

1 For operating temperatures as low as −50°F (−46°C)
   a. Killed low-carbon steel
   b. Three percent nickel low-carbon steel

2 For operating temperatures as low as −150°F (−101°C)
   a. Six percent nickel low-carbon steel
   b. Stainless steels with 8 percent nickel

3 For operating temperatures below −150°F (−101°C)
   a. Stainless steels with 8 percent nickel or more
   b. Nine percent nickel steel
   c. FCC metals such as aluminum or monel

Expansion and Conductivity of Metals (Physical Properties of Metals)

Metals conduct heat better than nonmetals. Silver conducts heat the best of all metals. This ability to conduct heat and the ability to conduct electricity are related. Since silver is the best heat conductor, it is also the best electrical conductor. Figure 24 compares the thermal conductivity of some common metals and alloys. Note that in all cases the pure metals are better conductors than their alloys. Pure copper would be a better choice for electrical

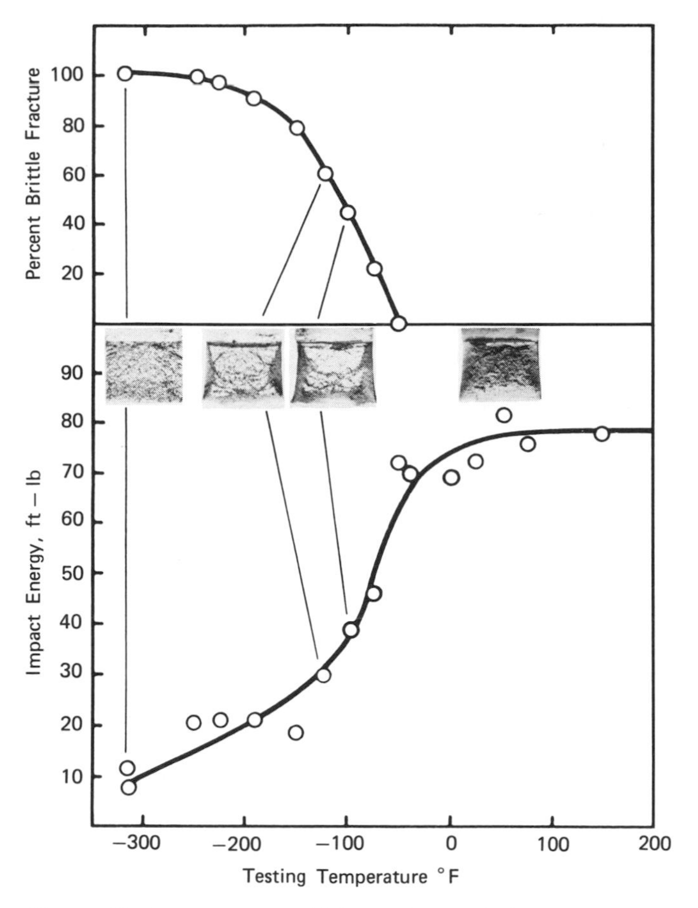

Figure 23. Appearance of Charpy V-notch fractures obtained at a series of testing temperatures with specimens of tempered martensite of hardness around Rockwell C 30 (Courtesy Republic Steel Corporation).

wiring than a copper alloy. A pure aluminum automobile radiator would conduct heat away from the water inside better than an aluminum alloy radiator.

Thermal Expansion

In almost all cases solids become larger when heated and smaller when cooled. Each substance expands and contracts at a different rate. This rate is expressed in inches per inch per degree Fahrenheit and is called the coefficient of thermal expansion. Figure 25 compares the coefficient of thermal expansion of some common metals and alloys.

Some Cases Where a Knowledge of Thermal Expansion Helps

Knowing the thermal expansion coefficient of steel allows the engineer to calculate the sizes of the expansion joints in bridges and other steel structures. Heat treaters must be aware of differing expansion and contraction rates when heating and quenching steels. Internal expansion rate is often lower than the external rate when a piece of steel is heated rapidly. The stresses caused by uneven heating can cause cracking in metals with low ductility.

If a mechanic must remove a bronze bushing from a housing, or if heat is inadvertently applied to a bushing area such as by welding in the vicinity of the bushing, it may be loosened as a result of applying heat (Figure 26). The thermal expansion coefficient of bronze is almost twice that of steel. If the bushing and housing are heated, the bronze will try to expand at almost twice the rate of steel for a given amount of heat. The steel restricts the bronze from expansion. The bronze then is stressed above its elastic limit and into the plastic range, where the bushing is deformed to a slightly smaller diameter. When both steel and bronze are

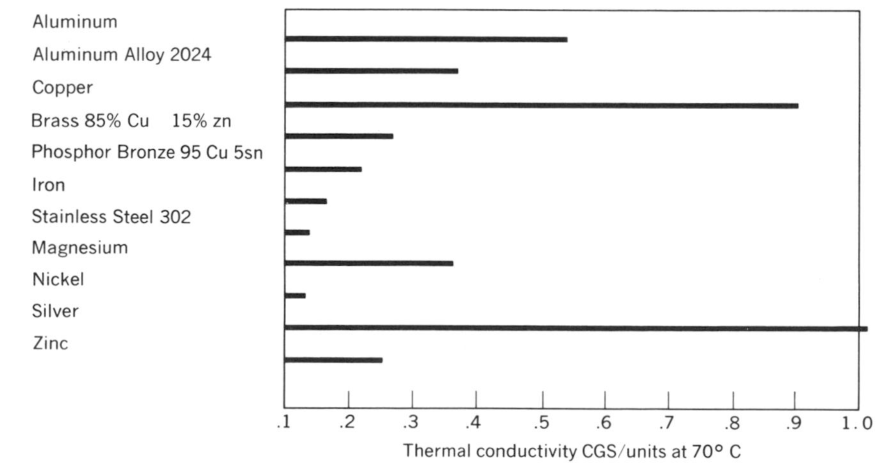

Figure 24. Comparison of thermal conductivity *(Machine Tools and Machining Practices).*

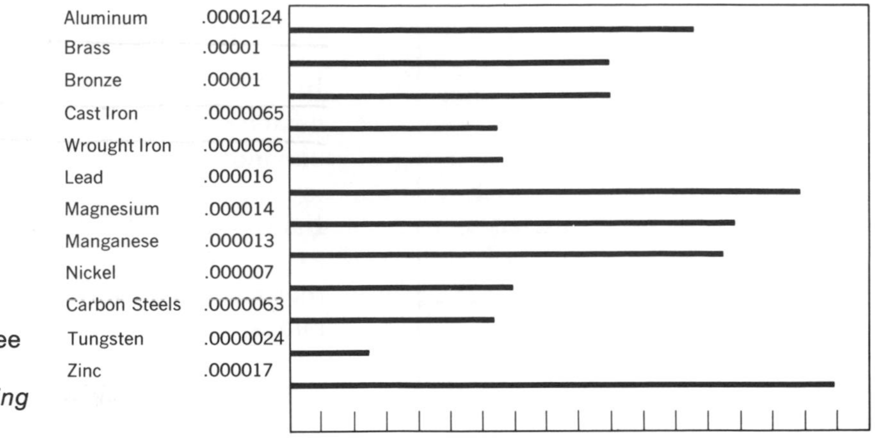

Figure 25. Coefficient of thermal expansion per degree Fahrenheit per unit length *(Machine Tools and Machining Practices).*

cool, the bushing is now smaller than the steel bore which makes it easy to remove from the hole.

A machinist has turned a bearing fit with a 0.0001 inch tolerance on a 4 inch steel shaft. The shaft is still hot from turning when the operator measures it. After a coffee break, he returns and checks his work to find that it is 0.0025 inch under size. What happened? The shaft had cooled down 100°F (37.8°C) to room temperature, causing it to shrink in size. The following formula is used to calculate the amount of contraction after cooling to room temperature.

Coefficient of expansion × diameter ×

rise in temperature F = expansion

or,

$$0.0000063 \times 4 \text{ inches diameter} \times 100°\text{F} = 0.0025 \text{ inch}$$

If a 1 inch diameter steel shaft would expand 0.0000063 inch for 1°F rise in temperature, it would expand 0.00063 inch for a 100°F rise. If the shaft were 4 inches diameter and had a 100°F rise, the expansion would be 0.0025 inch. If machined at that temperature, it would contract the same amount when cooled. All lathe operators are familiar with the lengthwise expansion of turned workpieces that causes the dead center to tighten and heat up.

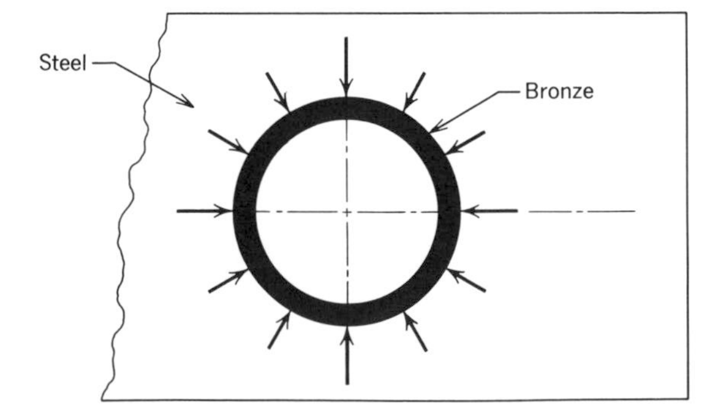

Figure 26. An application of thermal expansion of metals *(Machine Tools and Machining Practices).*

Temperature differences on workpieces, especially thin parts, can cause them to "crawl" on the milling machine table. The heat generated by milling with carbide cutters, when no coolant is used, is most likely to cause this problem.

Mechanics and machinists sometimes need to remove a sprocket or gear that has been pressed onto a shaft. Since the bore is often corroded and has no lubrication, the pressure required for removal is often many times greater than that required to press it on. When the available press is not capable of removing the gear or sprocket, heat is often applied to the hub in order to expand it so it will come off easily. The trouble with this process is that unless a high heat input rate is used, the shaft will heat along with the hub and it will not help. Even with a high heat input rate, if only the hub is heated on a large gear or sprocket with a solid flange or rim, the bore may not be expanded. In fact just the reverse may happen, the bore is squeezed tighter than ever because of the constraint placed upon the hub by the cold flange and rim (Figure 27). The expansion of the hub from the heat is deflected inward instead of outward. The solution is to heat the entire outer portion of the gear or sprocket gently and then to heat quickly the hub area with a high heat input while pressing on the shaft.

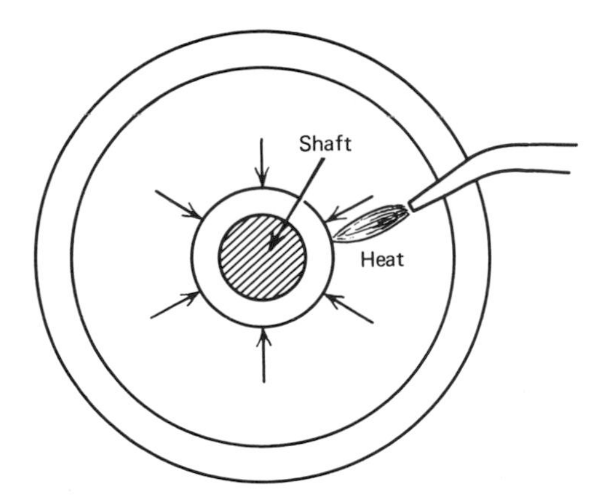

Figure 27. Heat applied to hub only causes it to expand. The restraint of the flange and rim makes the hub grip the shaft even more tightly.

Welders often experience expansion and contraction problems because of the heat involved in welding. In structural welding, large beams or weldments are sometimes preheated prior to welding. If a beam were 100 ft long and preheated to 300°F, the amount of expansion could cause it to buckle if no allowance were made for expansion. Its total expansion would be

$$0.0000063 \times 100 \times 12 \times 300°\text{F} = 2.268 \text{ in.}$$

Steel structures such as bridges have expansion joints to allow for differences in temperature that can be as much as 150°F.

QUESTIONS & ANSWERS

Self-Evaluation

1. Does creep occur within the elastic or plastic range of steels?
2. Explain the relative rate of failure in creep. Are creep failures sudden or do they take years to fail?
3. What happens below the transition temperature of a metal?
4. What happens to the properties of hardness, strength, and modulus of elasticity with a decrease in temperature?
5. Name an alloying element that can be added to steel to lower its transition temperature.
6. Describe the three categories of hardness and how they can be measured.
7. Name the three basic stresses.
8. A 2 in. square bar in tension has a load of 40,000 lb. What is the unit stress?
9. Explain ductility.
10. Explain malleability.
11. In what ways can fatigue strength be improved?
12. What is the correlation between metals for electrical and thermal conductivity?
13. In what state does a metal conduct heat best, alloyed or unalloyed?
14. How is the rate of thermal expansion for a particular material expressed?
15. Why should a machinist be very aware of thermal expansion in the metal he is using?

Worksheet 1

Objectives

1. Learn to use the tensile testing machine.
2. Calculate elongation, reduction in area, and unit stress of a pulled specimen.

Materials A tensile testing machine and a prepared specimen.

Procedure

1. Prepare a specimen of mild steel as shown in Figure 28. Punch the gage marks exactly 2 in. apart.

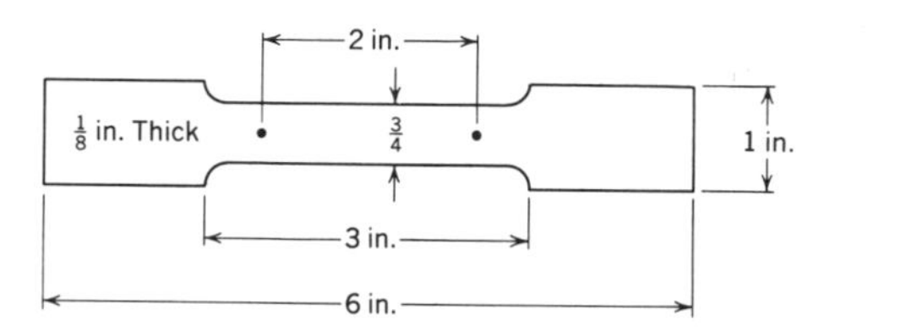

Figure 28.

2. With a micrometer measure the width at the narrowest point and the thickness.
3. Record this information.
4. Set up the tensile tester with flat gripping jaws and a 0 to 10,000 lb dial if it is the interchangeable type.
5. Place the specimen in the tensile testing machine and pull it to rupture.
6. Record the yield point.
7. Record the breaking load.
8. Remove the specimen and fit the broken pieces together. Measure the width and thickness of the specimen at the narrowest point; measure the length between gage marks and record the information.
9. Calculate the elongation and reduction in area using the formulas given in Figure 15.
10. Calculate the unit stress using the following formula:

$$\text{Unit stress} = \frac{\text{Load}}{\text{The original area}}$$

Note The original area is equal to the width × thickness at the narrowest point before pulling.

Conclusion Do you think this was a ductile metal? Why?

Data Tensile specimen

Length between gage marks	=
Thickness	=
Width	=
Area	=

Tensile specimen after pull

Length between gage marks	=
Breaking load	=
Yield point	=
Thickness	=
Width	=
Area	=

Results

Ultimate unit stress	=
Reduction in area, percent	=
Elongation, percent	=

Worksheet 2

Objective Determine the relative notch toughness of a carbon steel in its annealed and hardened state.

Materials An Izod-Charpy testing machine and prepared specimens.

Procedure

1. Prepare two Izod or Charpy specimens of annealed or as-rolled high carbon steel, SAE 1080 to 1095 according to specifications in Figures 19 or 20.
2. Harden one specimen in a water quench from 1500°F (815.6°C) and temper to 400°F (204.4°C).
3. Test both specimens for hardness. Record your results.
4. Test both specimens on the Izod-Charpy machine. Record your results.

Conclusion Which metal shows greater toughness? Which is most brittle?

Data

	Specimen 1 (soft)	Specimen 2 (hard)
Hardness		
Ft-lb		

Worksheet 3

Objective Demonstrate the difference in thermal conductivity between copper and stainless steel.

Materials A convenient heat source such as a bunsen burner or propane torch, a strip of copper, and an equally shaped strip of stainless steel.

Procedure

1 Set up the burner so the flame is at the ends of the strips arranged as in Figure 29.

2 Mark the opposite ends with a 200°F (93.3°C) temperature crayon. Instead of the crayon, a wooden match will work, as shown in Figure 29.

3 Note which crayon mark melts first or which match lights first.

Conclusion Which metal has the highest thermal conductivity?

Note You may demonstrate the difference between thermal conductivity of other metals such as steel and stainless steel if you use closely controlled conditions.

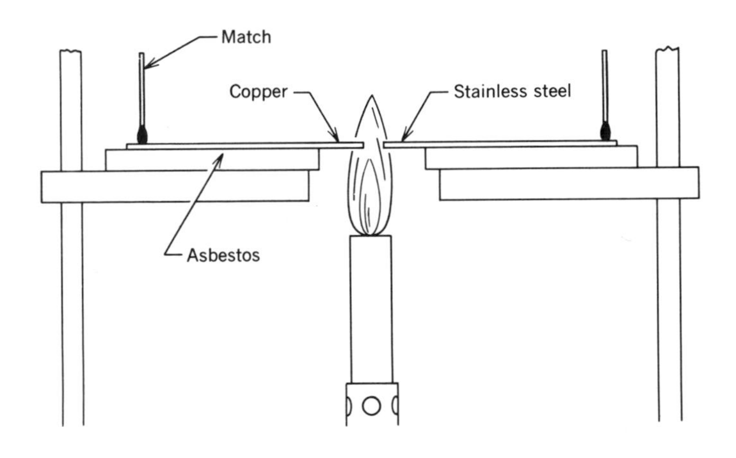

Figure 29.

Worksheet 4

Objective Determine the relative scaling characteristics of two metals.

Materials A furnace, one piece of mild steel about $\frac{1}{16} \times 1 \times 4$ in., and one piece of stainless steel about $\frac{1}{16} \times 1 \times 4$ in.

Procedure

1. Place the two specimens over a bunsen burner or in an electric furnace and allow them to remain at a yellow heat for one hour.
2. Check and record observations at 10 minute intervals.

Conclusion What happens to the mild steel?
What happens to the stainless steel?
Which one would you suggest for high temperature service?

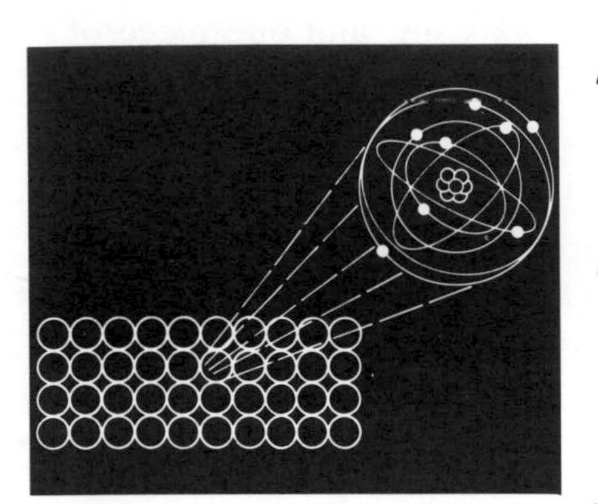

7 The Crystalline Structure of Metals

What are the forces that hold metals together? Why do metals behave as they do? These and other questions relating to the atomic and crystalline structure of metals will be discussed in this chapter.

OBJECTIVES

After completing this chapter, you will be able to:

1. Explain the various phases of crystalline structures of metals.
2. Explain the various aspects of solid solutions.
3. Conduct a Metcalf experiment and determine the approximate grain size in steel samples.
4. Prepare metal specimens for microscopic study by polishing and etching.

INFORMATION

The great utility of metals is due to their elastic behavior to a certain level of stress followed by a plastic behavior at higher levels of stress. Along with ceramic materials that are brittle, or polymers such as wood or leather, metals play a unique and useful role in the economy.

The physical world is composed of matter and energy. Matter is defined as anything that occupies space; it exists in three forms: solid, liquid, and gas. Energy is the ability to do work. It can be either potential or kinetic. Potential energy may be chemical or physical; it is "stored" energy, for example, as water in a reservoir or as the power in explosives. Kinetic energy is energy in motion that is doing work. The laws of conservation of matter and energy state that energy can neither be created nor destroyed, but may be changed from one form to another. This is true of the everyday world, but Einstein's discoveries leading to this atomic age revealed that matter and energy are related. A small amount of matter releases a tremendous amount of energy under certain conditions.

Matter is composed of atoms too small to be seen with the aid of ordinary microscopes, but the outline of molecules has been detected with such devices as the ion field emission microscope and the electron microscope (Figure 1). The **molecule** is defined as the smallest particle of any substance that can exist free and still exhibit all the chemical properties of that substance. A molecule can consist of one or more atoms. Atoms of different materials vary only in number and arrangement, but not the type of their parts. Matter composed of a single kind of atom is called an **element.** There are more than 100 elements and new ones are being discovered. Those that have atomic numbers greater than 92 are not found in nature, but are produced by atomic reactions. Elements are classed in two groups: metals and nonmetals. A **compound** is composed of two or more elements combined chemically. A **mixture** is two or more elements or compounds physically combined in the same way in which two fine powders are mixed together.

Atom

An atom resembles a miniature solar system that has orbits in many planes. Its chief parts are shown in Figure 2. The nucleus of the atom consists of protons and neutrons. The protons have a positive electrical charge. Neutrons weigh essentially the same as a proton, but are neutral in charge. Revolving at high speed around the nucleus are much smaller particles, called electrons. Electrons are negatively charged, which means that they are strongly attracted to the positively charged nucleus.

Each atom has preferred electron paths or orbits, called shells. The number, arrangement, and spin of the electrons in these shells in combination with the positively charged nucleus determine the kind of atom and its characteristics. Each shell contains a given number of electrons for each atom. The sum of the electrons in all shells equals the atomic number. Also the number of protons in the nucleus equals the atomic number. For example, the atomic number of hydrogen is 1, and it contains one proton and one electron.

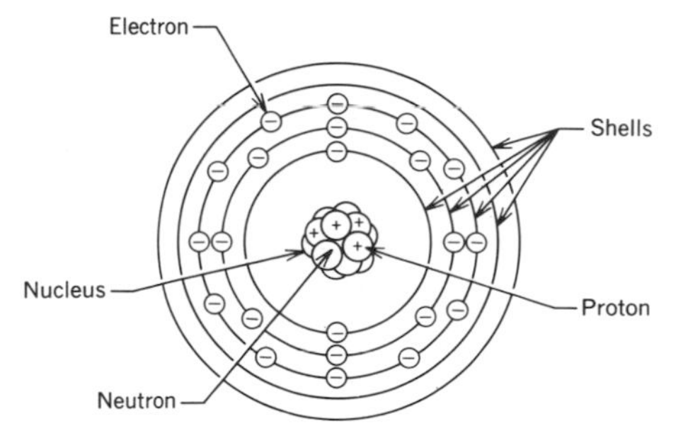

Figure 2. The atom (*Machine Tools and Machining Practices*).

The combining power of an atom is called its **valence.** The electrons on the outer (and sometimes the second) shell, called the valence electrons, are the most important in determining chemical and physical properties. When the valence shell in a free atom has a full complement of electrons, it is said to have zero valence. When an atom has more or less electrons in the outer shell than in its uncombined or free state, it is called an ion and possesses an electrical charge. The charge is negative when extra electrons are present and positive when some are missing. Metal atoms are easily stripped of their valence electrons and thus form positive **ions** when combined as a solid metal.

Each element is assigned a symbol that is derived from its name, such as O for oxygen and C for carbon. Some are derived from the Latin name. For example, the Latin for sodium is Natrium (Na), iron is Ferrum (Fe), silver is Argentum (Ag), lead is Plumbum (Pb), and gold is Aurum (Au).

The number of atoms or ions of each element in a compound is indicated by a subscript immediately following the symbol. H_2O indicates that in one molecule of pure water there are two atoms of hydrogen (H) and one of oxygen (O). A number placed in front of the symbol indicates the number of molecules of that substance. For example, $2CO_2$ indicates two molecules of carbon dioxide.

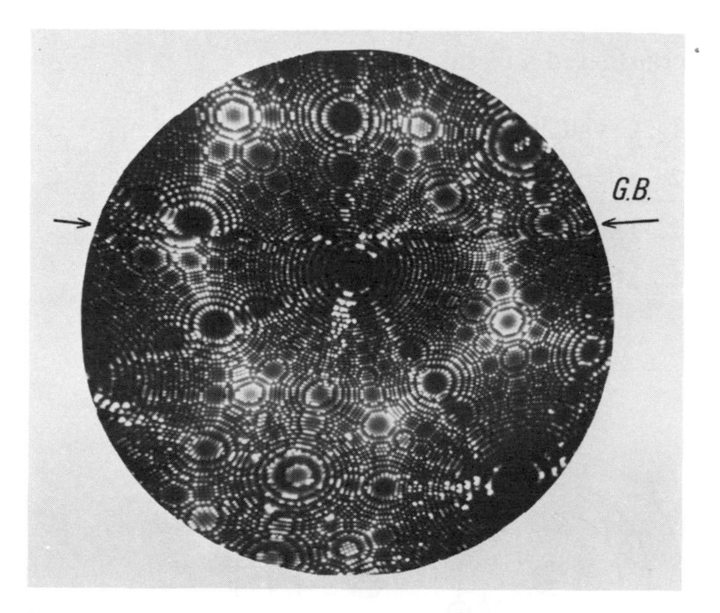

Figure 1. Grain boundary between metal crystals is only a few atoms wide, as seen in a field ion micrograph. Here the tip of a tungsten needle is enlarged several million diameters. Each bright dot represents a tungsten atom. The distance between dots is approximately 5 angstroms (John J. Hren).

Bonding

With this information, we can now go on to find out how metals are held together. There are four basic types of bonding arrangements that hold the atoms together. They are ionic, covalent, metallic, and Van der Waals forces. However, a molecule may

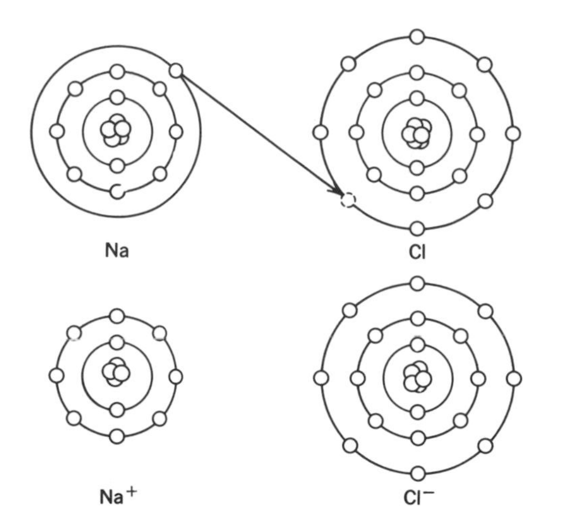

Figure 3. Ionic bonding (*Machine Tools and Machining Practices*).

be held together by a combination of several or all of these forces.

Ionic bonding is the attraction of negative and positive ions. Sodium chloride (NaCl) is an example of ionic bonding (Figure 3), where a metal (sodium) loses its single valence electron to a nonmetal (chlorine) to complete its valence shell. The sodium atom is now positively charged and the chlorine atom negatively charged. The resultant structure of salt (Figure 4) is rather weak since the electrostatic attractions are very selective and directional.

Covalent bonding, or shared valence electrons, is a very strong atomic bond whose strength depends on the number of shared electrons. Covalent bonding is found primarily between nonmetallic elements such as carbon (in diamonds, for instance). As in ionic bonding, the structure is rigid and directional.

For example, the oxygen atom has six electrons in the outer shell. This shell would like to be complete by having eight electrons. To make an oxygen molecule, two electrons are shared by each atom to provide a satisfactory arrangement (Figure 5).

The **metallic bond** is where some electrons in the valence shell separate from their atoms and exist in a cloud surrounding all the positively charged atoms. These positively charged atoms are arranged in a very orderly pattern. The atoms are held together because of their mutual attraction for the negative electron cloud (Figure 6). The free movement of electrons accounts for the high electrical and heat conductivity of metals and for their elasticity and plasticity. The metallic bond is very strong.

Van der Waals bonding is found in neutral atoms such as the inert gases. There is only a very weak attractive force, and it is of importance in the study of metals only at very low temperatures.

Metals and Nonmetals

Approximately three-quarters of all the elements are considered to be metals. Some of these are metalloids such as silicon or germanium. Some of the properties that an element must have to be considered a metal are:

1 Crystalline structure—grain structure.

2 High thermal and electrical conductivity.

3 The ability to be deformed plastically.

4 Metallic luster or reflectivity.

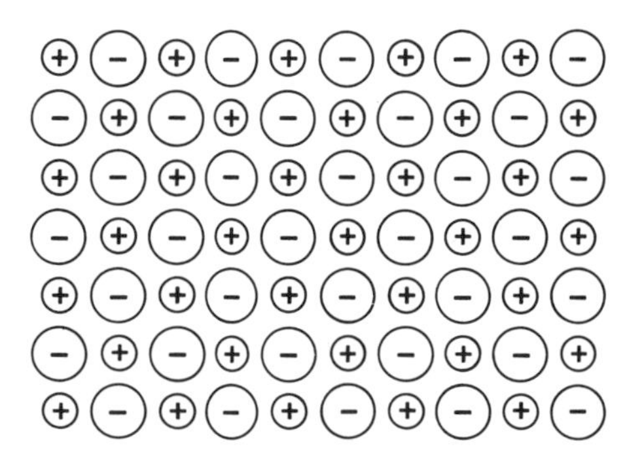

Figure 4. Lattice structure of salt (*Machine Tools and Machining Practices*).

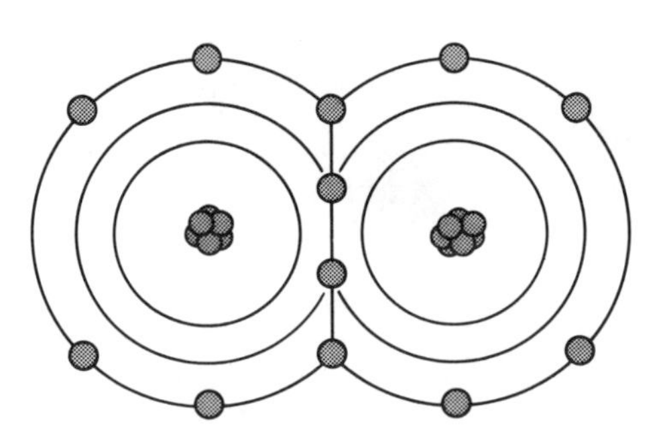

Figure 5. Oxygen molecule has a covalent bond (*Machine Tools and Machining Practices*).

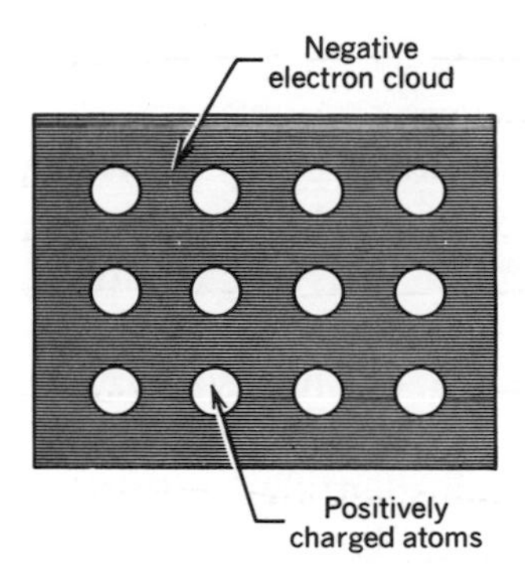

Figure 6. Metallic bond (*Machine Tools and Machining Practices*).

Metalloids, such as silicon, possess one or more of these properties, but they are not true metals unless they have all of the characteristics of metal.

Crystalline Unit Structures

Metals solidify into six main lattice structures.

1. Body-centered cubic (BCC)
2. Face-centered cubic (FCC)
3. Close-packed hexagonal (CPH)
4. Cubic
5. Body-centered tetragonal
6. Rhombohedral

As solidification is taking place, the arrangement of the crystalline lattice structure takes on a characteristic pattern. Each unit cell builds on another to form crystalline needle patterns that resemble small pine trees. These structures are called **dendrites** (Figure 7).

Figure 7. This is an actual dendrite growth on the surface of tin. Most dendrites are not as large as this example (*Machine Tools and Machining Practices*).

The crystalline lattice structures begin to grow first by the formation of seed crystals or nuclei as the metal solidifies. The number of nuclei or grain starts formed determines the fineness or coarseness of the metal grain crystal structure. Slow cooling promotes large grains and fast cooling promotes smaller grains. The grain grows outward to form the dendrite crystal until it meets another dendrite crystal that is also growing. The places where these grains meet are called grain boundaries.

Figure 8 represents the growth of the dendrite from the nucleus to the final grain when metal is solidifying from the melt. The nucleus can be a small impurity particle or a unit cell of the metal. Iron, copper, silver, and other metals are composed of these **tiny grain structures** that can be seen under a microscope when a specimen is polished and etched. This grain structure can also be seen with the naked eye as small crystals in the rough broken section of a piece of metal (Figure 9).

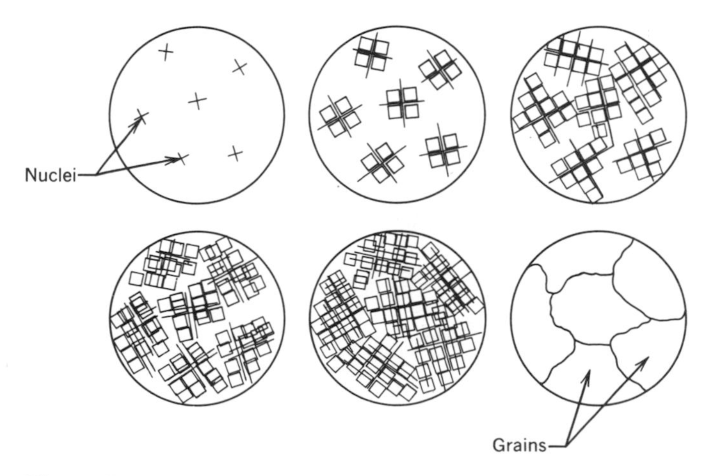

Figure 8. The formation of grains during solidification (*Machine Tools and Machining Practices*).

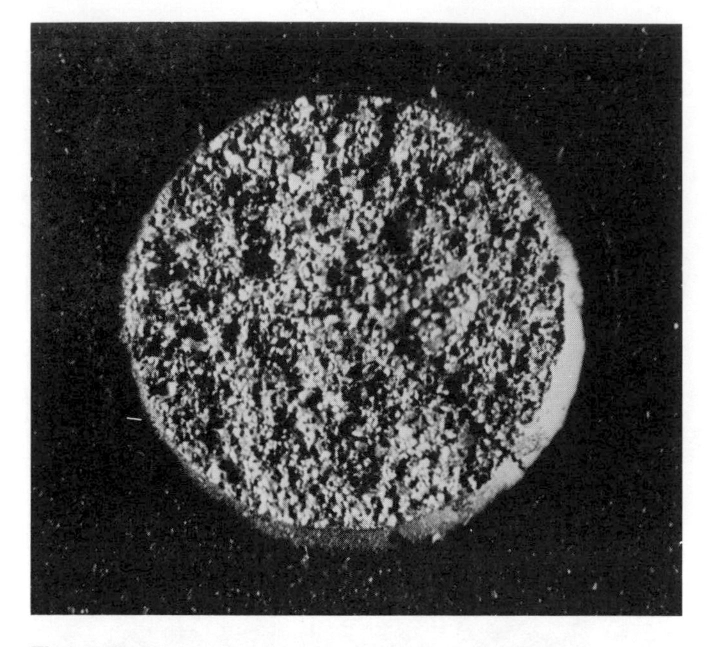

Figure 9. Broken section showing the crystalline structure (*Machine Tools and Machining Practices*).

Grain Boundary

As the crystal structures grow in different directions, it can be seen that at the **grain boundaries,** the atoms are jammed together in a misfit pattern (Figure 10). This strained condition also makes the grain boundaries stronger than the adjacent grain lattice structure at low temperatures (under red heat), but weaker at high temperatures (yellow or white heat). Grain boundaries are only about one or two atoms wide, but their strained condition causes them to etch differently; this means they may be observed with the aid of a microscope.

Body-Centered Cubic (BCC)

This cubic unit structure is made up of atoms at each corner of the cube and one in the very center. Steel under 1333°F (723°C) has this arrangement, and it is called alpha iron or **ferrite.** Other metals such as chromium, columbium, barium, vanadium, molybdenum, and tungsten crystallize into this lattice structure. These cubes are identified within the lattice structure as seen in Figurre 11. Body-centered cubic metals (Figure 12) show a lower ductility but higher yield strength than face-centered cubic metals.

Face-Centered Cubic (FCC)

Atoms of calcium, aluminum, copper, lead, nickel, gold, platinum, and some other metals arrange themselves with an atom in each corner of the cube, and one in the center of each cube face. When steel is above the upper critical temperature, it rearranges its atoms to this FCC structure and is called gamma iron or **austenite** (Figure 13).

Close-Packed Hexagonal (CPH) or Hexagonal Close-Packed (HCP)

The close-packed hexagonal structure (Figure 14) is found in many of the least common metals. Beryllium, zinc, cobalt, titanium, magnesium, and cadmium are examples of metals that crystallize into this structure. Because of the spacing of the lattice structure, rows of atoms do not easily slide over one another in CPH. For this reason, these metals have lower plasticity and ductility than cubic structures.

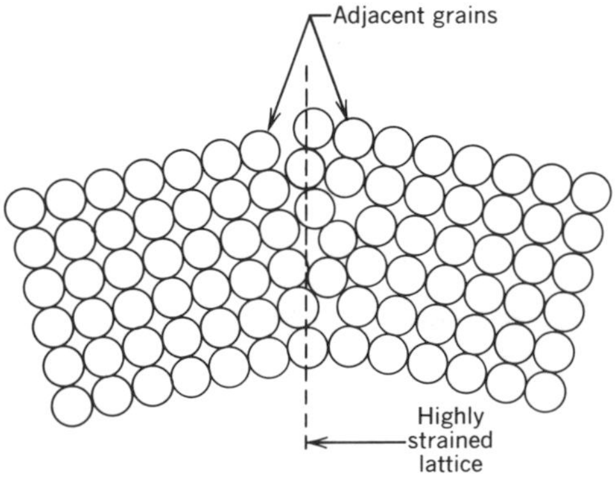

Adjacent grains

Highly strained lattice

Figure 10. Grain boundary is highly strained (*Machine Tools and Machining Practices*).

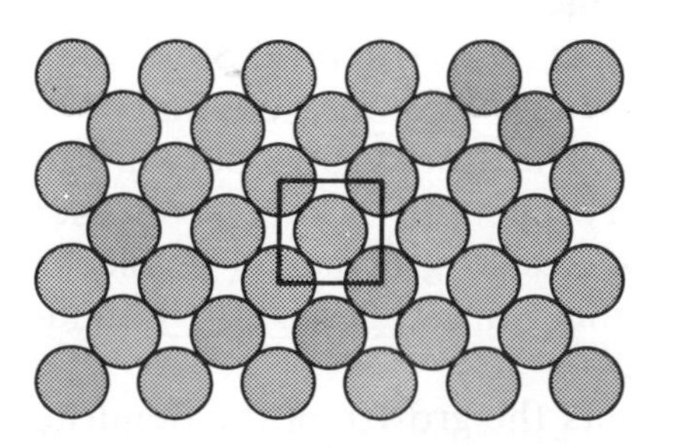

Figure 11. Lattice structure showing body-centered cubic formation (*Machine Tools and Machining Practices*).

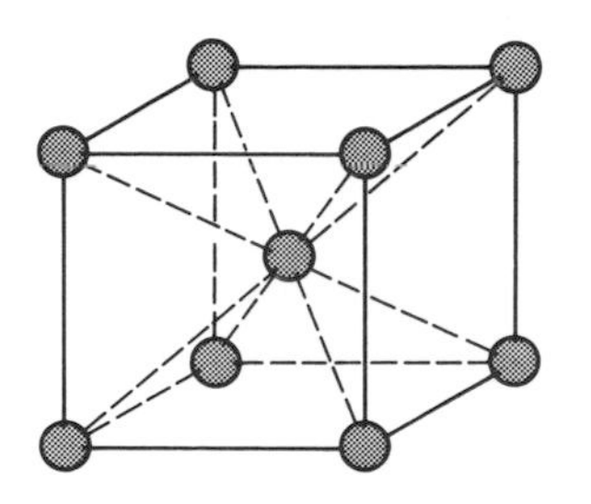

Figure 12. Idealized body-centered cubic structure (*Machine Tools and Machining Practices*).

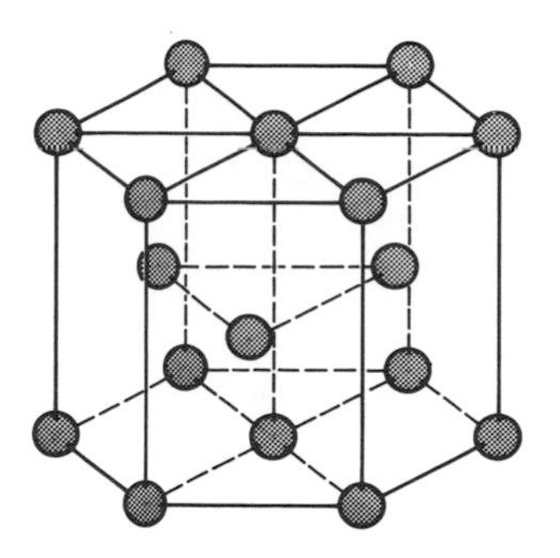

Figure 14. Close-packed hexagonal structure (*Machine Tools and Machining Practices*).

Other Structures

The metal manganese has a simple cubic structure. Manganese is used as an alloying element in steel. Antimony has a rhombohedral crystal structure and is used as an alloying element with zinc and tin.

When carbon steel above the upper critical temperature is quenched, the FCC structure attempts to change to the BCC structure. Since there is a solid solution of carbon and iron at the quenching temperature, the lattice contains the smaller carbon atoms in the interstices (spaces between the iron atoms) and complete conversion to BCC is not possible. This is because the carbon atoms interfere with the conversion since there is not enough room in the BCC structure to hold all of them. The result is a structure that ranges from an elongated body-centered cubic crystal to a body-centered tetragonal crystal. This distortion of the lattice is what causes the hardness of **martensite** that has a body-centered tetragonal unit structure (Figure 15). (Martensite is the hard structure that is formed when carbon steel is quenched from high temperatures.) Table 1 lists some common metals and their chemical symbols and crystalline structures. See the periodic table of the elements in Appendix I.

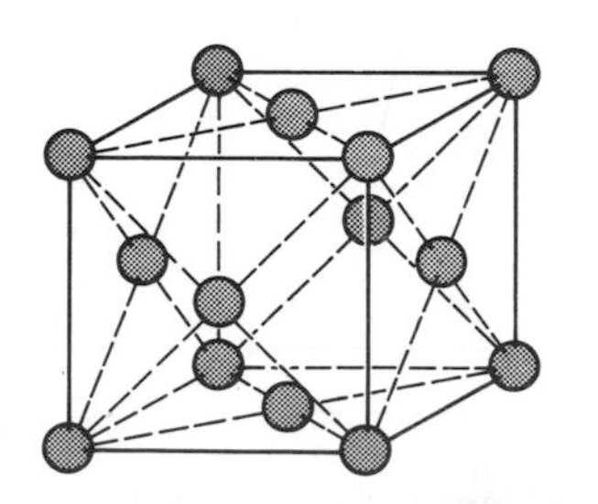

Figure 13. Idealized face-centered cubic structure (*Machine Tools and Machining Practices*).

Table 1 Crystal Structures of Some Common Metals

Symbol	Element	Crystal Structure
Al	Aluminum	FCC
Sb	Antimony	Rhombohedral
Be	Beryllium	CPH
Bi	Bismuth	Rhombohedral
Cd	Cadmium	CPH
C	Carbon (graphite)	Hexagonal
Cr	Chromium	BCC (above 26°C)
Co	Cobalt	CPH
Cu	Copper	FCC
Au	Gold	FCC
Fe	Iron (alpha)	BCC
Pb	Lead	FCC
Mg	Magnesium	CPH
Mn	Manganese	Cubic
Mo	Molybdenum	BCC
Ni	Nickel	FCC
Nb	Niobium (columbium)	BCC
Pt	Platinum	FCC
Si	Silicon	Cubic, diamond
Ag	Silver	FCC
Ta	Tantalum	BCC
Sn	Tin	Tetragonal
Ti	Titanium	CPH
W	Tungsten	BCC
V	Vanadium	BCC
Zn	Zinc	CPH
Zr	Zirconium	CPH

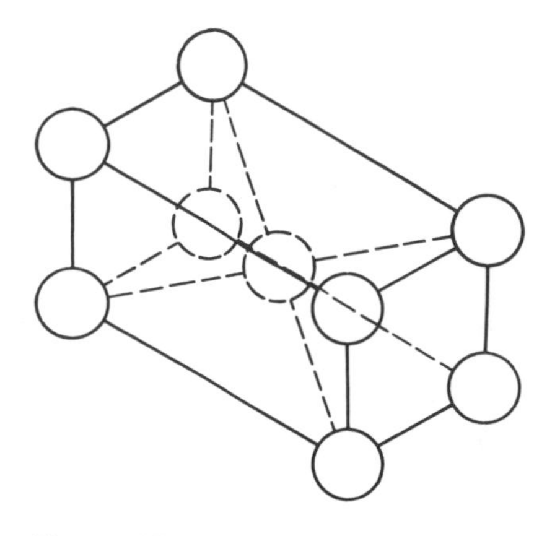

Figure 15. Body-centered tetragonal structure. This is the distorted cubic form of the unit cell.

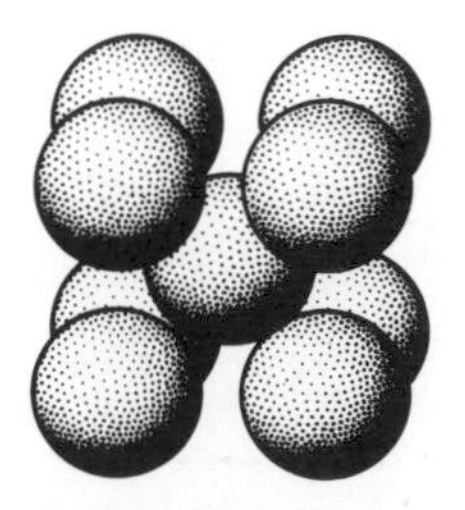

Figure 16. Body-centered cubic (BCC) (*Machine Tools and Machining Practices*).

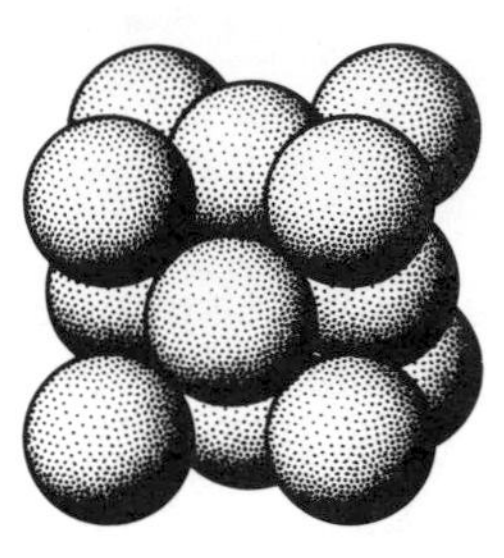

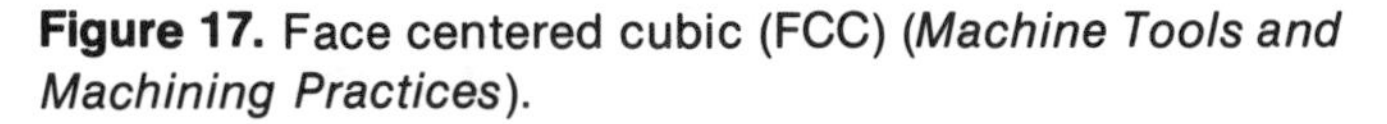

Figure 17. Face centered cubic (FCC) (*Machine Tools and Machining Practices*).

Crystalline Changes During Heating

When metals are heated slowly to their melting points, certain changes take place. Most nonferrous metals, such as aluminum, copper, and nickel, do not change their crystalline lattice structure before becoming a liquid. However, this is not the case with iron. (The term iron refers to elemental or pure iron unless specified as wrought or cast.)

Iron is a special type of metal that does undergo a crystalline change as it is heated to the liquid stage. Iron at room temperature is BCC (Figure 16) but, when it is heated to about 1700°F (927°C), it changes to FCC (Figure 17). A material that can exist in more than one crystalline lattice structure depending on temperature is called **allotropic.** An allotropic element is able to exist in two or more forms having various properties without a change in chemical composition. For example, carbon exists in three allotropic forms: amorphous (charcoal, soot, coal) graphite, and diamond. Iron also exists in three allotropic forms: BCC (below 1300°F or 704°C), FCC (above 1670°F or 911°C), and delta iron (between 2550 and 2800°F or between 1498 and 1371°C) (Figure 18).

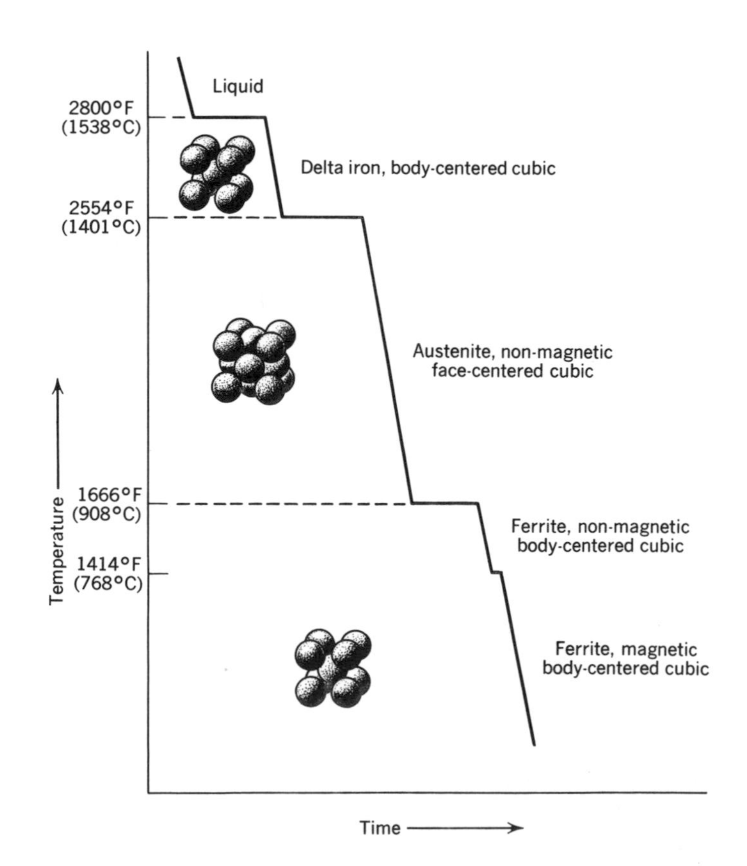

Figure 18. Cooling curve for pure iron. As iron is cooled slowly from the liquid phase, it undergoes these allotropic transformations (*Machine Tools and Machining Practices*).

Cold Forming and Plastic Deformation in Metals

As you learned in Chapter 7, the ability of a metal to undergo plastic deformation when stressed beyond its elastic limit is one of its most useful characteristics. Forging, drawing, forming, extruding, rolling, stamping, and pressing involve plastic deformation. A great deal of useful information regarding the behavior of metals in plastic deformation may be obtained by the study of the crystalline and lattice structures under stress. Deformation may occur by slip, twinning, or by a combination of these. Some of the factors that influence slip and twinning in metals are: precipitate particles or inclusions in the grains, impurity atoms added intentionally or unintentionally, vacancies where no atom exists in the lattice, atoms ordered and arranged in a pattern or disordered, interstitial atoms, dislocation or distortion of the lattice arrangement, and factors involving polycrystalline material (grains whose crystal axes are oriented at random).

In an amorphous material like tar, plastic deformation begins immediately when a load is applied and the rate is proportional to the load. In metals, however, the plastic deformation rate is essentially zero until a given stress is reached (the yield point).

It is theorized that commercial metals do not have perfectly arranged lattice structures. It is believed that the irregularities of vacancies, impurity atoms, and dislocations are responsible for the plasticity of metals. Metallurgists have grown "whiskers" of pure metal with perfectly ordered lattice structures. These "whiskers" have tensile strengths many times that of the commercial metal. This is because these pure metals have few irregularities such as dislocations to allow slip to take place.

When a dislocation is present, rows of atoms in the lattice can slip or slide until the vacancy or dislocation is blocked by impurity atoms or other irregularities (Figure 19). The dislocation moves one atom at a time across the lattice (Figure 20). When many planes of atoms are free to slide to a dislocation, a large percentage of elongation or deformation is possible. This is roughly analogous to a deck of cards, each card sliding a small amount, adding to the total movement (Figure 21).

Up to this point, the discussion about slip has been confined to the single crystal or grain (Figure 22). In the polycrystalline state of metal, there is no uniform order of lattice planes from one grain to another. Since each grain has a random orientation (direction of dendrite growth), no continuous slip line through the material is possible since resistance to slip is exhibited at the grain boundaries. This explains why large grain metals have a greater capacity to slip and have more plasticity than small grain metals. A fine grain steel, for instance, is stronger than a coarse grain steel. Slip takes place along certain crystallographic planes (Figure 23). BCC metals have four times as many possibilities or directions for slip than FCC metals and 16 times that of CPH metals such as zinc, magnesium, and cadmium.

Twinning takes place along certain planes and results in a new lattice orientation or direction along the twinning plane (Figure 24). The CPH metals are particularly susceptible to twinning when a shear stress is applied. Twinning forms a mirror image and occurs in pairs. It is easily observed by microscopic examination. If the surface steps are removed by additional polishing, the twin pattern can still be seen as wide bands. Slip lines, by contrast, are seen as lines that are changed or removed by additional polishing.

When any of these forms of deformation have reached their limit because of filled vacancies or

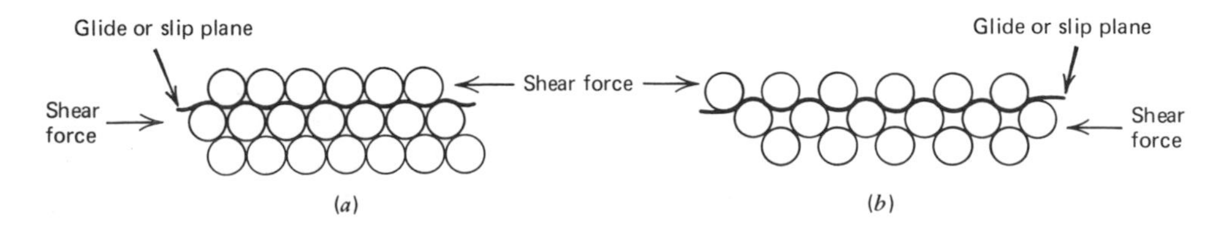

Figure 19. Rows of atoms can glide along slip planes as shown. BCC and FCC have rows of atoms closely spaced as in *(a)* making movement easier than in *(b)*. CPH type metals are less ductile since many slip planes as in *(b)* are spaced closer.

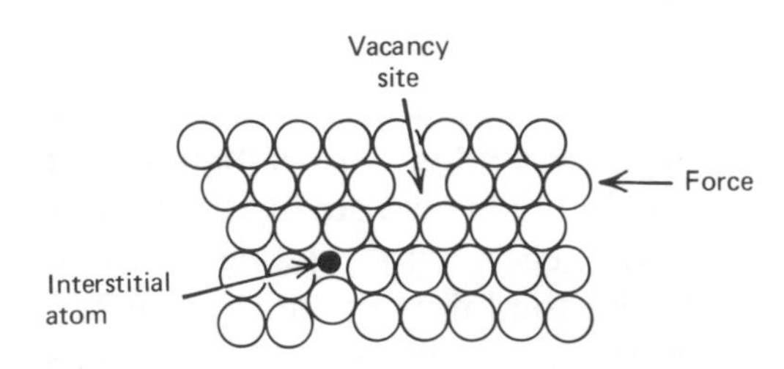

Figure 20. Rows of atoms move, one atom at a time, to fill in a vacancy or space created by dislocated atoms. Interstitial atoms jam up the slip planes of atoms, making the metal that contains them harder and less ductile.

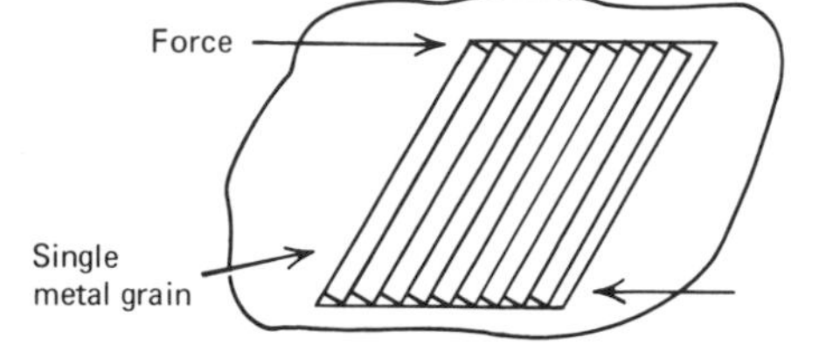

Figure 21. When stress is applied to a metal grain above its plastic range, slip occurs and the grain flattens and becomes enlongated. Slip planes are rotated when stress is applied.

jammed up dislocations resulting from impurity or alloying atoms, no more slip is possible and the material is no longer plastic. Continued stress results in fracture at this point.

Grain Size Consideration

The size of the grain has a great effect on the mechanical properties of the metal. The effects of **grain growth** caused by heat treatment are easily predictable. Temperatures, alloying elements, and soaking time all affect grain growth.

In metals, a small grain is generally preferable over a large grain. The small grain metals have more tensile strength, greater hardness, and will distort less during quenching, as well as be less susceptible to cracking. Fine grain is best for tools and dies. However, in steels a large grain has increased hardenability, which is often preferable for carburizing and for steel that will be subjected to extensive cold working.

All metals will experience grain growth at high temperatures (Figure 25). However, there are some steels that can reach relatively high temperatures (about 1800°F or 982°C) with very little grain growth but, as the temperature rises, experience a

Figure 22. Slip planes, as they appear in a micrograph, resulting from the cold working of anneal strip (250 ×) (By permission, *Metals Handbook* Volume 7, Copyright American Society for Metals, 1972).

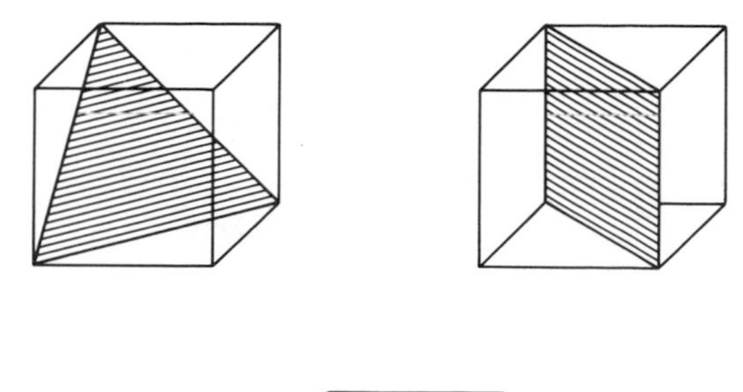

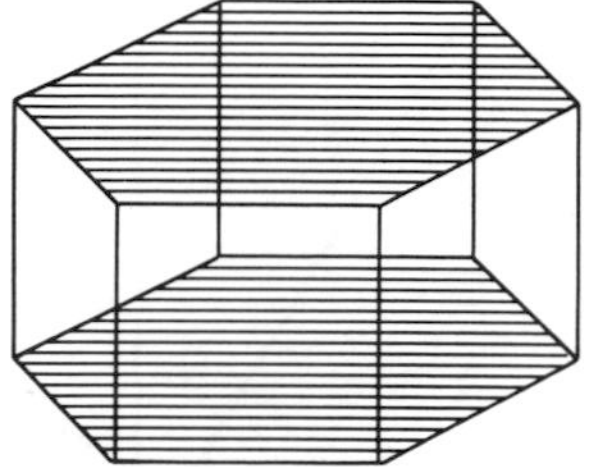

Figure 23. Slip is confined to certain crystallographic planes (darkened area).

rapid growth rate. These steels are referred to as fine grained. A wide range of grain sizes may be produced in the same steel.

Grain Size Classification

There are several methods of determining grain size, as seen under a microscope. The method explained here is one that is widely used by suppliers. The size of grain is determined by a count of the grains per square inch under 100× magnification. Figure 26 is a chart representing the actual size of the grains as they appear when magnified 100 times. Specified grain size is generally the austenitic grain size. A properly quenched steel should show a fine grain.

In describing a piece of steel, it is often necessary to specify the grain size. The actual method is done on a comparison between the specimen and grain size classification chart. The chart includes eight different grain sizes. A steel is considered fine grained if it is predominately 5 to 8 inclusive, and coarse grained if it is predominately 1 to 5 inclusive. If 70 percent of the grain size falls into the given range, it is considered acceptable. Two size classifications may be necessary if there is a wide variation in a section of metal. When austenitic grain size is specified, as it is in most mechanical property tables, the accepted method of determining it is the McQuaid-Ehn test. This test consists of carburizing a specimen at 1700°F (927°C) followed by slow cooling to develop a carbide network at the grain boundaries. The specimen is then polished and etched, and then compared to the grain size standards at 100× magnification.

Another method of determining grain size is by comparing a fractured surface of the metal to a fracture standard. A common standard is the Shepherd grain size standard (Figure 27). This comparison is made without any magnification.

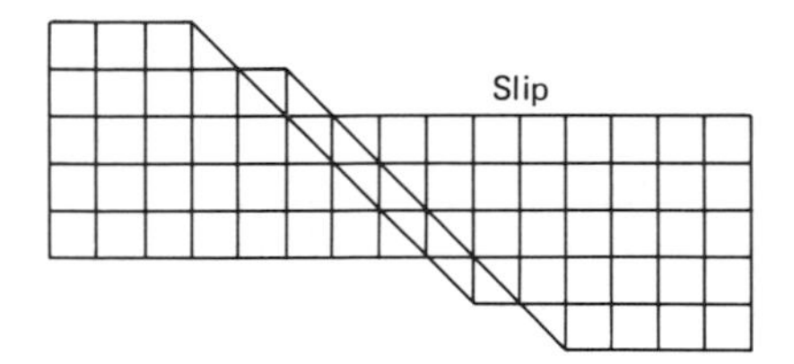

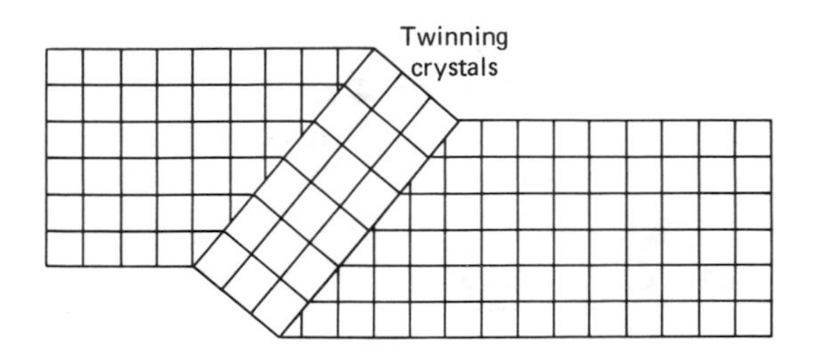

Figure 24. The difference between slip and twinning. Twinning is a kind of slip in which a uniform tilting of the twinning crystals results from shear force. Slip occurs along the planes of greatest atomic density.

Microscopic Examination of Metals

Details of the structure of metals are not readily visible to the naked eye, but grain structures in metals may be seen with the aid of the microscope. Metal characteristics, grain size, and carbon content may be determined by studying the micrograph (Figures 28*a* through 28*e*). The approximate percent of carbon may be estimated by the percentage of pearlite (dark areas) in annealed carbon steels. For this purpose, a metallurgical microscope (Figure 29) and associated techniques of photomicroscopy are used. The metallurgical reflected light microscope is similar to those used for other purposes, except that it contains an illumination system within the lens system to provide vertical illumination (Figure 30). Some microscopes

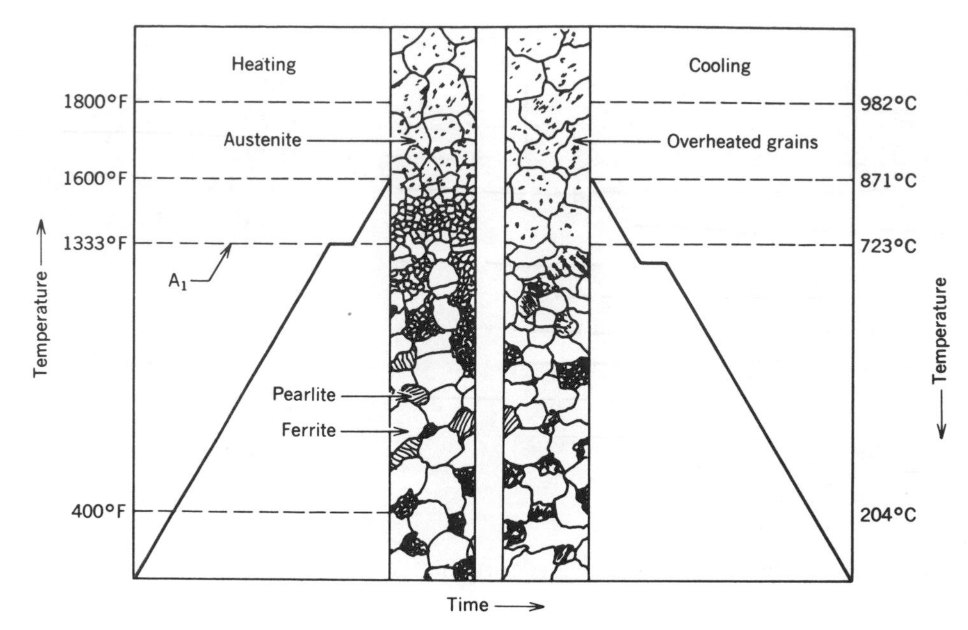

Figure 25. The effect of overheating carbon steel and the resultant grain growth (*Machine Tools and Machining Practices*).

are also equipped with a reticle and micrometer scale for measuring the magnified image. Another reticle used contains the various grain sizes at 100× magnification for the purpose of comparing or measuring relative grain size. Filters and polarizers are used in the illumination or optical system to reduce glare and to improve the definition of grain structures. The magnifying power of the microscope may be determined by multiplying the power of the objective lens by that of the eyepiece. Thus, a 40× objective lens with a 12.5× eyepiece would enlarge the image to 500× (500 diameters).

Inverted stage microscopes are a more modern design (Figure 31). The specimen is placed face down on the stage on this instrument. Metallograph instruments make possible a group observation of the metallurgical magnification. The image is projected on a ground glass screen (Figure 32). Large models are used in large metallurgical laboratories (Figure 33). Many metallograph instruments have the capability of making polaroid or standard microphotographs. Adaptors are available for most microscopes for making photographs. Simple sleeve adaptors may be used with a 35 mm SLR camera for taking microphotographs (Figure 34). With this simple arrangement, the shutter is opened and the light is then turned on for a few seconds (6 to 8 seconds with ASA 32 panatomic X film). Focus is made on the ground glass in the camera.

Preparation of the Specimen

The specimen should be selected from the area of the piece that needs to be examined and in the proper orientation. That is, if grain flow or distortion is important, a cross section of the part may not show elongated grains; only a slice parallel to the direction of rolling would adequately reveal elongated grains from rolling. Sometimes, more than one specimen is required. A weld is usually cross sectioned for examination.

Soft materials (under 35 Rc) may be sectioned by sawing, but harder materials must be cut off with an abrasive wheel. Metallurgical cut off saws with abrasive blades and coolant flow are used for this purpose (Figure 35). **The specimen must not be overheated** whether it is hard or soft. The grain structures may become altered with a high cutting temperature.

The specimen should be small $\frac{1}{4}$ to $\frac{3}{8}$ inch in width for ease of preparation. The specimen is

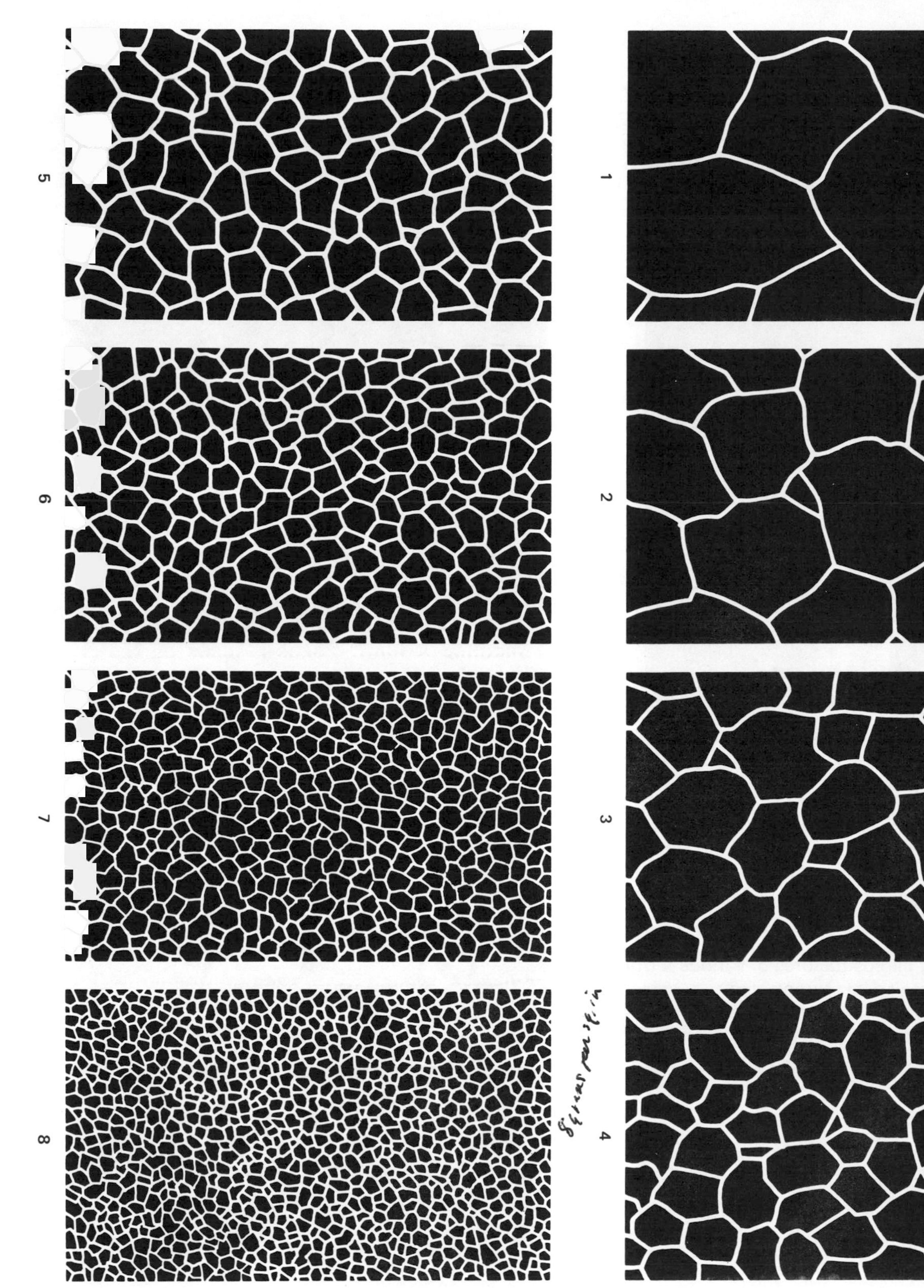

Figure 26. Standard grain size numbers. The grain size per square inch at 100 magnification (Bethlehem Steel Corporation).

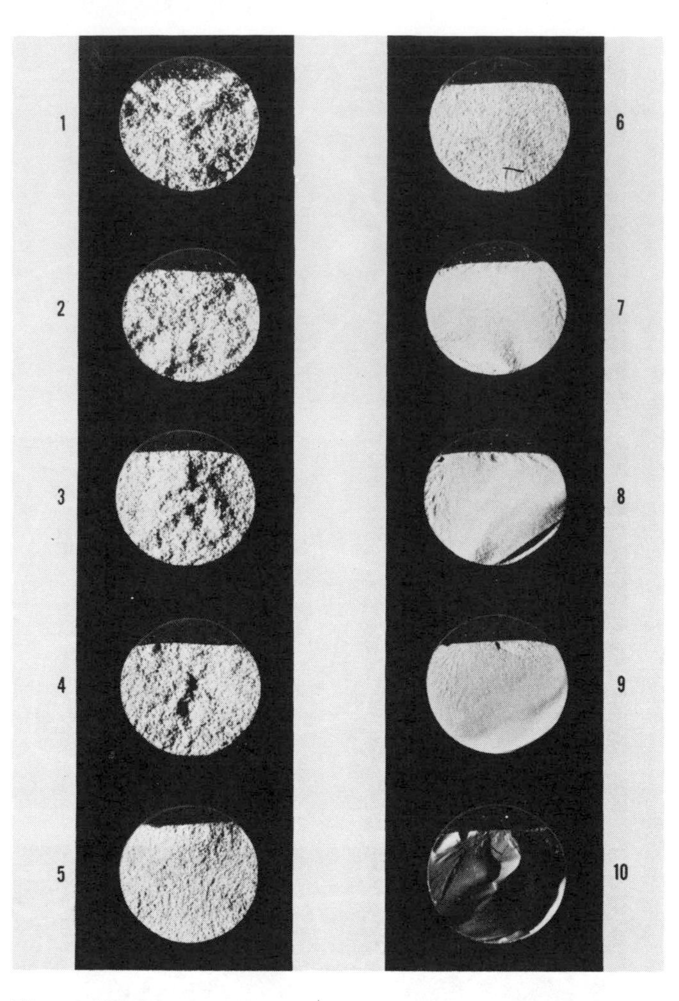

Figure 27. Shepherd grain size fracture standards. The fracture grain size test specimen can be compared visually with this series of ten standards. Number 1 constitutes the coarsest and Number 10 the finest fracture grain size (Courtesy Republic Steel Corporation).

(a)

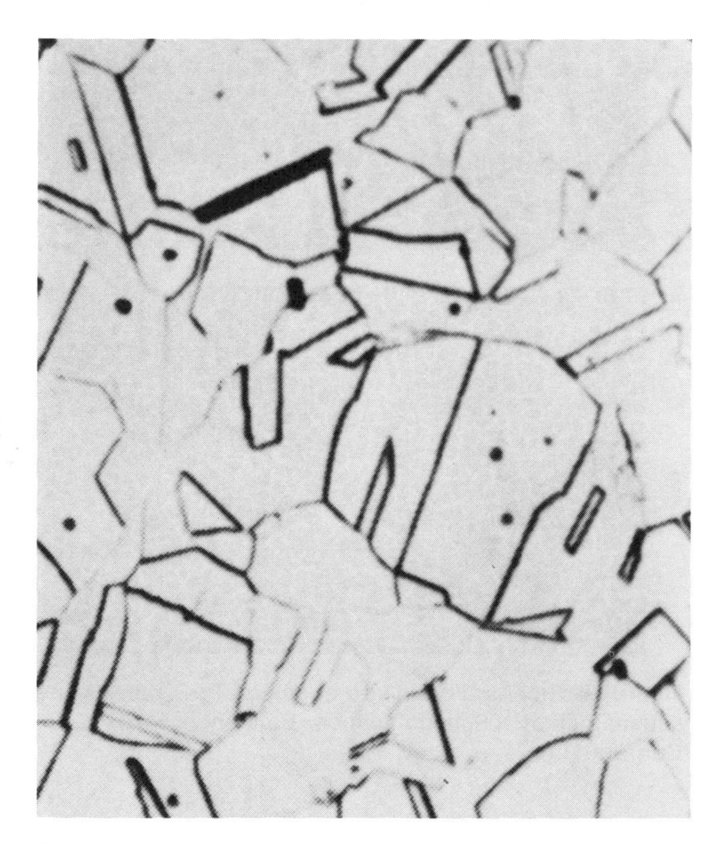

(b)

Figure 28. Identifying microstructures of various ferrous metals *(a)* ferrite (annealed low carbon steel), *(b)* austenite, *(c)* gray cast iron (etched), *(d)* martensite, and *(e)* pearlite. (*b* is by permission from *Metals Handbook* Volume 7, Copyright American Society for Metals, 1972).

(c)

(d)

usually mounted in plastic using a mounting press (Figure 36). Thermosetting plastic is often used and is formed by heat and pressure around the specimen (Figure 37). The mold must be kept hot through the curing period (3 to 5 minutes) for thermosetting materials. If thermoplastic materials such as lucite are used, the mold is first heated to soften the plastic and then cooled before the specimen is removed. Time for this operation is about 15 to 20 minutes.

Polishing the Specimen

Grains and other features of metal cannot be seen unless the specimen is ground and polished to remove all scratches. Various methods of polishing are used such as electrolytic rotating, or vibrating polishers. The most common procedure is to first rough grind the face of the specimen on a belt sander and then hand grind on several grades of abrasive paper from 240 to 600 grit. The Handimet® grinder (Figure 38) provides four grinding surfaces upon which water flows constantly. The water removes surface particles and maintains sharp cutting edges. The specimen is first moved

(e)

Figure 29. Metallurgical microscope (Lane Community College).

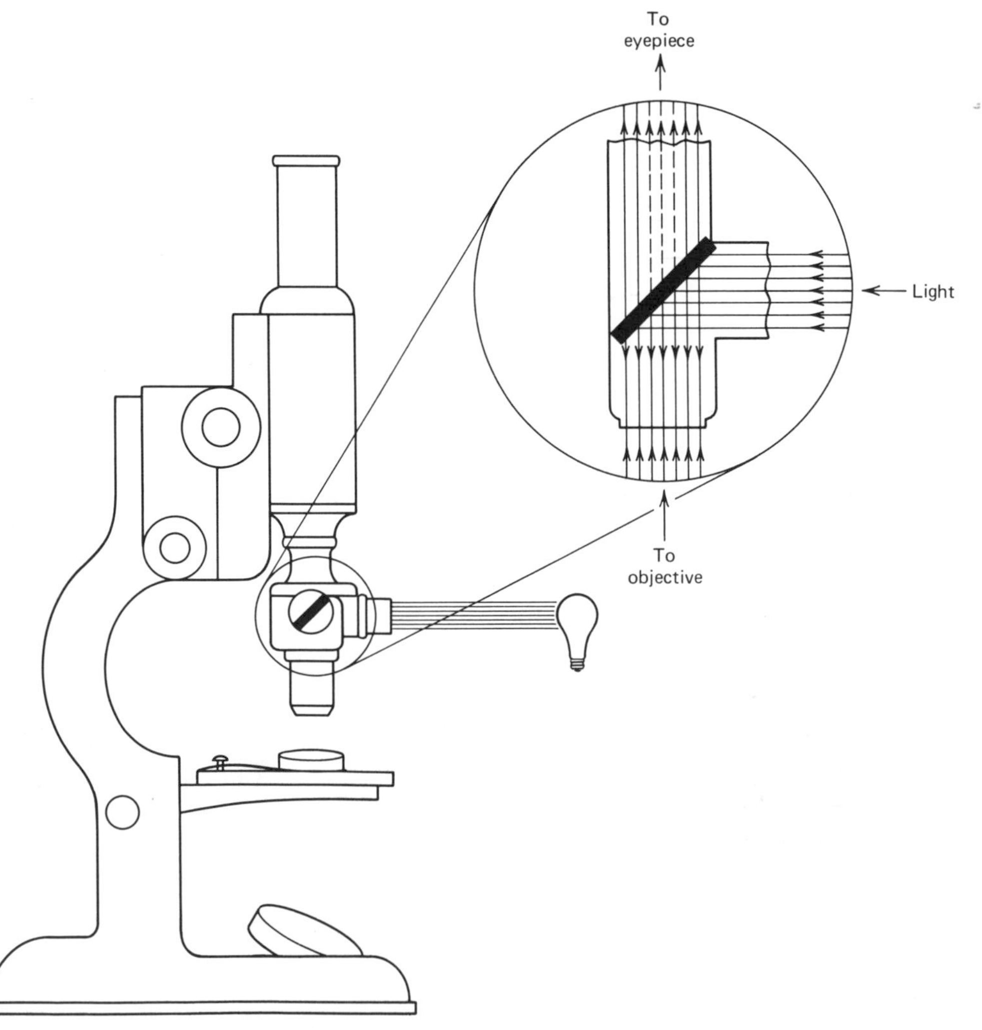

Figure 30. Illumination in a metallurgical microscope.

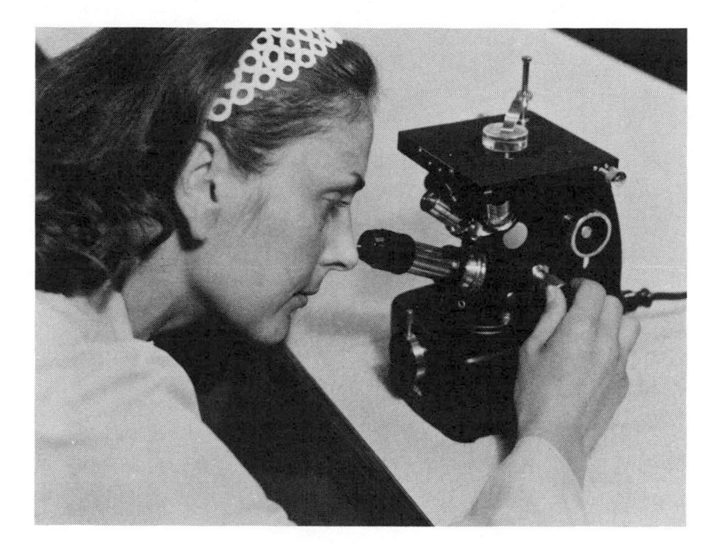

Figure 31. Inverted stage microscope (Photograph courtesy of Buehler Ltd., Evanston, Illinois).

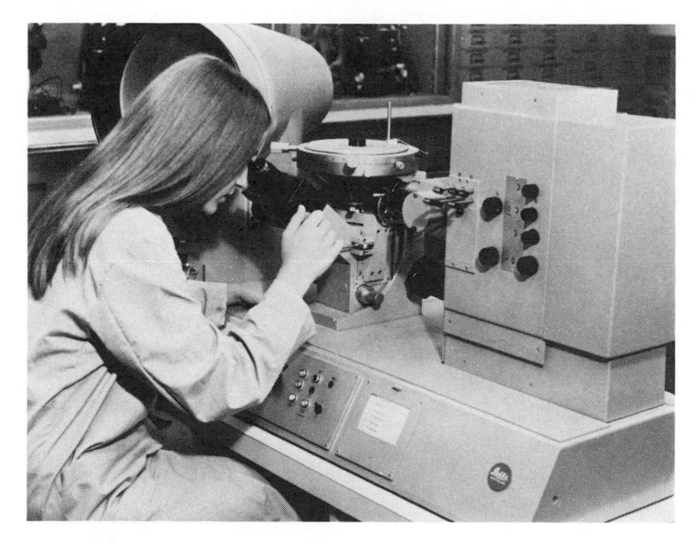

Figure 33. Metallograph used in a metallurgical laboratory (Photograph courtesy of Buehler, Ltd., Evanston, Illinois).

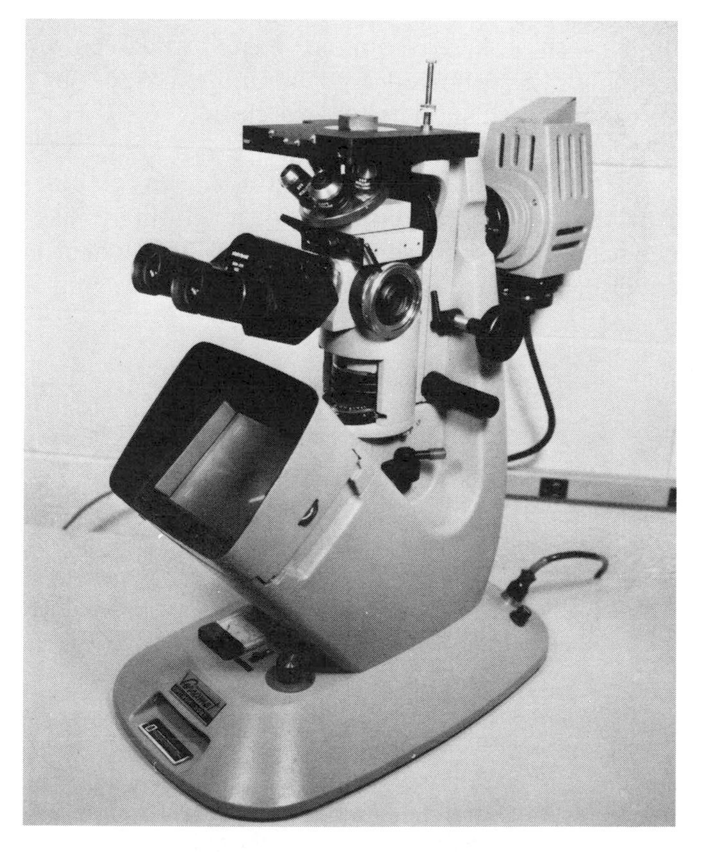

Figure 32. Metallograph with projection screen (Photograph courtesy of Buehler Ltd., Evanston, Illinois).

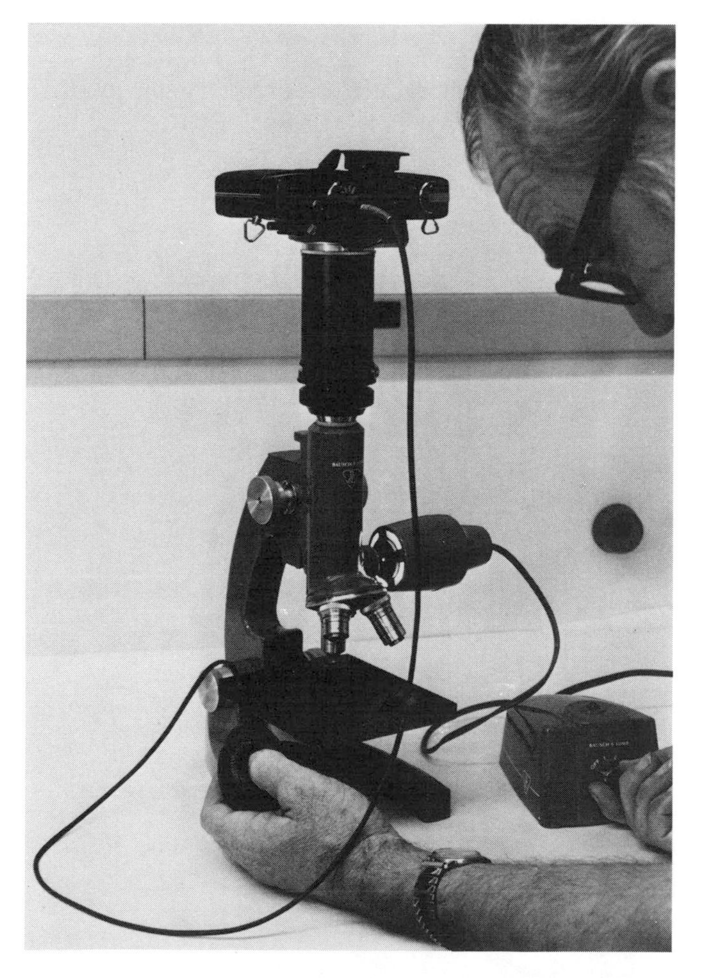

Figure 34. Method of making micrographs using a 35mm camera with adaptor on a small metallurgical microscope (Lane Community College).

Figure 35. Metallurgical cutoff saw (Photograph courtesy of Buehler Ltd., Evanston, Illinois).

Figure 36. Specimen mount press with mold (Photograph courtesy of Buehler Ltd., Evanston, Illinois).

Figure 37. Small specimen mounted in thermosetting plastic.

back and forth on the coarse grit paper until all the scratches go in one direction **and then the specimen must be thoroughly cleaned before moving to a finer grit.** The second step in grinding should be done so that the new grinding scratches or lines are 90 degrees to the previous lines. This process should continue through all the grits of paper. The specimen is now prefinished and is ready for polishing, which can be done in one operation or by rough polishing followed by finish polishing. This is best done with coarse diamond paste abrasives on nylon cloth, followed by aluminum oxide on a short nap cloth such as billiard cloth.

The specimen may be put in an electropolisher, a vibratory polisher, or a rotating polishing wheel may be used (Figure 39). The wheel is covered with a cloth such as billiard felt, and a

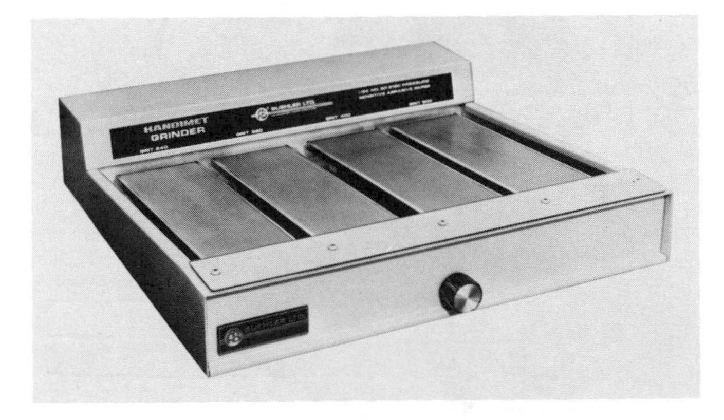

Figure 38. The Handimet® grinder is used to prepare specimens for polishing with either the rotating or electropolishers (Photograph courtesy of Buehler Ltd., Evanston, Illinois).

Figure 39. Two unit polishing table with polishers and lower cabinet (Photograph courtesy of Buehler, Ltd., Evanston, Illinois).

slurry of finely divided alumina is applied. The specimen is held face down on the wheel and slowly moved around in the opposite direction to the rotation (Figure 40). When the metal specimen is mirror bright and shows no scratches or lines, it should be cleaned in water. Methyl alcohol should be used for further cleaning and to help the drying operation. A small pyrex (petri) dish may be used for this purose or the specimen may be swabbed with cotton saturated in the alcohol. The alcohol should be dried with a flow of hot air. This can be done with an ordinary hair dryer.

Improper polishing techniques will produce a less than desirable outcome. Insufficient polishing will leave many scratches, which are magnified by the microscope (Figure 41). Too much pressure applied when polishing can deform the surface layer and make grain boundaries and structures indistinct (Figure 42).

Etching the Specimen

Enchants are composed of organic or inorganic acids and alkalis dissolved in alcohol, water, or other solvents. Some common enchants are given in Table 2. The specimen may now be etched for the required time by immersing it face down in a solution contained in a petri dish (Figure 43). An

Figure 40. Method of hand polishing a specimen (Lane Community College).

Figure 41. Etching the specimen.

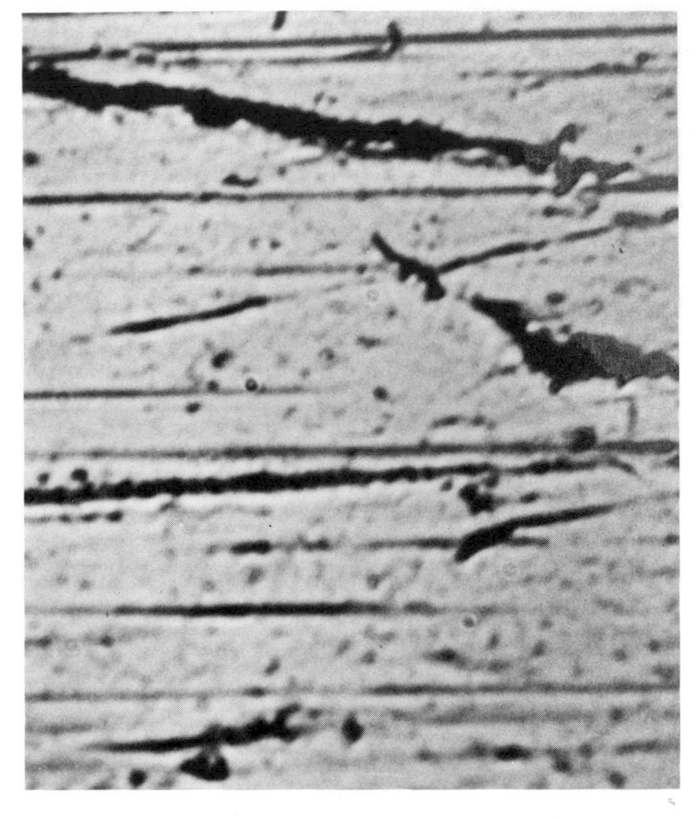

Figure 42. Micrograph showing scratches resulting from insufficient polishing or contamination from a coarser grit (500 ×).

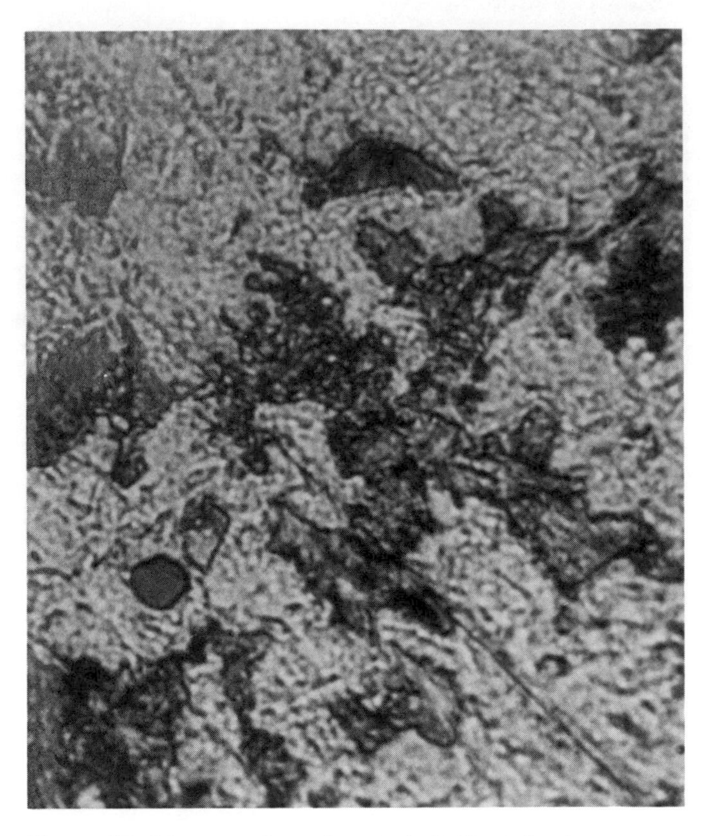

Figure 43. Micrograph surface is indistinct due to deformed surface (100 ×).

alternate method is to apply the enchant with an eye dropper. If the etching time is too short, the specimen will be under etched and the grain boundaries and other configurations will be faint and indistinct when viewed in the microscope. If the etching time is too long, the specimen will be overetched and will be very dark, having unusual colors. The time of etching must be controlled very carefully. Etching action is stopped by placing the specimen under a stream of water. Clean the specimen with alcohol and use a hair dryer to finish drying it. Do not rub the polished and etched specimen with a cloth or your finger as this will alter the surface condition of the metal.

Safety with Chemicals

Etching reagents, such as acids that are used on prepared and polished specimens of metals to bring out grain structures and other characteristics for viewing under a microscope, can be hazardous if they are mishandled. Since acids or alkaline solutions can cause severe burns to the skin and eyes, care must be exercised in their use. Small amounts of reagents are involved when etching for microscopic study, but much larger amounts are used for macroetching (for visual study) larger parts, which are often immersed in a heated bath.

Face masks and protective clothing (rubberized aprons and rubber gloves) should be used when using reagents. Rubber gloves and tongs should be used when handling the specimen. If any acid is accidentally spilled on your skin or splashed in your eyes, it should be **immediately** washed off in cold running water. Sodium bicarbonate solutions should always be available for neutralizing acids. Only water should be used for washing out eyes that are burned by acids, and it should be done as quickly as possible.

Red fuming nitric acid should not be used for testing purposes. If it is used on a sample of titanium, for example, a very dangerous explosion may follow. Only dilute acids may be used for testing purposes.

A ventilation system with a hood should be installed over the area where acids are used in

Table 2 Etching Reagents for Microscopic Examination of Metals

Metals	Enchant	Composition	Remarks
Iron and carbon steel	Nital	2 to 5 percent nitric acid in Methyl alcohol	Darkens pearlite in carbon steels. Differentiates ferrite from martensite; reveals ferrite grain boundaries. Shows core depth of nitrided steels. Time: 5 to 60 seconds.
	Picral	4 g picric acid 100 ml methyl alcohol	For annealed and quench-hardened carbon and low alloy steel. Not as good as Nital for revealing ferrite grain boundaries. Time: 5 to 120 seconds.
	Hydrochloric and picric acid	5 g hydrochloric acid 1 g picric acid 100 ml methyl alcohol	Reveals austenitic grains in quenched and quenched and tempered steels.
Alloy and stainless steels	Ferric chloride and hydrochloric acids	5 g ferric chloride 20 ml hydrochloric acid 100 ml distilled water	Reveals structures in iron-chromium-nickel and steels. Reveals structure of stainless and austenitic nickel steels.
High speed steels	Hydrochloric and nitric acid	9 ml hydrochloric acid 9 ml nitric acid 100 ml methyl alcohol	Reveals grain size of quenched and tempered high speed steels. Time: 15 seconds to 5 minutes.
Aluminum and aluminum alloys	Sodium hydroxide	10 g sodium hydroxide 90 ml distilled water	General enchant. Can be used for both micro- and macro-etching. Time: 5 seconds.
Magnesium and magnesium alloys	Glycol	75 ml ethylene glycol 24 ml distilled water 1 ml concentrated nitric acid	Can be used to reveal grain structures of most magnesium alloys.
Nickel and nickel alloys	Acetic and nitric acid	50 ml glacial acetic acid 50 ml concentrated nitric acid	Nickel-nickel copper alloys. Time: 5 to 20 seconds.

order to carry away the toxic fumes. An eyewash spray should also be available in the immediate vicinity.

Note. When mixing acids with water or methanol, **always** pour the acid into the solvent. When in doubt about the safe mixing procedures for chemicals, consult a chemist or a comprehensive chemistry handbook.

Self-Evaluation

1 Briefly describe the structure of an atom and the importance of the valence electrons.

2 How does the metallic bond work, and what effect does it have on metals?

3 Name five crystalline unit structures found in metals.

4 The growth from the nucleus of the grain that resembles small pine trees is called a ________________.

5 Are the grain boundaries a continuation of the regular lattice structure from one grain to another? Explain.

6 When austenitized carbon steel is quenched, why is the BCC crystal elongated into a body-centered tetragonal crystal structure?

7 What is meant by the term "allotropic" when applied to iron or steel?

8 Explain the relative advantages of fine grained steel as compared to coarse grained steel.

9 Briefly describe the theoretical mechanism of plastic deformation in metals.

10 How can a hard specimen be removed from a large piece of metal without damaging it and be prepared for metallurgical microscopic inspection?

Worksheet 1

Objectives 1 Show the effect of heat on fine and coarse grain materials.

2 Determine approximate grain size of steel samples used.

Material A piece of SAE 1095 $\frac{3}{8}$ inch diameter × 4 inches, and a set of number stamps.

The Metcalf Experiment When high carbon steels (about 1 percent carbon) are overheated to 1800°F (1024°C) and quenched in water, the grain structure becomes very coarse and weak. If the specimen is quenched from 1300°F (704°C) to 1500°F (815.6°C), however, the grain structure is very fine and much stronger. In their normal state the grains are fairly coarse, but the steels are soft and tough. This experiment will allow you to visually observe varying grain sizes.

By using the metal selected, you will be able to compare the grain growth of a fine grain to a coarse grain structured metal.

Procedure 1 Take the material and make notches or vee-grooves in it every $\frac{1}{2}$ inch, and about 0.050 inch deep. See Figure 44. File or machine a flat $\frac{1}{8}$ inch wide on one side and stamp a number on each section.

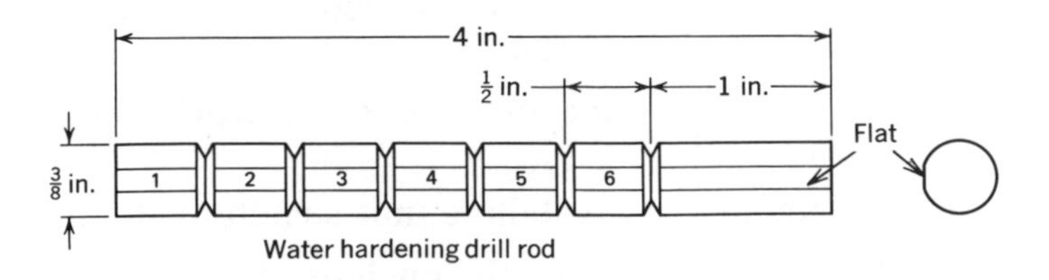

Figure 44. Metcalf experiment.

2 Heat one end (#1) with a torch to 1800°F (1024°C), a bright orange color, and make each section progressively cooler so #6 is gray or black and #4 is a dull red.

3 Quench in water until cool.

4 Place the sample in a vise with one section extending into the jaws.

5 Use a pipe or tube and break off each section. **Wear safety glasses.**

6 Compare the grain size of your specimens to the standard grain size chart in the chapter. Use a 100× power magnifier or microscope when viewing the specimen.

7 If a hardness tester is available, test the hardness of each section and list on the chart. Take the results of your experiment to your instructor for evaluation.

SAE 1095 steel

Grain size	#1	#2	#3	#4	#5	#6
Approximate temperature	______	______	______	______	______	______
Rc	______	______	______	______	______	______

Conclusions

1 Which piece of metal can be considered fine grained, coarse grained? Explain why. (Compare them with the Shepherd grain size standards.)

2 How does grain size affect tensile strength?

3 What conclusion can you draw from the experiment relating grain size to hardness? To toughness?

4 What type of fractures do you see in the broken sections?

Worksheet 2

Objectives

1 Learn to polish and etch specimens for microscopic viewing.

2 Learn to use a metallurgical microscope.

3 Learn to identify several kinds of microstuctures.

Material Metallurgical microscope, nital enchant (five percent nitric acid in alcohol), wash bottle of alcohol, electric polishing wheel and polishing compound, and four mounted specimens of unidentified carbon steel.

Procedure

1 Encapsulate the specimen in a mounting press and grind or sand any rough surfaces lightly (do not overheat) to make them level. Be sure to permanently mark each plastic capsule immediately.

2 Sand the surface of the hand-held specimen on three or four progressively finer grit papers to 600 grit placed on a flat surface.

3 When the lines are all one direction, change to a finer grit and rotate the

specimen to show the new lines. Be sure to clean the specimen thoroughly between each grit change and carefully avoid contaminating the felt on the polishing wheel.

4 Take a mounted specimen to the polishing wheel. Turn on the wheel and add a little slurry (polishing compound and water) to the felt surface. Gently press the specimen to the wheel surface and move it in the opposite direction of wheel rotation. Move the specimen around the wheel several times, then check the surface. When it becomes mirror bright, it is ready for etching.

5 Apply the nital on the surface of the specimen and count about 2 or 3 seconds. Rinse with water and immediately follow with an alcohol rinse from the wash bottle. This will remove the water and prevent rusting. Allow the alcohol to dry.

6 Place the specimen under the microscope objective on the stage, and carefully lower the objective to the surface of the specimen. It should not quite touch. Use the fine adjustment button to raise the microscope until the specimen comes into focus.

7 If the microstructure is not visible, apply the nital again for a few seconds, rinse with water and then with alcohol wash.

8 Observe each specimen and determine its microstructure.

Conclusion Draw four circles on a piece of paper to represent each specimen and draw in each circle the lines and shapes that you see in the microscope. Write the names of the structures and extend lines to point them out as shown in Figure 45. Write your conclusions for each microstructure.

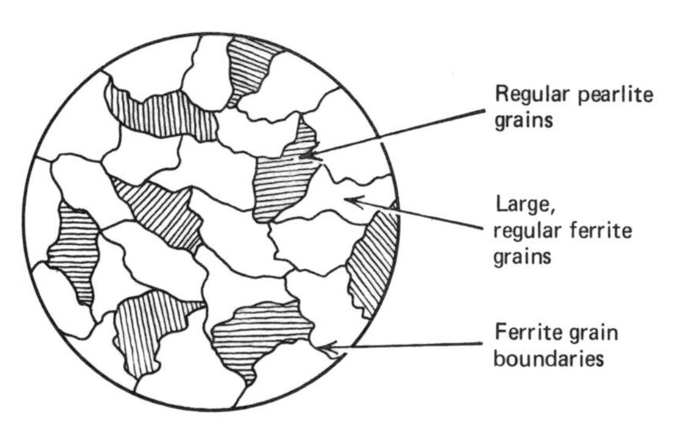

Figure 45. Sketch of etched and polished specimen, as seen in the microscope.

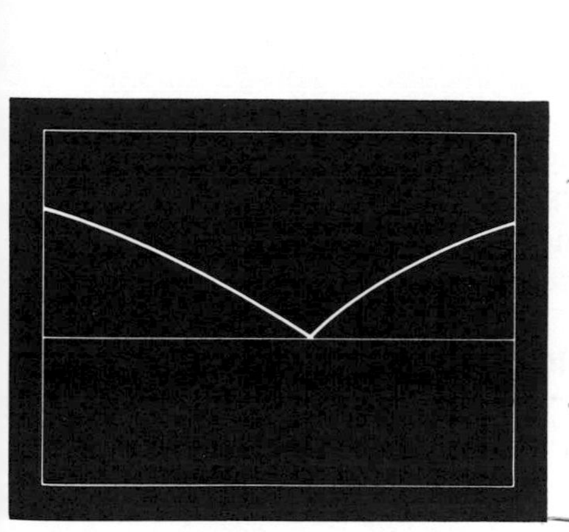

8 Phase Diagrams and the Iron-Carbon Diagram

Constitutional diagrams (sometimes called alloy or phase diagrams) are a useful means of explaining and understanding the behavior of metals. Many very complex diagrams for various alloys are used by metallurgists, but only several of the simple types will be used in this chapter. These alloy diagrams are the "road maps" of the metallurgist that help him or her to develop new alloys. The iron-carbon diagram is basic to an understanding of heat treating iron and steel.

OBJECTIVES

After completing this chapter, you will be able to:

1 Demonstrate your understanding of phase diagrams by recognizing their parts.
2 Establish relative carbon content by microscopic evaluation.
3 Identify various cast iron compositions by microscopic examination.

INFORMATION

Matter may exist in three states: solid, liquid, or gaseous. Some substances are capable of changing within the solid state to other phases or crystalline structures. The ability to change into different phases is called allotropy. Iron is an allotropic element and changes from face-centered cubic to body-centered cubic during cooling. When a substance changes phases while cooling, there is a release of heat, which appears on the cooling curve graph as a straight or curved horizontal line (Figure 1). As the temperature rises, more energy is required to make the transformation from one structure to another. This means that here too there is an elapsed time at the point of transformation, shown in Figure 1 as a straight line. Heating and cooling transformation points are at slightly different temperatures.

When two metals are alloyed together, the temperature of phase changes will be different for every combination of the two metals. The tempera-

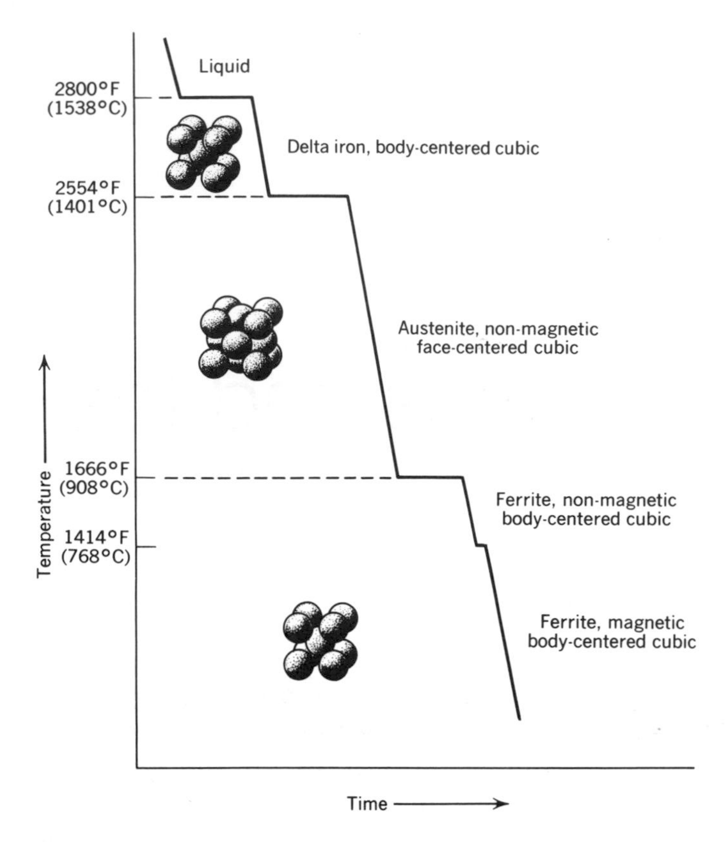

Figure 1. Cooling curve for pure iron (*Machine Tools and Machining Practices*).

tures and compositions of phase changes can be graphed so that all possible combinations of two pure metals are represented.

Solutions, Liquid, and Solid

When two or more metals are heated to or above their melting points and combined, they usually become a solution that is considered an alloy. The metal composing the greatest percentage is the solvent and the metal composing the smaller percentage is the solute. Some molten metals will not dissolve at all in other molten metals, but separate or form a mixture. We are accustomed to think of solutions in terms of liquids such as salt or sugar solutions. There are also limits to solubility; water will dissolve only a certain amount of salt or sugar. Oil *will dissolve* in water, but to a very limited extent. There is a similarity between liquid metal solutions and other liquid solutions as some metals will only partially dissolve other metals.

Solutions may also be found in solid metals, but the changes are made within the lattice and grain structure in solids. The atoms are not quite so free to move about as they are in liquids. Atoms move only to a limited extent and much slower in solids. The rate of movement (diffusion) is dependent upon the temperature.

Types of Solution

The dissolving of one material into another can take place in two ways: **substitutional** (replacement of atoms by others) and **interstitial** (spaces between the atoms in the lattice), as illustrated in Figure 2.

Substitutional Solid Solutions

A substitutional solid solution is a solution of two or more elements with atoms that are of relatively the same size. This requirement is necessary in that the alloying atoms need to replace the regular atoms in the lattice structure and not just fit in the spaces between the regular atoms as it does in interstitial solutions.

Type I alloys are completely soluble in both the liquid and solid states, and their solid phase has a substitutional lattice (Figure 3). **Type II alloys** are soluble in the liquid state but insoluble in the solid state (Figure 4). Alloys that are not solid solutions are usually mixtures of various forms. Compounds may also form between two metals (for example, copper and aluminum to form copper aluminide) and between metals and nonmetals like iron and carbon to form iron carbide (Fe_3C). The following are some examples of mixtures in metals.

1 Solid solution and solid solution.
2 Solid solution and compound.
3 Compounds and compounds.
4 Compound and pure material.

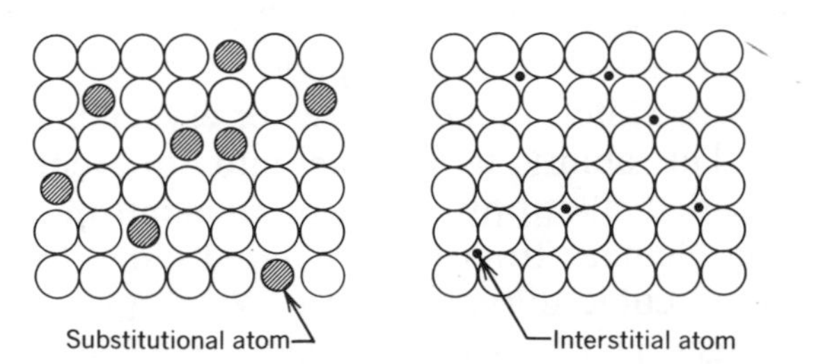

Figure 2. Substitutional solid solution and interstitial solid solution (*Machine Tools and Machining Practices*).

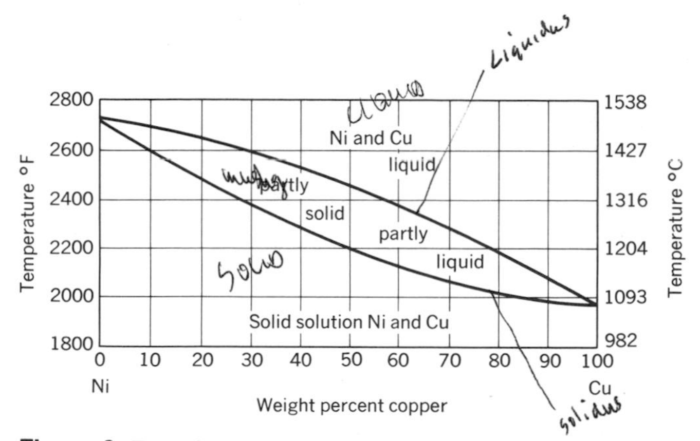

Figure 3. Type I, copper-nickel phase diagram (*Machine Tools and Machining Practices*).

When one metal is *completely soluble* in another, such as copper and nickel, both metals must have the same lattice structure (distance between the atoms), atoms of the same relative size, and a chemical desire to combine. This type of solution is called a *continuous solid solution*. Figure 3 is a diagram showing how nickel and copper combine into a continuous solid solution. When the above factors vary, metals take on varying degrees of solubility. Copper-silver is an example of this. Their atom size differs somewhat more than copper–nickel, and their chemical desire to combine is much less.

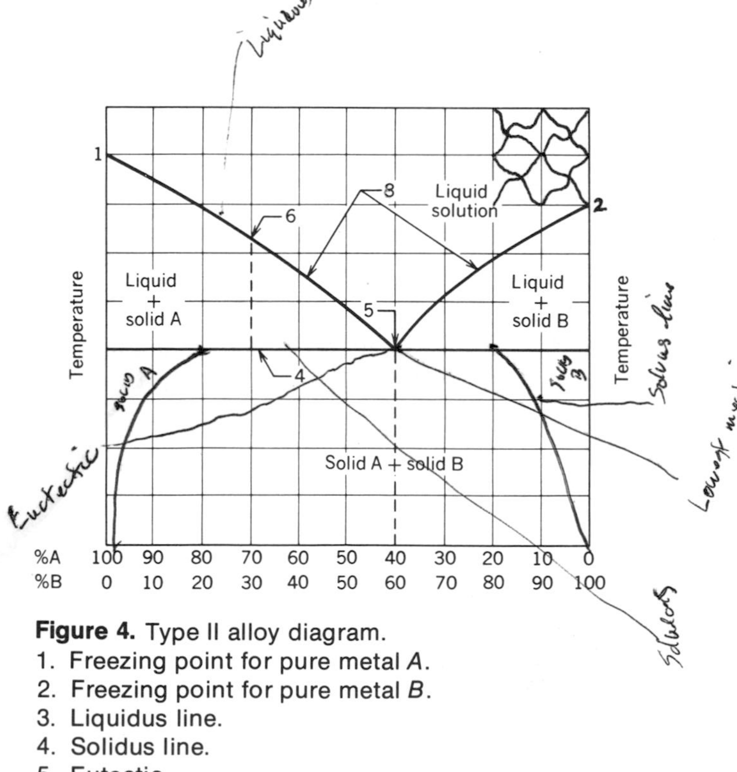

Figure 4. Type II alloy diagram.
1. Freezing point for pure metal *A*.
2. Freezing point for pure metal *B*.
3. Liquidus line.
4. Solidus line.
5. Eutectic.
6. Freezing point for 70 to 30 percent alloy.

To summarize: as a general rule, the more alike the two metals are chemically and physically, the more they tend to form continuous solid solutions. The following are necessary for a continuous solid solution to form.

1. The size of the atom of the alloying metals cannot differ by more than 15 percent.
2. The chemical characteristics should be similar.
3. The metals must crystallize in the same pattern, such as BCC, FCC, or CPH.

It can be seen from Figure 5 that solubility is *continuous* at each end, but *insoluble* (a mixture) in the middle. This type of solution is called *terminal solid solution,* **Type III.** These terminal solid solutions are found in solution heat treating and in the precipitation type of hardening used for some kinds of stainless steels and some nonferrous metals. This is further discussed in Chapter 14 "Heat Treating of Nonferrous Metals."

Interstitial Solid Solutions

Solid solutions that become insoluble below the transformation temperature often separate into a **lamellar** (platelike) structure, which is a mixture (Figure 6). A 100 percent lamellar structure is found in these mixtures in the **eutectic composition** (lowest melting point); other percentages in a two-alloy system produce alloy A plus eutectic, or alloy B plus eutectic. See Figure 4.

Interstitial solid solutions are made up of alloying elements or atoms that differ greatly in size, as Figure 2 demonstrates. The alloying atoms must be small enough to fit within the lattice structure of the base material. It has been determined that the alloying atom should be about one-half the size of the base atom. Common elements that are able to form interstitially with iron are carbon, nitrogen, oxygen, hydrogen, and boron. Some of these elements are also important in their ability to combine chemically with the base metal to form compounds such as the combining of iron (Fe) and carbon (C) to form iron carbide (cementite) (Fe_3C). Iron carbide and other compounds such as iron nitrides and chromium nitrides have the potential for greatly increasing hardness, strength, heat resistance, and abrasion resistance of metals. The development of the carbides, nitrides, and borides in

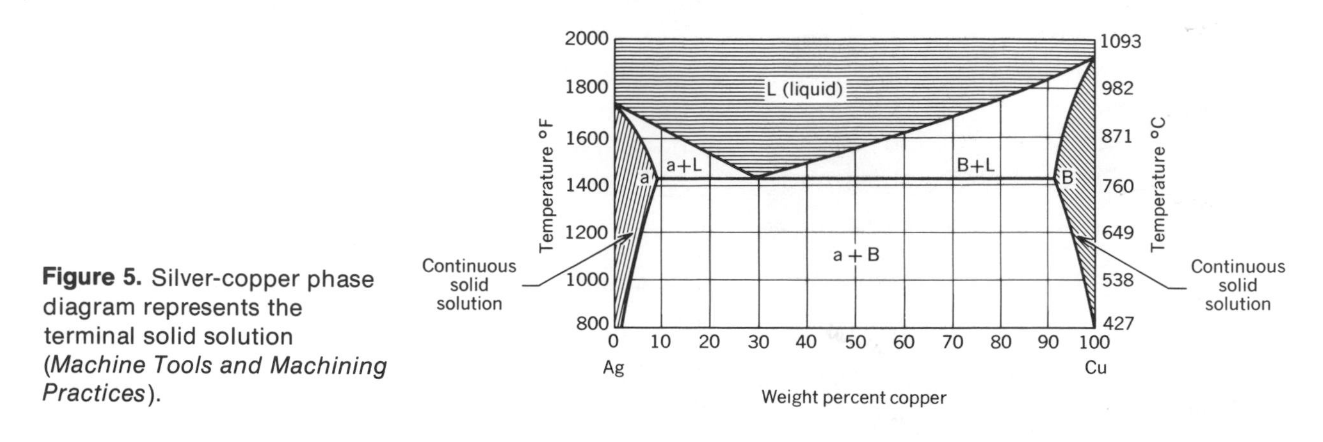

Figure 5. Silver-copper phase diagram represents the terminal solid solution (*Machine Tools and Machining Practices*).

metals has greatly aided the aircraft and space industry in their building programs.

A third condition is the combining of substitutional and interstitial atoms with the same base metal. Figure 7 illustrates how this would be done. Many alloys are formed in this way. It makes possible the strength, hardness, and heat treatment advantages of both types of lattice structures within the same material.

In both the substitutional and interstitial solid solutions, the strengthening of the material is accomplished through the distortion of the lattice structure caused by the alloyed atoms. The lattice distortion creates a strain along the slip planes and grains of the material, which results in the increase of strength and hardness.

Phase Changes of Iron

Iron, being an allotropic element, can exist in more than one lattice unit structure, depending on temperature. When a substance goes through a phase change while cooling, it releases heat. When a phase change is reached while heating, it absorbs heat. This characteristic is used to construct cooling curve graphs. If a continuous record is kept of the temperature of cooling iron, we can construct a graph that will resemble Figure 1.

Each flat segment of the cooling curve represents a phase change. These flat portions are caused by the release of heat during phase changes. At 2800°F (1538°C), iron changes to a solid body-centered cubic structure (delta iron). All of the remaining changes involve a solid of one lattice structure transforming into a solid of another lattice structure. At 2554°F (1401°C) BCC delta iron changes to FCC austenite. Austenite transforms to BCC ferrite at 1666°F (908°C). The next change is not a phase change at all, but it does give off heat. This is the change from nonmagnetic ferrite to magnetic ferrite.

The Iron-Carbon Diagram

In the preceding phase diagrams of alloy systems, none of the metals went through any solid state phase changes (transformation). There are some new lines that must be added to account for solid phase transformations (Figure 8). Many of the lines, the liquidus, solidus, and the eutectic point, are similar. This diagram differs from previous ones in that the diagram ends on the right at 6.67

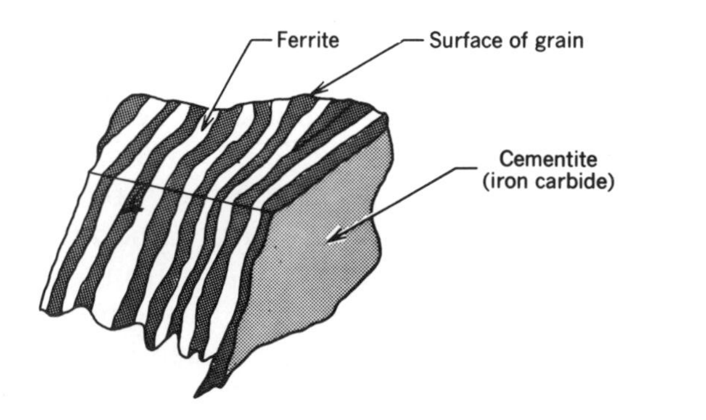

Figure 6. The lamellar pearlite microstructure (*Machine Tools and Machining Practices*).

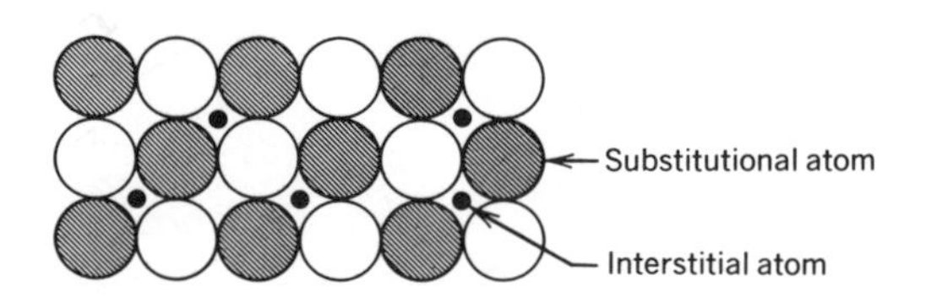

Figure 7. Substitutional and interstitial solid solution (*Machine Tools and Machining Practices*).

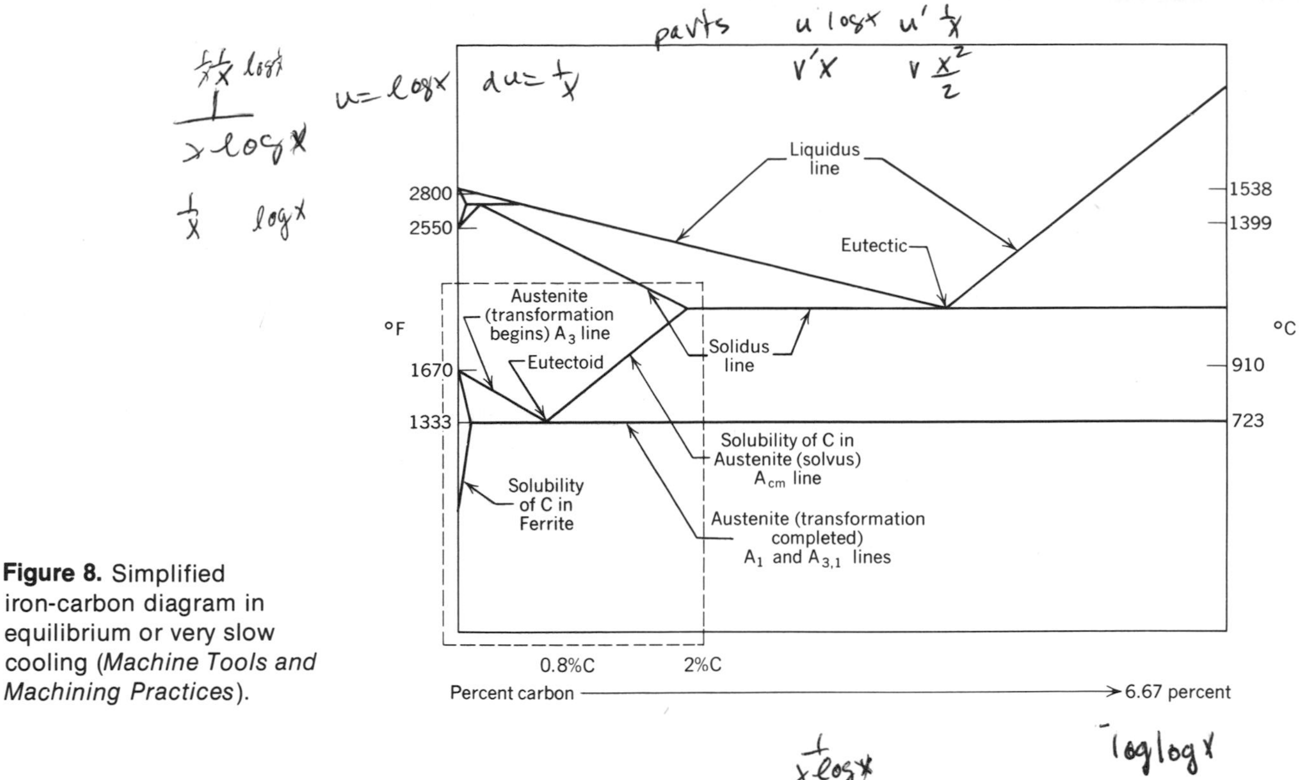

Figure 8. Simplified iron-carbon diagram in equilibrium or very slow cooling (*Machine Tools and Machining Practices*).

percent carbon (C) instead of 100 percent carbon. The rest of the diagram from 6.67 to 100 percent carbon would give no useful information about steels and cast irons. But a more important reason for ending the diagram at this point is that 6.67 is the percent of carbon by weight in the compound iron carbide (Fe_3C). The iron-carbon diagram is referred to as an equilibrium or phase diagram. A phase diagram indicates the transformations that take place during very slow cooling or equilibrium conditions.

The Delta Iron Region

The area at the left of the diagram between 2800°F (1538°C) and 2554°F (1410°C) in Figure 8 describes the solidification and transformation of delta iron. This area has no commercial value in heat treating and therefore is only of passing interest.

The Steel Portion

Study the portion of the diagram outlined by dashed lines in Figure 8 and you will see how it resembles phase diagrams that you have already studied. However, a little different terminology is used for the lines. Since the metal is now a solid, the terms liquidus and solidus do not apply. Since eutectic means low melting point, this term is not correct, either. The suffix "-oid" means similar but not the same. The **eutectoid** point appears like a eutectic on the diagram, but it is the lowest temperature transformation point of solid phases, while eutectic is the lowest freezing point of a liquid phase. Instead of a liquidus line, we have the line that shows the transformation from austenite to ferrite, called the A_3 line, and the line showing the amount of carbon that is soluble in austenite, called the A_{cm} line (Figure 9). Instead of a solidus line, we have the line that shows where austenite completes its transformation to ferrite and where pearlite is formed. This line is called A_1 to the left of the eutectoid point and $A_{3,1}$ to the right of the eutectoid point.

Microstructures

When carbon steel is viewed through a metallurgical microscope, the microstructure may be identified as **pearlite, ferrite, cementite** (iron carbide), **austenite,** or **bainite** with some variations, depending on heat treatment. See Chapter 7 for illustrations of microstructures.

Cementite, or iron carbide, is the hardest structure on the iron-carbon diagram. It is 6.67 percent carbon by weight. The iron-carbon diagram shows the composition of iron carbide or cementite on the right. Massive cementite in grain

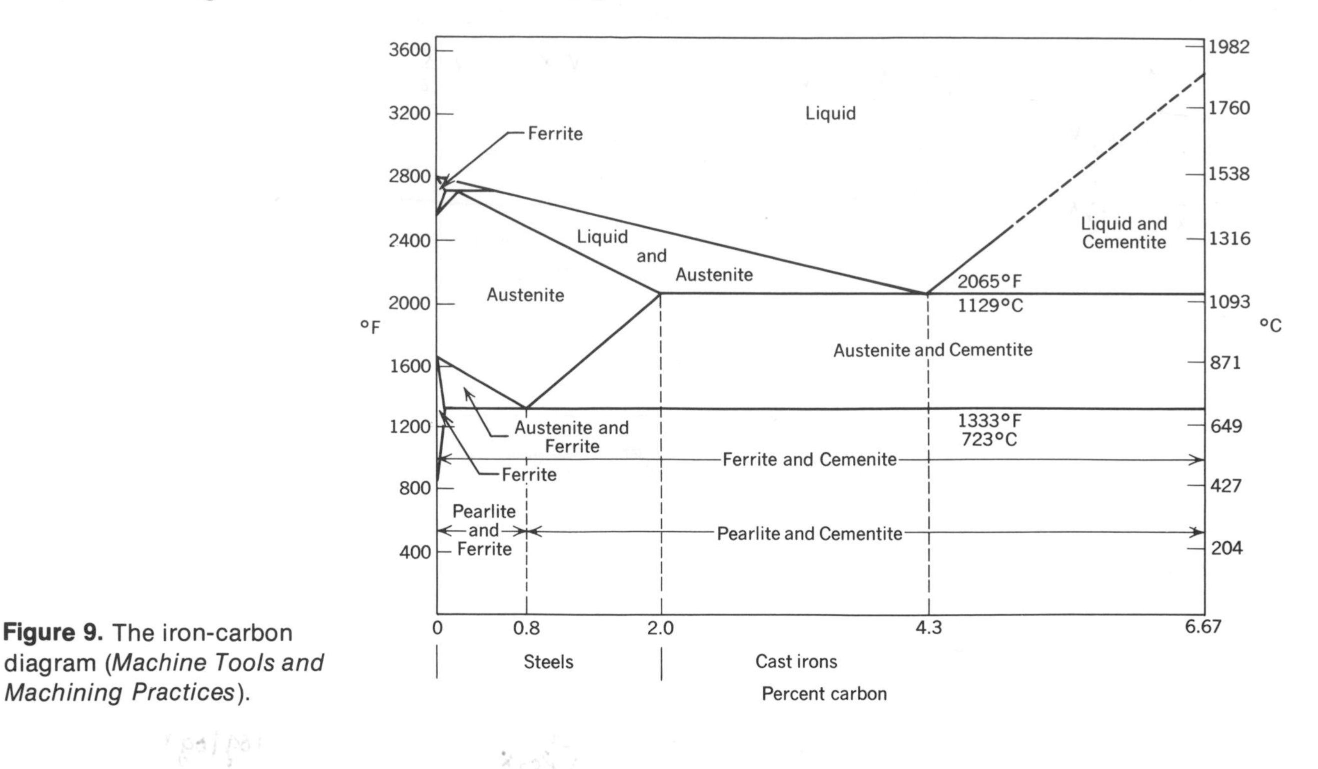

Figure 9. The iron-carbon diagram (*Machine Tools and Machining Practices*).

boundaries and some cast irons etch white (Figure 10). This is found in steels containing more than 0.8 percent C.

Austenite is a face-centered cubic iron. It has the ability to dissolve carbon interstitially to a maximum of 2 percent at 2065°F (1129°C). See Figure 8. The iron-carbon diagram shows the solubility range of carbon in austenite. Austenite is not stable at room temperature except in some alloy steels, but decomposes into pearlite, ferrite, and/or cementite.

Ferrite is a body-centered cubic iron. It will dissolve only 0.008 percent carbon at room temperature and a maximum of 0.025 percent carbon at 1330°F (721°C). Locate the narrow solid solution area at the left of the diagram as seen in Figure 9. To a limited extent in this area ferrite forms an interstitial solid solution with carbon. Ferrite is the softest structure that appears on the diagram. It appears light gray or white through the microscope (Figure 11).

Pearlite is the eutectoid mixture with 0.8 percent carbon (Figures 12 and 13). It appears like a fingerprint and is actually composed of thin alternating layers of ferrite and cementite. The cementite etches dark and the ferrite etches light. Pearlite forms at 1330°F (721°C) when cooling; 0.8 percent carbon produces 100 percent pearlite, the iron-carbon eutectoid.

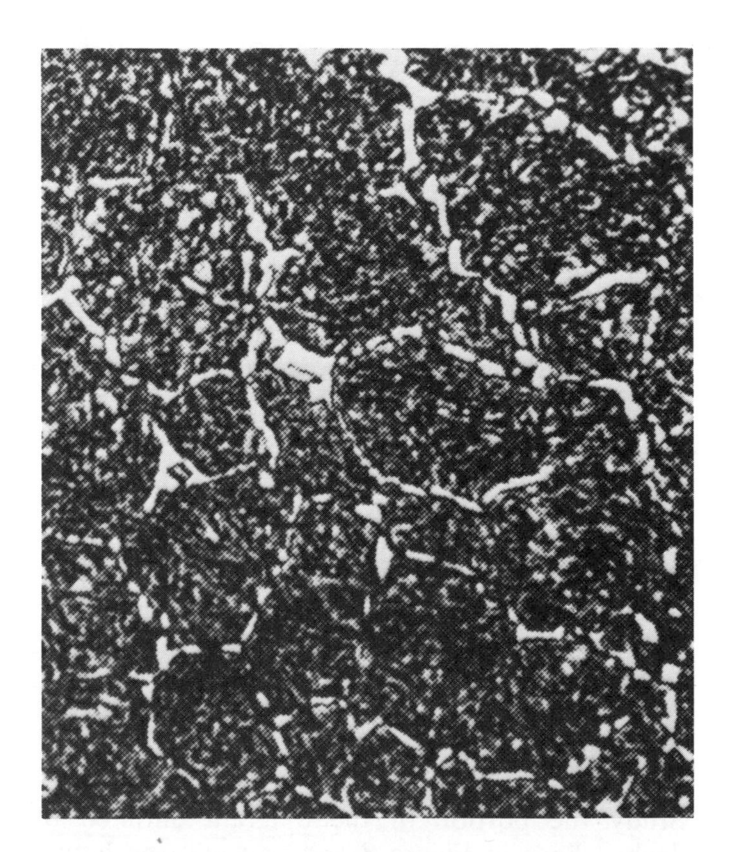

Figure 10. Microstructure of 1 percent carbon steel showing cementite in the grain boundaries (500 ×) (by permission, from *Metals Handbook* Volume 7, Copyright American Society for Metals, 1972).

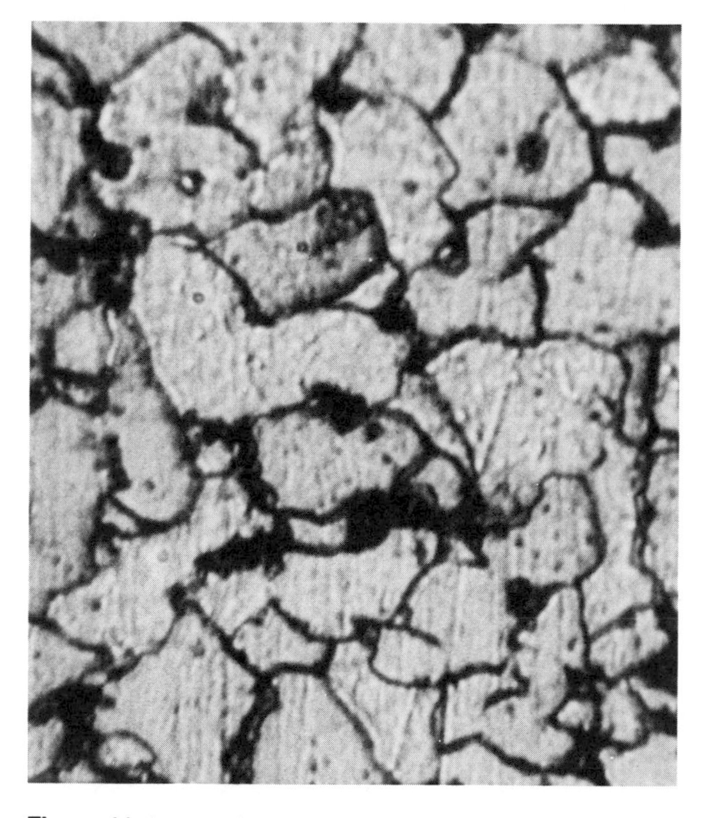

Figure 11. Low carbon steel showing the grain boundaries of mostly ferrite grains with isolated grains of fine pearlite (250 ×).

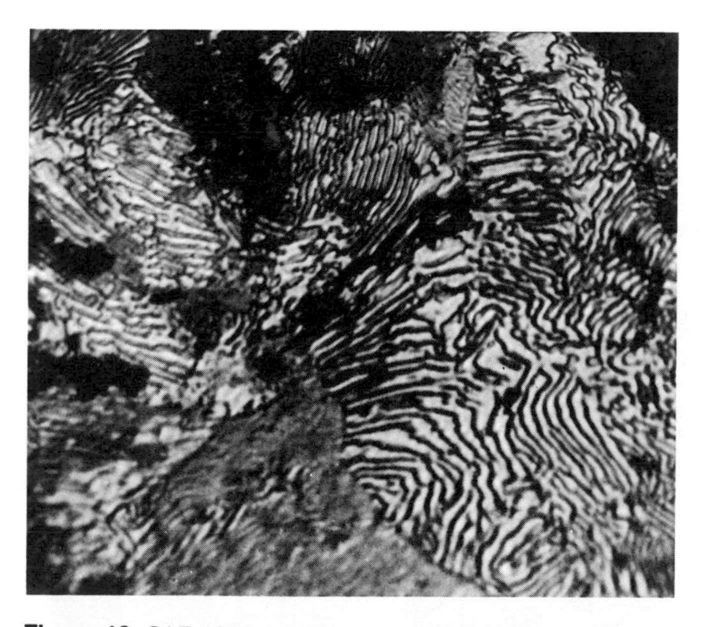

Figure 12. SAE 1090 steel slowly cooled (100 percent very coarse pearlite (500 ×) (*Machine Tools and Machining Practices*).

Figure 13. SAE 1030 steel (fine pearlite and ferrite) 100 × (*Machine Tools and Machining Practices*).

The Slow Cooling of 1020 Carbon Steel

For a plain carbon steel less than eutectoid, such as SAE 1020 (or 0.20 percent carbon) steel during cooling, it can be seen in Figure 14, line 1 at point *a* that the entire microstructure is austenite. After crossing the A_3 line (point *b*), ferrite begins to form along the austenite grain boundaries (Figure 15). Further cooling causes more ferrite to form. Because the solubility of carbon in ferrite is so low, excess carbon must be dissolved by the remaining austenite. At point *c*, most of the austenite has become ferrite. The remaining austenite has a carbon content of 0.8 percent carbon. As the steel crosses the A_1 line, the remaining carbon-rich austenite transforms to the eutectoid or pearlite at point *d*.

Slow Cooling of a 1095 Carbon Steel

At point *a* (Figure 16) the SAE 1095 steel is entirely austenite and interstitially dissolved carbon. When the steel cools to point *b*, it has crossed the A_{cm} line. The A_{cm} line indicates the limit of carbon solubility in austenite. Since the austenite can no longer hold the entire amount of carbon in solution, cementite begins to appear along the grain boundaries. At point *c* the solubility of the austen-

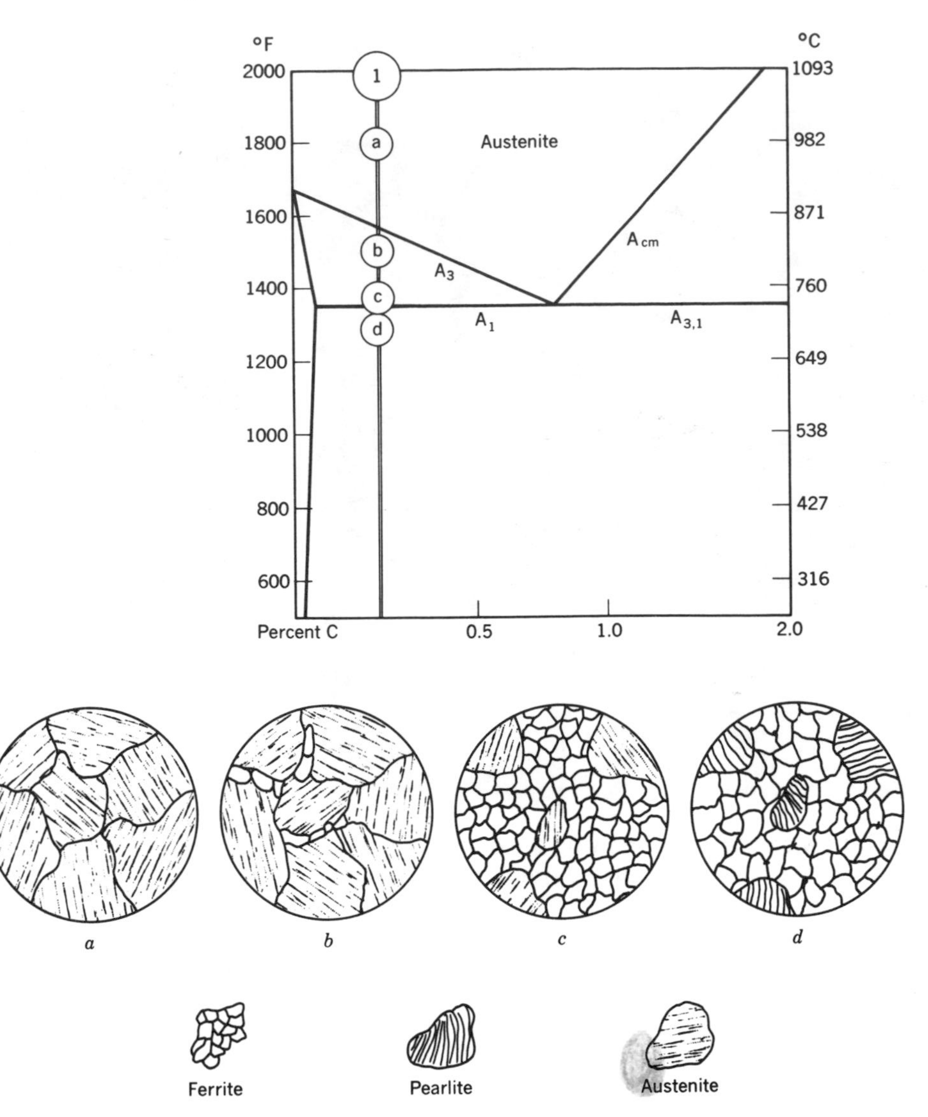

Figure 14. Microstructures at various temperatures of very slow cooling SAE 1020 steel (*Machine Tools and Machining Practices*).

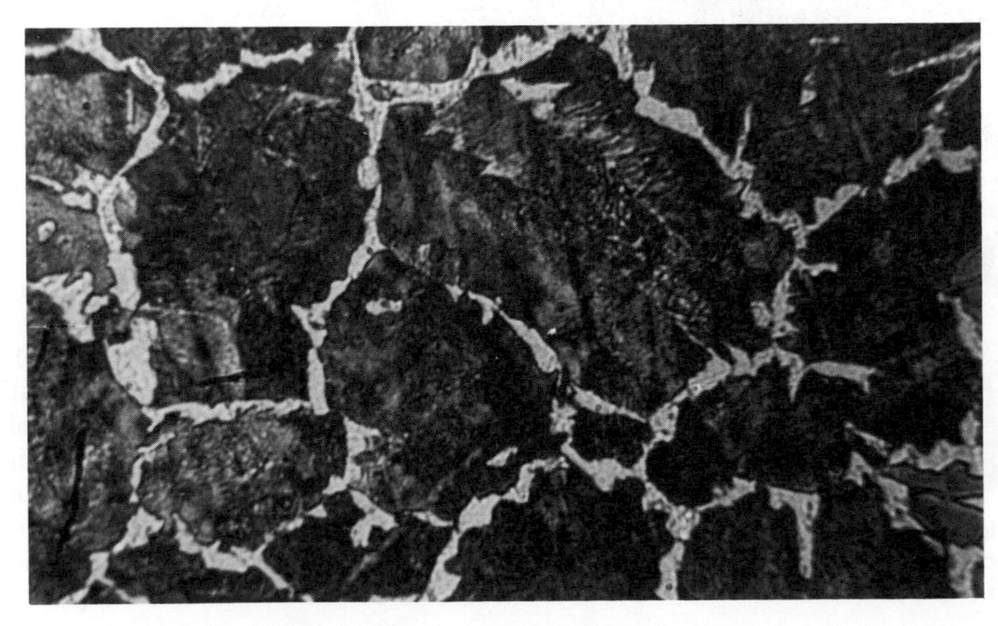

Figure 15. This SAE 1045 carbon steel micrograph shows the ferrite at the grain boundaries where it is first formed while cooling. The ferrite, in this case, remained at the grain boundaries instead of forming ferrite grains because of the relatively low percentage of ferrite and fairly rapid cooling rate (500 ×).

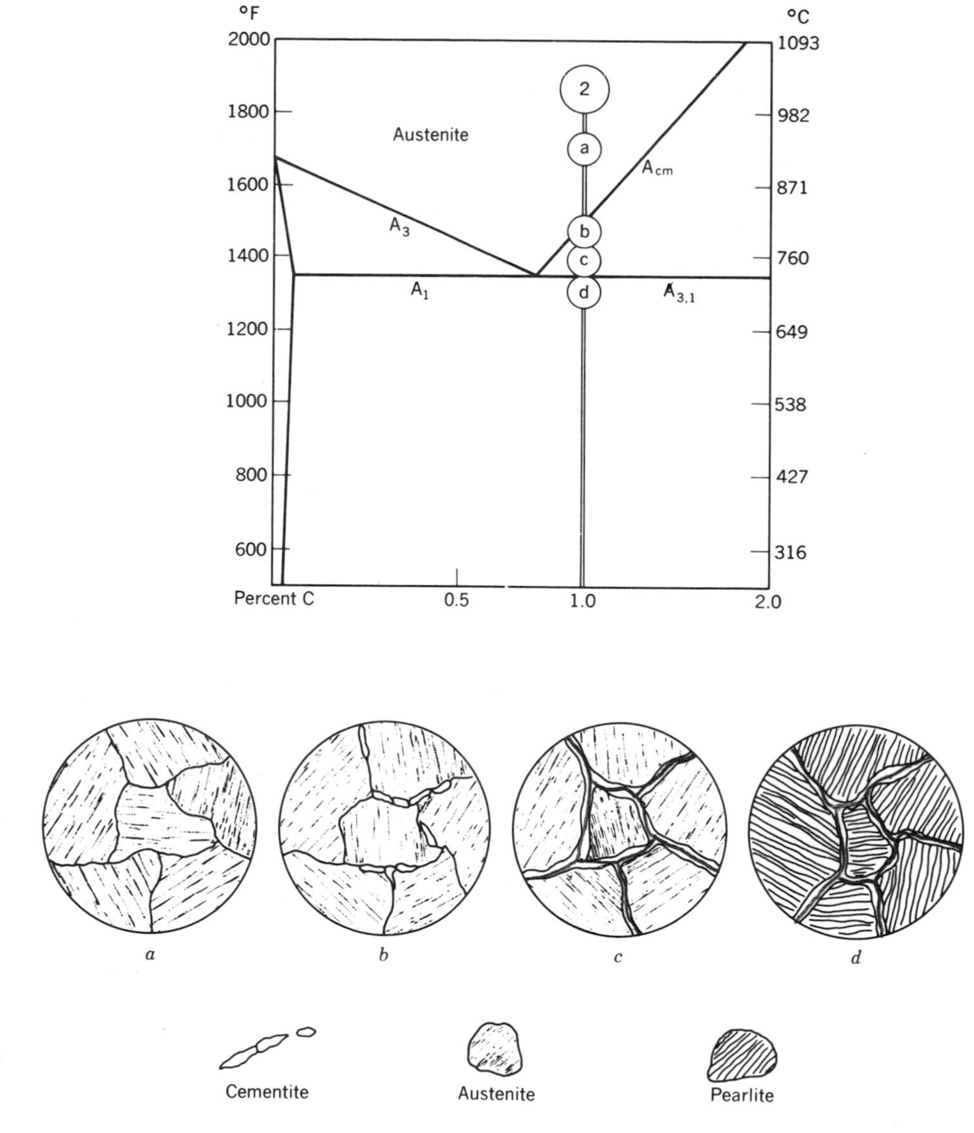

Figure 16. Microstructures at various temperatures when cooling SAE 1095 steel (*Machine Tools and Machining Practices*).

ite has dropped even further, so that the additional cementite has formed a fairly continuous network between the austenite grains. The carbon content of the remaining austenite is now the eutectoid composition of 0.8 percent carbon. At point *d* the steel has crossed the $A_{3,1}$ line and the remaining austenite transforms to pearlite.

Heating 1020 Steel

There are some very important differences in the transformation of steels during cooling and heating. If in Figure 14 the 1020 steel is raised from room temperature to 1800°F (982°C), the following changes take place. At room temperature, the microstructure is ferrite and pearlite; no change takes place until the A_1 line is reached. Now the pearlite grains transform to austenite, but the remaining ferrite grains are not affected until the A_3 line is reached. At the A_3 line the remaining ferrite grains recrystallize to austenite. The structure becomes fine grained at this temperature because grain growth is interrupted by the diffusion of carbon. The austenite that was formed at a lower temperature from pearlite is richer in carbon than the new austenite. Consequently, carbon must diffuse into the new austenite until the carbon content is uniform throughout. As the temperature increases beyond the A_3 line, grain size also increases dramatically.

Grain size has a marked effect on the strength, toughness, and plasticity of steels at room temperature. Large grain steels have lower strength and toughness and should be avoided when these properties are important. If the steel is to be cold formed, the increased plasticity of large grain size may be desirable, but not always; in some cold forming applications an "orange peel" effect develops from large grains.

When steels are overheated into the area of large grain growth, the grains remain large until transformation takes place. Cooling steel that has been overheated will not restore small grain size for heat treating. A reheating from normal temperature is necessary, and often a normalizing treatment is needed to restore a fine grain size.

Alloying Elements and the Iron-Carbon Diagram

Alloying elements move the transformation lines of the iron-carbon diagram. A common alloying element is chromium. Considering the austenite area of the iron-carbon diagram, it can be seen that the effect of increasing chromium is to decrease the austenite range (Figure 17). This will increase the ferrite range. Many other alloying elements are ferrite promoters also, such as molybdenum, silicon, and titanium.

Nickel and manganese tend to enlarge the austenite range and lower the transformation temperature (austenite to ferrite). A large percentage of these metals will cause steels to remain austenitic at room temperature. Examples are 18-8 stainless steel and 14 percent manganese steel. The A_1-$A_{3,1}$ line is also lowered by rapid cooling. Any alloy addition will move the eutectoid point to the left (with a few exceptions such as copper) or, in other words, lower the carbon content of the eutectoid composition.

Cast Iron Alloys

Cast iron, essentially an alloy of iron, carbon, and silicon, is composed of iron and from 2 to 6.67 percent carbon, plus manganese, sulfur, and phosphorus. Commercial cast iron contains no more than 4 percent carbon. Cast iron is often alloyed with such elements as nickel, chromium, molybdenum, vanadium, copper, and titanium. Alloying elements toughen and strengthen cast irons.

Since the maximum amount of carbon that iron can contain in an austenitic solution is 2 percent, any amount of carbon in excess is in a supersaturated condition. The solubility of carbon in iron varies with the temperature. The supersaturated or excess carbon will either form graphite flakes as seen in gray cast iron (Figure 18), or iron carbide as often seen in white cast iron. Graphite may also take the form of spheres or nodules as found in nodular cast iron (Figure 19) and malleable cast iron (Figure 20).

The types of cast irons are:

1 White cast iron. Most of the carbon is in the combined form of cementite (iron carbide) (Figure 21). This is a very hard, brittle material that often contains pearlite grains.

2 Gray cast iron. Most or all of the carbon is in the form of graphite. Other microstructures are pearlite and ferrite.

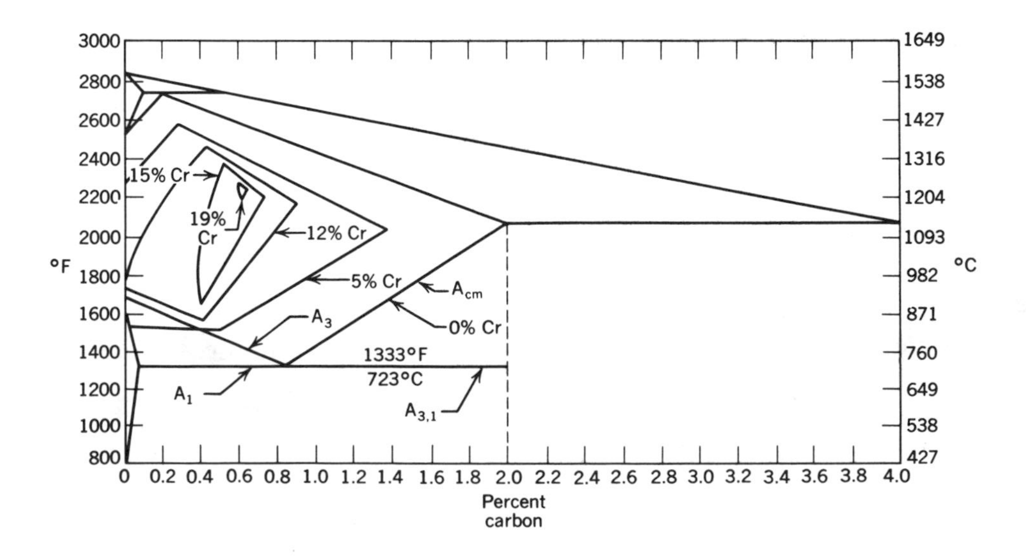

Figure 17. The effect of chromium on the austenite range of steel (*Machine Tools and Machining Practices*).

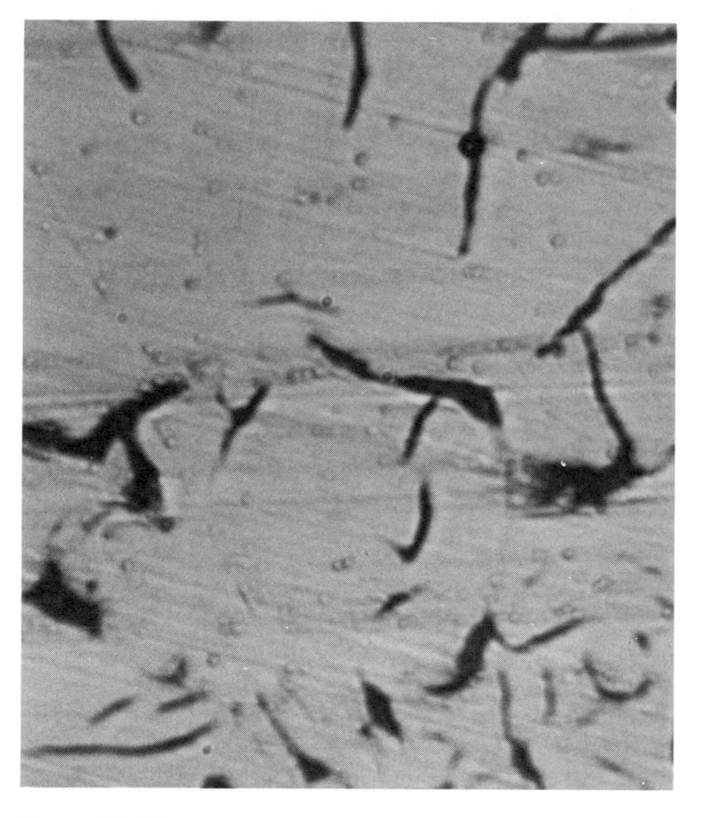

Figure 18. Micrograph showing the graphite flakes in gray cast iron (500 ×).

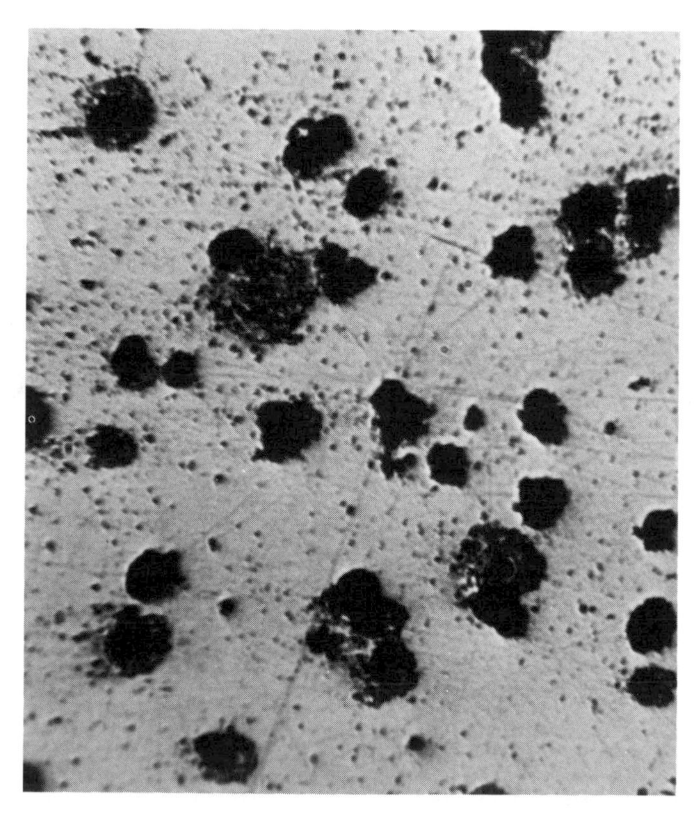

Figure 19. Nodular cast iron, unetched (500 ×).

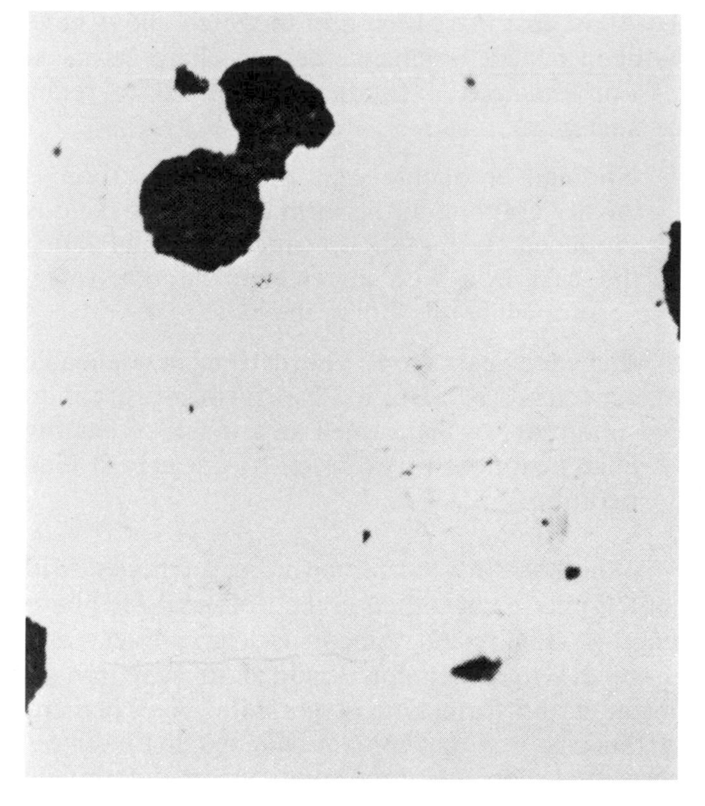

Figure 20. Micrograph of malleable cast iron, unetched (100 ×) (By permission, from *Metals Handbook* Volume 7, Copyright American Society for Metals, 1972).

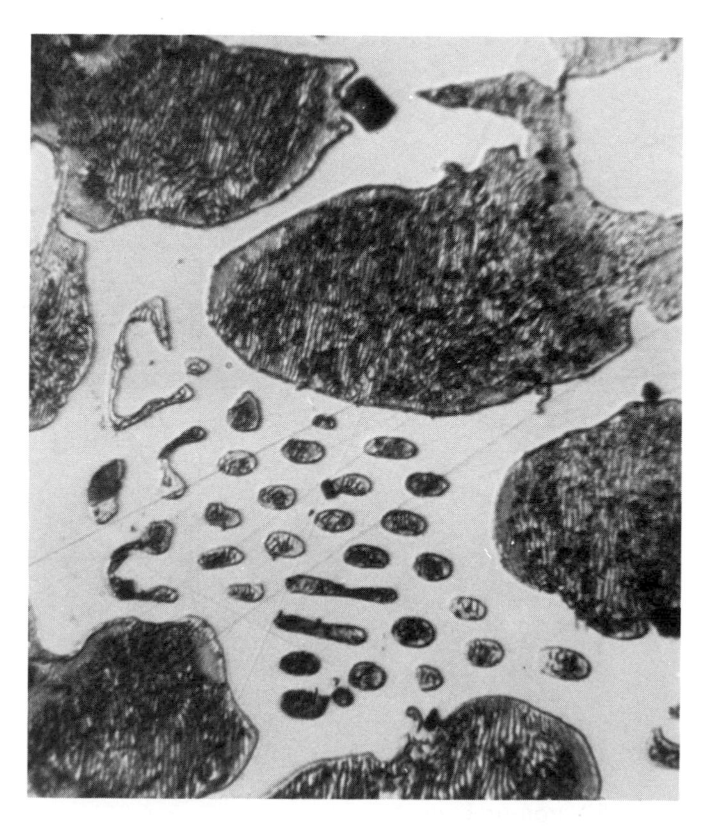

Figure 21. White cast iron, unetched (500 ×).

3 Alloy cast iron. Cast iron to which alloys have been added to enhance certain characteristics. For example, an addition of nickel to retain austenite.

4 Nodular or ductile cast iron. The carbon is mostly graphite in the form of spheroids and is produced during solidification by inoculating the cast iron with an element such as magnesium while it is still in the ladle.

5 Malleable cast iron. The carbon in malleable cast iron is also in the form of graphite spheroids, but is formed as a result of lengthy heat treatment of white cast iron at high temperatures.

The solubility of carbon in iron when slowly cooled may be observed in the iron-graphite phase diagram (Figure 22). Silicon is a graphitizer and about 2.5 to 3 percent is added to cast iron to promote the formation of graphite. A 3 percent carbon cast iron that contains silicon can dissolve 2 percent carbon. The remaining 1 percent carbon is free to change to graphite when solidification occurs at 2075°F (1135°C). The solubility of carbon decreases as the temperature drops and at 1360°F (738°C) the solubility of carbon has dropped to 0.5 percent. Now a total of 1.5 percent carbon has been precipitated (pushed out) to form graphite because of the silicon content. The final structure, therefore, will be ferrite and graphite flakes (ferritic gray cast iron). If less silicon were added, the final structure would probably be pearlite and graphite flakes (pearlitic gray cast iron). The rate of cooling also determines the type of final structure. Rapid cooling promotes cementite or massive white areas to form in combination with gray iron (mottled cast iron). Very rapid cooling can produce white cast iron having a composition of iron carbide (cementite) and pearlite. Figure 23 shows the composition limit for gray, mottled, and white cast irons. It can be seen in the graph that as the carbon increases, a smaller amount of silicon is required to form gray cast iron.

Pearlitic cast irons may be heat treated just as steel can. It can also be locally or surface hardened by flame or induction heating. Hardening operations do not affect the graphite constituent of the casting; it only changes the pearlite to martensite.

Gray Cast Iron

Gray cast irons are classified according to their tensile strength as Table 5 in Chapter 3 indicates. The American Society for Testing Materials (ASTM) classes run from 20 to 60. The number correlates to the tensile strength of the material in thousand pounds per square inch.

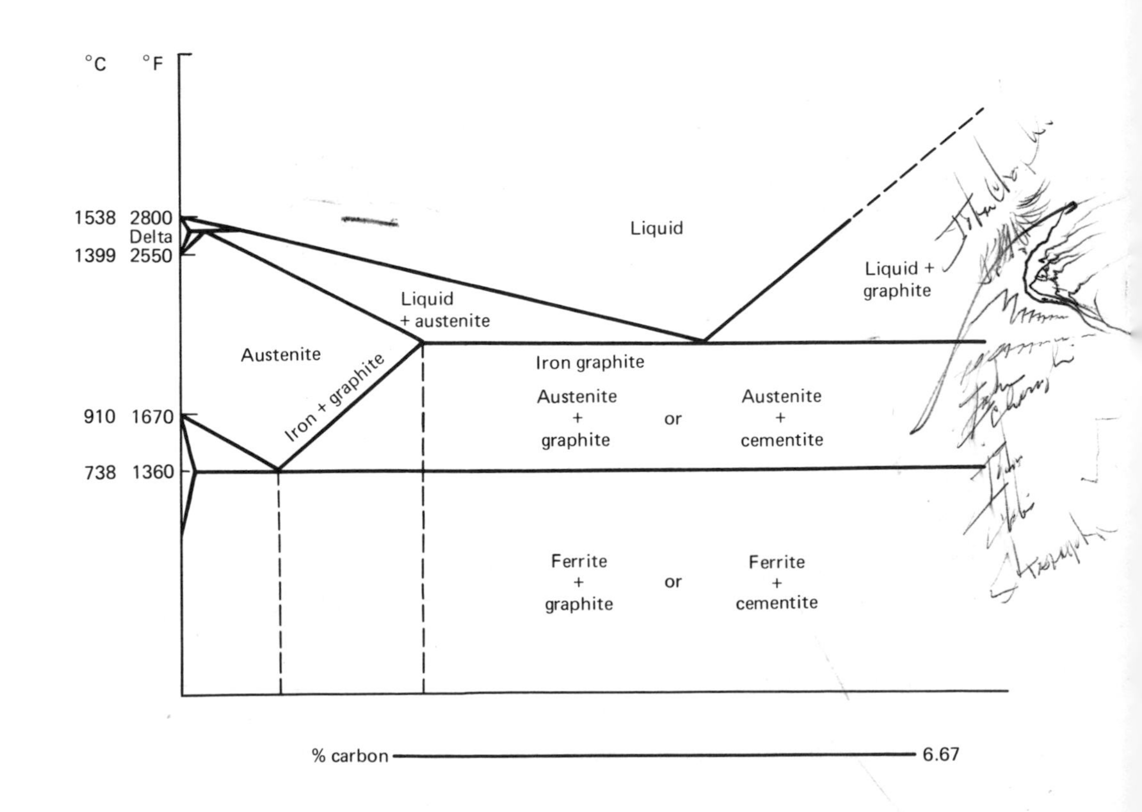

Figure 22. Iron-graphite equilibrium diagram.

Gray cast iron is a relatively brittle material, mainly because of its long thin graphite flakes that are very weak. Gray cast iron is a metal that will stand large compressive loads but small tensile loads.

White Cast Iron

White cast iron is very hard, brittle, and virtually nonmachinable. In some cases it is used where there is a need for resistance to abrasion. White cast iron is often found in combination with other cast iron, such as gray cast iron, to improve the hardness and wear resistant properties.

There are basically two ways of obtaining white cast iron. One way is by lowering the iron's silicon content; the second is by rapid cooling, which in this case yields what is called chilled cast iron. When cooled at a rapid rate, the excess carbon forms iron carbide and not graphite, thus making white cast iron.

Many times it is advantageous to have a hard wear resistant surface on the part, such as a bearing surface or outer rim. This is easily accomplished be putting chill plates in the mold so that the molten iron will cool faster in these localized areas, creating white cast iron.

Malleable Cast Iron

Malleable cast iron is noted for its strength, toughness, ductility, and machinability. In the process of making malleable cast iron, it is first necessary to begin with white cast iron. The white cast iron is then heat treated as follows:

1. Heat to about 1700°F (927°C).
2. Hold at this temperature for about 15 hours. This breaks down the iron carbide to austenite and graphite.
3. Slow cool to about 1300°F (704°C).
4. Hold at this temperature for approximately 15 hours.
5. Air cool to room temperature.

The above process breaks down the iron carbide into additional austenite and graphite. Upon cooling the graphite will form into clusters or balls. The austenite will take on any one of the transformation products depending on the cooling rate.

Typical properties of malleable cast iron are:

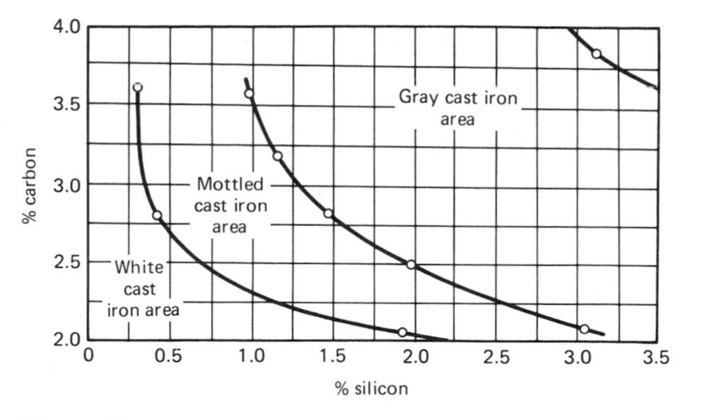

Figure 23. Composition limits for white, gray, and mottled cast irons.

Tensile strength	50,000 psi
Yield strength	35,000 psi
Elongation, 2 in.	15 percent
Brinell hardness	120

For pearlitic malleable cast iron typical properties are:

Tensile strength	70,000 psi
Yield strength	50,000 psi
Elongation, 2 in.	20 percent
Brinell hardness	180

The microstructure of the graphite formed will appear as shown in Figure 20, in a ball-like form. Graphite in this form makes a much more ductile metal than the metal would be if the graphite were in flake form.

Nodular Cast Iron

Nodular cast iron is known by several names: nodular iron, ductile iron, and spheroidal graphite iron. It gets the names from the ball-like form of the graphite in the metal and the very ductile property it exhibits. Nodular cast iron combines many of the advantages of cast iron and steel. Its advantages include good castability, toughness, machinability, good wear resistance, weldability, low melting point, and hardenability.

The formation of the graphite into a ball form is accomplished by adding certain elements such as magnesium and cerium to the melt just prior to casting. The vigorous mixing reaction caused by adding these elements results in a homogenous spheroidal or ball-like structure of the graphite in the cast iron. The iron matrix or background material can be heat treated to form any one of the microstructures associated with steels such as ferrite, pearlite, or martensite.

Self-Evaluation

1 Locate the following on the phase diagram (Figure 24):
a. Eutectic point
b. Liquidus line
c. Solidus line
d. Area of liquid and solid
e. Area of 100 percent solubility

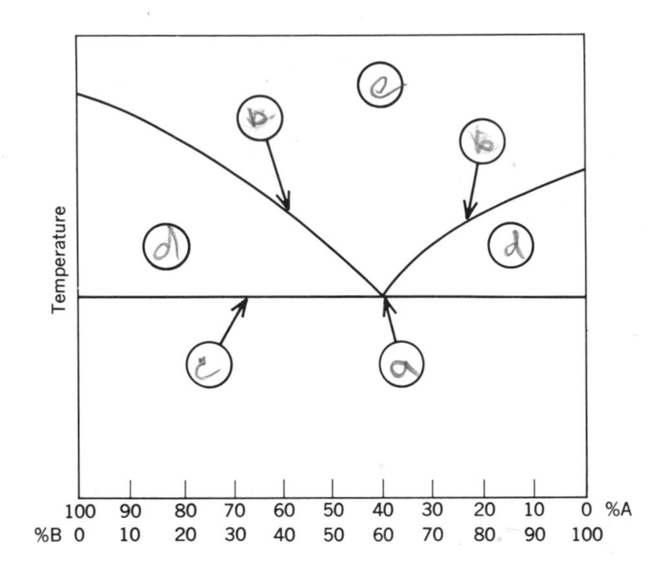

Figure 24.

2 When a substance changes phases while cooling, there is a release of heat. How does this appear on a cooling curve graph?

3 What does eutectoid mean? How does this differ from eutectic?

4 What does the A_3 line indicate? The A_1? The A_{cm}?

5 What is the hardest structure found in the iron-carbon alloy system? How would it appear in pearlite under the microscope?

6 Give a definition of austenite.

7 What is the softest structure found in the iron-carbon alloy system? How would it appear under the microscope? How much carbon can this structure dissolve at normal temperatures and at 1330°F (721°C)?

8 What is the name of the eutectoid composition of carbon steel slowly cooled to room temperature? How would it appear under the microscope? What is it made of?

9 What phase do steels with less than 0.8 percent carbon begin to form at the A_3 line after slow cooling? At what temperature will this transformation be complete?

10 When steels are heated, at which line does pearlite form austenite and at which line does ferrite form austenite?

11 Grain size increases as temperature goes above the A_3 line. Do these grains decrease in size when steel is cooled toward the A_3 line?

12 What effect does the addition of alloying elements have on the eutectoid composition?

13 How can you identify each of the four cast iron types by microscopic examination?

14 Name one alloying element that might make steel austenitic at room temperature.

15 How can white cast iron be formed?

Worksheet 1

Objective Determine carbon content and heat treatment by metallurgical observation.

Materials Metallurgical equipment for encapsulating, polishing, and etching; two unknown samples of carbon steel, one of relatively low carbon and one of very high carbon content; metallurgical microscope (100× to 500× magnification).

Procedure

1 Heat both specimens to above 1650°F (899°C) and slowly cool them in the furnace (anneal).

2 Prepare the two specimens for microscopic study as explained in Chapter 8. Identify the capsule by marking on the bottom or side.

3 Observe each specimen. Estimate the percentage of pearlite and ferrite. Compare your observations with the following:
100 percent pearlite = 0.8 percent C
75 percent pearlite = 0.6 percent C
50 percent pearlite = 0.4 percent C
25 percent pearlite = 0.2 percent C

Conclusion

1 How much carbon do you estimate for each sample?

2 Draw an iron-carbon diagram and locate each numbered specimen on the diagram for carbon content by drawing a vertical line.

3 Turn in your drawings and conclusions to your instructor.

Worksheet 2

Objective Learn to identify the microstructures of various forms of cast iron.

Materials Metallurgical equipment for microscopic study; encapsulated and polished specimens of gray, white, malleable, and nodular cast irons.

Procedure

1. Polish the specimens only. Do not etch.
2. View each specimen and make a drawing of each showing what you see. Identify each microstructure with an arrow and label.
3. Etch the specimens and view each in the microscope. Draw four more sketches and identify the new features that are revealed by the etch.

Conclusion

1. What are the differences between the unetched and etched specimens?
2. Are the cast irons ferritic or pearlitic?
3. How can you identify gray cast iron?
4. Turn in your sketches and conclusions to your instructor.

9 Hardening and Tempering of Plain Carbon Steel

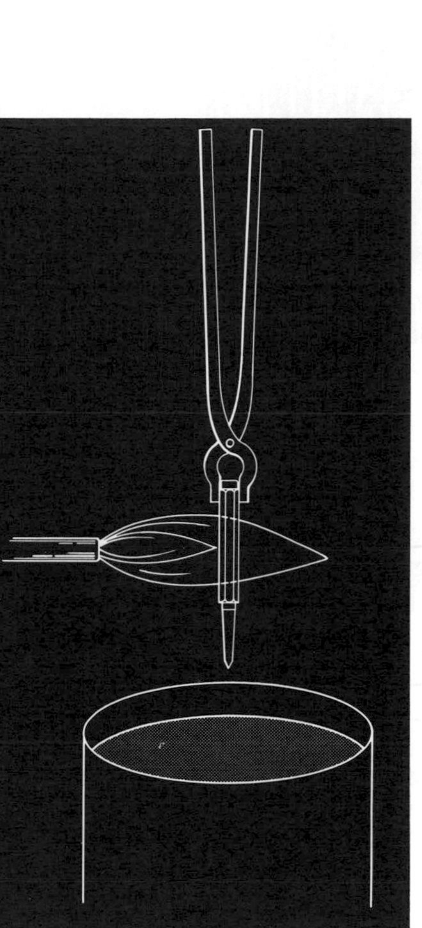

Plain carbon steel has been valued from early times because of certain properties. This soft silver-gray metal could be converted into a superhard substance that would cut glass and many other substances, including itself when soft. Furthermore, its hardness could be controlled. This converting of carbon steel into a steel of useful hardness is done with different heat treatments, two of the most important of which are hardening and tempering (drawing), which you will investigate in this chapter.

OBJECTIVES

After completing this chapter, you will be able to:

1. Correctly harden a piece of tool steel and evaluate your work.
2. Correctly temper the hardened piece of tool steel and evaluate your work.

INFORMATION

Most metals (except copper when used for electric wire) are not used commercially in their pure states because they are too soft and ductile and have a low tensile strength. When they are alloyed with other elements such as carbon or other metals, they become harder and stronger as well as more useful. A small amount of carbon (1 percent) greatly affects pure iron when alloyed with it. When heat treated, this alloy metal (carbon and iron) becomes a familiar tool steel used for cutting tools, files, and punches. Iron with 2 to 4½ percent carbon content yields cast iron, as you learned in Chapter 8.

In the last two chapters you also learned something about grain structures in metals and crystalline changes during heating and and cooling. These transformations from ferrite to austenite on heating and from austenite to ferrite when cooling are called phase changes. The temperatures where one phase changes to another are called **critical points or temperatures** by heat treaters and **transformation temperatures** by metallurgists. For example, the critical points of water are its boiling point (212°F or 100°C) and freezing point (32°F or 0°C). Figure 1 shows a critical temperature diagram for a carbon steel. The lower critical point in carbon steels is always about 1330°F (721°C), but the upper critical temperature varies as the carbon content changes, as can be seen in Figure 2.

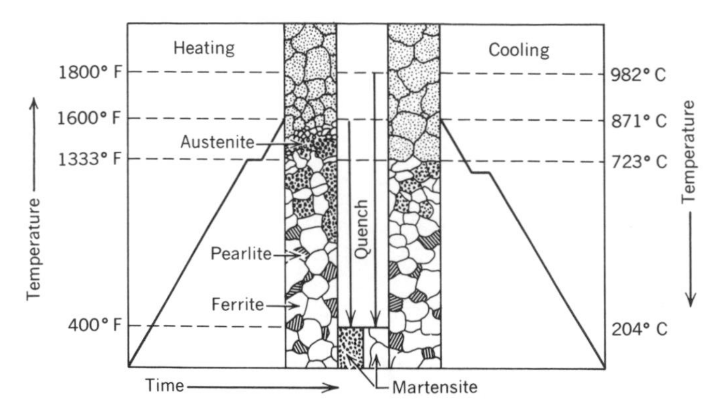

Figure 1. Critical temperature diagram of 0.83 percent carbon steel showing grain structures in heating and cooling cycles. Center section shows quenching from different temperatures and the resultant grain structure *(Machine Tools and Machining Practices).*

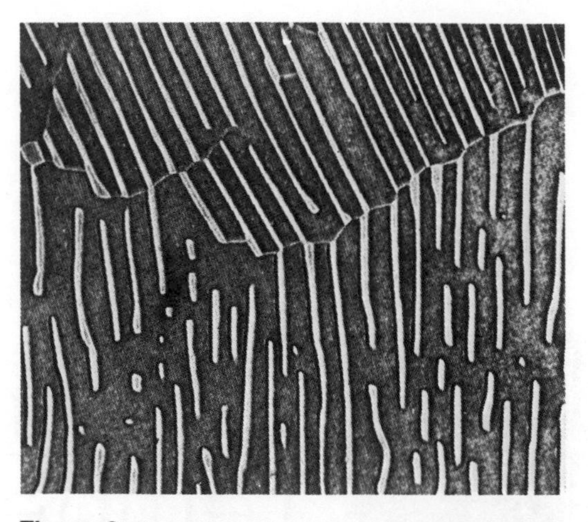

Figure 3. A replica electron micrograph. Structure consists of lamellar pearlite (11,000×) (By permission, from *Metals Handbook,* Volume 7, Copyright American Society for Metals, 1972).

Hardening by Quenching

As steel is heated above the lower critical temperature of 1330°F (721°C), the carbon that was in the form of layers of iron carbide in pearlite (Figure 3) begins to dissolve in the iron and form a solid solution called austenite (Figure 4). When this solution of iron and carbon is suddenly cooled or quenched, a new microstructure is formed. This is called martensite (Figure 5). Martensite is very hard and brittle, having a much higher tensile strength than the steel with a pearlite microstructure. It is quite unstable, however, and must be tempered (drawn) to relieve internal stresses, in order to have the ductility and toughness needed to be useful. AISI-C1095, commonly known as water hardening tool (W1) steel, will begin to show hardness when quenched from a temperature just over 1330°F (721°C), but will not harden at all if quenched from a temperature lower than 1330°F (721°C). This steel will become as hard as it can get

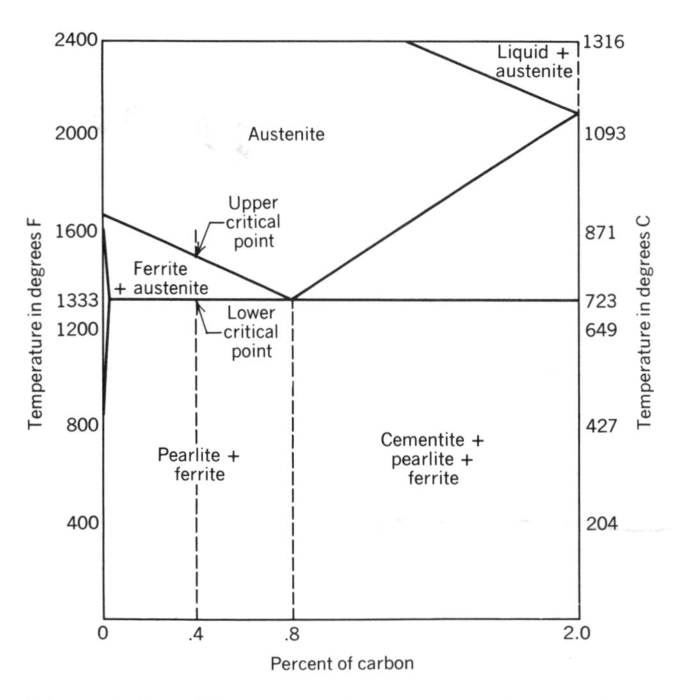

Figure 2. Simplified phase diagram for carbon steel *(Machine Tools and Machining Practices).*

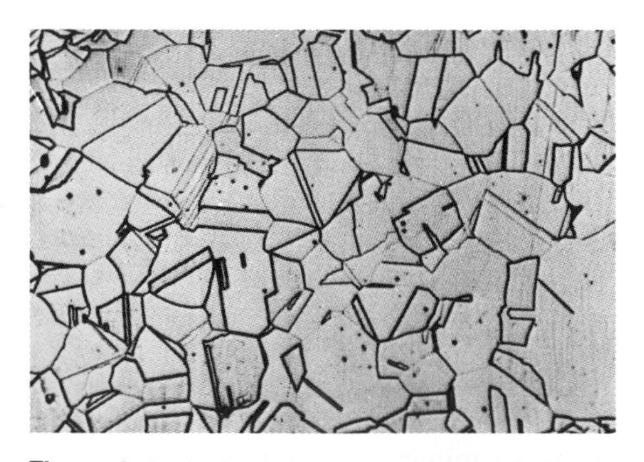

Figure 4. A microstructure of annealed 304 stainless steel, which is austenitic at ordinary temperatures (250×) (By permission, from *Metals Handbook,* Volume 7, Copyright American Society for Metals, 1972).

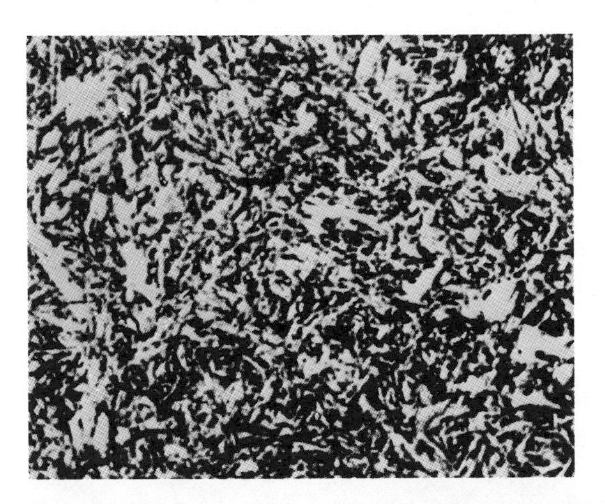

Figure 5. 1095 steel, water quenched from 1500°F (816°C) (1000×). The needlelike structure shows a pattern of fine martensite (By permission, from *Metals Handbook,* Volume 7, Copyright American Society for Metals, 1972).

when heated to 1450°F (788°C) and quenched in water. This quenching temperature changes as the carbon content changes. It should be 50°F (10°C) above the upper critical temperature for carbon steels containing less than 0.83 percent carbon (Figure 6). The reason carbon steel, less than eutectoid, should be heated above the upper critical temperature is that the ferrite is not all transformed into austenite below this point, and, when quenched, is retained in the martensitic structure. The retained ferrite causes brittleness even after tempering.

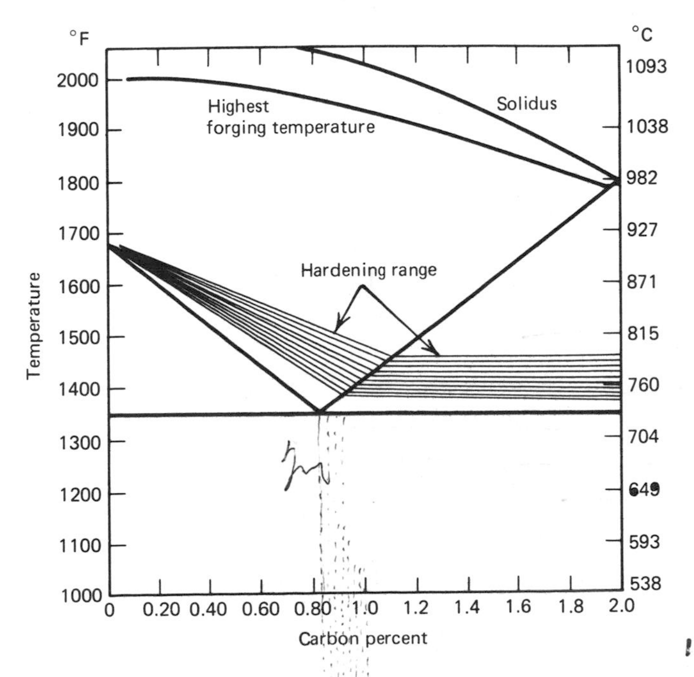

Figure 6. Temperature ranges used for hardening carbon steel.

Low carbon steels such as AISI 1020 will not, for all practical purposes, harden when they are heated and quenched. Oil and air hardening steels have a higher hardenability and do not have to be quenched as rapidly as plain carbon steels. Consequently, they are deeper hardening than water hardening types, which must be cooled to 200°F (93°C) within one or two seconds. (You will learn more about deep hardening steels in later chapters.) As you can see, it is quite important to know the carbon content and alloying element so that the correct temperature and quenching medium can be used. Fine grained tool steels are much stronger than coarse grained tool steels (Figure 7). If a piece of tool steel is heated above the correct temperature for its specific carbon content, a phenomenon called grain growth will occur and a coarse, weak grain structure develops. The grain growth will remain when the part is hardened by quenching and, if used for a tool such as a punch or chisel, the end may simply drop off when the first hammer blow is struck (Figure 8). Tempering (drawing) will not remove the coarse grain structure. If the part has been overheated, simply cooling back to the quenching temperature will not help, as the coarse grain persists well down into the hardening range (see Figure 1). The part should be air cooled and then reheated to the correct quenching temperature. Plain carbon steels containing 0.83 percent carbon can get as hard (RC67) as any plain carbon steel containing more carbon.

Transformation to martensite depends on three factors: (1) mass of the part, (2) severity of the quench, and (3) hardenability of the material. AISI-C1095, or water hardening tool steel (W1), can be quenched in oil, depending on the size of the part. For example, since a piece of small diameter W1 drill rod or thin sheet would have a high cooling rate, oil should be used as a quenching medium. Oil is not as "severe" as water because it conducts heat less rapidly than water and thus prevents quench cracking. Larger sections, however, would not be fully transformed into martensite if they were oil quenched since they have a slower cooling rate, but would instead contain some softer transformation structures. Water quenching should be used in this case, but re-

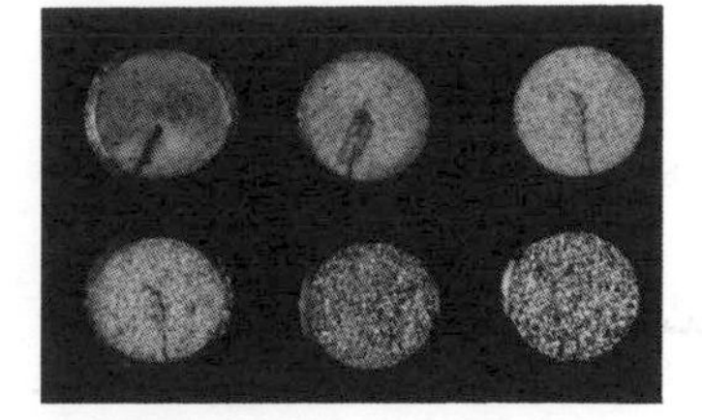

Figure 7. Fractured ends of 1095 water quenched tool steel ranging from fine grain, quenched from 1475°F (802°C), to coarse grain, quenched from 1800°F (982°C) *(Machine Tools and Machining Practices).*

member that W1 is shallow hardening and will only harden about $\frac{1}{8}$ inch deep.

If a torch is used to heat a part for quenching, heat colors are used to determine the approximate temperatures. When using a furnace to heat W1 for quenching, the temperature control would be set for 1450°F (788°C). See Table 1. If the part is small, a preheat is not necessary; but if it is thick, it should be heated slowly. If the part is left in a furnace without a controlled atmosphere for any length of time, the metal will form an oxide scale as the carbon leaves the surface. This decarburization of the surface will cause it to remain soft when quenched, while the metal directly under the surface will become hard. This loss of surface carbon can be avoided by using a carburizing compound to replace lost carbon, or by burying the part in cast iron chips. Wearing gloves, face shield, and a long sleeve shirt for protection, place the part in the furnace using tongs (Figure 9). When the part has become the same color as the furnace bricks, remove it by grasping one end with the tongs and immediately plunge it into the quenching bath. If

Figure 8. This chisel failed in service after being hardened and tempered because it was overheated during the hardening process.

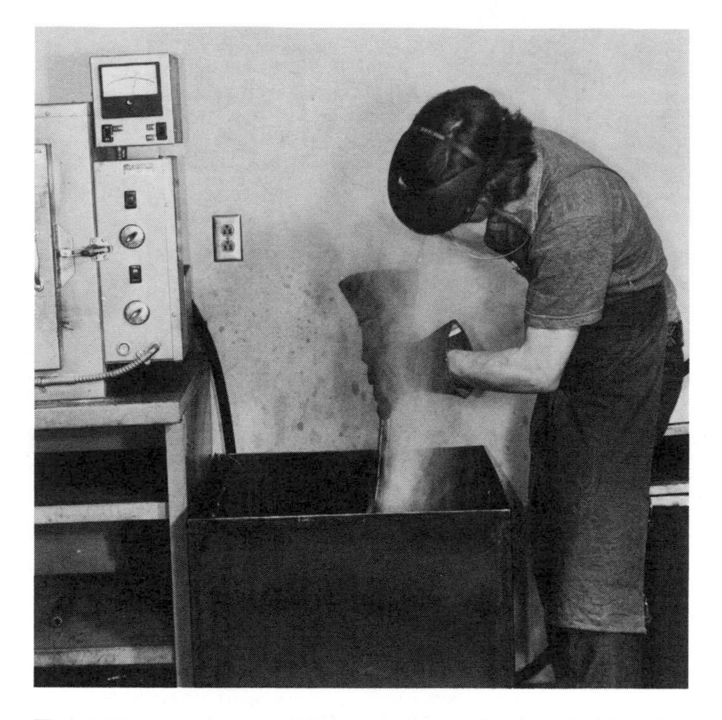

Figure 9. A long part is being quenched vertically. In this case, agitation of the part should be an up-and-down motion (Lane Community College).

the part is long, like a chisel or punch, it should be inserted into the quench vertically (straight up-and-down) not at a slant. Quenching at any angle can cause unequal cooling rates and bending of the part. Also, agitate the part in an up-and-down or a figure-8 motion to remove any gases or bubbles that might cause uneven quenching.

Oil quenching is used for oil hardening steels (01). When hardening various tool steels, a manufacturer's catalog should be consulted to get the correct temperature, time periods, and quenching media.

Tempering

Tempering, or drawing, is a process of reheating a steel part that has been previously hardened to transform some of the hard martensite into softer structures. The higher the tempering temperature used, the more martensite is transformed, and the softer and tougher (less brittle) the piece becomes. Therefore, tempering temperatures are specified according to the strength and ductility desired. Mechanical properties charts, which may be found in steel manufacturers' handbooks and catalogs, give this data for each type of alloy steel. Figure 10 is an example of a mechanical properties chart for

Table 1 Temperatures and Colors for Heating and Tempering Steel

	Colors	Fahrenheit	Process	
	White	2500°		
		2400°	High speed steel hardening (2250°–2400°F)	
	Yellow white	2300°		
		2200°		
		2100°		
	Yellow	2000°		
		1900°		
	Orange red	1800°	Alloy steel hardening (1450°–1950°F)	
Heat colors		1700°		
		1600°		
	Light cherry red	—		
		1500°		
	Cherry red	1400°	Carbon steel hardening (1350°–1550°F)	
		1300°		
	Dark red	—		
		1200°		
		1100°		
	Very dark red	1000°		
		900°		
	Black red in dull light or darkness	800°		High speed steel tempering (350°-1100°F)
		700°	Carbon steel tempering (300°-1050°F)	
	Pale blue (590°F)	600°		
	Violet (545°F)	—		
Temper colors	Purple (525°F)	500°		
	Yellowish brown (490°F)	—		
	Straw (465°F)	400°		
	Light straw (425°F)	—		
		300°		
		200°		
		100°		
		0°		

Source. Pacific Quality Steels, "Stock List and Reference Book," No. 75, Pacific Machinery and Tool Steel Company, 1971.

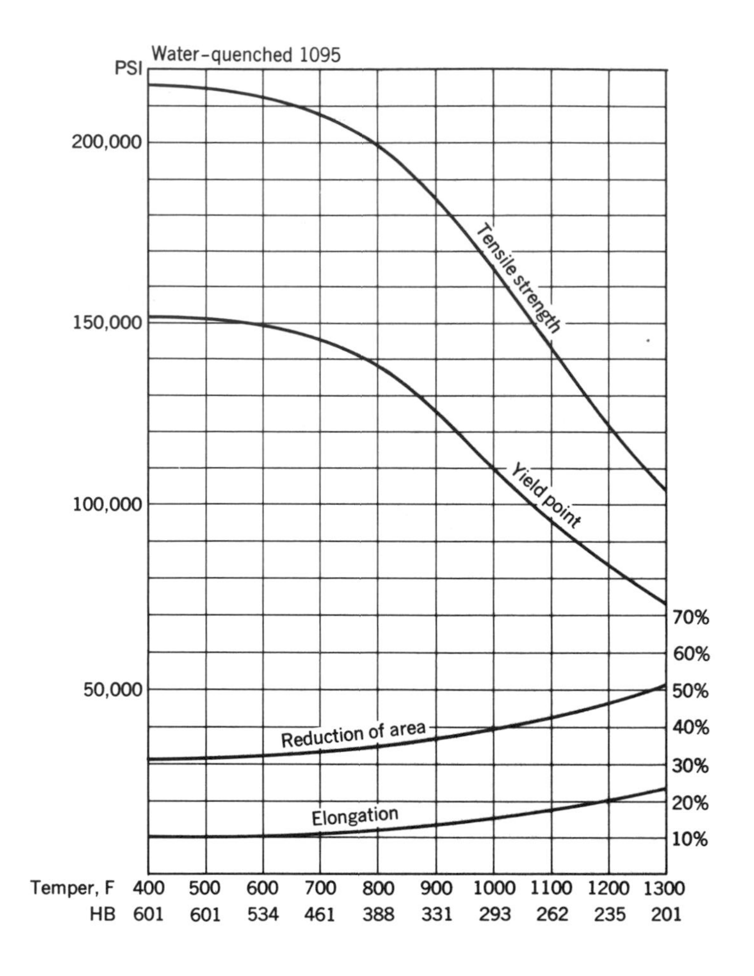

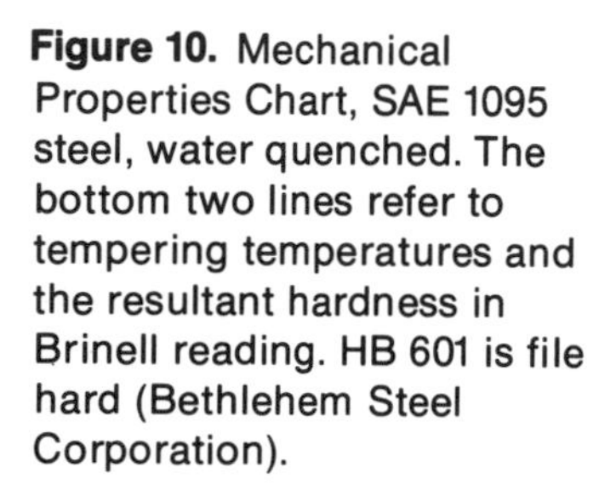
Figure 10. Mechanical Properties Chart, SAE 1095 steel, water quenched. The bottom two lines refer to tempering temperatures and the resultant hardness in Brinell reading. HB 601 is file hard (Bethlehem Steel Corporation).

water quenched 1095 steel. A tempering chart for tool steels is also available in Appendix 1.

A part can be tempered in a furnace or oven by bringing it to the required temperature and holding it there for a length of time, then cooling it in air or water. Some tool steels should be cooled rapidly after tempering to avoid temper brittleness. Small parts are often tempered in liquid baths such as oil, salt, or metals. Specially prepared oils that do not ignite easily can be heated to the tempering temperature. Lead and various salts are used for tempering since they have a low melting temperature. These methods are covered in more detail in Chapters 12 and 13.

When there are no facilities to harden and temper a tool with controlled temperatures, tempering by color is done. The oxide color used as a guide in such tempering will form correctly on steel only if it is polished to the bare metal and is free from any oil or fingerprints. An oxy-acetylene torch, steel hot plate, or an electric hot plate can be used. If the part is quite small, a steel plate is heated from the underside, while the part is placed on top. Larger parts such as chisels and punches can be heated on an electric plate (Figure 11) until the needed color shows, then cooled in water. See Table 1 for oxide colors and temperatures.

When grinding carbon steel tools, if the edge is heated enough to produce a color, you have in effect retempered the edge. If the temperature

Figure 11. Tempering a punch on an electric hot plate *(Machine Tools and Machining Practices).*

Table 2 Temper Color Chart

Degrees C°	Degrees F°	Oxide Color	Suggested Uses for Carbon Tool Steels
220	425	Light straw	Steel cutting tools, files, and paper cutters
240	462	Dark straw	Punches, dies
258	490	Gold	Shear blades, hammer faces, center punches, and cold chisels
260	500	Purple	Axes, wood cutting tools, and striking faces of tools
282	540	Violet	Springs, screwdrivers
304	580	Pale blue	Springs
327	620	Steel gray	Cannot be used for cutting tools

Harder——Softer

reached was above that of the original temper, the tool has become softer than it was before you began sharpening it. Table 2 gives the hardnesses of various tools as related to their oxide colors and the temperature at which they form.

Tempering should be done as soon as possible after hardening. The part should not be allowed to cool completely, since untempered it contains very high internal stresses and tends to split or crack. Tempering will relieve the internal stresses. A hardened part left overnight without tempering may develop cracks by itself.

Forging

Forging temperatures should be well below the solidus or freezing point of steel (refer to Figure 6). If a small tool, such as a chisel made of carbon steel, is being forged (hammered while hot), the temperature should be in the range that produces an orange red to yellow color (1800 to 1950°F or 982 to 1066°C). If this temperature is exceeded, there is a danger of "burning" the steel and ruining it for any use. If excessive sparking is evident, it is too hot. Carbon steels are somewhat "hot short" above forging temperatures; that is, they tend to split when hammered upon. Excess sulfur also causes hot short at any temperature above a red heat. If a previously hardened tool is to be reshaped by forging, it should first be annealed. See Chapter 10. Forging temperatures vary with other types of steels. Consult a reference book when forging tool steels other than plain carbon tool steels.

Self-Evaluation

1. If you heated AISI-C1080 steel to 1200°F (649°C) and quenched it in water, what would be the result?
2. If you heated AISI-C1020 steel to 1500°F (815°C) and quenched it in water, what would happen?
3. List as many problems encountered with water hardening steels as you can think of.
4. Name some advantages of using air and oil hardening tool steels.
5. What is the correct temperature for quenching AISI-C1095 tool steel? For any carbon steel?

6 Why is steel tempered after it is hardened?

7 What factors should you consider when you choose the tempering temperature for a tool?

8 The approximate temperature for tempering a center punch should be ______________. The oxide color would be ______________.

9 If a cold chisel became blue when the edge was ground on an abrasive wheel, to approximately what temperature was it raised? How would this temperature affect the tool?

10 How soon after hardening should you temper a part?

Worksheet 1

Objective Harden a piece of tool steel in the form of any tool or part, such as a punch or a chisel that has been previously forged.

Materials An electric or gas furnace, tongs, quenching media (oil or water), and safety equipment (face shield and gloves).

Procedure

1 Assuming the part is a small tool such as a center punch, it should be placed in a furnace that has already been brought up to the correct temperature.

2 Determine the best place to grip the part with the tongs so that it will not be damaged where it is red hot. Use the properly shaped tongs.

3 Make sure you can grasp the part in the furnace to remove it in the proper orientation to enable you to quench it straight in.

4 Heat the end of the tongs so they will not remove heat from the part.

5 When the piece has become the same color as the furnace bricks, remove it and immediately quench it **completely under** in the bath.

6 Agitate it up and down or in a figure-8 motion.

7 Be sure it has cooled below 200°F (93°C) before you remove it from the quench.

Note: If no furnace is available, a torch may be used with a temperature chart. See Table 1.

Conclusion Do you think the part got hard? Test with an old file in an inconspicuous place. If a hardness tester is available, a quick test could be taken. For information on hardness testing, see the chapter on Rockwell and Brinell hardness testers.

Is the part as hard as it should be? If not, check with your instructor.

Worksheet 2

Objective Temper a part that has just been hardened.

Materials An electric or gas furnace or oxy-acetylene torch, fine emery cloth, tongs and safety equipment.

Procedure for Tempering in the Furnace

1 Polish all smooth surfaces of the part with emery cloth and remove all oil. A cold furnace should be brought up to the correct temperature.

2 Small parts are then placed in the furnace for about 15 mintues.

3 Remove and cool in air.

4 When this method is used, the striking end of punches, chisels, and other striking tools should be further heated with a torch until they are a blue color on that end. This is done to ensure the safety of the user by preventing the struck end from shattering, since it is softer when tempered to blue.

Procedure for Tempering with a Torch or Hot Plate

1 When using a torch or hot plate, make sure that heat is applied to the body of a punch or a part of the tool that can be softer. Allow the heat to travel slowly out to the cutting edge. This way the colors may be observed.

2 See that the striking end of a tool is blue or gray before the proper color arrives at the cutting end.

3 When the proper color has arrived, quickly cool the piece in water. **Do not** delay or it will be overtempered.

Conclusion

1 Are the colors right according to Table 2?

2 If possible, recheck the hardness and compare with Figure 10.

3 Leave the temper colors on your project so your instructor can evaluate it. Turn it in for grading.

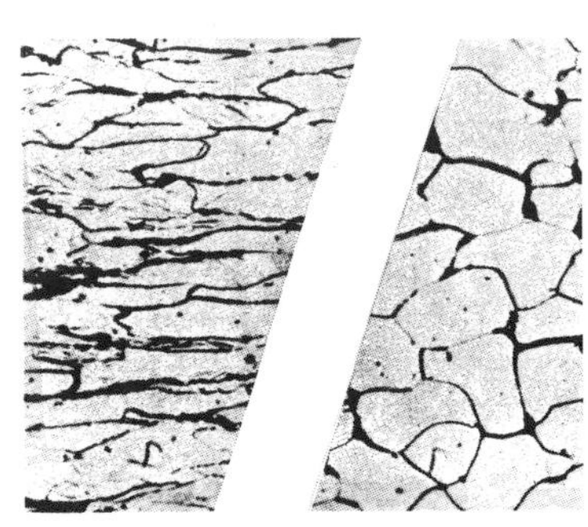

10 Annealing, Normalizing, and Stress Relieving

Since the machinability and welding of metals are so greatly affected by heat treatments, the processes of annealing, normalizing, and stress relieving are important to the welder and machinist. You will learn about these processes in this chapter.

OBJECTIVES

After completing this chapter, you will be able to:

1. Explain the principles of and differences between the various kinds of annealing processes.
2. Test various steels with annealing, normalizing, and stress relieving heat treatments in order to determine their effect on machinability and welding.

INFORMATION

Annealing

The heat treatment for iron and steel that is generally called annealing can be divided into several different processes: full anneal, normalizing, spheroidize anneal, stress relief (anneal), and process anneal.

Full Anneal

The full anneal is used to completely soften hardened steel, usually for easier machining of tool steels that have more than 0.8 percent carbon content. Lower carbon steels are full annealed for other purposes. When welding or prior heat treating has been done on a medium to high carbon steel that must be machined, a full anneal is needed. Full annealing is done by heating the part in a furnace to 50°F (10°C) above the upper critical temperature (Figure 1), and then cooling very slowly in the furnace or in an insulating material. By this process, the microstructure becomes coarse pearlite and ferrite, which is soft enough to machine. It is necessary to heat above the critical temperature, as in full annealing, in order to recrystallize the grains containing iron carbides (pearlite or martensite) in low carbon steels and to reform new soft whole grains from the old hard ones (Figures 2*a* and 2*b*). However, the stressed and deformed ferrite grains will recrystallize below the critical temperature at about 900°F (482°C) and reform into soft whole grains.

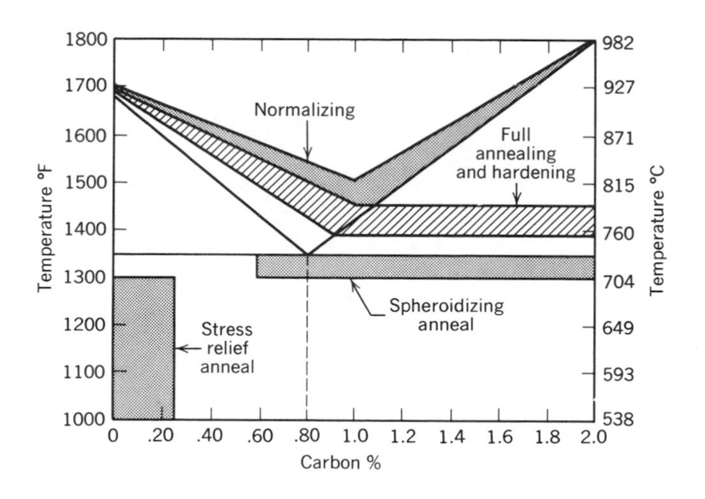

Figure 1. Temperature ranges used for heat treating carbon steel *(Machine Tools and Machining Practices).*

Normalizing

Normalizing is somewhat similar to annealing, but it is done for several different purposes. Medium carbon steels are often normalized to give them better machining qualities. Medium (0.3 to 0.6 percent) carbon steel may be "gummy" when machined after a full anneal, but can be made sufficiently soft for machining by normalizing. The finer, but harder, microstructure produced by normalizing gives the piece a better surface finish. The piece is heated to 100°F (38°C) above the upper critical line, and cooled in still air. When the carbon content is above or below 0.8 percent, higher temperatures are required. See Figure 1.

Forgings and castings that have unusually large and irregular grain structures are corrected by using a normalizing heat treatment (Figures 3*a*

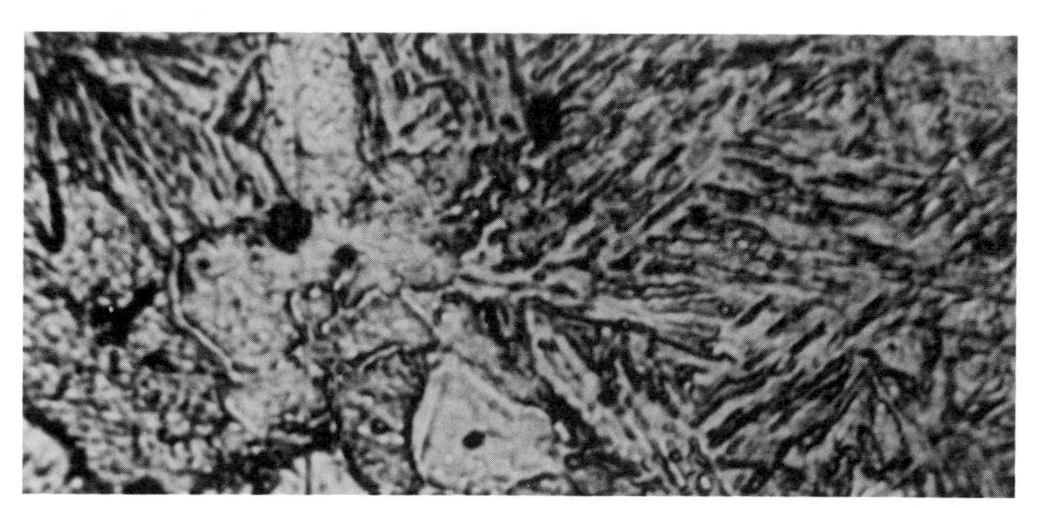

Figure 2*a*. Very hard martensite structures are evident in this micrograph of SAE 1090 steel. This material is nonmachinable (500×).

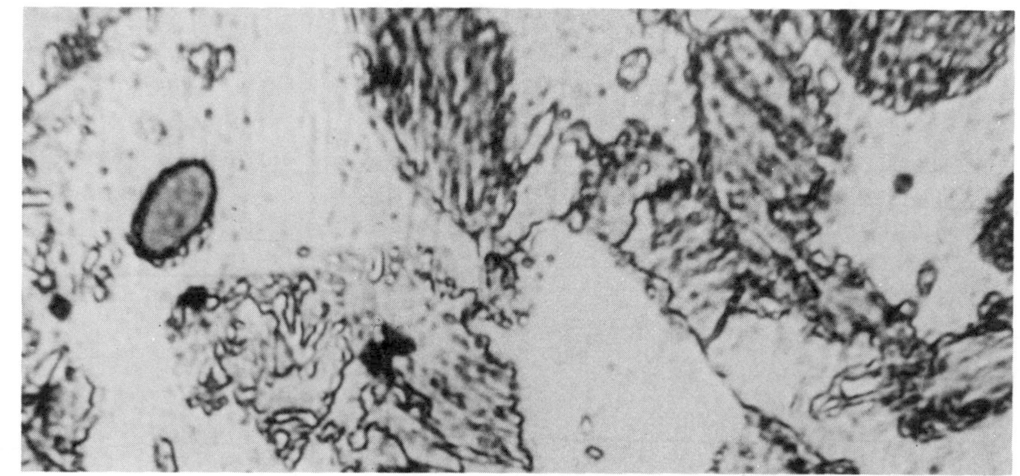

Figure 2*b*. The same material as shown in Figure 2*a* has been fully annealed and is now machinable. The martensite has been recrystallized into large grains of pearlite and ferrite (500×).

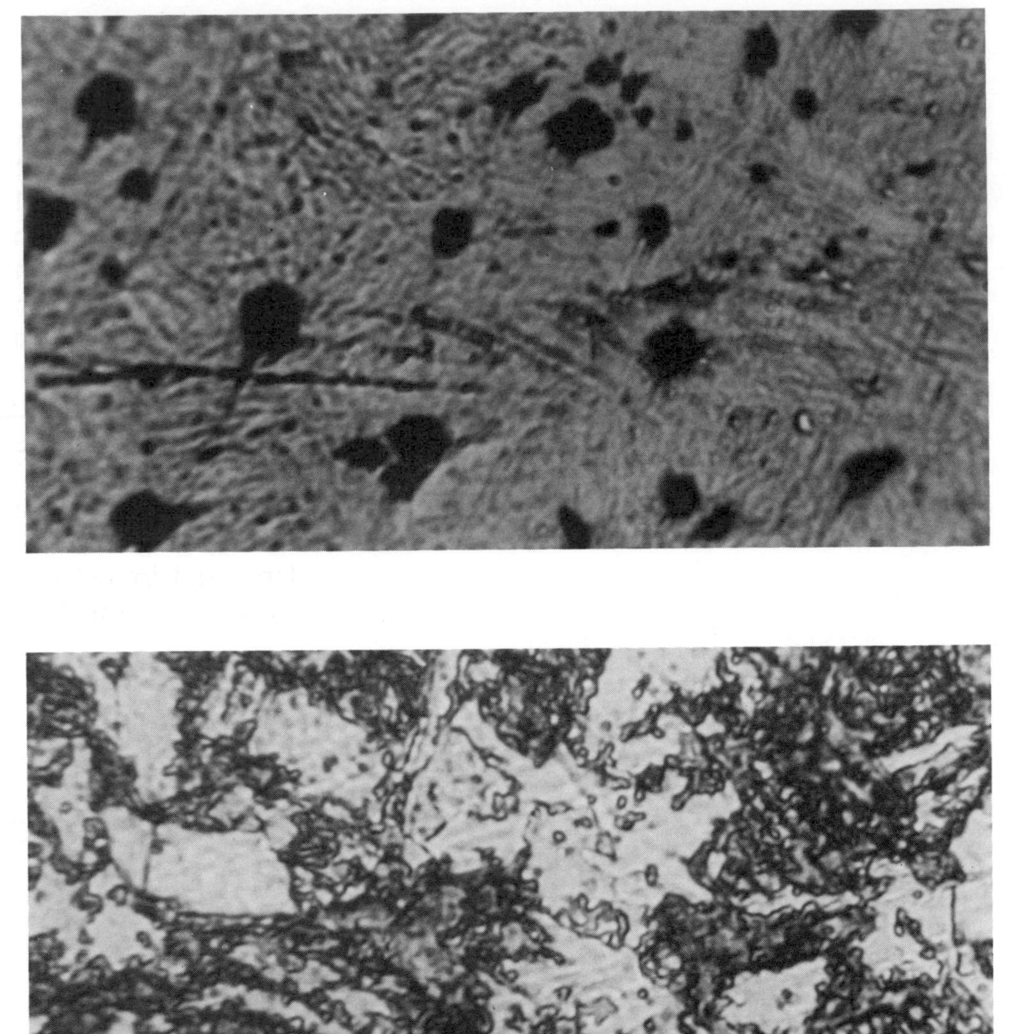

Figure 3*a*. This carbon steel contains small concentrated areas of iron carbide (dark patches) in a matrix of martensite. Normalizing will homogenize this steel and give it a regular grain structure (500×).

Figure 3*b*. Same material as in Figure 3*a* except that it has been normalized. A more homogenous structure can be seen with large grains of pearlite (black) and partial spheroidization (500×).

and 3*b*). Stresses are removed, but the metal is not as soft as with full annealing. The resultant microstructure is a uniform fine grained pearlite and ferrite, including other microstructures depending on the alloy and carbon content. Normalizing also is used to prepare steel for other forms of heat treatment such as hardening and tempering. Weldments are sometimes normalized to remove welding stresses that develop in the structure as well as in the weld.

Spheroidizing

Spheroidizing is used to improve the machinability of high carbon steels (0.8 to 1.7 percent carbon). It produces a spherical or globular carbide grain structure in the steel rather than a lameller (platelike) structure of pearlite (Figure 4). Low carbon steels (0.08 to 0.3 percent carbon) can be spheroidized, but their machinability gets poorer since they become gummy and soft, causing tool edge build up and poor finish. The spheroidization temperature is close to 1300°F (704°C). The steel is held at this temperature about four hours. Hard carbides that develop from welding on medium carbon steels, causing them to be brittle can be changed to the more ductile iron carbide spheroids by the process of spheroidization (Figure 5). Maximum ductility is obtained in steel by this method.

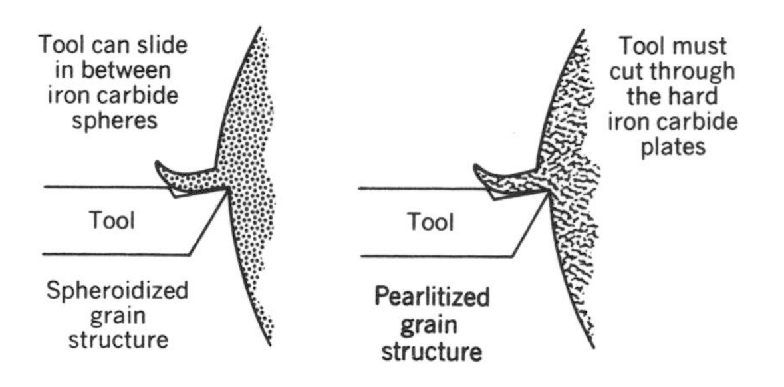

Figure 4. Comparison of cutting action between spheroidized and normal carbon steels *(Machine Tools and Machining Practices).*

Stress Relief Anneal

Stress relief annealing is a process of reheating low carbon steels to 950°F (510°C). Stresses in the ferrite (mostly pure iron) grains caused by cold working steel such as rolling, pressing, welding, forming, or drawing are relieved by this process. The distorted grains reform or recrystallize into new softer ones (Figures 6 and 7).

The pearlite grains and some other forms of iron carbide remain unaffected by this treatment, unless done at the spheroidizing temperature and held long enough to effect spheroidization. Stress relief is often used on weldments as the lower temperature limits the amount of distortion caused by heating. Full anneal, for example, can cause considerable distortion in steel.

Process Anneal

Process annealing is essentially the same as stress relief annealing. It is done at the same temperatures and with low and medium carbon steels. In the wire and sheet steel industry, the term is used for the annealing processes used in cold rolling or wire drawing processes and those used to remove the final residual stresses when necessary. Wire and other metal products that must be continuously formed and reformed would become too brittle to continue after a certain amount of forming. The anneal, between a series of cold working operations, reforms the grains to the original soft, ductile condition so that cold working can continue. Process anneal is sometimes referred to as bright annealing, and is usually carried out in a closed container with inert gas to prevent oxidation of the surface.

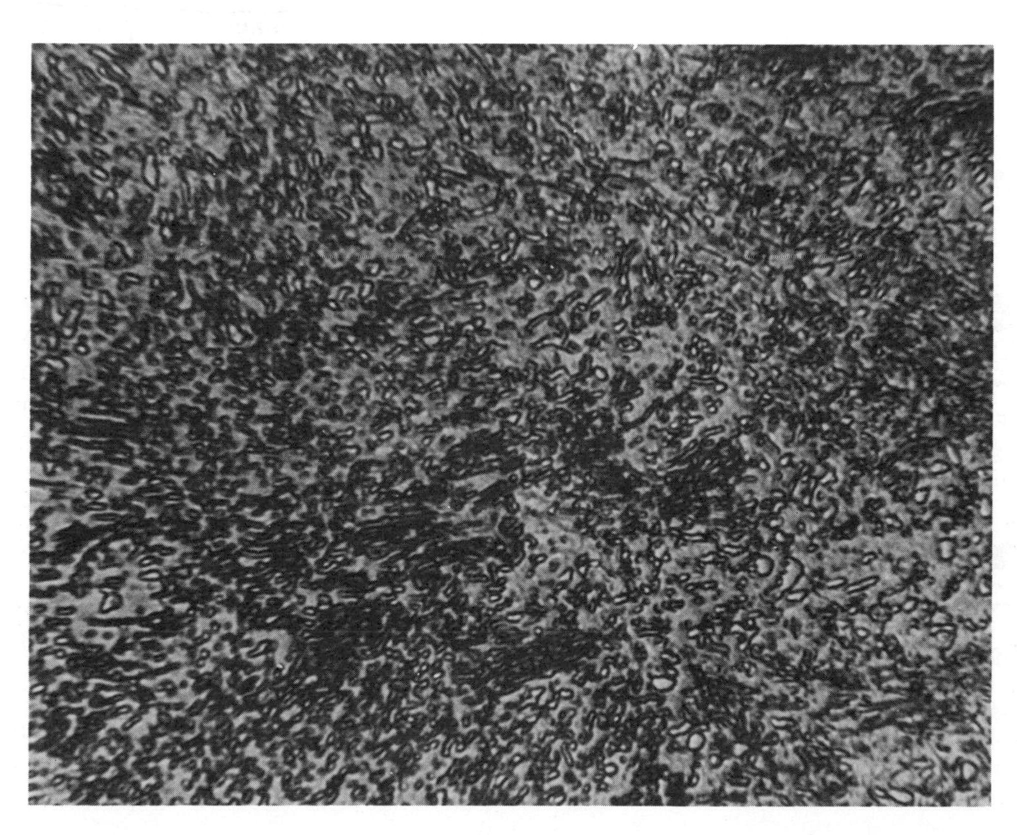

Figure 5. SAE 1090 spheroidized steel. Machining is much easier in this condition than it is when the steel is normalized or full annealed (500×).

Figure 6. Microstructure of flattened grains of 0.10 percent carbon steel, cold rolled (1000×) (By permission, from *Metals Handbook,* Volume 7, Copyright American Society for Metals, 1972).

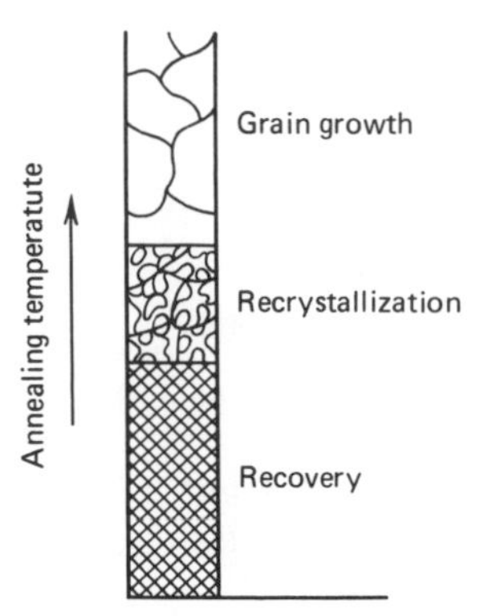

Figure 8. Changes in metal structures that take place during the annealing process.

Recovery, Recrystallization, and Grain Growth

When metals are heated to temperatures less than the recrystallization temperature, a reduction in internal stress takes place. This is done by relieving elastic stresses in the lattice planes and not by reforming the distorted grains. Recovery in annealing processes used on cold worked metals is usually not sufficient stress relief for further extensive cold working (Figure 8), yet it is used for some purposes and is called stress relief anneal. Most often, recrystallization is required to reform the distorted grains sufficiently for further cold work.

Recovery is a low temperature effect in which there is little or no visible change in the microstructure. Electrical conductivity is increased and often a decrease in hardness is noted. It is difficult to make a sharp distinction between recovery and recrystallization. Recrystallization releases much larger amounts of energy than does recovery. The flattened, distorted grains are sometimes reformed to some extent during recovery into polygonal grains, while some rearrangement of defects such as dislocations takes place.

Recrystallization not only releases much larger amounts of stored energy but new, larger grains are formed by the nucleation of stressed grains and the joining of several grains to form larger ones. To accomplish this joining of adjacent grains, grain boundaries migrate to new positions, which changes the orientation of the crystal structure. This is called grain growth.

The following factors affect recrystallization.

1. A minimum amount of deformation is necessary for recrystallization to occur.
2. The larger the original grain size, the greater amount of cold deformation is required to give an equal amount of recrystallization with the same temperature and time.

Figure 7. The same 0.10 percent carbon steel as in Figure 6, but annealed at 1025°F (552°C). Ferrite grains are mostly reformed to their original state, but the pearlite grains are still distorted (1000×) (By permission, from *Metals Handbook*, Volume 7, Copyright American Society for Metals, 1972).

3 Increasing the time of anneal decreases the temperature necessary for recrystallization.

4 The recrystallized grain size depends mostly on the degree of deformation and, to some extent, on the annealing temperature.

5 Continued heating, after recrystallization (reformed grains) is complete, increases the grain size.

6 The higher the cold working temperature, the greater amount of cold work is required to give equivalent deformation.

Nonferrous Metals

Annealing of most nonferrous metals consists of heating them to the recrystallization temperatures or grain growth range and cooling them to room temperature (Table 1). The rate of cooling has no effect on most nonferrous metals such as copper or brass, but quenching in cold water is sometimes beneficial. Annealing temperatures and procedures are very critical with some metals such as stainless steels and precipitation hardening nonferrous metals. See Chapter 14 "Heat Treating of Nonferrous Metals" for further information. The phenomenon of grain growth, as discussed in Chapter 7, is found at higher temperatures. When a large amount of deformation is required in one operation, large grains are sometimes preferred although a surface defect, called orange peel (Figures 9*a* and 9*b*), is sometimes seen on formed metals having large grains. In this case, a stress relief anneal could be used; that is, recovery without grain growth.

Table 1 Recrystallization Temperatures of Some Metals

Metal	Recrystallization Temperature °F
99.999% aluminum	175
Aluminum bronze	660
Beryllium copper	900
Cartridge brass	660
99.999% copper	250
Lead	25
99.999% magnesium	150
Magnesium alloys	350
Monel	100
99.999% nickel	700
Low carbon steel	1000
Tin	25
Zinc	50

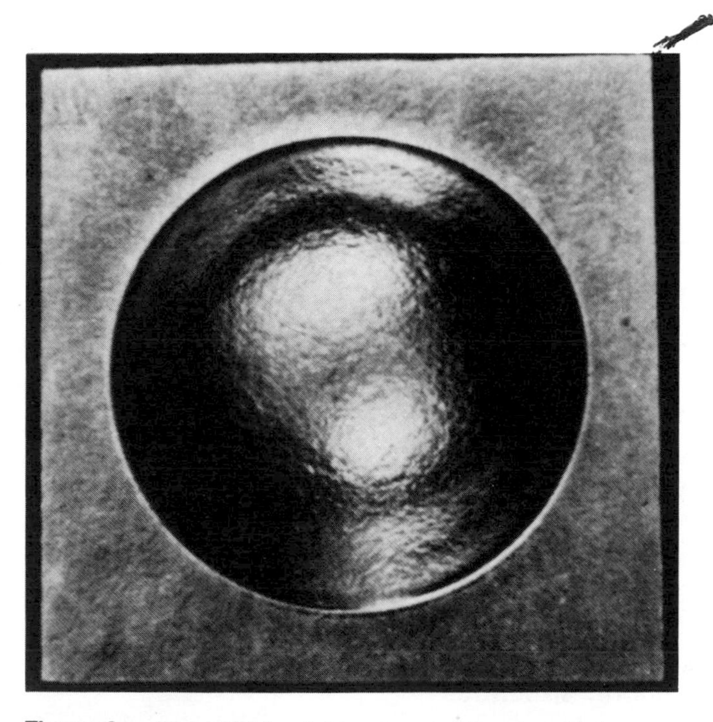

Figure 9*a*. Alloy 260 (cartridge brass, 70 percent) drawn cup showing rough surface or "orange peel" (actual size) (By permission, from *Metals Handbook* Volume 7, Copyright American Society for Metals, 1972).

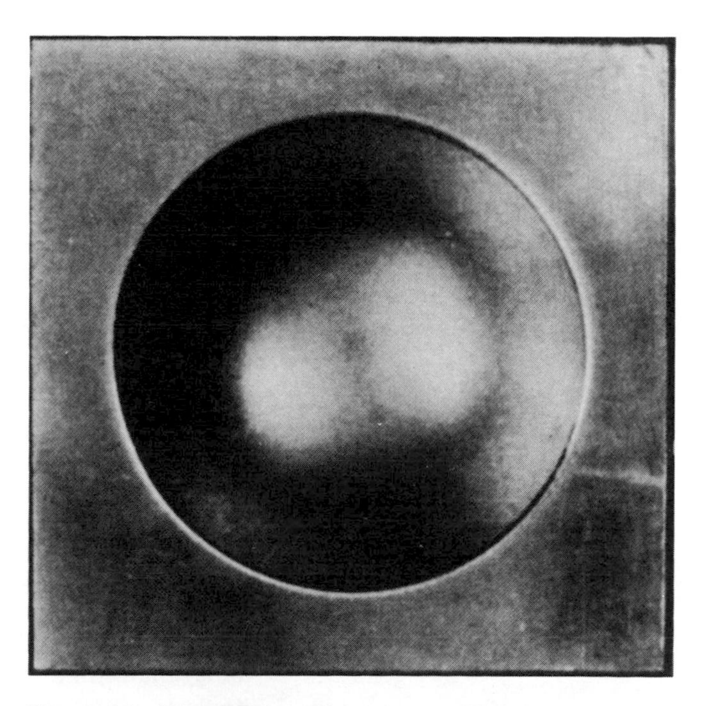

Figure 9*b*. Alloy 260 (cartridge brass, 70 percent) drawn cup with a smooth surface, no "orange peel" (actual size) (By permission, from *Metals Handbook,* Volume 7, Copyright American Society for Metals, 1972).

Self-Evaluation

1. When might normalizing be necessary?
2. At what approximate temperature should you normalize 0.4 percent carbon steel?
3. What is the spheroidizing temperature of 0.8 percent carbon steel?
4. What is the essential difference between full anneal and stress relieving?
5. When should you use stress relieving?
6. What kind of carbon steels would need to be spheroidized to give them free machining qualities?
7. Explain process annealing.
8. How should the piece be cooled for a normalizing heat treatment?
9. How should the piece be cooled for the full anneal?
10. What happens to machinability in low carbon steels that are spheroidized?

Worksheet 1

Objective Determine the changes in microstructures as a result of annealing heat treatments.

Materials Furnace and metallurgical equipment for microscopic examination, samples of SAE 1010 to 1020 carbon steel and of SAE 1090 carbon steel

Procedure

1. Take one small specimen of each type of steel and full anneal both.
2. Normalize two specimens, one of each type.
3. Stress relieve two specimens.
4. Spheroidize anneal one of each specimen by holding near 1300°F (704°C) for several hours in the furnace.
5. Mount, polish, and etch each specimen. Be sure to permanently mark each plastic capsule immediately.
6. Study each specimen with a microscope and write down your results.

Conclusion

1. Make circles and draw in each microstructure as you see it. Note the magnification and parts of the structure.
2. Note the changes that have taken place as the result of each heat treatment under each drawing.
3. Turn in your sketches and conclusions to your instructor.

Worksheet 2

Objective Determine relative tensile strengths and ductility as a result of heat treatment.

Materials Furnace, tensile testing machine, and four samples of SAE 1040 steel prepared for tensile testing as described in Chapter 6 "The Mechanical and Physical Properties of Steel," Worksheet 1.

Procedure

1. Identify each sample with a different number.
2. Heat the furnace to the quenching (hardening) temperature (about 1650°F or 899°C) and put three of the samples in the furnace.
3. Quench one sample in water (brine if available) and temper it to 400°F (204°C). Record the number and heat treatment.
4. Raise the temperature 50°F (10°C) and remove one sample to air cool (normalize); record the number.
5. Lower the temperature 50°F (10°C) and then let the furnace cool slowly. Remove the remaining sample and record the number.

Conclusion

1. Draw a stress strain diagram as seen in Chapter 6 and graph the elastic ranges, yield points, and plastic ranges of the three heat treated samples and the remaining as-rolled sample after testing.
2. Test all four samples in a small tensile tester and record the percent of elongation and percent of reduction of area for each on a sheet of paper.
3. Turn in your results to your instructor.

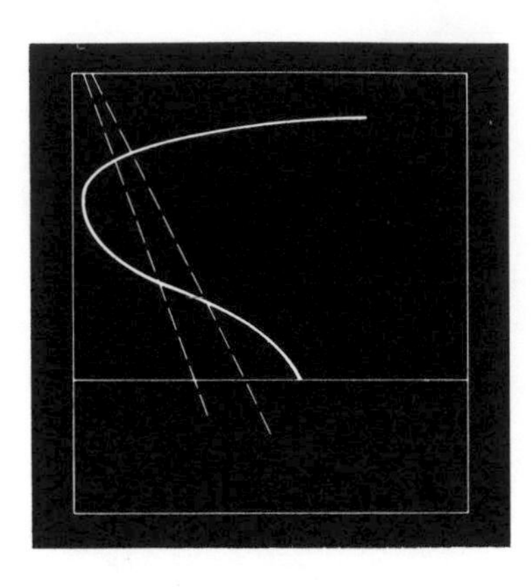

11 I-T Diagrams and Cooling Curves

When steel is held in the austenite phase at a constant temperature below its minimum stable temperature, it will transform into various transformation products. A graph is used to show these results. It is called an Isothermal Transformation (I-T) diagram. Hardening procedures, rates of cooling, and various transformation products and how they are obtained are all explained by the study of this diagram. A working knowledge of transformation products is very important to the heat treater.

OBJECTIVES

After completing this chapter, you will be able to:

1. Determine the hardenability of steels and their quenching rates by using information gained from the I-T diagrams.
2. Recognize certain microstructures of transformation products produced at various temperatures by the use of the metallurgical microscope.
3. Estimate the hardness of a quenched steel by using the I-T diagram and microscope.

INFORMATION

Hardening Process

The process of hardening steel is carried out by performing two operations. The first step is to heat the steel to the austenite range (austenitizing), which is heating to a temperature above the upper critical temperature. The second step is that of and rapid cooling or quenching near to room temperature.

Austenitizing produces the solid solution of carbon in the face-centered cubic structure. The temperature used for austenitization is usually about 50°F (10°C) above the A_3 or $A_{3,1}$ lines (Figure 1). Alloys with 0.8 percent carbon or less will become 100 percent austenite at this temperature, while steel with more than 0.8 percent carbon will be austenite with some free cementite.

Higher carbon steels that contain carbide forming elements such as chromium, molybdenum, tungsten, or vanadium require more soaking time at the austenitizing temperature as the complex carbides are relatively slow to dissolve. If the temperature is too low, there may be incomplete solution of carbides and the steel may still

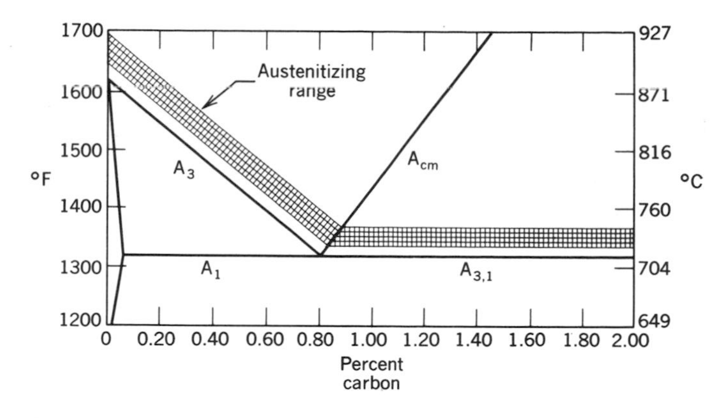

Figure 1. Diagram showing hardening temperature (austenitizing range) *(Machine Tools and Machining Practices).*

contain undissolved ferrite grains that are not beneficial in a hardened tool steel. If the temperature is too high, large grains may form and cause cracking during heat treatment resulting in failure of the part. Most steel producers publish data sheets containing correct austenitizing temperatures for various alloys.

Quenching undercools the austenite to form a new structure below the Ms temperature. This structure is called martensite. The martensite is an extremely hard acicular or needlelike structure that, for most purposes, is too brittle to be of any use, and a tempering process is needed as an additional operation to toughen it.

Isothermal-Transformation (I-T) Diagrams

Isothermal means the same or constant temperature. I-T diagrams are also called the Bain S-Curve or T-T-T (time-temperature-transformation) diagrams. An isothermal-transformation diagram is plotted on a graph by heating small steel specimens of a specific kind of steel to the austenitization temperature (Figure 2). They are next quenched in a liquid salt bath that is held at a constant temperature, such as 1200°F (649°C). At regular time intervals a specimen is removed and rapidly quenched. Microscopic examination will then show martensite if transformation has not yet started, but martensite and pearlite if transformation has started, and only pearlite if transformation is complete. A mark is placed on the graph indicating the time and temperature. This procedure is repeated at other temperatures until the entire graph is plotted for that particular steel (Figure 3).

The vertical scale on the left represents temperature and the horizontal scale on the bottom represents time. It is plotted on a log scale that corresponds to one minute, one hour, one day, and one week. The letters Ms can be found at a specific temperature for each kind of steel. Ms represents the temperature at which austenite begins to transform to martensite during cooling. The Mf temperature is the point at which the transformation of austenite to martensite is completed or near 100 percent during cooling. This is sometimes replaced by a percentage of transformation showing the area

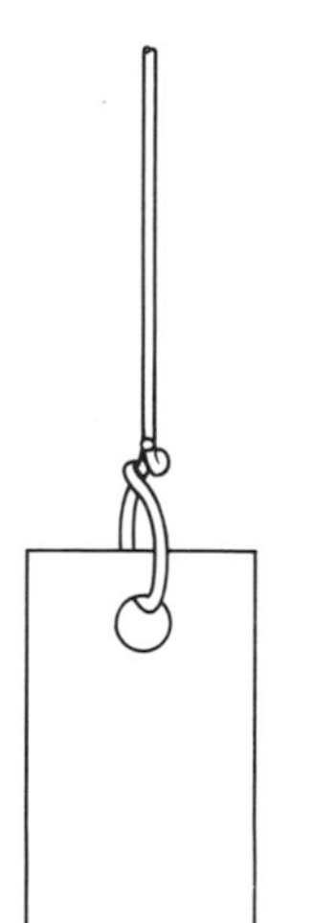

Figure 2. Isothermal test specimen *(Machine Tools and Machining Practices).*

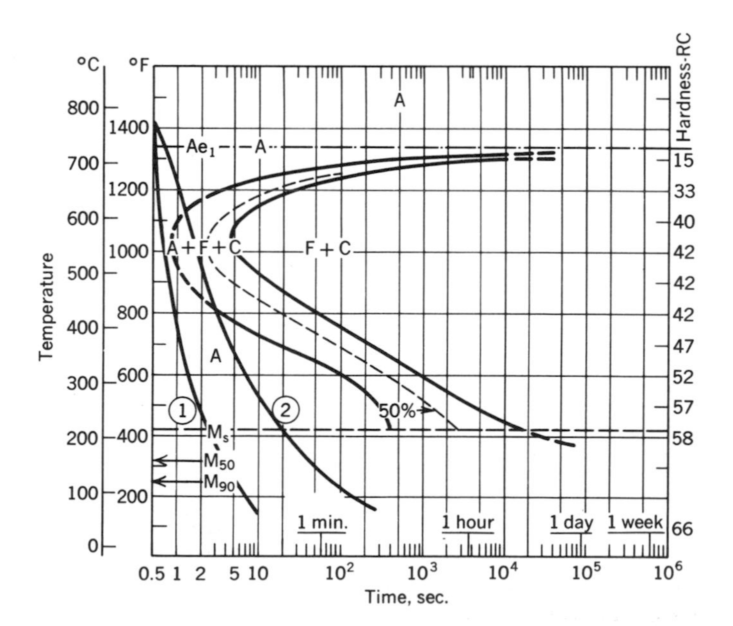

Figure 3. Method of plotting an isothermal diagram (Copyright 1951 by United States Steel Corporation).

of retained austenite. Some tool steels will retain untransformed austenite even below the Mf temperature. Suitable tempering or subzero treatments will usually fully transform the austenite to martensite. Retained austenite can cause serious distortion in tool steels at a later time if its transformations result from external stresses or heat.

Transformation Products

Austenite, when cooled below the transformation temperature and held at a constant temperature, decomposes into various transformation products such as pearlite, ferrite, or bainite. Austenite containing 0.8 percent carbon that is cooled quickly and held at 1300°F (704°C) does not begin to decompose or transform until after about 15 minutes has elapsed and does not completely decompose until it is at that temperature for more than 5 hours (Figure 4). A very coarse pearlite structure has developed at this temperature and the material is very soft. If the austenite is quickly cooled to and held at a lower temperature of 1200°F (649°C), decomposition begins in about 5 seconds and is completed after about 30 seconds. The resultant pearlite is coarse grained and slightly harder. At a

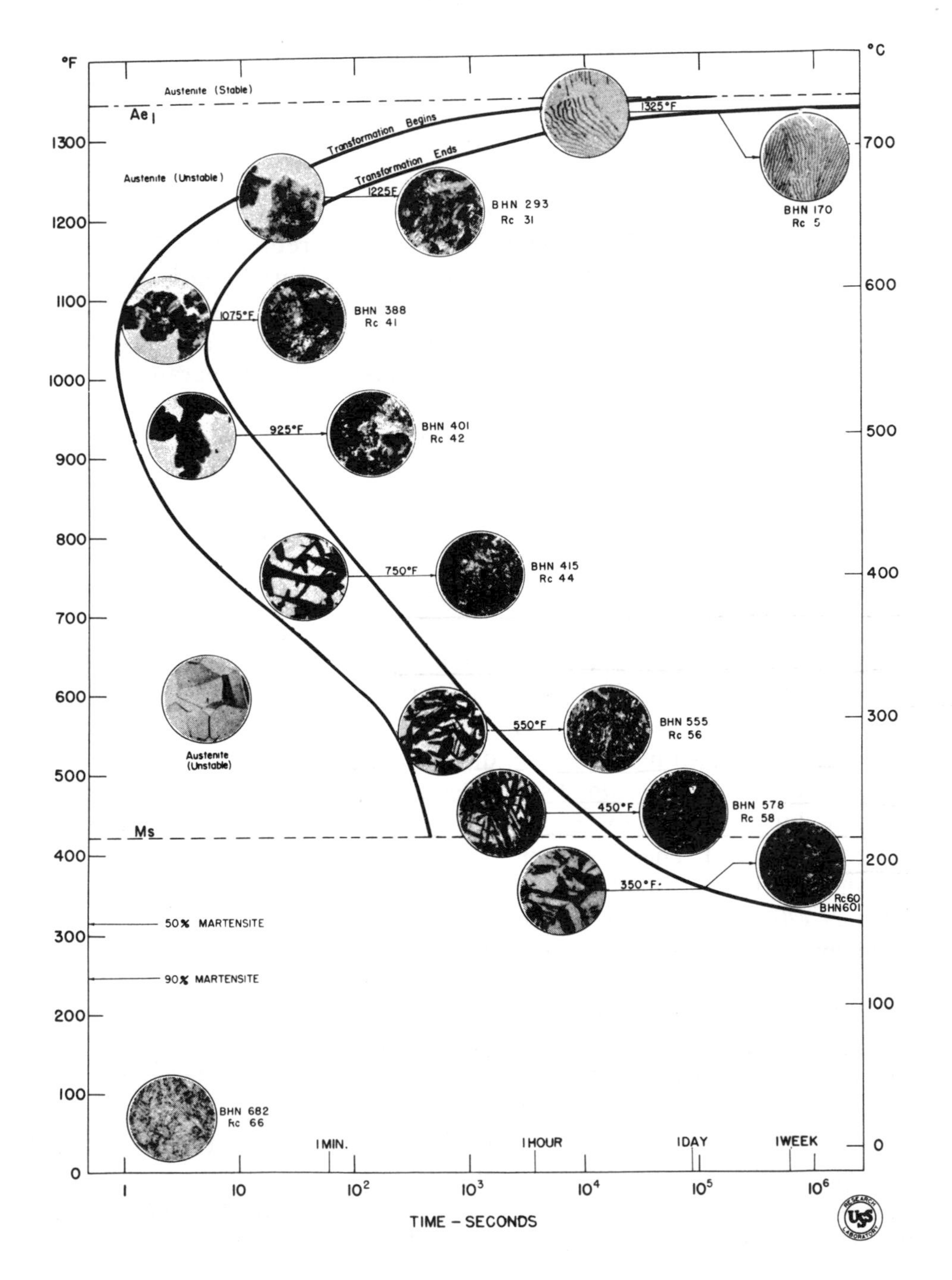

Figure 4. I-T diagram of 0.89 percent carbon steel (Copyright 1951 by United States Steel Corporation).

temperature of about 1000°F (538°C), the austenite decomposes extremely rapidly. It takes only about 1 second before transformation begins and 5 seconds to complete it. The resultant pearlite is extremely fine and its hardness is relatively high. This region of the S-curve, when decomposition of austenite occurs, is called the nose of the curve on an isothermal-transformation (I-T) diagram.

Cooling Curves

If the austenite is cooled to temperatures below the nose of the curve and held at these temperatures for a sufficient length of time, a transformation would take place, producing a product that would be called bainite. If the austenite were cooled quickly to a temperature below the Ms line, the product would then be called martensite. As noted, this transformation to martensite is complete at the Mf temperature. These temperatures vary considerably in steels and are a function of carbon content. The Ms and Mf temperatures are lower for high carbon steels than for low carbon steels (Figure 5).

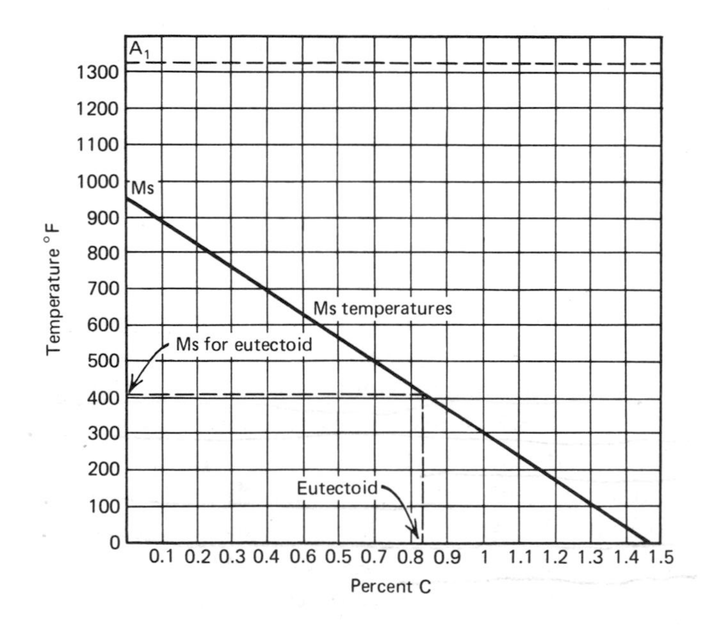

Figure 5. The Ms temperature is a function of the carbon content and will be further lowered when alloying elements are added. After finding the Ms temperature for steel containing a particular carbon content, subtract the following: 70 times the percentage of chromium, 70 times the percentage of manganese, 50 times the percentage of molybdenum, and 35 times the percentage of nickel (based on Fahrenheit temperatures).

If a cooling curve (Figure 6) is superimposed on the I-T diagram, it can be seen that it must pass to the left (1) of the nose of the diagram to transform to martensite. If the cooling rate is too slow, however,the cooling curve will cut into the nose of the diagram, showing that a partial or complete transformation has taken place at that point (2) and that fine pearlite plus martensite developed in the material instead of the desired martensite. Therefore, the rate of cooling for a hardening quench must be such that the nose of the diagram is to the right of the cooling curve. However, the *formation* of martensite is temperature, not time, dependent.

The Critical Cooling Rate

Different alloys can affect the shape of the I-T diagrams. An increase in carbon content moves the S-curve to the right (increases the time before transformation takes place). Grain size also has an effect on hardenability (the property that determines the depth and distribution of hardness induced by quenching a ferrous alloy). Larger grain carbon steels also take more time to transform. This also moves the S-curve to the right. The addition of alloy to the steel also moves the S-curve to the right.

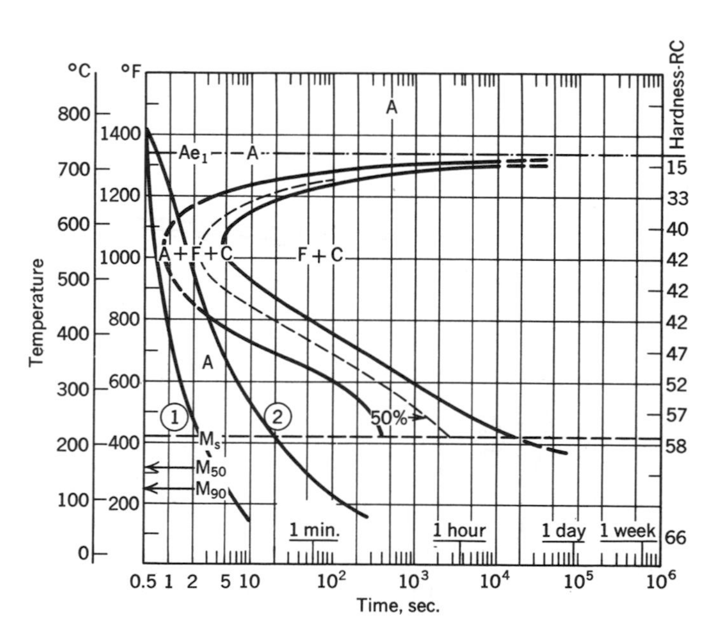

Figure 6. Two cooling curves shown on a SAE 1095 I-T diagram (Copyright 1951 by United States Steel Corporation).

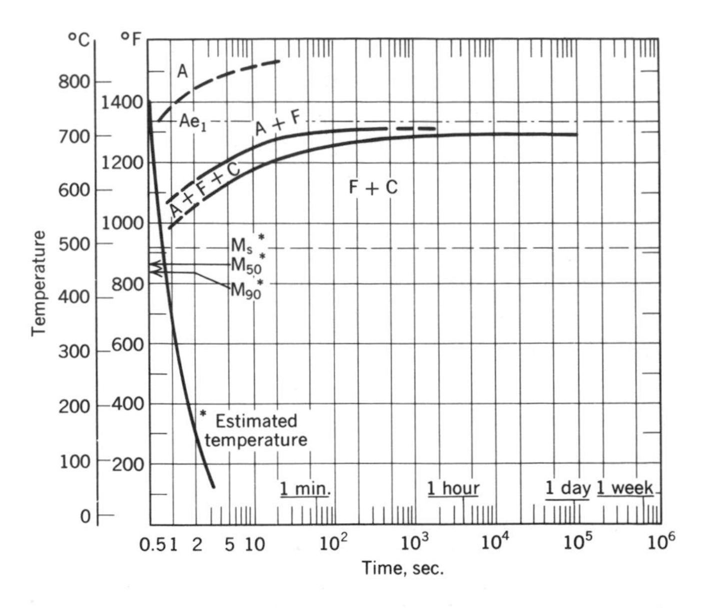

Figure 7. Cooling curve for SAE 1008 carbon steel (Copyright 1951 by United States Steel Corporation).

A plain low carbon steel cannot be hardened for practical purposes because the nose of the diagram is at or falls short of the zero time line, and it would be impossible to avoid cutting into it with the quenching or cooling curve (Figure 7). However, with carbon steels above 0.30 percent, it is possible to quench rapidly enough to effect a transformation to martensite (Figure 8). Plain carbon steel of 0.83 percent must be quenched in water to make the quench rapid enough so that it would take place within the 1 or 2 seconds needed to avoid cutting into the nose on the diagram. The critical cooling rate, then, is the rate of cooling that avoids cutting into the nose of the S-curve.

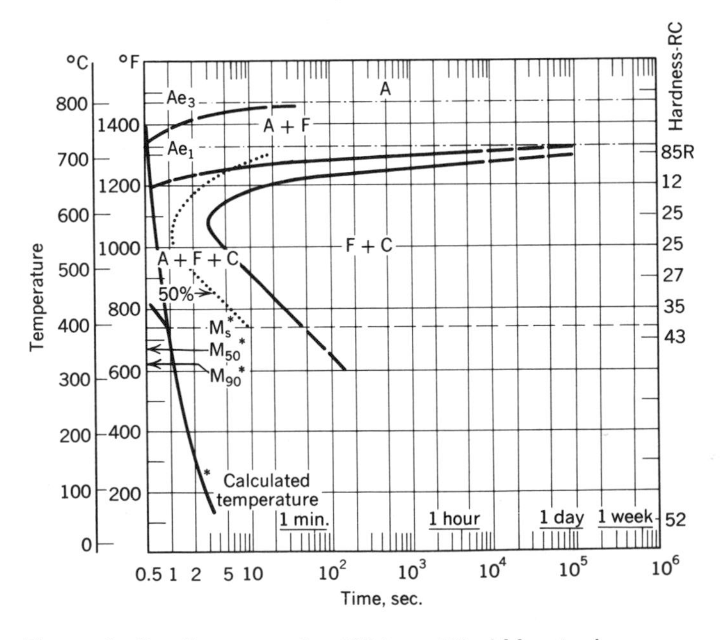

Figure 8. Cooling curve for 1034 modified Mn steel (Copyright 1951 by United States Steel Corporation).

Oil hardening steels with alloying elements, such as chromium and molybdenum, cause the nose of the diagram to move toward the right, thus increasing the time in which hardening can take place. Also the shape of the nose is often changed on the diagram. These changes often allow a great deal of time for the quench to take place. It is easy to see on the I-T diagram how oil hardening and air hardening (deep hardening) steels are affected by cooling rates (Figure 9).

Transformation to the martensite depends on three factors:

1 Mass of part.
2 Severity of quench.
3 Hardenability of the material.

The surface area of the part and the thickness have a considerable effect on the cooling rates. A very thin part, such as a razor blade, with a large surface area would have a cooling rate that would be many times greater than a cube of steel 2 or 3 inches square. Therefore, a normally water hardened steel, when extremely small or thin, should be quenched in oil to achieve the proper cooling rate, while the steel block being of the same material would still not achieve critical cooling even in a severe quench of cold salt water.

Water quenched steels normally will harden only to a depth of approximately $\frac{1}{8}$ inch, while the core is left quite soft. These are termed shallow hardening steels. A 1 or 2 inch thick, air cooled steel may harden completely to the core. This is deep hardening steel. As the time increases in which quenching can take place, the depth of hardening increases.

When quenching rates are drastic, such as in brine or water quench, the stresses in the part caused by the different cooling rates from the interior and exterior of the part can cause warping and cracking (Figure 10). Water quenched steels are particularly prone to this problem. The slower cooling rates of oil and air allow more uniform

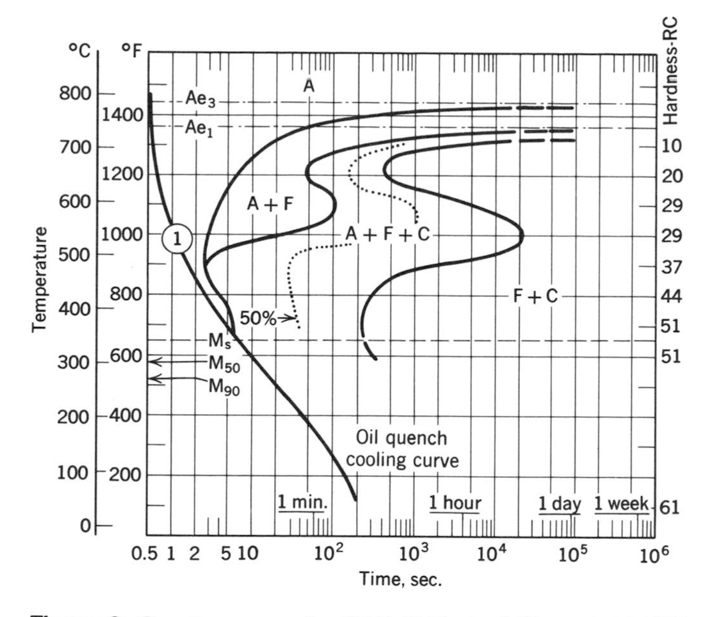

Figure 9. Cooling curve for SAE 4140 steel (Copyright 1951 by United States Steel Corporation).

cooling and, because of this, these steels suffer less cracking and warping. For this reason, when large or heavy sections must be heat treated, an alloy steel that can be oil or air hardened should be selected (Figure 11).

Figure 11. The design of this forming die, made of Type W1 tool steel, presents an almost impossible problem to the heat treater. Because of the blind holes and the thin section between them, the die cracked during heat treatment. Unless the die can be totally redesigned, the use of an air hardening steel is imperative (Bethlehem Steel Corporation).

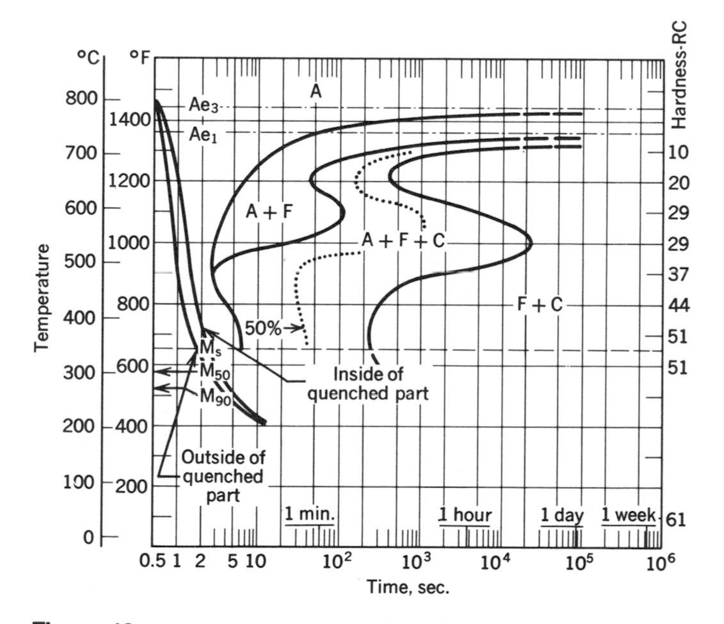

Figure 10 Internal and external cooling curves on the same part on this I-T diagram show how two different cooling rates can cause the high stresses to build up that sometimes result in quench cracks (Copyright 1951 by United States Steel Corporation).

Self-Evaluation

1 What is the austenitizing temperature for a carbon steel?

2 How is the hard structure, martensite, produced? What is the major consideration when martensite is produced?

3 What are the Ms and Mf temperatures?

4 What is the critical cooling rate?

5 If the cooling curve of a 0.8 percent carbon steel cut partly through the nose of the I-T diagram, what would the resultant microstructure of the metal be?

6 How does the carbon content affect the position of the S-curve?

7 What are the two steps needed in hardening steel to produce useful articles?

8 A thick section of W1 steel developed a quench crack when it was hardened. What do you think caused this? What steps can be taken to correct the problem?

9 How can you determine from studying an I-T diagram of very low carbon steel why little or no martensite can be produced when the steel is quenched from an austenitizing temperature?

10 How can you tell by studying an I-T diagram whether an alloy steel has deep hardening capabilities?

Worksheet

Objectives

1 Harden the two specimens by quenching them from the austenitizing temperature. One specimen in water and one in oil.

2 Determine their respective grain structures by microscopic observation.

3 Draw the approximate cooling curves for each quench.

Materials

Metallurgical equipment such as a specimen mounting press, polishing equipment, abrasive cutoff saw, microscope, and two pieces of $\frac{3}{8}$ inch diameter SAE 1095 water hardening steel 3 inches long.

Procedure

1 Heat both specimens to the austenitizing temperature and quench one in oil and one in water. Do not temper them. Make an identifying mark on each one immediately.

2 Use the abrasive cutoff saw with coolant to cut a small section out of each specimen near the outer edge.

3 Encapsulate in plastic and identify the specimen with a permanent mark.

4 Polish and etch $\frac{1}{2}$ to 2 seconds with nital.

5 Set up the metallurgical microscope and observe the specimen that was quenched in water. What is its microstructure?

6 Observe the specimen quenched in oil. What is the microstructure?

7 Draw a cooling curve in the isothermal diagram (Figure 12) for each specimen based on the microstructures that you see and compare with those in Figure 2.

Conclusion

1 Do you think the cooling rates were different? How did you determine this?

2 What effect do different cooling rates have on some steel?

3 Show your instructor your cooling curve and conclusions.

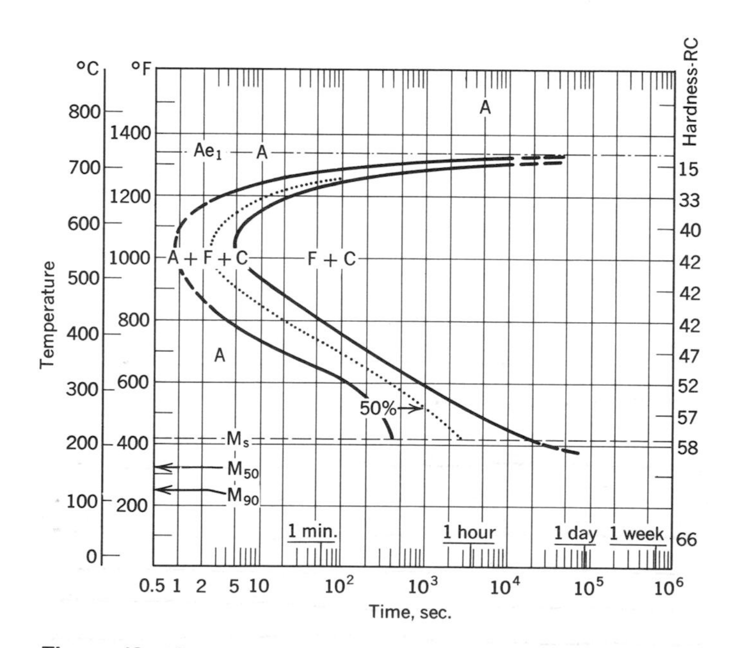

Figure 12. (Copyright 1951 by United States Steel Corporation.)

12 Hardenability of Steels and Tempered Martensite

If all hardened parts were less than $\frac{1}{2}$ inch thick, then water hardening plain carbon steels would be sufficient for most purposes. However, many design applications call for the use of large sections of high strength steels. This requires the designer to specify deep hardening or high hardenability steels. The difference in hardenability of tool steels is explained and demonstrated in this chapter.

OBJECTIVES

After completing this chapter, you will be able to:

1. Explain the methods of determining and evaluating the depth of hardening (hardenability) of various steels.
2. Demonstrate and measure the hardenability of a shallow hardening steel.
3. Demonstrate the use of a mechanical properites chart for predicting the hardness and strength of a hardened and tempered specimen.

INFORMATION

Hardenability Testing and the Jominy End-Quench Test

This test is used to determine the depth of hardening or hardenability of various types of steels. In conducting this test, a one inch round specimen approximately four inches long is heated uniformly to the correct austenitizing temperature for the length of time needed to effect complete austenitization (diffusion of carbon in the austenite). The specimen is quickly removed and placed in a bracket in such a way that a jet of water (or other quenching medium) at room temperature impinges on the bottom face of the hot specimen without wetting the sides (Figure 1). It is allowed to remain in the water jet until the entire specimen has cooled. After it has cooled, longitudinal flat surfaces are ground on the side to remove decarburization, and Rockwell C scale readings are taken at $\frac{1}{16}$ inch intervals from the quenched end. Since the quenching effect is concentrated on the end surface and the cooling rate diminishes with the distance from the end, the measurement of hardness at each location from the end corresponds to a

Figure 1. A jominy end-quench hardenability test being performed
(Courtesy of Pacific Machinery & Tool Steel Company).

certain cooling rate and hardness penetration at that depth of the particular type of metal being tested. The data secured by this means are plotted on a graph.

From a study of the curves, it becomes apparent that initial surface hardness is a function largely of the carbon content and that **hardenability** (depth hardness) **depends on the amount of carbon present, the alloy content, and the grain size.** Manganese, boron, chromium, and molybdenum are the chief elements that promote depth hardness, while nickel and silicon help to a lesser degree.

Effect of Mass on Heat Treated Steel

Figures 2 and 3 compare the depth of hardening between eutectoid (0.83 percent) carbon steel and SAE 4140. Note that the rates of cooling become slower as the distance from the quenched end increases. These differing rates of cooling can be plotted on an I-T diagram and will show cooling curves in the nose area of the S-curve. It can readily be seen that transformation products other than martensite will be formed at certain distances back from the quenched end (or near the center of an equally thick section). The effect of various types of cooling media on the hardenability or depth of hardening is shown in Figure 4.

It is also true that the mechanical properties of quenched steel depend on the mass of the piece being quenched. If a series of different diameter cylinders of steel were quenched from the same temperature in the same quenching medium and by the same procedure, the mechanical properties will vary depending on the diameters. Figure 5 shows the hardness penetration curves of round stock that survey the hardness as it varies from the outside diameter to the center for six different diameters in two steels, C1040 and A4142.

The temperature at which martensite begins to form is called the Ms temperature. This temperature can be lowered considerably by increasing the carbon content. When 0.83 percent carbon steel is quenched to below the Ms temperature, or approximately 400°F (204°C), martensite begins to form. Just above 300°F (149°C), about 50 percent transformation to martensite (Figure 6, line 1) has

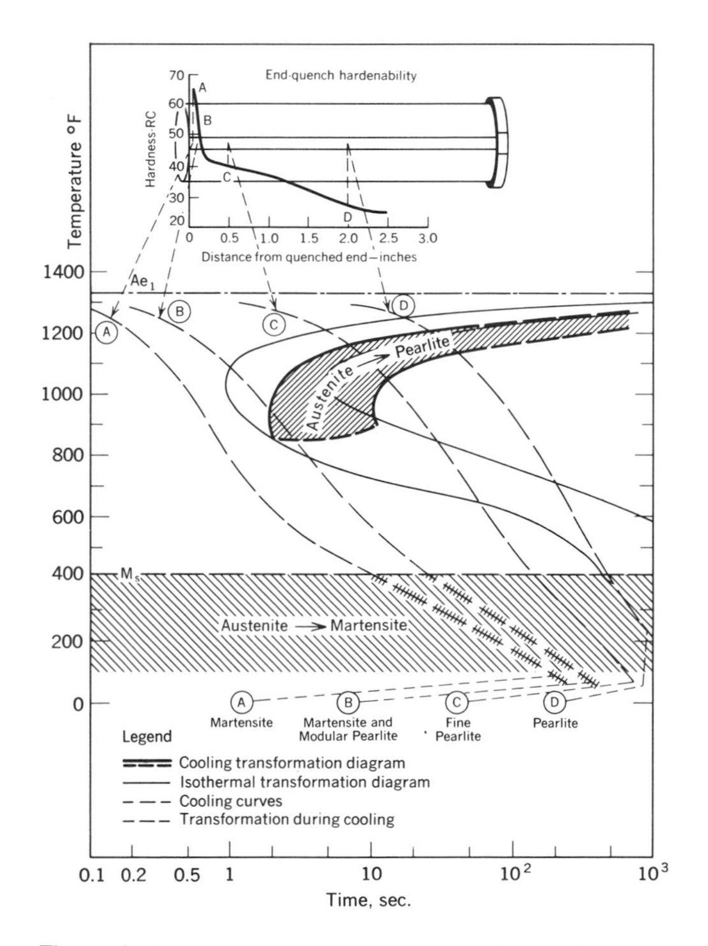

Figure 2. Correlation of continuous cooling and isothermal transformation diagrams with end-quench hardenability test data for eutectoid carbon steel (Copyright 1951 by United States Steel Corporation).

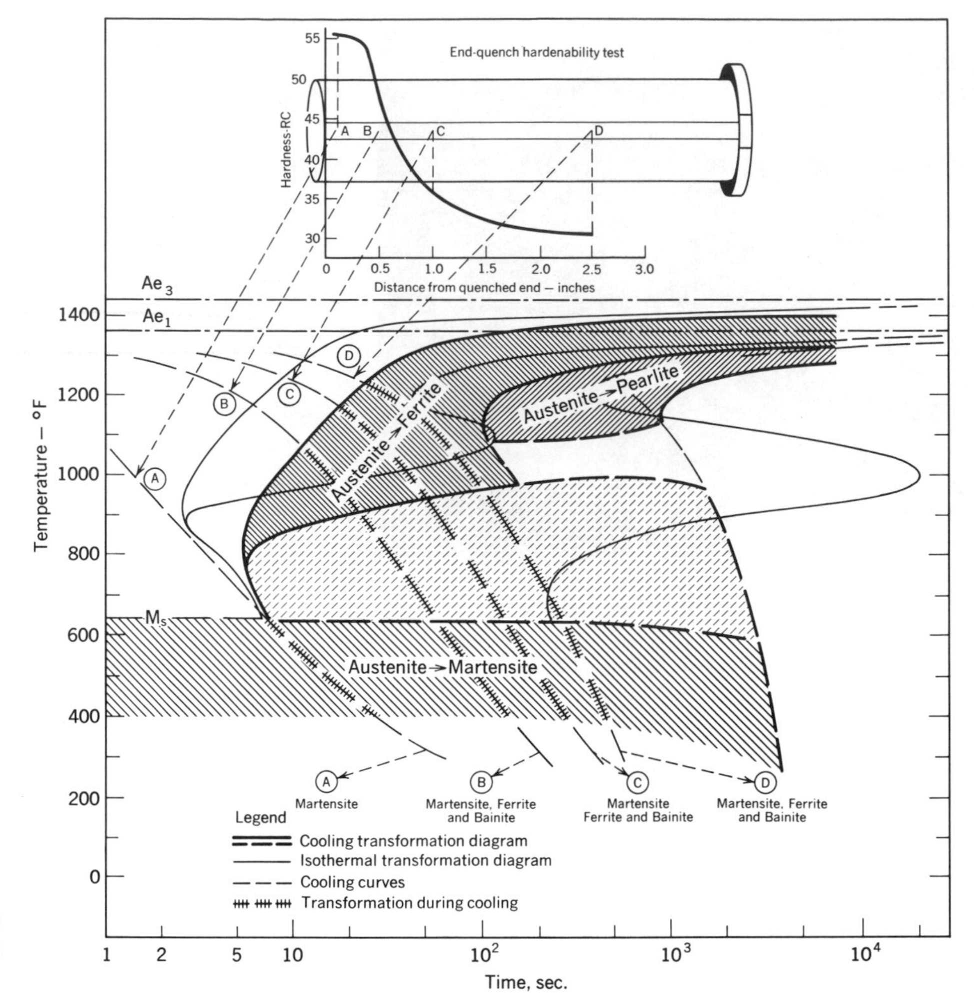

Figure 3. Correlation for continuous cooling and isothermal transformation diagrams with end-quench hardenability test data for 4140 steel (Copyright 1951 by United States Steel Corporation).

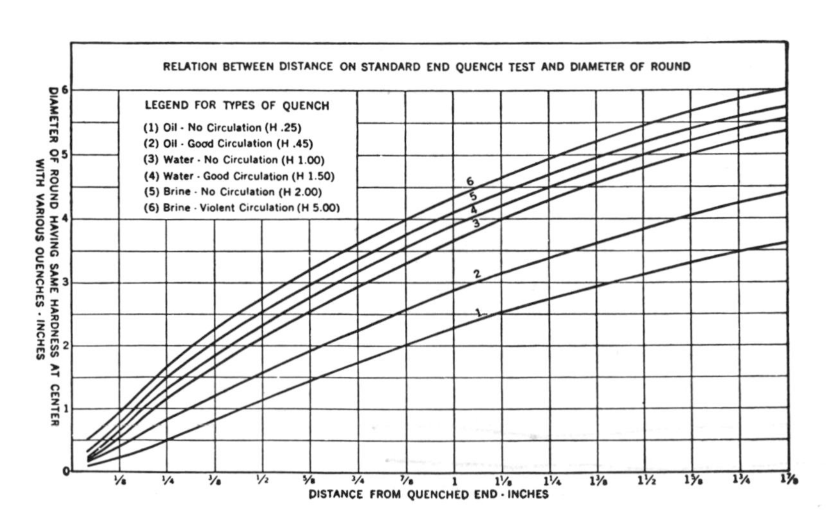

Figure 4. Cooling curves for various quenching media (Courtesy of Pacific Machinery & Tool Steel Company).

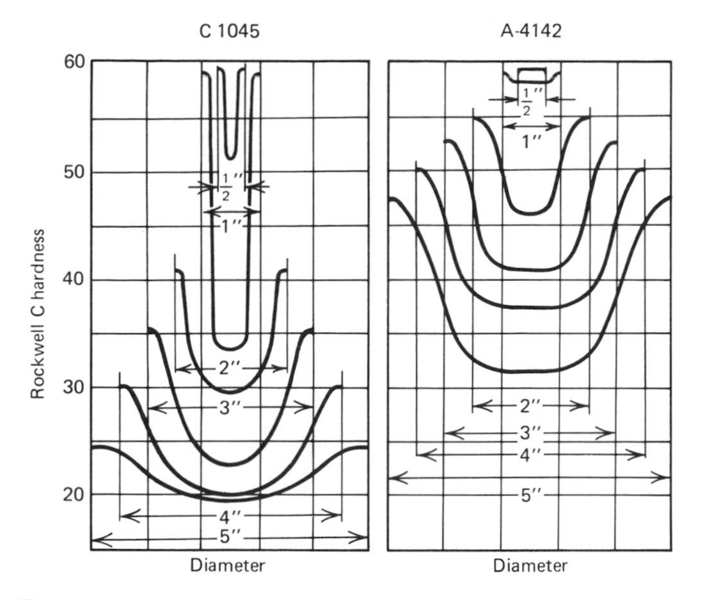

Figure 5. The effect of mass on the hardness of several cylindrical (round) quenched specimens of different diameters (Courtesy of Pacific Machinery & Tool Steel Company).

transpired; just above the Mf temperature, about 90 percent transformation has transpired; and at the Mf temperature, or approximately 200°F (93°C), about 100 percent transformation has taken place. When austenitized carbon steel is quenched to below the Mf temperature, it is at its maximum hardness unless there is retained austenite.

Isothermal Heat Treatments

When cooling curves are such that they cut across the S-curve on the I-T diagram in various places, certain microstructures are formed. A soft, coarse pearlite develops when a very slow cooling takes place at approximately 1200°F (649°C). This would be the case when a part is furnace annealed (Figure 6, line 2). When a part is air cooled after heating in the furnace to 100°F (38°C) above its upper critical temperature, the process is known as normalizing. The cooling curve for normalizing would be approximately through a medium pearlite or upper bainite section (Figure 6, line 3) of the S-curve in eutectoid steels, forming smaller, more uniform grains that leaves a stronger structure than full anneal will produce.

Another method of hardening and tempering is a form of isothermal quenching called **austempering** (Figure 7), in which a part is austenitized and quenched into a lead or salt bath held at a temperature of approximately 600°F (316°C) to produce a desired microstructure of lower bainite. It is held at this temperature for several hours until a complete transformation has taken place. This type of hardening eliminates the need for tempering. Au-

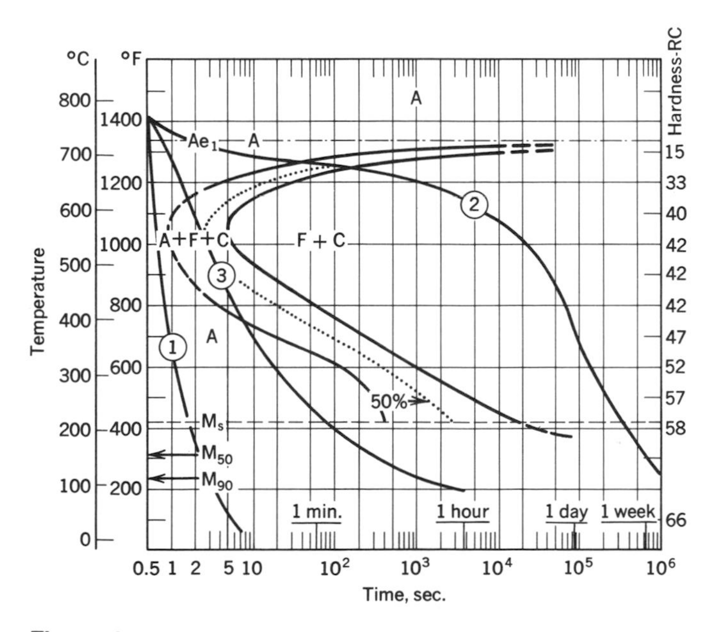

Figure 6. Cooling curves on an I-T diagram for eutectoid steel. Line 1 shows the quench for undercooling to produce martensite; line 2 shows an annealing curve; line 3 is a normalizing curve (Copyright 1951 by United States Steel Corporation).

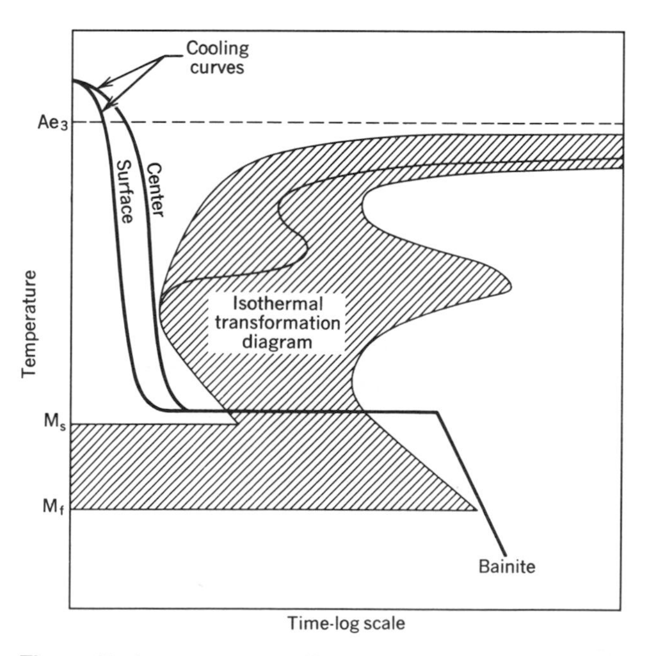

Figure 7. Austempering (Bethlehem Steel Corporation).

stempering produces a superior product that is much tougher than that developed in the conventional hardening and tempering process. There is one drawback, however; it is confined mostly to small or thin sections. Large, heavy sections of carbon steel cannot be austempered.

Isothermal quenching is also used for **martempering** (Figure 8) in which the austenitized part is brought to slightly over the Ms temperature and held for a few minutes in order to bring the interior and exterior temperatures to an equalized temperature to avoid stresses. Then the quench is continued to the Mf temperature, followed by conventional tempering. **Isothermal annealing** is done by quenching from above the critical range to the desired annealing temperature in the upper portion of the I-T diagram and holding at the anneal temperature for a length of time sufficient to produce complete transformation (Figure 9). This method produces a more uniform microstructure than conventional annealing in which the steel is very slowly cooled.

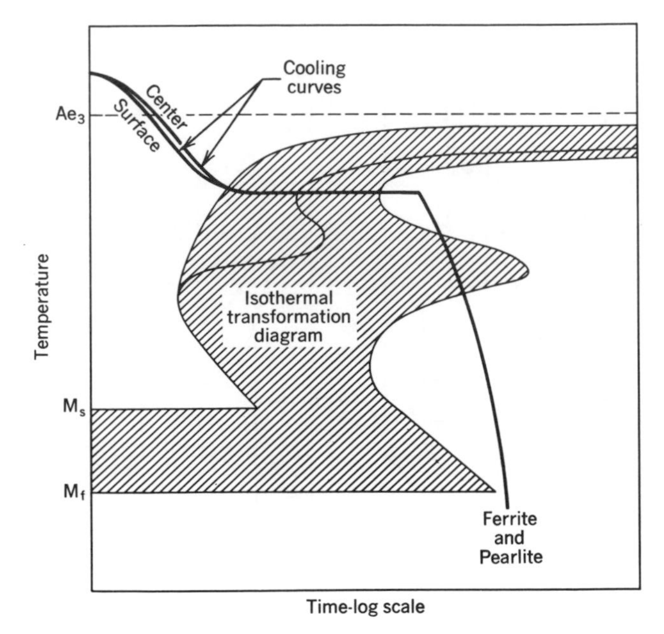

Figure 9. Isothermal annealing (Bethlehem Steel Corporation).

Tempering Procedures

Tempering, sometimes called drawing or temper drawing, is the process of reheating hardened martensitic steels to some temperature below the lower critical or Ac_1 line. Proper tempering of a hardened steel requires soaking at the tempering temperature for a certain length of time. With any selected tempering temperature, the hardness drops rapidly at first and gradually decreases over a period of time. For instance, within a half hour of

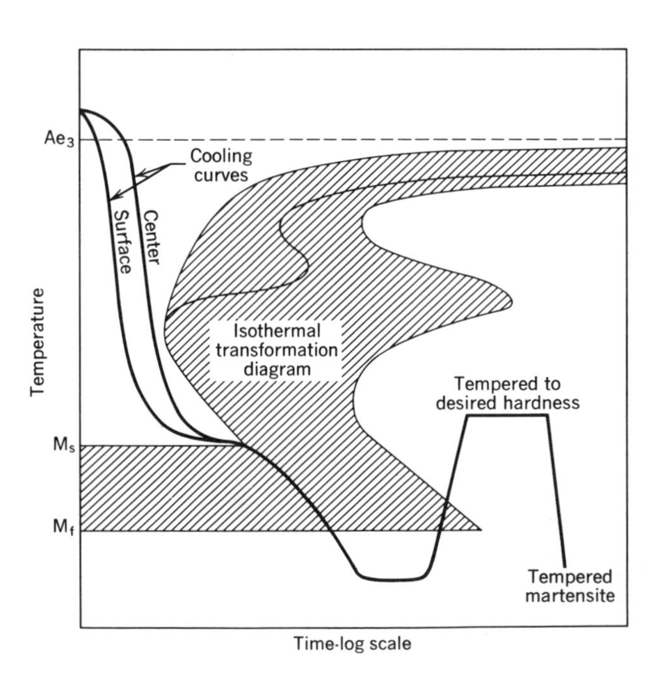

Figure 8. Martempering (Bethlehem Steel Corporation).

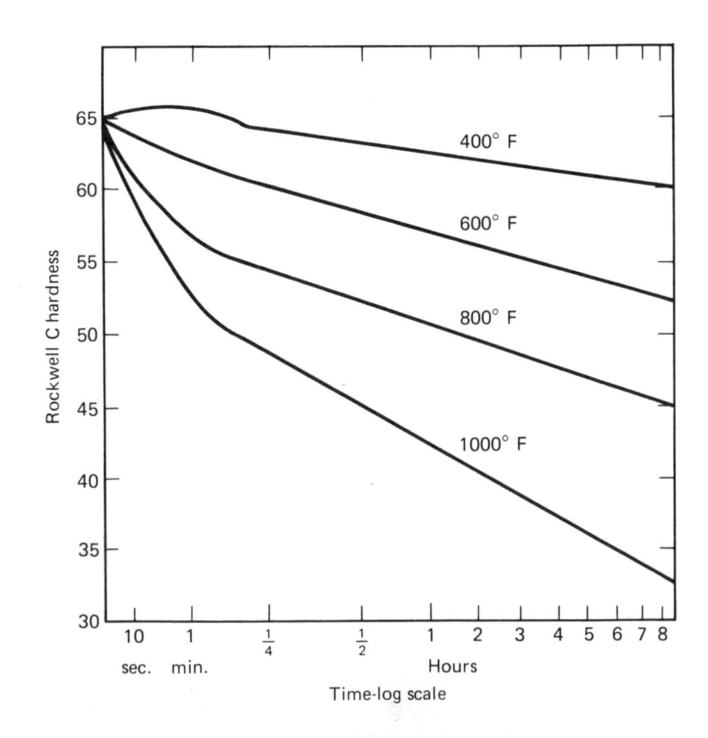

Figure 10. The effect of tempering time at four different tempering temperatures on 0.83 percent carbon steel hardened to 65 RC by quenching.

tempering at 600°F (316°C), a part is reduced in hardness about 5 or 6 points on the Rockwell scale (Figure 10). Some heat treaters prefer to temper by use of color and, in this case, they must stop the tempering process when the proper color has been reached by cooling in water, thus causing the tempering time to be very short. This is not the best practice. Carbon steels and most of the low alloy steels should be tempered as soon after quench hardening as possible. Carbon steels should not be tempered before they cool to room temperature because in some steels the Mf temperature is quite low and untransformed austenite might be present. Part or all of this residual austenite will transform to martensite or bainite on cooling from the tempering temperature so that the final structure will consist of tempered and untempered martensite and perhaps some bainite depending on the tempering temperature. The brittle untempered martensite can easily cause failure of the heat treated part.

In most cases, toughness increases as the hardness decreases due to the temperature increase of tempering but, if the Izod-Charpy notch bar test is used as a measure of toughness, it has been found that a part tempered between 400 and 800°F (204 and 427°C) has a reduced notch toughness (Figure 11). It is true that ductility increases in this tempering range so that it may be used for most purposes; but, if parts and designs include stress raisers, this tempering range should be avoided. This is sometimes called the **blue brittle tempering range.**

Some alloy steels show a loss in notch toughness when tempered between 1000 and 1250°F (538 and 677°C) followed by slow cooling. This is called **temper brittleness** and can be avoided by quenching from the tempering temperature. Steels high in manganese, phosphorus, and chromium suffer from temper brittleness, while the addition of molybdenum will retard it.

Tempered Martensite

Martensite is a supersaturated solution of carbon and iron. As heat energy is added, carbon gets the mobility to escape the iron lattice structure and form fine iron carbide particles called **transition carbides.** Tempering from 100 to 400°F (38 to 204°C) leaves low carbon martensite and submicroscopic iron carbide particles. This martensite etches dark but still retains the needlelike formation. It is sometimes called black martensite (Figure 12).

Higher tempering temperatures cause transformation of the low carbon martensite, 450 to 700°F (232 to 371°C), to lower bainite plus ferrite

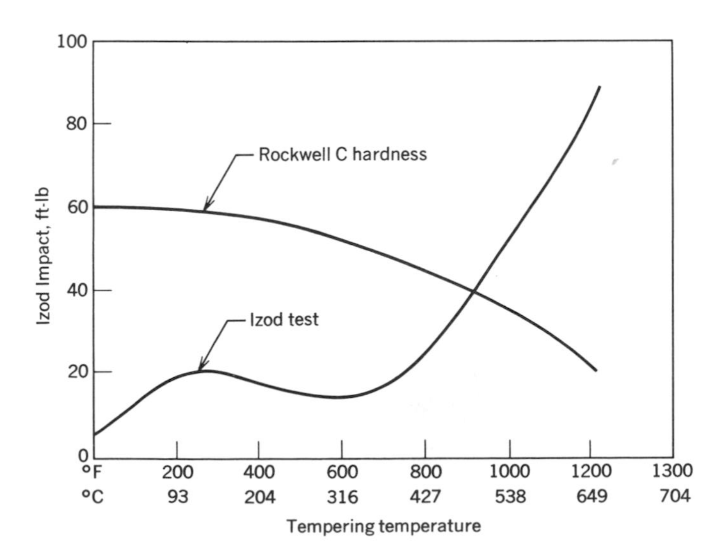

Figure 11. Graph showing the blue brittle tempering range *(Machine Tools and Machining Practices).*

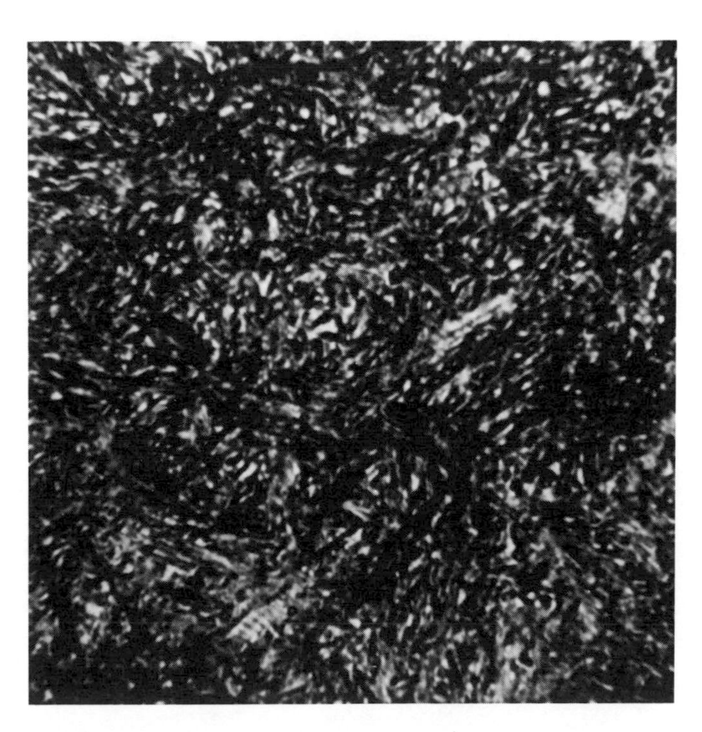

Figure 12. Microstructure of black martensite (500×) *(Machine Tools and Machining Practices).*

and an increase in carbide particle size. This etches to a black mass once known as trootsite. At a temperature between 700 and 1200°F (371 and 649°C), the carbide particles become visible and the ferrite becomes a more continuous network. This change is responsible for the marked increase in its toughness and ductility. Martensite tempered in this range begins to etch light (Figure 13). This structure was once known as sorbite. Many metallurgists use only the term "tempered martensite" for all these conditions. Martensite and carbide particles (also pearlite), when held in a 1200 to 1300°F (649 to 704°C) temperature range, tend to form spheroidite (Figure 14).

Many alloy steels retain a great deal of austenite after quenching and consequently are not fully hardened. The air cool and oil quench steels have this characteristic. After tempering, the retained austenite transforms to martensite, causing an increase in hardness. This tendency is so pronounced in the high alloy steels that double tempering is sometimes necessary to develop full hardness. Subzero treatments are also used to transform retained austenite to martensite. The steel is cooled to −70°F or below using dry ice or refrigeration. Since there is some danger of cracking the hardened steel when it is cooled to subzero temperatures, it is usually tempered at 300 to 350°F (149 to 177°C) before the cold treatment.

Mechanical properties charts are available in metals handbooks. These charts are useful to the heat treater for determining temperatures for heat treating. Tempering temperatures versus hardness, for instance, are readily determined for any carbon steel by using the mechanical properties chart. See Figure 16 in Worksheet 2. See also tempering tables for various steels in Appendix 1.

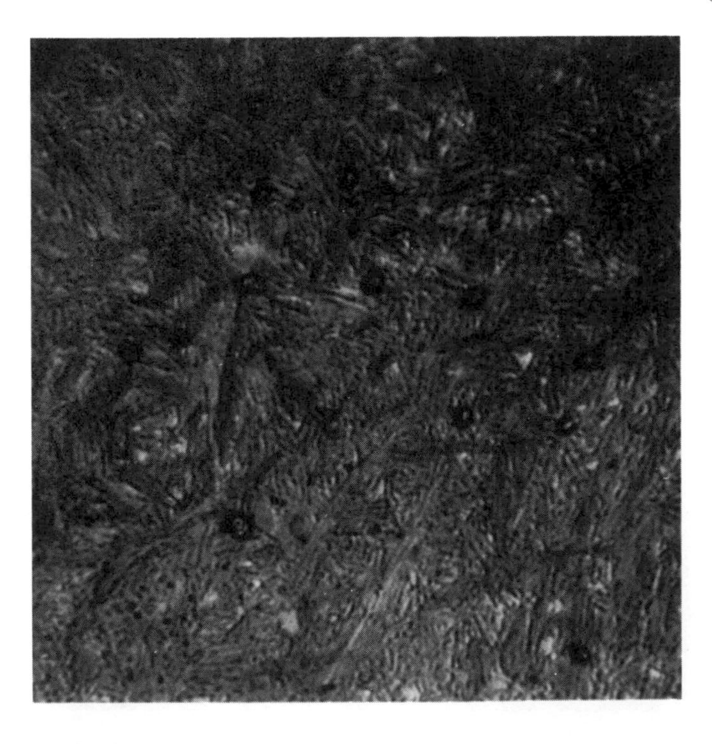

Figure 13. Microstructure of light martensite (500×) *(Machine Tools and Machining Practices).*

Figure 14. Microstructure of spheroidite (500×) *(Machine Tools and Machining Practices).*

Self-Evaluation

1. What test is used to determine hardenability?
2. Briefly explain how the test in question 1 is carried out.
3. What relationship does the test referred to in questions 1 and 2 have to the S-curve in the I-T diagram?
4. Refer to the graph in Figure 4. What effect does circulation of the quench seem to have on hardenability?
5. Approximately what is the maximum hardness of an austenitized steel of 1.50 percent carbon when quenched to the Mf temperature?
6. What type of microstructure develops in eutectoid steel when it is furnace annealed?
7. What is austempering? Name one advantage.
8. When is the best time to temper? Explain.
9. Explain the difference between the blue brittle tempering range and temper brittleness in some steels.
10. How can you predict the final tempered hardness of a hardened carbon steel that you are preparing to temper?

Worksheet 1

Objectives

1. Show the difference in hardness and cooling rate between the surface and center of a $\frac{1}{2}$ inch diameter steel specimen.
2. Demonstrate hardenability depth on a shallow hardening steel.

Material

One $\frac{1}{2}$ inch diameter 1040 (0.40 percent carbon) piece of steel 3 inches long, Rockwell hardness tester, furnace, tongs, and a quenching bath of water.

Procedure

1. Place the specimen in the furnace and set for 1550°F (843°C). Allow the specimen to reach the same color as the furnace bricks. Heat the end of the tongs so that they will not quench the specimen when you remove it from the furnace.
2. Remove the specimen and quench it in still water.
3. Cut the specimen in half using the metallurgical cutoff saw. Be careful not to overheat the specimen. Cut a short length, about $\frac{3}{8}$ inch long.
4. Next mark $\frac{1}{16}$ inch increments across the diameter of the freshly cut surfaces.

5 Rockwell test at the $\frac{1}{16}$ inch increments on center (Figure 15).

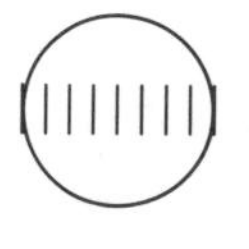

Figure 15.

6 Record your results.

Note If your lab has a Jominy end-quench hardenability tester, make up a standard sample and make tests as explained in this chapter.

Conclusion

1 Why is there a difference in hardness between the surface and center?
2 How deep does full hardening extend?
3 Show your sample and conclusions to your instructor.

Worksheet 2

Objectives

1 Learn the effect of tempering on SAE 1095 carbon steels.
2 Learn to use the mechanical properties chart to estimate the physical properties resulting from tempering.

Material Three specimens of $\frac{1}{2}$ inch round × 1 inch SAE 1095 steel, Rockwell hardness tester, heat treat furnace, water quenching tub, and tongs.

Procedure

1 Test the as-rolled specimens for hardness and compare them with the chart (Figure 16).
2 Next place the specimens in a furnace heated to 1550°F (843°C). Allow the specimens to reach the same color as the furnace bricks. The steel should be soaked at the austenitizing temperature for a few minutes before quenching.
3 Heat the end of the tongs that will come in contact with the specimens. This will avoid cooling of the specimens before they can be quenched.
4 Remove the austenitized specimens and quench in water. Agitate the specimens for better quenching.

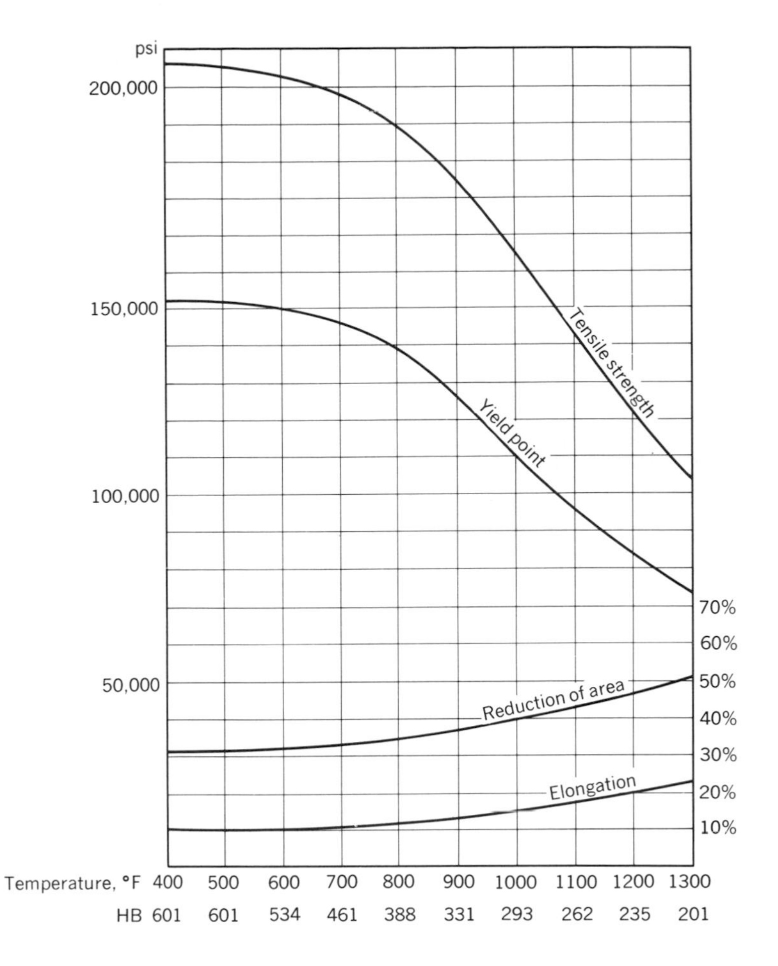

Figure 16. Water-quenched 1095 steel Mechanical Properties Chart (Bethlehem Steel Corporation).

5 Check the hardness of the as-quenched steel. The mechanical properties chart gives the as-quenched hardness of 1095 steel to be 601 Brinell. Use conversion tables if your hardness tests are Rockwell.

6 Place a hardened specimen in a furnace at 400°F (204°C) for 20 minutes.

7 Remove, cool, and test for hardness. Repeat this procedure with the remaining specimens, one at 800°F (427°C) and one at 1000°F (538°C).

8 Compare your results with the mechanical properties chart provided.

Conclusion

1 Did the results you acquired correspond with the chart? If not, how can you account for the difference?

2 Show your conclusions to your instructor.

13 Heat Treating Equipment and Procedures

The heat treating of steels is a process that requires some very critical furnace operations. The proper steps must be taken for a particular grade of steel or a failure will almost surely be the result. Study these procedures and follow them so that you will be able to harden and temper tool steels successfully.

OBJECTIVES

After completing this chapter, you will be able to:

1. Describe the proper heat treating procedures for most tool steels.
2. Correctly harden a SAE 4140 steel part.
3. Correctly draw temper the SAE 4140 steel to a predetermined hardness.

INFORMATION

Furnaces

Electric, gas, or oil fueled furnaces are used for heat treating steels (Figure 1). They use various types of controls for temperature adjustment. These controls make use of the thermocouple for controlling and recording furnace temperatures (Figure 2). Temperatures generally range up to 2500°F (1371°C). One of the disadvantages of some small electric furnaces is that they allow the atmosphere to enter the furnace, and the oxygen forms oxide on the heated metal and also combines with the surface carbon to form carbon dioxide. This oxide formation causes scale and **decarburization** of the surface of the metal. A decarburized surface will not harden. Many electric furnaces, however, use a cover (inert) gas or a slightly carbonizing atmosphere to prevent decarburization. The part may also be wrapped in stainless steel foil or packed in cast iron chips. A thick solution of boric acid in water may be applied as a paste to the metal (when it is warmed to 400°F) to prevent decarburization.

High temperature salt baths are also used for heating metals for hardening or annealing. These are sometimes called pot furnaces (Figure 3). Metal baskets are used to hold a number of small parts that are lowered into the molten bath for hardening or carburizing purposes (Figure 4). Salt baths such as molten sodium chloride (not a salt-water solution) have the advantage of limiting thermal shock. When a piece of cold metal is placed in the bath, a ''cocoon'' of solid salt immediately encases the part preventing rapid heating and consequent distortion.

One of the most important factors when heating steel is the rate at which heat is applied. When steel is first heated, it expands. If cold steel in a

Figure 1 *a*. Gas fired heat treating furnace. The gas must *not* be turned on before lighting it. Accumulated gas could cause an explosion (Mt. Hood Community College).

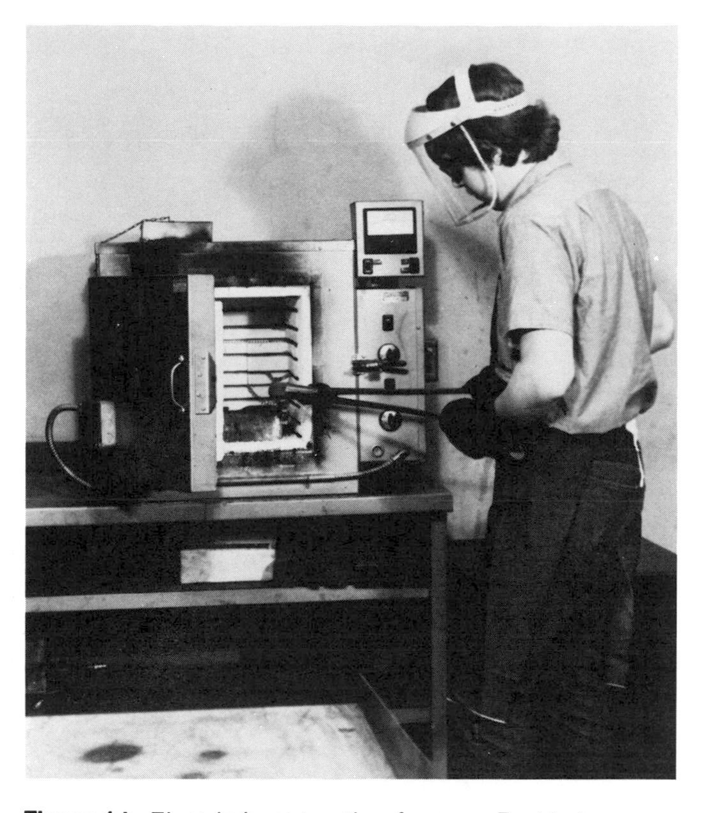

Figure 1 *b*. Electric heat treating furnace. Part being placed in furnace by heat treater wearing correct attire and using tongs (Lane Community College).

highly stressed condition is placed in a hot furnace, the surface expands more rapidly than the still cool core. The surface will then have a tendency to pull away from the center, thus inducing internal stress. This can cause cracking and distortion in tool steels. If the metal is in a ductile condition or is a low alloy steel, the rate of heating is not particularly important. Most furnaces can be adjusted for the proper rate of heat input (Figure 5) to bring the part up to the soaking temperature (Figure 6). Soaking means holding the part for a given length of time at a specified temperature.

The soaking time is usually at the austenitizing temperature to allow complete diffusion of the carbon atoms in the solid solution of austenite. Preheating a tool steel at a lower temperature is often necessary. Another factor is the time of soaking required for a piece of steel of a certain size. According to an old rule of thumb, the steel should be soaked in the furnace for 1 hour for each inch of thickness, but there are considerable variations to this rule since some steels require much more soaking time than others. This is necessary because complex carbides of chromium, molybdenum, vanadium, tungsten, and others are somewhat slow to dissolve. The correct soaking period for any specific tool steel may be found in tool steel reference books. If an incomplete soaking period is used for a tool steel, it may not harden when quenched even after the correct austenitizing tem-

Figure 2. Thermocouple. These twisted ends of dissimilar metal wires develop a feeble electric current when heated. The differences in current levels are measured by a control system and registered on a dial face as Fahrenheit or Celsius temperatures (Lane Community College).

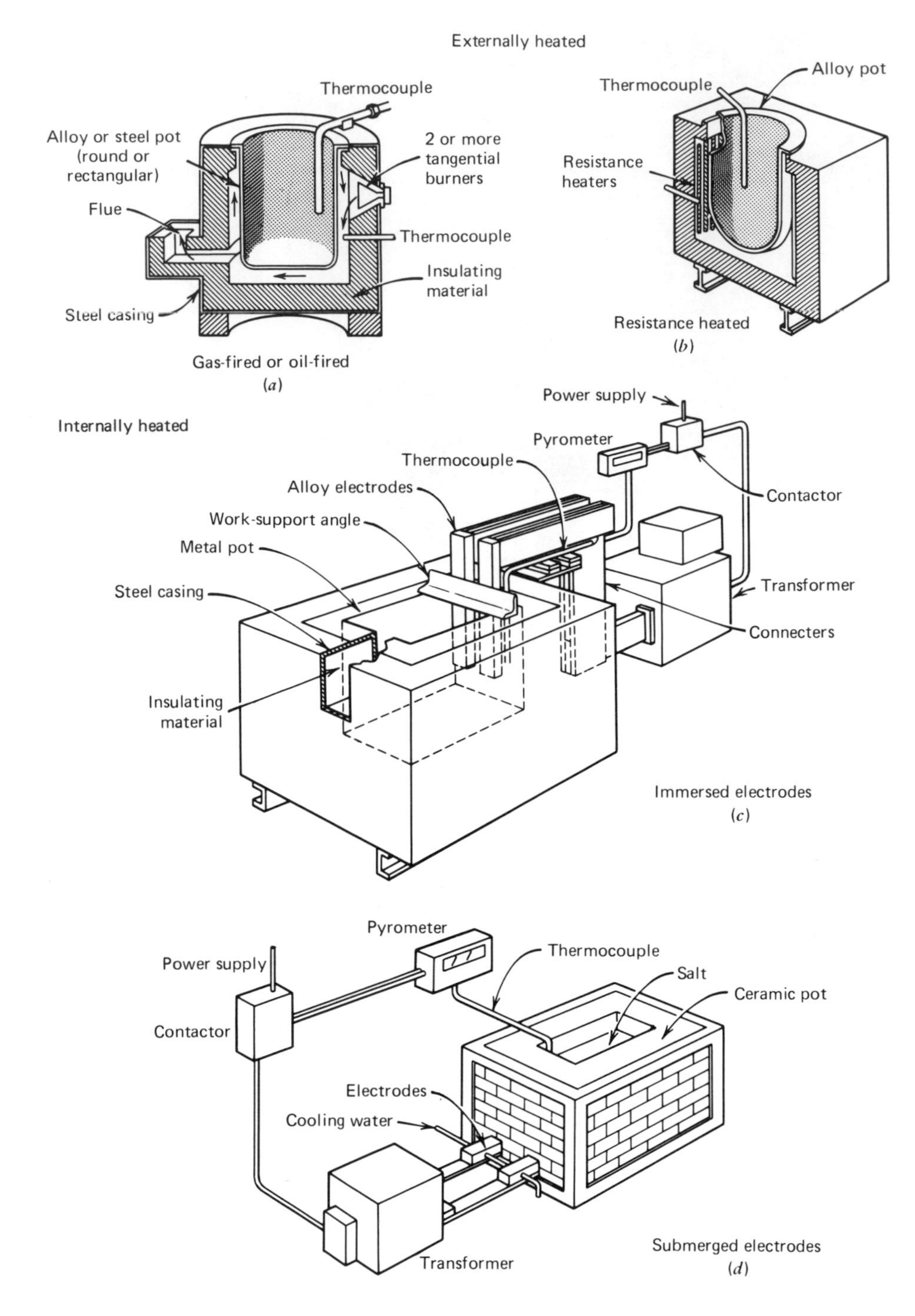

Figure 3. Pot furnace. Principal types of externally and internally heated salt bath furnaces used for liquid carburizing (By permission, from *Metals Handbook,* Volume 2, Copyright American Society for Metals, 1963).

perature is applied. Plain carbon steel, however, need only to be heated evenly throughout the part before it is removed from the furnace for quenching.

Quenching Media

In general, six media are used to quench metals, They are listed here in order of decreasing severity or speed of quenching.

1 Brine (water and salt); that is, sodium chloride or sodium hydroxide.

2 Tap water.

3 Fused or liquid salts.

4 Soluble oil and water.

5 Oil.

6 Air.

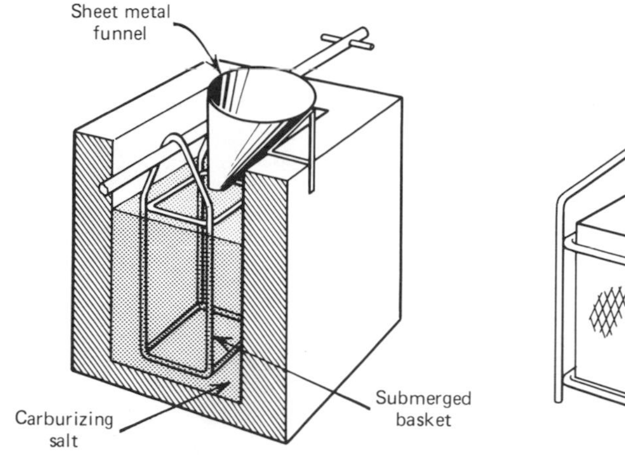

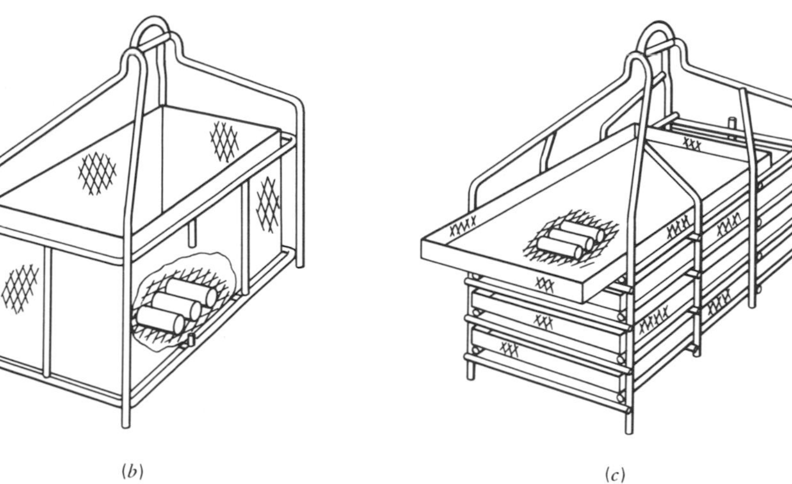

Figure 4. *(a)* Typical holding basket for small parts, equipped with a funnel for loading parts into the basket without splashing. Funnel, which is made of sheet metal, also insures that parts are coated with salt before nesting together. Basket may be made of carbon or alloy steel rod and steel wire mesh. Work must be *free* from oil, or the parts will stick together. Parts must be dry. *(b)* Inconel basket of simple design. Upper loop of the handle is for lifting; lower loop accommodates a rod that supports the basket over the furnace. *(c)* Simple basket with trays, intended for small parts. Trays provide a maximum of loading space without adversely affecting circulation. Entire fixture is made of Inconel. (By permission, from *Metals Handbook,* Volume 2, Copyright American Society for Metals, 1963).

Figure 5. Input controls on furnace (Lane Community College).

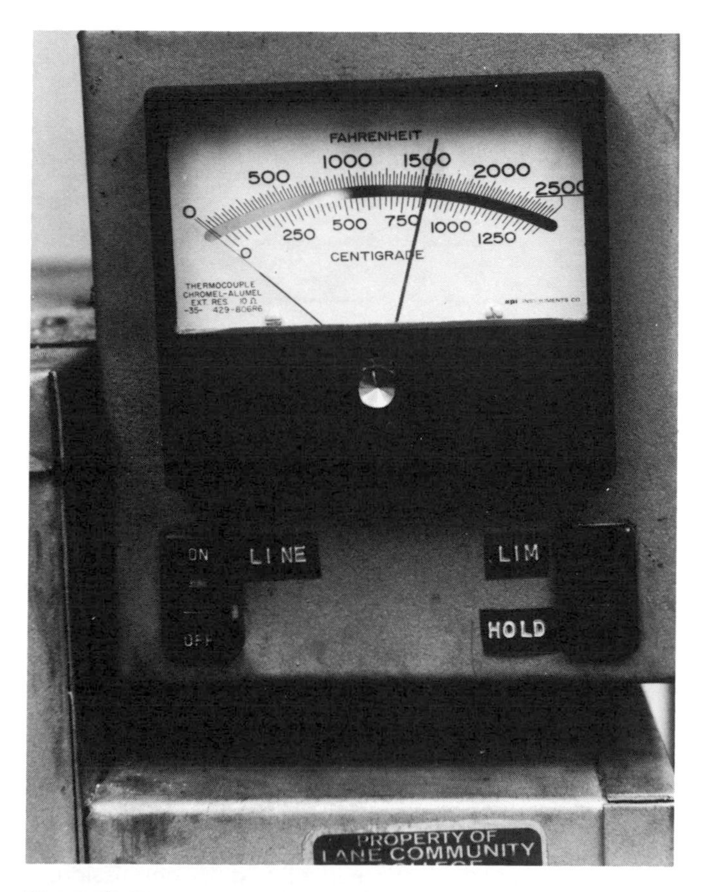

Figure 6. Temperature control (Lane Community College).

Liquid quenching media goes through three stages. The **vapor blanket** stage occurs first because the metal is so hot that it vaporizes the liquid. This envelops the metal with vapor, which insulates it from the cold liquid bath. This causes the cooling rate to be relatively slow during this stage. The **vapor transport cooling** stage begins when the vapor blanket collapses, allowing the liquid medium to contact the surface of the metal. The cooling rate is much higher during this stage. The **liquid cooling** stage begins when the metal surface reaches the boiling point of the quenching medium. Since boiling no longer occurs at this stage, heat must be removed by conduction and convection. This is the slowest stage of cooling (Figure 7).

It is important in liquid quenching baths that either the quenching medium or the steel being quenched should be agitated (Figure 8). The vapor that forms around the part being quenched acts as an insulator and slows down the cooling rate. This can result in incomplete or spotty hardening of the part. Agitating the part breaks up the vapor barrier. An up and down motion works best for long, slender parts held vertically in the quench. A figure eight motion is sometimes used for heavier parts.

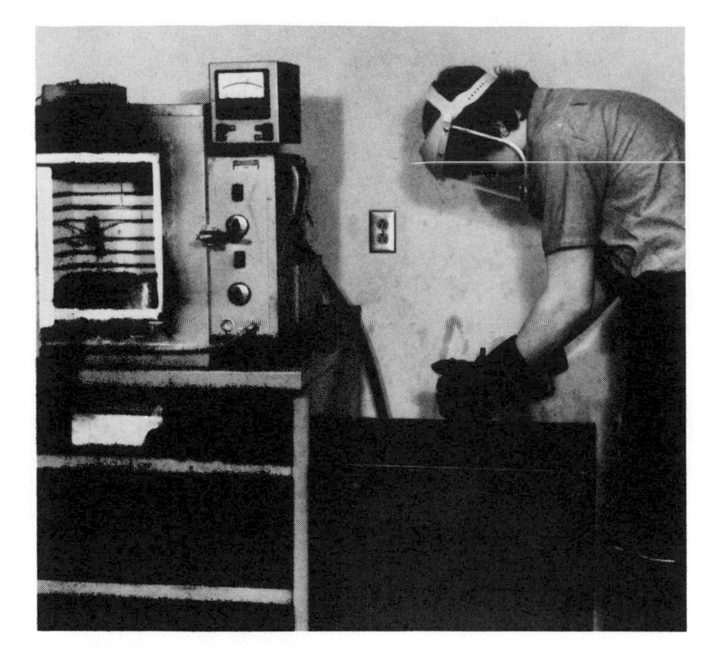

Figure 8. Heat treater is agitating part during quench (Lane Community College).

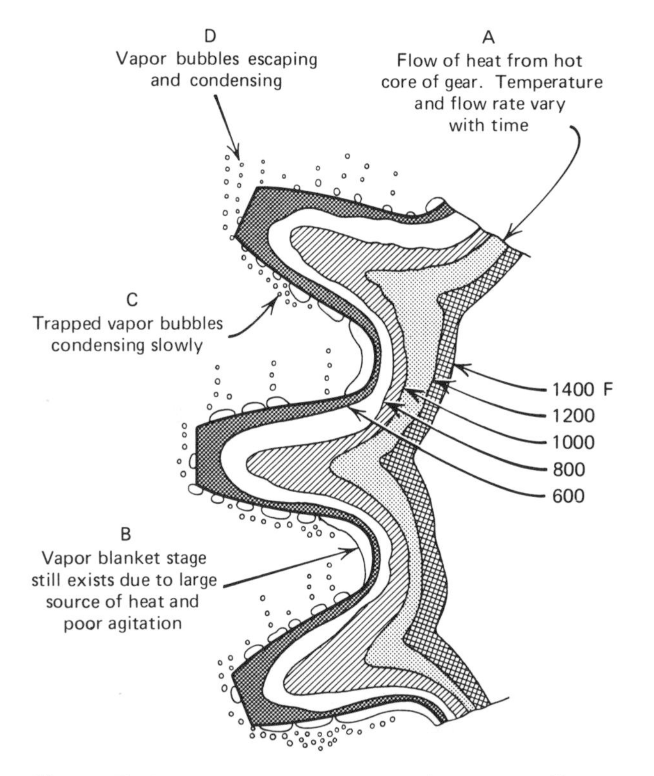

Figure 7. Stages of cooling. Temperature gradients and other major factors affecting the quenching of a gear. The gear was quenched edgewise in a liquid (By permission, from *Metals Handbook,* Volume 2, Copyright American Society for Metals, 1963).

Gloves and face protection must be used in this operation for safety (Figure 9). Hot oil could splash up and burn the heat treater's face if he or she is not wearing a face shield.

Molten salt or lead is often used for isothermal quenching. This is the method of quenching used for austempering as mentioned in chapter 12. The part is quenched at the austenitizing temperature in a molten bath, held about 500 to 600°F (260 to 315.5°C), and kept there until it is completely transformed. Austempered parts (Figure 10) are superior in strength and quality than those produced by the two-stage process of quenching and tempering. The final austempered part is essentially a fine, lower bainite microstructure (Figure 11). As a rule, only parts that are thin in cross section are austempered; a common example is the power lawn mower blade.

In martempering the part is quenched in a lead or salt bath about 400°F (204°C) until the outer and inner parts of the material are brought to the same uniform temperature. The part is next quenched below 200°F (93°C) to transform all of the austenite to martensite. Tempering is then carried out in the conventional manner.

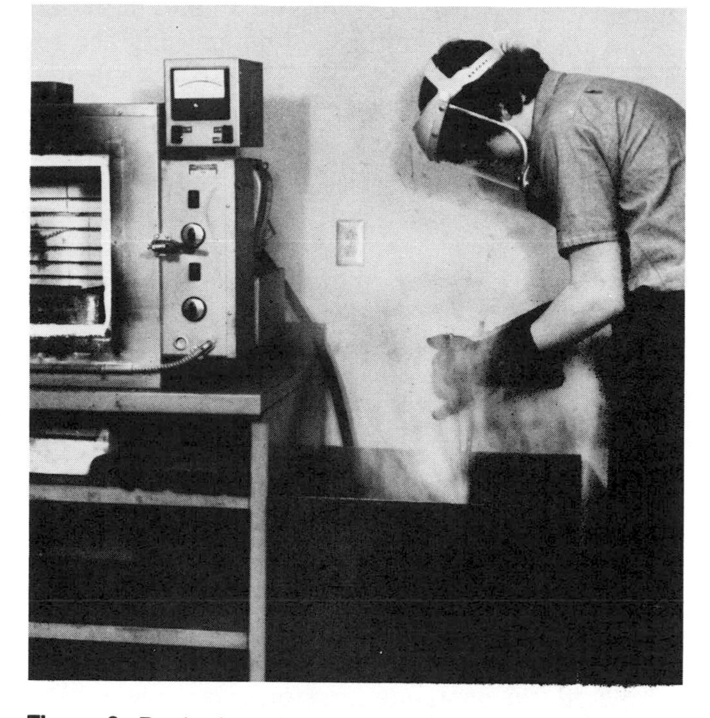

Figure 9. Beginning of quench. At this stage heat treater could be burned by hot oil if he is not adequately protected with gloves and face shield (Lane Community College).

Figure 11. Lower bainite microstructure (500×) *(Machine Tools and Machining Practices).*

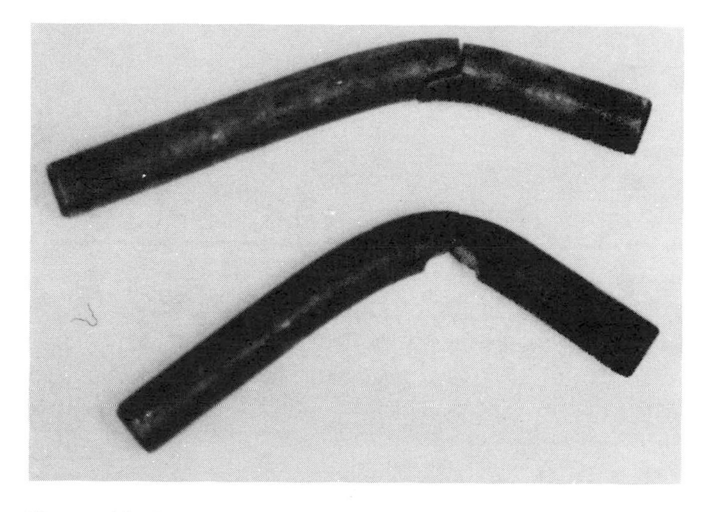

Figure 10. Austempered parts compared to the same kind of part hardened and tempered by the conventional method (Lane Community College).

Tool steels are often classified by the type of quenching medium used to meet the requirements of the critical cooling rate. See Table 1. For example, water quenched steels, which are the plain carbon steels, must be rapidly quenched. Oil quenched steels are alloy steels, and they must be hardened in oil. The air cooled steels are alloy steels that will harden when allowed to cool from the austenitizing temperature in still air. Air is the slowest quenching medium; however, its cooling rate may be increased by movement (the use of fans, for example). Tool steels are also classified by their uses such as shock or heat resisting steels.

Step or multiple quenching is sometimes used when the part consists of both thick and thin sections. A severe quench will harden the thin section before the thick section has had a chance to cool. The resulting uneven contraction often results in cracking. With the step method, the part is quenched for a few seconds in a rapid quenching medium, such as water followed by a slower quench in oil. The surface is first hardened uniformly in the water quench, and time is provided by the slower quench to relieve stresses. A suitable tank for liquid quenching is shown in Figure 12.

Tempering

Furnace tempering is one of the best methods of producing a tempered martensite of the correct hardness and toughness that the part requires. Tempering should be accomplished immediately after hardening. While still warm, the part should be put into the furnace (Figure 13). If it is left at room temperature for even a few minutes, it may develop a quench crack (Figure 14).

A soaking time should also be used when tempering, and the length of time is based on the type

Figure 12. Tank for liquid quenching showing basket for retrieving parts that have been dropped into the tank (Lane Community College).

of tool steel used. A cold furnace should be raised to the correct temperature for tempering. The residual heat in the bricks of a previously heated furnace may overheat the part, even though the furnace has been cooled down to the right temperature. The hardened part is placed in the furnace, using tongs, in an easily retrievable position, as on a firebrick to raise it above the floor of the furnace.

Some heat treaters, however, prefer to use the color system to determine the required temperature. With this system, the tempering process must cease when the part has come to the correct temperature by color, and the part must be dropped in water to stop further heating of the critical areas. There is no possible soaking time when this method is used.

Double tempering is used for some alloy steels such as high speed steels that have incomplete transformation of the austenite when they are tempered for the first time. The second time they are tempered, the austenite transforms completely into the martensite structure.

Table 1 Tool Steel Uses

Tool Steel		Uses
Water Hardening	W1, W2	Woodworking tools, files, saws, cold chisels, cutting tools
Cold work tool and die steels	01	Riveting tools, bushings
	06	Gages
	A2	Cold forming dies
	A4, A5	Shear blades, woodworking knives
	D2	Slitters
Shock Resisting	S1	Riveting tools, anvil facing, drill bushings
	S5	Blacksmith tools, cold chisels, veneer knives
	S7	Chuck jaws, swaging dies, hammers, mauls, pneumatic tools, screwdriver blades
Hot work	H1, H11	Hot riveting tools, cutting tools (hot)
	H12	Die casting molds, hot blanking dies
	H13	Hot forging dies
High speed	T1	End mills, flat drills, counterbores, countersinks, milling cutters, broaches, stone cutting tools
	M1	Twist drills, center drills, rock drills
	M2	Threading tools, broaches, woodworking tools, reamers, blanking punches, and dies

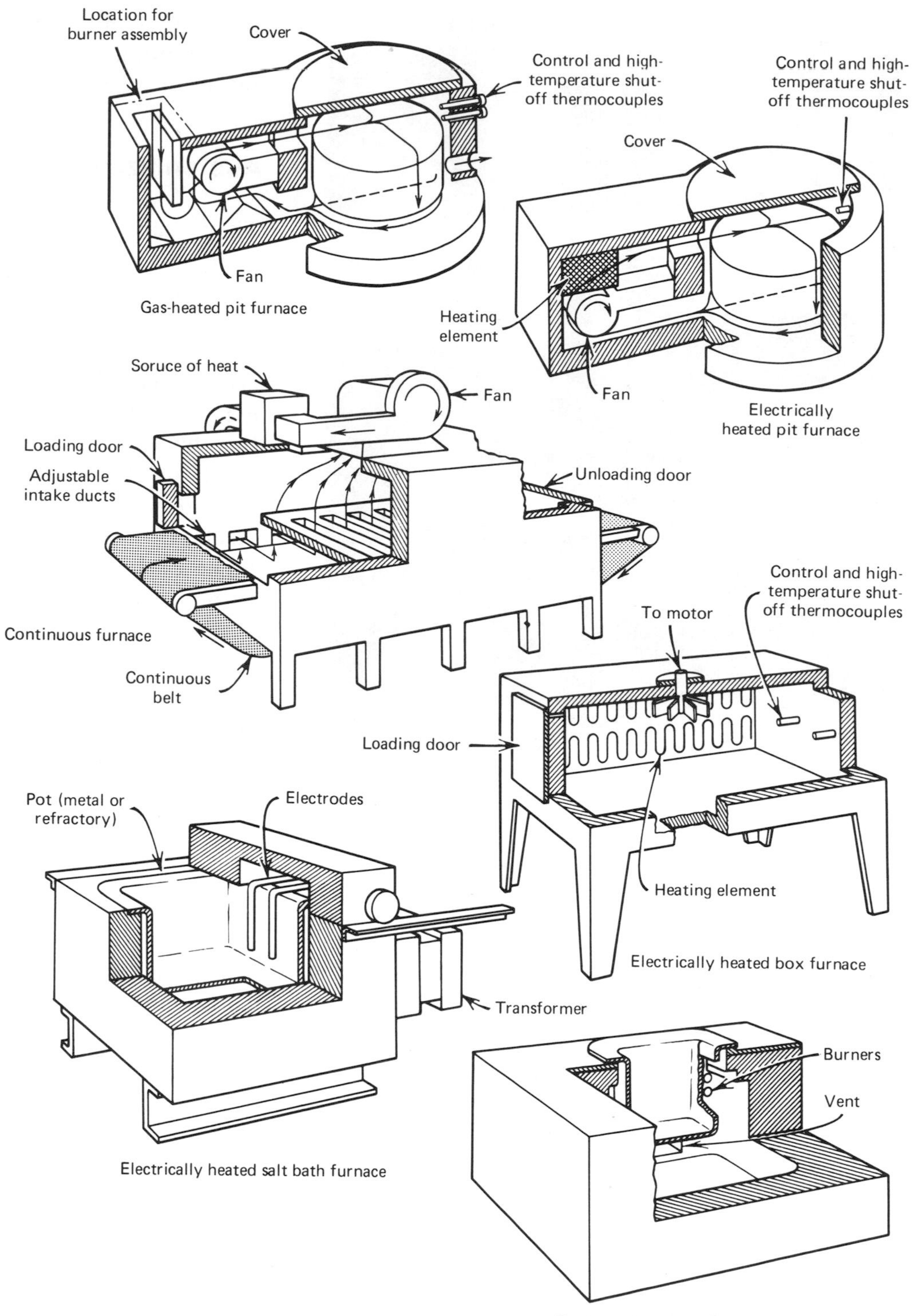

Figure 13. Type of furnaces used for tempering steel. Tempering may also be done in small electric furnaces that have automatic temperature controls (By permission, from *Metals Handbook,* Volume 2, Copyright American Society for Metals, 1963).

Conditions and Preparations Needed Prior to Hardening Steels

Before the hardening process of austenitization and quenching, normalizing is often necessary. Normalizing requires slow and uniform heating above the transformation temperatures (austenitizing) and a slow cooling in still air. Normalizing breaks up nonuniform structures, relieves residual stresses, and produces greater uniformity in grain size. Steel is conditioned by normalizing for processes such as spheroidization, annealing, or hardening.

Some tool steels harden in still air and for

Figure 14. These two breech plugs were made of Type 16 tool steel. Plug # 1 cracked in the quench through a sharp corner and was therefore not tempered. Plug # 2 was redesigned to incorporate a radius in the corners of the slot and a soft steel plug was inserted in the slot to protect it from the quenching oil. Plug # 2 was oil quenched and checked for hardness (Rockwell C 62) and after tempering at 900°F (482°C) was found to be cracked. The fact that the as-quenched hardness was measured proves that there was a delay between the quench and the temper that was responsible for the cracking. The proper practice would be to temper immediately at a low temper, check hardness, and then retemper to desired hardness (Bethlehem Steel Corporation).

these tool steels normalizing is not recommended. The tool steels that should not be normalized include all high speed steels, shock resisting steels, hot work steels, and cold work steels, types A (except for A10) and D. See Chapter 3, which is concerned with the identification of tool steels.

Tool steels when purchased from the supplier are in the annealed condition, but they usually have a decarburized surface skin. Some companies, however, produce steel that has been premachined to remove the decarburized surface so that the steel can be heat treated immediately in the condition that it comes from the factory without prior machining.

Austenitizing Procedures

The full anneal or stress relief anneal is often used to remove stresses prior to hardening. A preheat step before austenitizing the part safeguards it from thermal shock that could cause cracking even before the tool has been fully heated. But the most critical heating operation of all is that of austenitizing. Austenitizing temperatures that are too high or that have an excessively long holding time may result in abnormal grain growth, distortion, and low strength. Preheating of carbon and low alloy steels, if it is used, is normally carried out in the range of 1200 to 1300°F (649 to 704°C), and high carbon chromium steels and other highly alloyed steels at 1200 to 1450°F (649 to 788°C).

Carbon and low alloy steels are austenitized at fairly low temperatures from 1300 to 1650°F (704 to 899°C). The exact temperature depends entirely on the percent of carbon dissolved in the austenite. Considerable latitude is therefore permissible with respect to time and temperature in heat treating these steels. See Table 2.

The soaking times for these steels are very short, approximately 5 to 10 minutes for each inch of thickness. Longer times are not harmful to the steel when at the correct temperature, but in some cases can cause excessive decarburization.

The die steels with high carbon and chromium are more sluggish in dissolving carbides. The soaking time should be increased over that of the low alloy steels to avoid the danger of undersoaking. The high speed steels depend on the solution of various complex alloy carbides during austenization to develop their outstanding heat resisting qualities. These alloy carbides do not dissolve to an appreciable extent unless the steel is heated extremely close to its melting point. Exceedingly accurate temperature control is mandatory for austenitizing these steels.

Hardening and Tempering High Speed Steels

The high speed tool steels are classed in two types: **tungsten** and **molybdenum.** High speed tool steels are used primarily for cutting tools such as drills, milling machine cutters, reamers, taps, and hobs. These steels have a quality known as red hardness; that is, they can be heated by machining friction to a fairly high temperature and still not lose their hardness. High speed tool steel must be fully annealed after forging or when rehardening is required. Normalizing is not recommended. Annealing of high speed tools may be done by packing them in closed containers containing dry sand, cast iron chips, or lime so that the tool is insulated from the container. When the pack has reached the annealing temperature, it should be held one hour for each inch of thickness of the container and then slowly cooled in the furnace. Types T1 through T15 can be annealed from 1600°F (871°C), and the cooling rate is less than 40°F (4°C) per hour.

When high speed steels are brought to the austenitizing temperature from the preheat, they must not be held more than 2 to 5 minutes at that temperature. This is a very critical part of the heat treating operation for these steels and, if it is not done correctly, the tool will fail in use due to grain growth, oxidation, and thermal checking (fine cracking). As a general rule, preheating is carried out for about twice the length of time that the tool is held at the austenitizing temperature.

High speed tools must be multiple tempered for complete hardening to take place. Most of these tools undergo secondary hardening when reheated to temperatures above 700°F (371°C) because of retained austenite after quenching. For this reason they need to be tempered more than once. Tempering two or three times tends to complete the transformation of the austenite into martensite. Slow cooling is required from the tempering temperature.

Hardening Cast Irons

Cast irons may also be hardened by heat treatment. Cast iron, especially gray iron, is likely to crack if it is not heated uniformly. The heating rate must be very slow to avoid cracking. Oil or molten salt is most often used for quenching cast iron parts. Cast iron is also induction or flame hardened on surfaces of gears or on the ways of lathes and other machine tools; in this case, the quenching bath may be water rather than oil.

Gray irons should have a ferritizing anneal at a temperature between 1300 and 1400°F (704 and 760°C) prior to hardening heat treatments. As with steels, gray cast iron must be tempered at a temperature below the austenitizing range to improve toughness. Gray iron may also be austempered or martempered. Other cast irons such as nodular and pearlitic malleable iron may also be hardened and tempered in much the same manner as gray iron.

Hardening Stainless Steels

The martensitic types (cutlery grades) of stainless steel may also be hardened by quenching from temperatures above the upper critical point, 1700 to 1900°F (927 to 1038°C). As with the other tool steels, the elimination of stresses is important prior to hardening. The quenching medium is oil or air depending on the type. Preheating is unnecessary in most cases for types 410, 410Cb, 416, 414, and 431. Preheating types 420 and 440 is recommended because of their high carbon content; they should be brought to 1450°F (788°C) and held from 1 to 2 hours. The hardening temperatures of the lower carbon types are from 1700 to 1850°F (927 to 1010°C), but for the high carbon types such as 440A, 440B, 440C, and 440F the austenizing temperature ranges from 1850 to 1950°F (1010 to 1065.5°C). In the latter cases, the Rockwell hardness ranges up to RC 60. Air quenching is recommended for the higher carbon type, but they can be oil quenched to give them a slightly greater hardness.

Case Hardening

Low carbon steels (0.08 to 0.30 percent carbon) do not harden to any great extent even when combined with other alloying elements. Therefore,

Table 2 Heat Treating Temperatures and Procedures for Various Tool Steels

Steel	Normalizing Degrees F	Preheat Degrees F	Austenitize Degrees F	Light Sections Holding Time in Minutes	Quenching Medium	Approximate Hardness Rc
Water hardening						
W1 through W5	1500 (0.6 to 0.9 C) 1600-1900 (0.9 to 1.5 C)	Stress relief for cold worked parts, large sections 1200	Uniformly heated through 1450–1525	15	Water (or oil for thin sections)	65 to 67
Oil hardening						
01 through 07	1600-1650	1200	1475–1575	15	Oil	63 to 66
Shock resistant tool steel						
S1	Not recommended	1200	1650–1750	15 to 45	Oil	60 to 61
S2, S3, S4	Not recommended	1200–1400	1650–1750	5 to 20	Brine, water	60 to 62
S5, S6	Not recommended	1200–1400	1650–1750	5 to 20	Brine, oil, water	60 to 62
S7	Not recommended	1200	1650–1750	15 to 45	Air, oil	60 to 61
Air hardening cold worked steels						
A2	Not recommended	1200–1450	1700–1850	20 to 40	Air	62 to 65
A4	Not recommended	1200–1450	1500–1600	15 to 60	Air	62 to 65
A5	Not recommended	1200–1450	1450–1550	15 to 60	Air	62 to 65
A7	Not recommended	1500	1750–1800	30 to 60	Air	62 to 65
A8, A9	Not recommended	1500	1800–1875	20 to 40	Air	62 to 65
A10	1450	1200	1450–1500	30 to 60	Air	62 to 65
High carbon, High Chromium cold work steel						
D1, D2, D4, D5, D6	Not recommended	1500	1700–1850	15 to 40	Air	61 to 65
D3	Not recommended	1500	1700–1850	15 to 40	Oil	61 to 65
D7	Not recommended	1500	1850–1950	15 to 40	Air	61 to 65
Chromium hot work steels						
H10, H11, H12, H13, H14	Not recommended	1550–1650	1825–1900	15 to 40	Air	55 to 59
H16, H19	Not recommended	1550–1650	2000–2200	5	Air	55 to 59
Tungsten hot work steels						
H20 through H25	Not recommended	1500	2000–2300	5	Air, oil	48 to 57
H26	Not recommended	1500	2000–2300	5	Air, oil	63 to 64

Table 2 (continued)

Molybdenum hot work steel						
H41 through H43	Not recommended	1350–1550	2000–2200	5	Air, oil Molten salt	54 to 66
Tungsten high speed tools						
T1 through T9	Not recommended	1500–1600	2200–2300	2 to 5	Air, oil, molten salt	63 to 67
T15	Not recommended	1500–1600	2200–2300	2 to 5	Air, oil, molten salt	63 to 67
Molybdenum high speed steels						
M1 through M7, M10, M15, M30, M33, M34, M35, M36, M41, M42, M43, M44	Not recommended	1350–1550	2150–2250	2 to 5	Air, oil, molten salt	63 to 66

Note. Preheat and austenitizing temperatures vary slightly for each numbered tool steel. Manufacturers' recommendations should also be consulted prior to heat treating the steels.

when a soft, tough core and an extremely hard outside surface are needed, one of several case hardening techniques is used. It should be noted that **surface hardening** is not necessarily the same as **carburizing** case hardening. Flame hardening and induction hardening on the surfaces of gears, lathe ways, and many products depend on the carbon that is already contained in the ferrous metal. On the other hand, carburizing causes carbon from an outside source to penetrate the surface of the steel by the process of diffusion when it is at a high temperature. Because the diffusion raises the carbon content, it also raises the hardenability of the steel (Figure 15).

Flame hardening is mostly used for production hardening where long pieces or small parts are passed through a flame and immediately quenched in a spray of water (Figures 16*a* to 16*d*). The surface of the part is generally hardened from $\frac{1}{32}$ to $\frac{1}{8}$ inch deep. In this process, since there is no addition or absorption of other elements, the hardenability determines the depth and hardness of case. Good hardness on the surfaces can be obtained with the core unaffected. There is relative freedom from scaling and pitting. This type of case gives good wearing qualities with a soft ductile core. The induction hardening process is similar to flame hardening except that the heat is generated by electromagnetic induction (Figure 17).

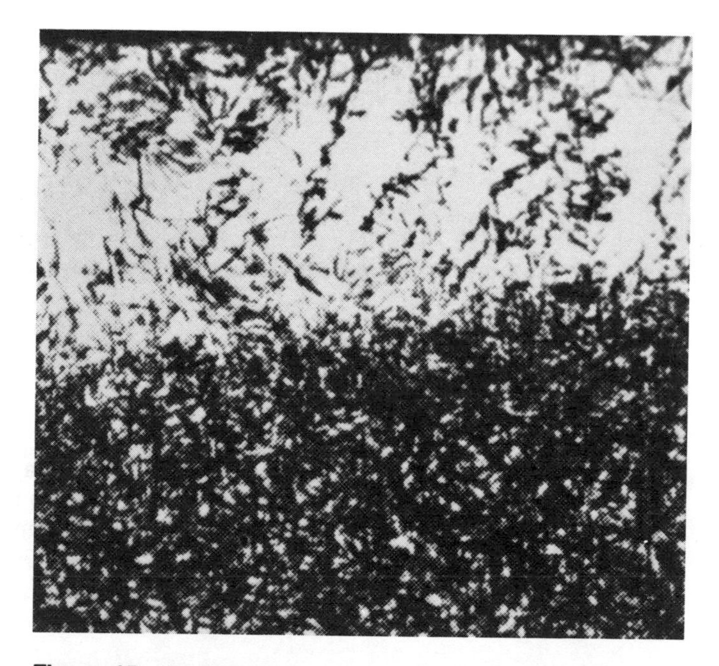

Figure 15. Micrograph of carburized case (250×) (By permission, from *Metals Handbook,* Volume 7, Copyright American Society for Metals, 1972).

Figure 16*a*. A cam lobe. This gray cast iron cam shaft was sawed halfway through and then broken, showing the flame hardened lobe of white cast iron.

Figure 16*b*. Micrograph showing the chilled iron area of cam lobe (100×).

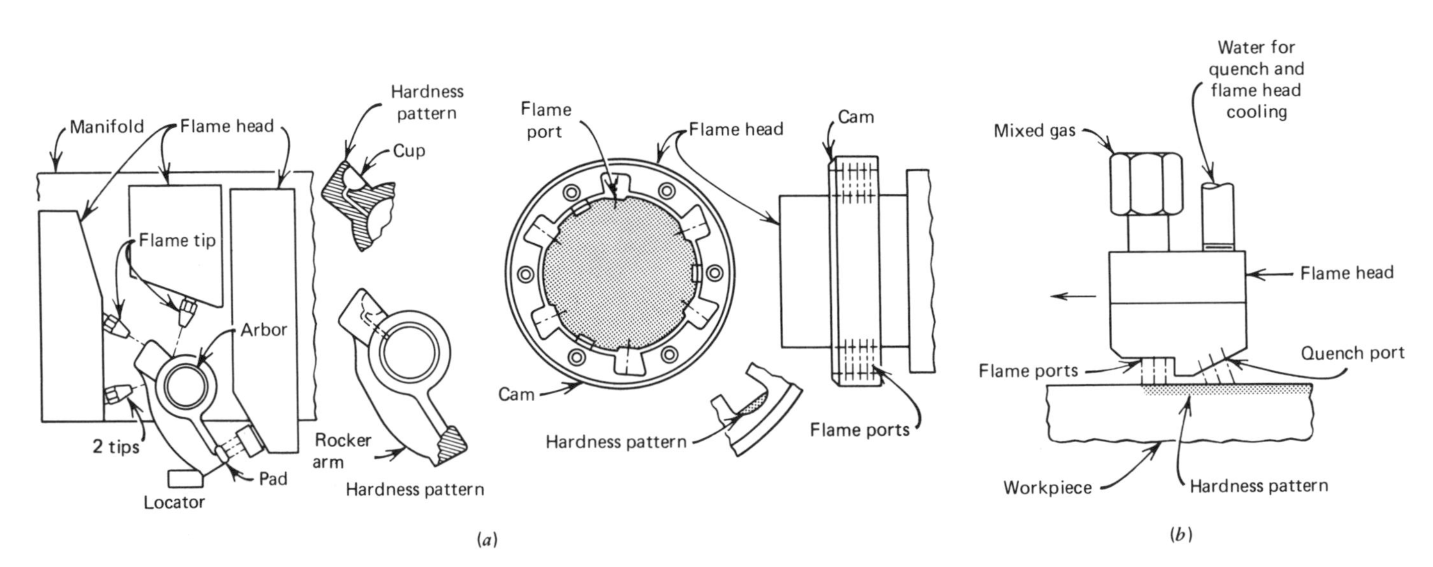

Figure 16c. *(a)* Spot (stationary) method of flame heating a rocker arm and the internal lobes of a cam. Quench is not shown. *(b)* Progressive method (By permission, from *Metals Handbook,* Volume 2, Copyright American Society for Metals, 1963).

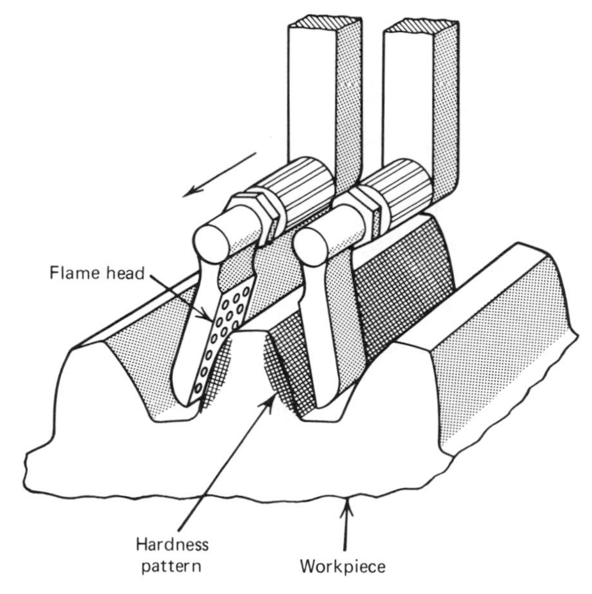

Figure 16*d*. Hardness pattern developed in sprocket teeth when standard flame tips were used for heating. When space permits this method, hardening one tooth at a time results in low distortion (By permission, from *Metals Handbook,* Volume 2, Copyright American Society for Metals, 1963).

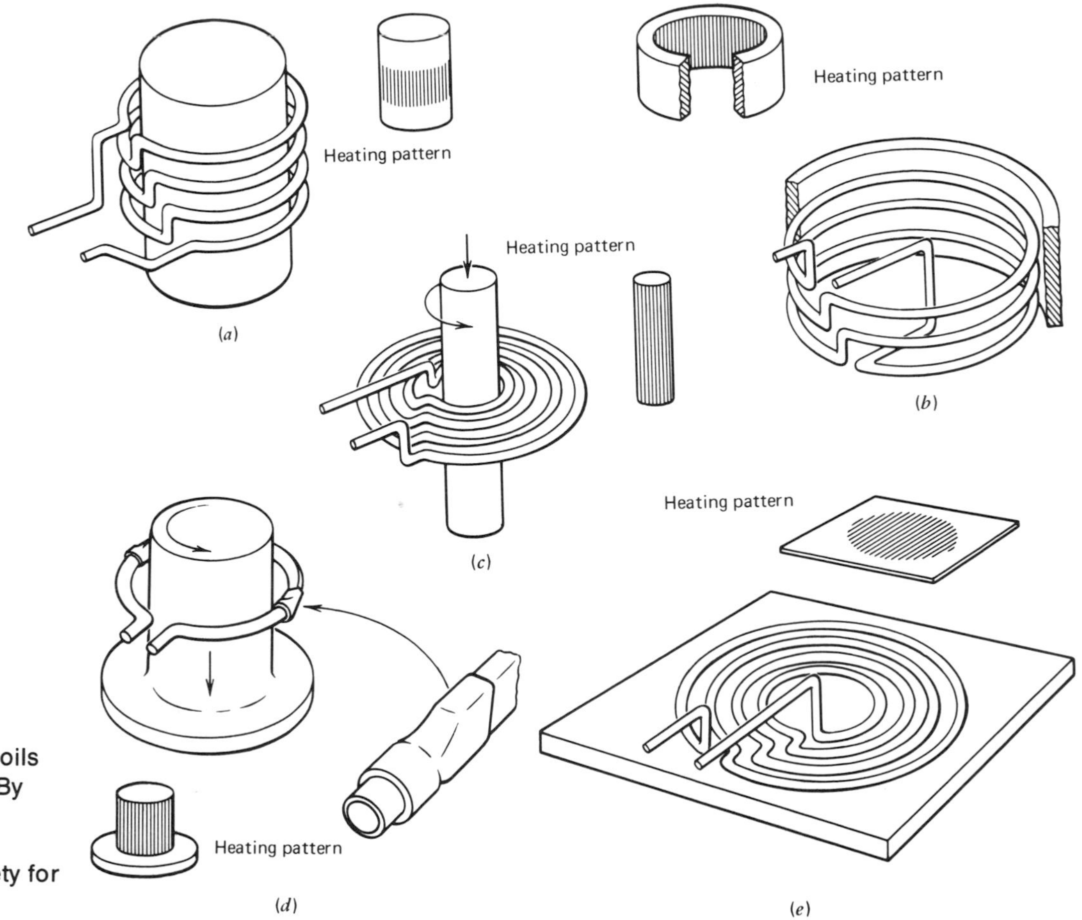

Figure 17. Induction hardening. Typical work coils for high-frequency units (By permission, from *Metals Handbook,* Volume 2, Copyright American Society for Metals, 1963).

Case Carburizing

Carburizing with a case hardening compound can be done by either of two methods. If only a shallow hardened case is needed, roll carburizing may be used. This consists of heating the part to 1650°F (899°C), rolling it in a carburizing compound, reheating, and quenching in water. In roll carburizing, use only a nontoxic compound such as Kasenite® unless special ventilation systems are used. Roll carburizing produces a maximum case of about 0.003 inch, but pack carburizing (Figure 18) can produce a case of $\frac{1}{16}$-inch in eight hours at 1700°F (925°C). The part is packed in carburizing compound in a metal box and placed in a furnace long enough to harden the case to the required depth. The part is then removed and quenched in water. After case hardening, tempering is not usually necessary since the core is still soft and tough. Therefore, unlike a piece hardened completely through that is softened by tempering, the surface of a case hardened piece remains hard, usually RC 65 (as hard as a file) or above.

Gas carburizing utilizes a high concentration of a hydrocarbon gas, such as propane or natural gas at high temperatures, to create a carbonizing atmosphere. These gases are usually piped into the furnace.

Gas nitriding produces an extremely hard case by bringing a ferrous metal into contact with a nitrogenous gas, usually ammonia. Since the heating temperature is relatively low (925 to 1050°F or 496 to 565.5°C), no quench is required for hardening. Little distortion is produced in the hardened part. Steels containing aluminum and molybdenum work best for this process. Carbo-nitriding is a modified gas carburizing process in which steels are held at high temperatures (about 1500°F or 815.5°C) in a gas containing both carbon and nitrogen (ammonia). Higher hardenability is obtained by this method than by other carburizing or nitriding processes.

Liquid carburizing is an industrial method in which the parts are bathed in carbonate cyanide and chloride salts and held at a temperature between 1500 and 1700°F (815.5 and 927°C). A pot furnace is used for this process. The part is quenched in oil or water after sufficient time is allowed for diffusion into the surface of the metal. Cyanide salts are extremely poisonous and adequate worker protection is essential.

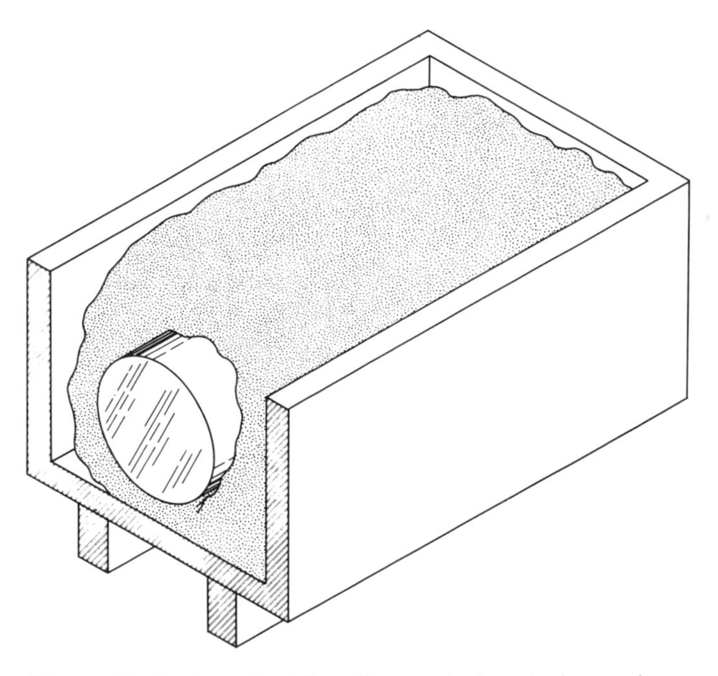

Figure 18. Pack carburizing. The workpiece to be pack carburized should be completely covered with carburizing compound. The metal box should have a close fitting lid *(Machine Tools and Machining Practices)*.

Problems in Heat Treating

Overheating of steels should always be avoided. In previous chapters, you have seen that if the furnace is set at a temperature that is too high for a particular type of steel, a coarse grain can develop. The result is often a poor quality tool, quench cracking, or failure of the tool in use. Extreme overheating causes burning of the steel and damage to the grain boundaries, which cannot be salvaged by heat treatment (Figure 19); the part must be scrapped. The shape of the part itself can be a contributing factor to quench failure and quench cracking. If there is a hole, sharp shoulder, or small extension from a larger cross section, a crack can develop in these areas (Figure 20).

A part of the tool being held by tongs may be cooled to the extent that it may not harden. The tongs should therefore be heated prior to grasping the part for quenching (Figure 21). As mentioned before, decarburization is a problem in furnaces that do not have controlled atmospheres (Figure 22). This can be avoided in other ways, such as wrapping the part in stainless steel foil, by cover-

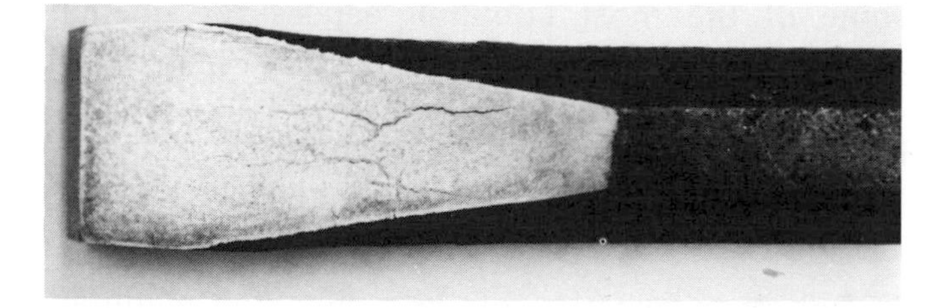

Figure 19. This tool has been overheated and the typical "chicken wire" surface markings are evident. The tool must be discarded (Lane Community College).

Figure 20. Drawing die made of Type W1 tool steel shows characteristic cracking when water quenching is done without packing the bolt holes (Bethlehem Steel Corporation).

ing it with cast iron chips, or by using an antiscaling compound.

A proper selection of tool steels is necessary to avoid failures in a particular application. If there is shock load on the tool being used, shock resisting tool steel must be selected. If there is to be heat applied in the use of the tool, a hot work type of tool steel is selected. If distortion must be kept to a minimum, an air hardening steel should be used. See Table 1.

Quench cracks have several characteristics that are easily recognized.

1 In general, the fractures run from the surface toward the center in a relatively straight line. The crack tends to spread open.

Figure 21. Heating the tongs prior to quenching a part (Lane Community College).

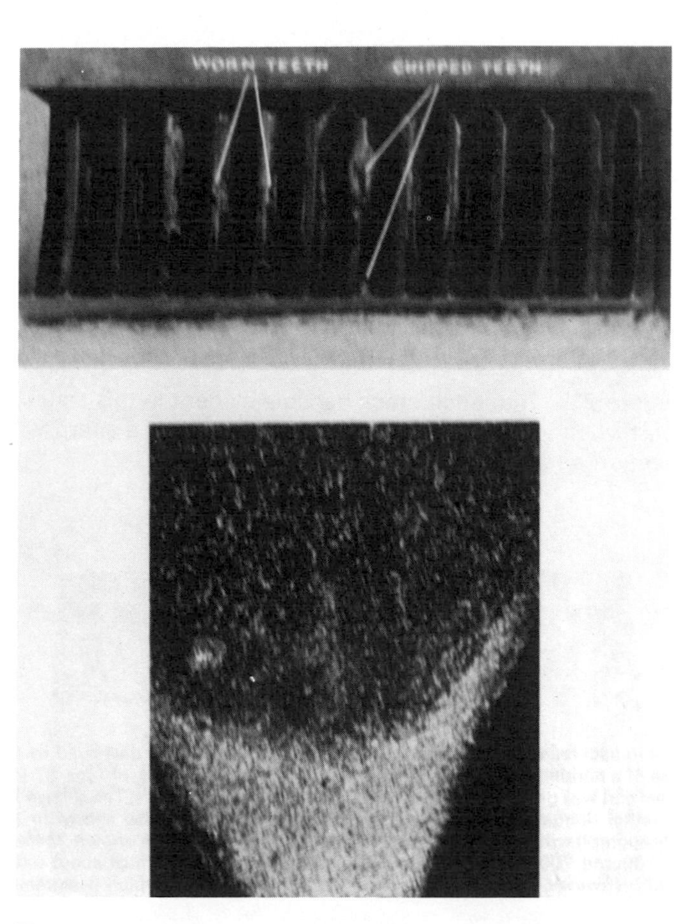

Figure 22. (At top) this thread chaser made of 18-4-1 high-speed steel failed in service because of heavy decarburization on the teeth. (Below) Structure of one tooth at 150× magnification shows decarburized structure on the point of the tooth and normal structure below the decarburized zone (Bethlehem Steel Corporation).

2 Since quench cracking occurs at relatively low temperatures, the crack will not show any decarburization.

3 The fracture surfaces will exhibit a fine crystalline structure when tempered after quenching. The fractured surfaces may be blackened by tempering scale (Figures 23*a* and 23*b*).

Figure 23*a*. A quench crack became evident in this small part when it was rough ground. The cause was a sharp V section and a quench that was too severe.

Figure 23*b*. After the part was separated, it could be seen that only a small area was holding it together (the small crystalline area). The quench crack was blackened by tempering scale.

Some of the most common causes for quench cracks are:

1 Overheating during the austenitizing cycle, causing the normally fine grained steel to become coarse.

2 Improper selection of the quenching medium; for example, the use of water or brine instead of oil for an oil hardening steel.

3 The improper selection of steel.

4 Time delays between quenching and tempering.

5 Improper design. Sharp changes of section such as holes and keyways (Figure 24).

6 Improper angle of the work into the quenching bath with respect to the shape of the part, causing nonuniform cooling.

It is sometimes desirable to stress relieve the part before hardening it. This is particularly appropriate for parts and tools that have been highly stressed by heavy machining or by prior heat treatment. If they are left unrelieved, the residual stresses from such operations may add to the thermal stress produced in the heating cycle and cause the part to crack even before it has reached the quenching temperature.

When a part is hardened but not tempered before it is ground, it is extremely liable to stress cracking (Figure 25). Faulty grinding procedures can also cause grinding cracks. Improper grinding operations can cause tools that have been properly hardened to fail. Sufficient stock should be allowed for a part to be heat treated so that grinding will remove any decarburized surface on all sides to a depth of 0.010 to 0.015 inch.

Heat Treating Safety

When heat treating, always wear a face shield, leather gloves, and long sleeves. There is a definite hazard to the face and eyes when cooling the tool steel by oil quenching, that is, submerging it in oil. The oil, hot from the steel, tends to fly upward, so you should stand to one side of the oil tank and not lean over it.

Always work in pairs during heat treatment. One person can open and close the furnace door, while the other handles the hot part. The heat

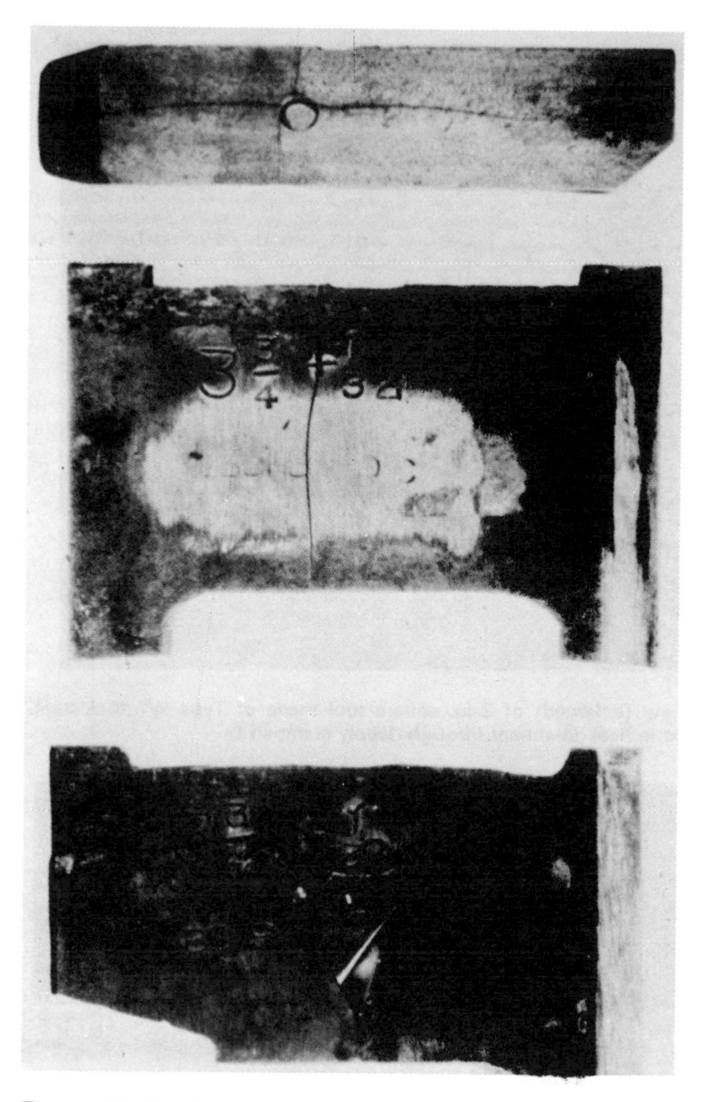

Figure 24. (Top) Letter stamp made of Type S5 tool steel, which cracked in hardening through the stamped O. The other two form tools, made of Type T1 high speed steel, cracked in heat treatment through deeply stamped + marks. Stress raisers such as these deep stamp marks should be avoided. Although characters with straight lines are most likely to crack, even those with rounded lines are susceptible (Bethlehem Steel Corporation).

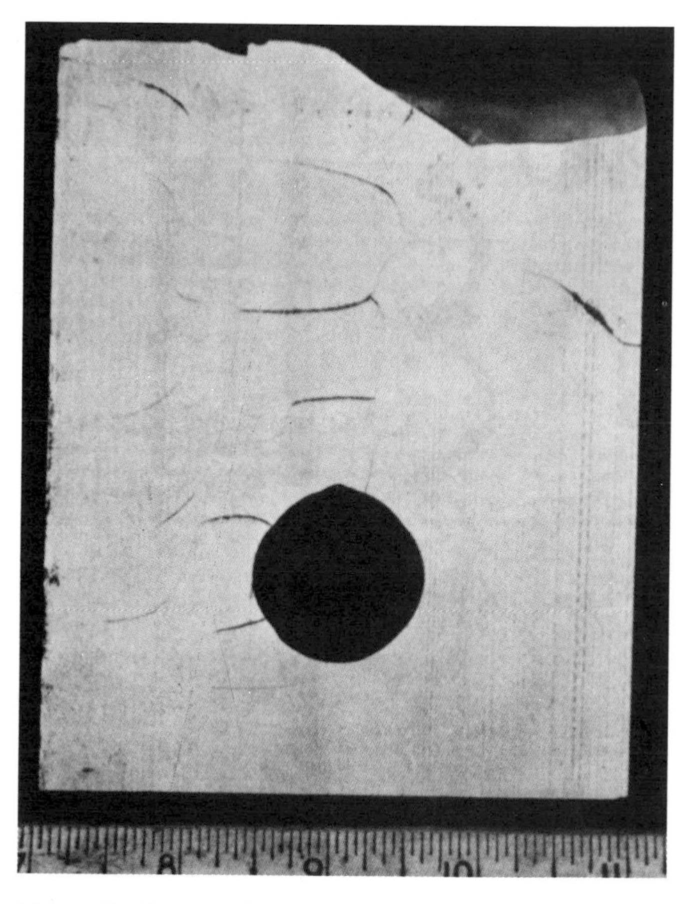

Figure 25. Severe grinding cracks in a shear blade made of Type A4 tool steel developed because the part was not tempered after quenching. Hardness was Rockwell C 64 and the cracks were exaggerated by magnetic particle test. Note the geometric scorch pattern on the surface and the fracture that developed from enlargement of the grinding cracks (Bethlehem Steel Corporation).

treated part should be positioned in the furnace so it can be conveniently removed. This will prevent the heat treater from dropping hot parts and help to insure successful heat treatment. Atmospheric furnaces should never be opened until the gas supply is turned off. Failure to do so could result in an explosion.

Very toxic fumes are present when parts are being carburized with compounds containing potassium cyanide. These cyanogen compounds are deadly poisonous and every precaution should be taken when using them. Kasenite®, a trade name for a carburizing compound that is not toxic, is often found in school and machine shops.

Self-Evaluation

1 Name three kinds of furnaces used for heat treating steels.

2 What can happen to a carbon steel when it is heated to high temperatures in the presence of air (oxygen)?

3 Why is it necessary to allow a soaking period for a length of time (which varies according to the kind of steels) before quenching the piece of steel?

4 Why should the part or the quenching medium be agitated when you are hardening steel?

5 Which method of tempering gives the heat treater the most control of the final product: by color or by furnace?

6 Describe two characteristics of quench cracking that you can recognize.

7 Name four or more causes of quench cracks.

8 In what ways can decarburization of a part be avoided when it is heated in a furnace?

9 Name two types of high speed tool steels.

10 When distortion must be kept to a minimum, which type of tool steel should be used?

11 What is the advantage of using low carbon steel for parts that are to be case hardened?

12 By which methods of carburizing can a deep case be made?

13 Are parts that are surface hardened always case hardened?

14 Name three methods by which carbon may be diffused into the surface of heated steel.

15 What method of case hardening uses ammonia gas?

Worksheet 1

Objectives 1 Correctly harden a steel part.

2 Correctly temper a steel part to Rc 48 to 52 (HB 470 to 514).

Materials Two heat treating furnaces, an oil quenching bath and accessories, plus a previously machined part of SAE 4140 steel.

Procedure 1 Determine the correct hardening temperature from the mass effect data in Table 3.

Table 3 Mass Effect Data for SAE 4140 Steel

SINGLE HEAT RESULTS

	C	Mn	P	S	Si	Ni	Cr	Mo	
Grade	.38/.43	.75/1.00	—	—	.20/.35	—	.80/1.10	.15/.25	Grain Size
Ladle	.40	.83	.012	.009	.26	.11	.94	.21	7–8

MASS EFFECT

Size Round in.	Tensile Strenth psi	Yield Point psi	Elongation % 2 in.	Reduction of Area, %	Hardness HB
Annealed (heated to 1500°F, furnace-cooled 20°F per hour to 1230°F, cooled in air)					
1	95,000	60,500	25.7	56.9	197
Normalized (heated to 1600°F, cooled in air)					
½	148,500	98,500	17.8	48.2	302
1	148,000	95,000	17.7	46.8	302
2	140,750	91,750	16.5	48.1	285
4	117,500	69,500	22.2	57.4	241
Oil quenched from 1550°F, tempered at 1000°F.					
½	171,500	161,000	15.4	55.7	341
1	156,000	143,250	15.5	56.9	311
2	139,750	115,750	17.5	59.8	285
4	137,750	99,250	19.2	60.4	277
Oil quenched from 1550°F, tempered at 1100°F.					
½	157,500	148,750	18.1	59.4	321
1	140,250	135,000	19.5	62.3	285
2	127,500	102,750	21.7	65.0	262
4	116,750	87,000	21.5	62.1	235
Oil quenched from 1550°F, tempered at 1200°F.					
½	136,500	128,750	19.9	62.3	277
1	132,750	122,500	21.0	65.0	269
2	121,500	98,250	23.2	65.8	241
4	112,500	83,500	23.2	64.9	229

As quenched Hardness (oil)

Size Round	Surface	½ Radius	Center
½	HRC 57	HRC 56	HRC 55
1	HRC 55	HRC 55	HRC 50
2	HRC 49	HRC 43	HRC 38
4	HRC 36	HRC 34.5	HRC 34

Source. Modern Steels and Their Properties, Seventh Edition, Handbook 2757, Bethlehem Steel Corporation, 1972.

2 Set the furnace thermocouple control to the correct temperature and turn it on.

3 With the tongs, place the part in the furnace on a fire brick.

4 Set the second furnace to the correct tempering temperature and turn it on. Consult the SAE 4140 mechanical properties chart (Figure 26) to

determine this temperature to obtain a draw temper hardness of Rc 48 to 52 (HB 470 to 514).

5 After the part is the same color as the furnace, allow a soaking time of 1 hour per inch of the narrowest cross section (about $\frac{3}{4}$ to 1 inch or $\frac{3}{4}$ to 1 hour).

6 Using face shield and gloves, heat the tongs on the gripping end. Remove the part from the furnace, close the door, and quickly plunge the part into the oil bath, agitating the part until it has cooled. It should still be warm to the touch.

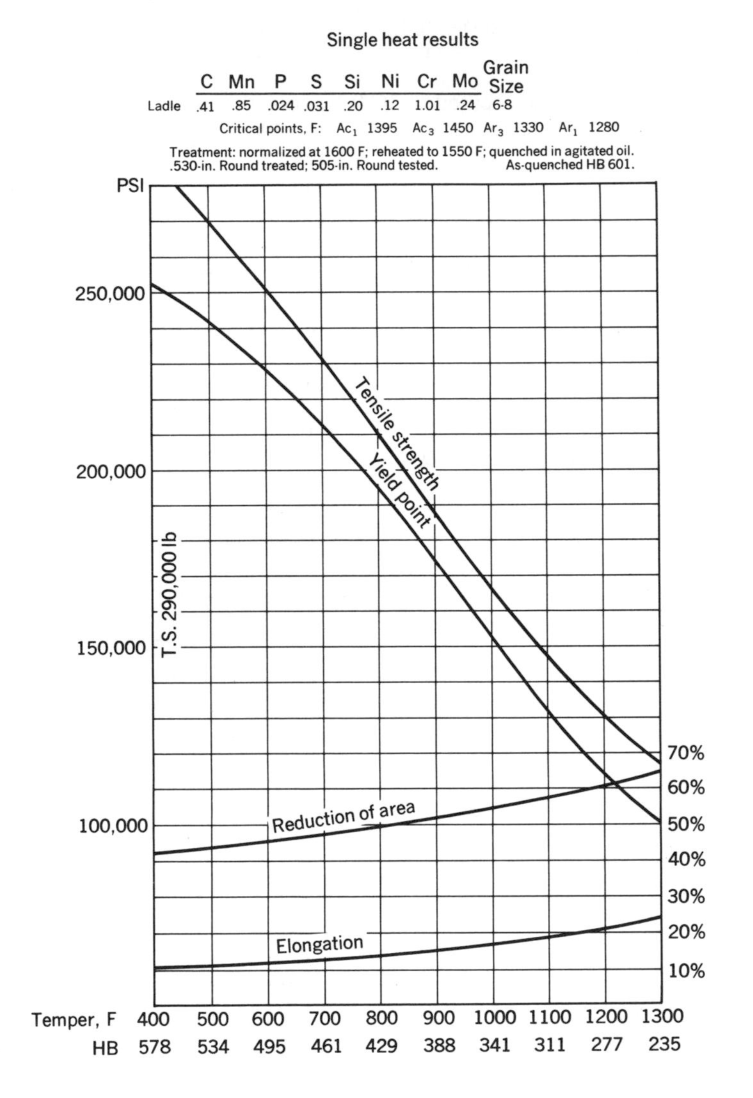

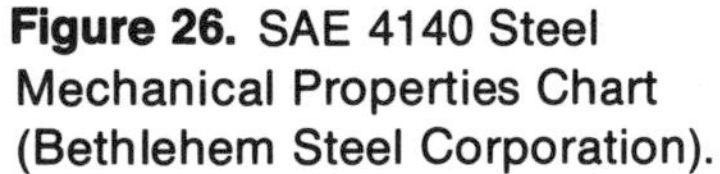

Figure 26. SAE 4140 Steel Mechanical Properties Chart (Bethlehem Steel Corporation).

7 Place the warm part immediately into the tempering furnace and hold it at that temperature for $\frac{1}{2}$ hour.

8 Remove the part and allow it to cool in air.

9 Check for hardness. Because of possible decarburization, the readings may be low. Check again after surface grinding.

Conclusion Is your steel part the same hardness that the mechanical properties chart indicated it would be at your selected tempering temperature? If not, what reason can you give for the discrepancy?

Worksheet 2

Objectives

1. Case harden a piece of mild steel using a furnace or heating torch.
2. Pack carburize a piece of mild steel in a furnace.

Materials A furnace, tongs, carburizing compound, protective clothing (face shield and gloves), and a small piece of low carbon steel.

Procedure for Case Hardening by the Roll Method

1. Heat the part to 1650°F (899°C) and remove from the furnace with tongs.
2. Roll part in carburizing compound.
3. Reheat to 1650°F (899°C).
4. Quench in cool water.

Procedure for Pack Carburizing

1. Place the part in a steel box containing the carburizing compound.
2. Place in furnace set at temperature of 1700°F (927°C). Leave it in the furnace for several hours.
3. Remove the part from furnace and quench in water.

Conclusion

1. Did the piece become hard on the surface? Check with a file.
2. Grind off a small amount and check again. How deep do you think the case is on the part hardened by the roll method? By the pack carburizing method?

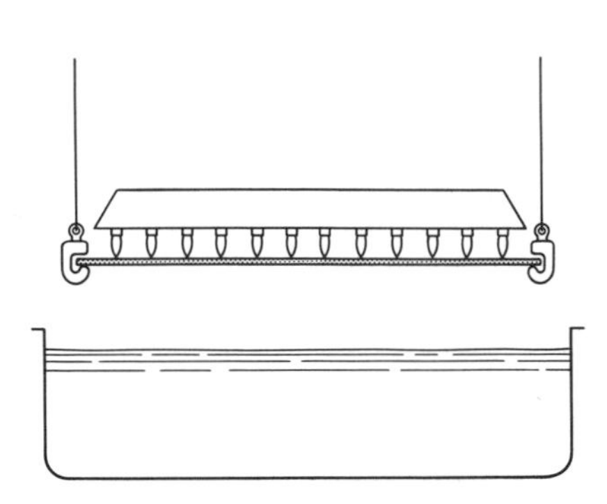

14 Heat Treating of Nonferrous Metals

Prior to 1900 nonferrous metals were not considered to be hardenable in the way that carbon steel could be hardened. Hardening by cold working to a limited extent was the only known way to strengthen nonferrous metals. It was believed that the ancient Egyptians had possessed an art for hardening copper for tools used in building the pyramids. This knowledge was however lost to later generations. Today, beryllium copper is used for tools in potentially explosive areas such as powder factories because of its nonsparking quality. It can be hardened to the range of Rockwell C 40 to 50 by precipitation hardening. This may very well be the "lost art" of hardening copper.

A method of hardening aluminum was discovered in Germany by Alfred Wilm in 1906. This new, light, and very strong metal alloy was called duralumin. One of its first uses was for the rigid skeletal structure of the zeppelins of the German navy. These lighter-than-air craft played a part in the air war in World War I. Modern aircraft and space vehicles could not be constructed without these strong, light alloys that can be hardened to exact specifications. This chapter will introduce you to the principles of heat treating nonferrous alloys by solution heat treatment and precipitation hardening.

OBJECTIVES

After completing this chapter, you will be able to:

1. Explain the reasons underlying the processes of solution heat treatment and precipitation hardening in which hardening takes place.
2. Demonstrate the process of hardening an aluminum alloy.

INFORMATION

In Chapter 8, eutectic and solid solution alloys were discussed. In order to understand precipitation hardening, the phase diagram for terminal solid solutions is used. Type III alloys are partly soluble at certain combinations in the solid phase. Phase diagrams for such alloys are essentially the same as Type II, the eutectic alloys. The only difference is a curve that shows the amount of solid phase solubility. This line is called the solvus curve (Figure 1).

The temperature at which metal *A* has the greatest solubility for metal *B* is at line 1 on the graph. At this temperature metal *A* will dissolve about 18 percent of metal *B* (seen on the left side of the graph). The situation is similar on the right side of the graph, where about 15 percent of metal *A* will dissolve in metal *B*. As the temperature drops, the ability for either metal *A* or *B* to dissolve the other metal decreases. This is shown by the slant of the solvus line toward the pure metal. Note that the eutectic mixture will no longer be pure metal *A* and pure metal *B*, but rather a mixture of solid solution *a* and solid solution *b*. This decrease in solid solubility is a necessary characteristic of precipitation hardening.

Solution Heat Treatment and Precipitation Hardening of Aluminum

In order to harden aluminum by the precipitation process, certain conditions must be met. First, there must be an element or compound present in significant amounts that has decreasing solid solubility in aluminum with decreasing temperature. Elements and compounds found in aluminum as precipitates are copper, zinc, silicon, magnesium, Mg_2Si, and $MgZn_2$.

To explain this first requirement, consider Figure 2, the aluminum-copper phase diagram. Copper is the major element that causes precipitation hardening of the 2000 aluminum alloys. Aluminum will hold more copper in solid solution at 900 to 1000°F (482 to 538°C) than it will at room temperature.

It can be seen from the phase diagram that at 1018 °F (548 °C) the aluminum can hold up to 5.65 percent copper in a saturated condition. If you follow the solubility line *A-B*, it can be seen that with decreasing temperature, the solubility of copper in aluminum also decreases. Successful **solution heat treatment** depends on putting the copper into solid solution and then trapping it there. This is done by quenching in water after heating to the correct temperature. Cooling a 4 percent copper-aluminum alloy is shown in Figure 2. At point 1, the copper is in solid solution, and at point 2 the alloy has been quickly cooled to form a supersaturated solution. The copper (after quenching the alloy) is in the form of highly dispersed globules of copper aluminide. Care must be exercised in solution heat treatment that the recommended soaking temperature of the alloy is not exceeded. If this happens, grain boundary melting may occur and the material will be ruined for further use as it

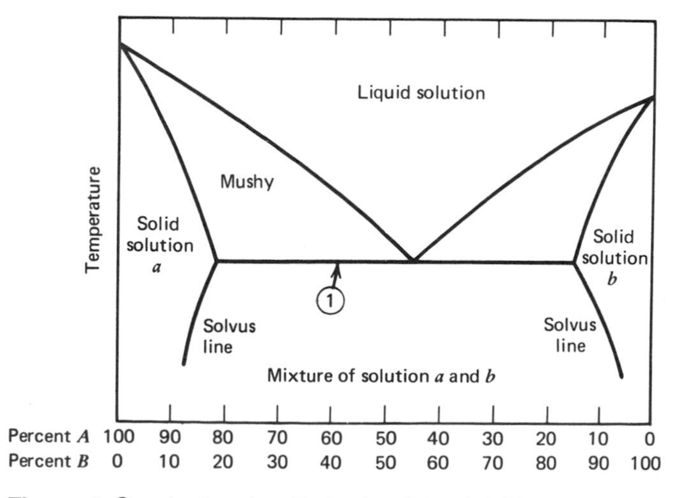

Figure 1. Graph showing limited solid solubility.

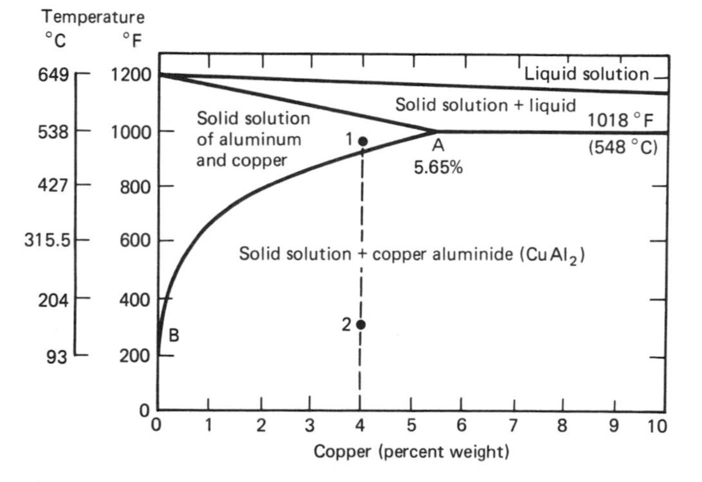

Figure 2. Aluminum-copper equilibrium diagram showing the aluminum-rich end. Line *A-B* represents the increase in solubility of copper with increasing temperture in aluminum in the solid state.

becomes brittle and cannot be salvaged. The surface is often blistered when overheating occurs. The soaking temperature and time vary with each alloy.

The second requirement is that the element, when placed into a supersaturated condition in the aluminum, will precipitate very fine particles of copper aluminide at the grain boundaries and along crystal planes, producing strains in the aluminum. This process is called **aging or precipitation heat treatment** (Figure 3). Aging causes these very fine particles of copper aluminide to act as keys to lock up the slip planes, causing a reduction of ductility and an increase in hardness and stress (Figures 4*a* and 4*b*). Very little forming is possible after aging is completed. As the term aging implies, hardening takes place over a period of time at room temperature. This process often requires many hours or days. **Artificial aging** speeds up this process and also increases strength, but lowers corrosion resistance in some alloys such as 2024. Artificial aging consists of heating the solution heat treated and quenched part for several hours at 250 to 360°F (121 to 182°C) depending on the alloy, and then cooling the part to below 100°F (38°C). This is sometimes repeated before cooling to room temperature, and then is called interrupted heat treatment. **Overaging** time or temperatures that are too high cause a loss of strength and corrosion resistance since this enlarges the copper aluminide particles. However, even with normal aging, corrosion resistance is lower in these hardenable alloys than in pure aluminum.

Cold working of aluminum is often done immediately after solution heat treatment and before aging begins. This process assures an even greater hardness and tensile strength of the aged part. An example of this procedure is in the use of aluminum alloy rivets for aircraft. They must be used before they become aged or the heads will split as a result of cold work when riveting. Table 1 shows solution heat treatment and aging times for some aluminum alloys.

Since pure aluminum is more corrosion resistant than any alloy of aluminum, a thin covering of pure aluminum is sometimes applied to sheets of aluminum alloy that can be hardened. Pure aluminum cannot be heat treated, but clad aluminum alloy can be heat treated in the same way as other heat treatable aluminum alloys because of the alloy part of the "sandwich." Care must be taken to avoid oversoaking as diffusion of alloying elements can occur in the clad surface. This process would eventually convert the clad surface of pure aluminum into an alloy that has low corrosion resistance (Figure 5).

Annealing

Aluminum may be stress relieved to remove the stresses of cold working by simply heating to 650°F (343°C) and allowed to cool. Since this temperature will not dissolve the copper aluminide, the alloy will remain in the annealed condition. Aluminum grain structure can be reformed or recrystallized by this stress relief process. However, in order for the heating cycle to recrystallize the

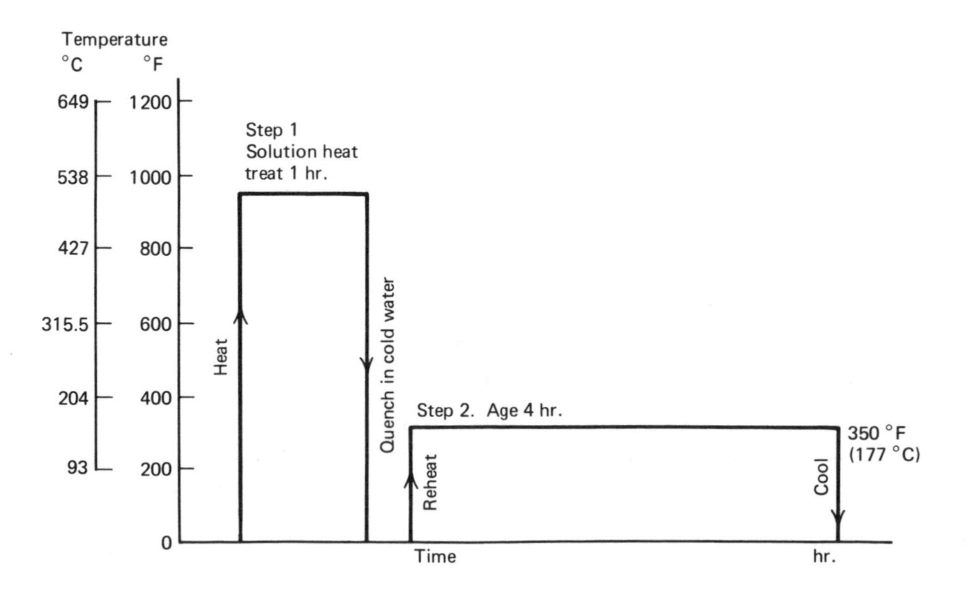

Figure 3. Process of solution heat treating and artificial aging of 2014-T6 aluminum alloys $\frac{1}{4}$ inch thick.

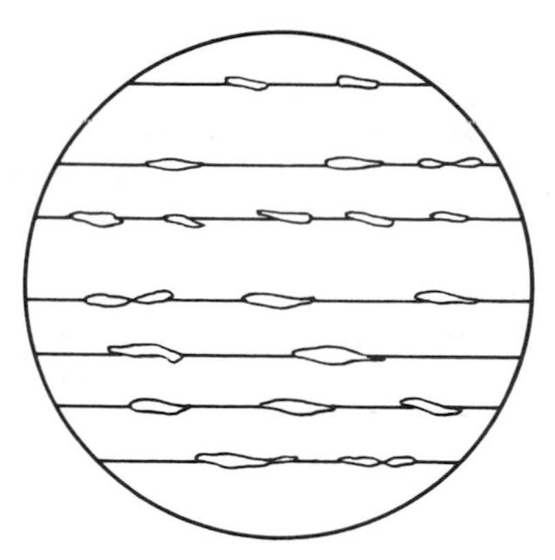

Figure 4*a*. Copper-aluminide particles lock up the slip planes by acting like keys to prevent plastic flow in the metal.

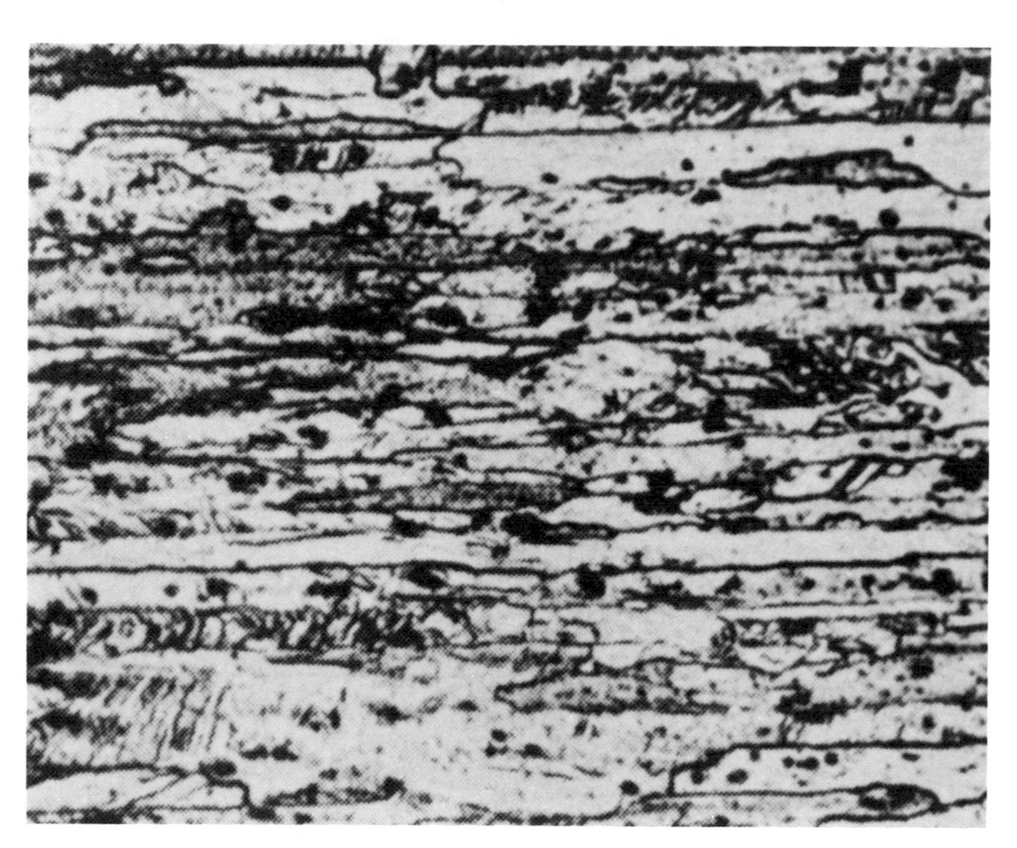

Figure 4*b*. The strain or slip lines in 2024-T6 aluminum sheet may be easily seen in this micrograph (100 ×) (By permission, from *Metals Handbook,* Volume 7, Copyright American Society for Metals, 1972).

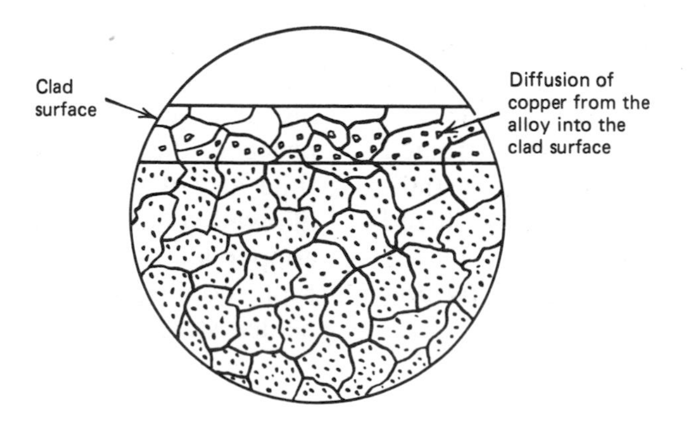

Figure 5. Diffusion of copper into the pure aluminum-clad surface will reduce its corrosion resistance. Excessive heating time can cause this transition zone of diffusion.

Table 1 Solution Heat Treatment with Aging Times and Temperatures for Some Commercial Hardenable Aluminum Alloys

Designation	Soaking Temperature in °F	Soaking Time for Various Thicknesses in Minutes			
		Up to 0.032 inch	Over 0.032 to 0.125 inch	Over 0.125 to 0.025 inch	Over 0.250 inch
2014-T6	925–950	20	20	30	60
2017	925–950	20	20	30	60
2117	890–950	20	20	30	60
2024	910–930	30	30	40	60
6061-T6	960–1010	20	30	40	60
7075	860-960	25	30	40	60

Note: Soaking time begins after the part has reached temperature.

Table 1 (continued)

Designation	Aging Temperature in °F	Aging Time in Hours
2014-T6	345–355	2 to 4
	355–375	½ to 1
2017	Room temperature	96
2117	Room temperature	96
2024	Room temperature	96
6061-T6	315–325	50 to 100
	345–355	8 to 10
7075	245–255	24
	315–325	1 to 2

grains, the piece must first be cold worked or strained to a minimum of 2 percent elongation. Nucleation, or the starting of new grains, begins at points of high stress on the crystal lattice structure. Therefore, a piece that has been highly strained will produce a fine grain structure after recrystallization. Thus, the grain size of the recrystallized metal can be controlled to a certain degree by the amount of cold work performed (Figures 6*a* and 6*b*).

Annealing to remove the effects of hardening in hardenable alloys caused by rapid cooling from welding, heat treating, or hot working is done at a higher temperature, about 970°F (521°C). In this process, the aluminum is soaked for a period of time at 770 to 825°F (410 to 440.5°C). It is then allowed to cool to room temperature. Since the copper aluminide is in solution in this operation, rapid cooling would not put the metal in an annealed state.

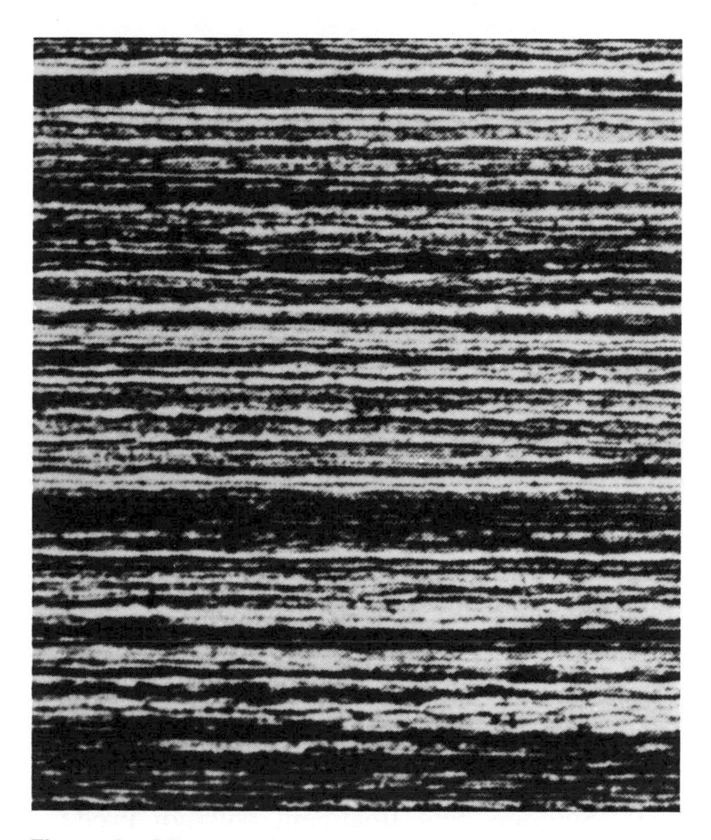

Figure 6a. Micrograph of highly strained cold worked aluminum (100×) (By permission, from *Metals Handbook*, Volume 7, Copyright American Society for Metals, 1972).

Figure 6b. Recrystallization of the same cold worked aluminum was produced by stress relief anneal (100×) (By permission, from *Metals Handbook*, Volume 7, Copyright American Society for Metals, 1972).

Heat Treating Copper Alloys

Some copper alloys may be hardened by solution heat treatment and precipitation (aging) and then can be stress relieved or annealed. The important copper alloys that can be age-hardened by precipitation are beryllium copper, aluminum bronze, copper-nickel-silicon, copper-nickel-phosphorus, chromium copper, and zirconium copper.

Aluminum bronzes containing more than 10 percent aluminum are hardened by quenching from a high temperature about 1200°F (649°C) to produce a martensitic type of structure similar to that of hardened steel. This is followed by tempering at a lower temperature.

Berryllium copper and other hardenable alloys are solution heat treated and precipitation hardened. Beryllium copper containing nickel or cobalt is solution treated at 1425 to 1475°F (774 to 802°C) for 1 to 3 hours and quenched in water. Aging time is from 2 to 3 hours at 575 to 650°F (302 to 343°C). The tensile strength of these hardened alloys ranges from 150,000 to 215,000 psi.

Beryllium copper is usually supplied solution heat treated, aged, and cold worked. It can be machined in this condition with proper tooling. It is, however, sometimes necessary to anneal beryllium copper for further cold working.

Safety note. Beryllium is a toxic metal and beryllium compounds are very toxic. Adequate protection should be used to avoid any fumes caused by any incipient melting or burning of the metal.

The annealing temperature for beryllium copper alloy is much higher than that of other copper alloys and it is usually held at 1425 to 1900°F (774 to 1038°C) for 3 hours and then quenched in water. Other hardenable alloys are annealed at temperatures above 1200°F (649°C). Cooling rates are not particularly important for annealing copper and copper alloys that are not hardenable and they are usually annealed at temperatures below 1400°F (760°C).

Stress relieving temperatures for nonhardenable copper alloys range from 400 to 475°F (204 to 246°C). Stress relief is used when further cold work is necessary where extensive change in mechanical properties is not desirable. The full anneal produces large, coarse grains that leaves the metal at a very soft temper and unsuitable for many uses.

Heat Treating Magnesium Alloys

Some magnesium alloys can be solution heat treated and aged, but there is a very great danger of magnesium fires when improper heat treating procedures are used. Extensive study and experience are required to heat treat this material safely. Manufacturer's catalogs and reference books such as the *Metals Handbook* from the American Society for Metals may be consulted for heat treating procedures.

Heat Treating Nickel and Nickel Alloys

Nickel and its alloys may be annealed, stress relieved, and, in some cases, solution heat treated and aged. Among the hardenable nickel alloys are permanickel 300, duranickel 301, monel 501, inconel 718, and hastelloy R-235. Solution temperatures for these alloys are from 1800 to 2000°F (982 to 1093°C) except for monel 501, which is 1525°F (829°C). Aging temperature for permanickel is about 900°F (482°C). The other hardenable alloys must be aged for 16 hours at 1100°F (593°C) and then at 1000°F (538°C) for 6 hours followed by 8 hours at 900°F (482°C) and air cooled. Annealing is carried out by heating to a predetermined temperature for a period of time and then quenching in water. As with most metals, a scale may be formed at high temperatures. This can be controlled by using a muffled (carbon rich) furnace atmosphere or by bright annealing in a closed container with an inert gas.

Heat Treating Titanium

Titanium alloys may also be heat treated but there is a danger of contamination of the metal during the heating cycle. Oxygen, hydrogen, and nitrogen have a detrimental effect on titanium, but it is particularly sensitive to chlorides such as salt. Even salt from fingerprints can cause stress corrosion when heated to temperatures above 600°F (315.5°C).

Pure titanium is allotropic, having a close-packed hexagonal (CPH) structure (alpha) below 1625°F (885°C), which changes to body-centered cubic (BCC) structure (beta) above this temperature. Alloying elements tend to affect these two crystal structures to make titanium either CPH or BCC. Solution heat treatment will not produce high strength in either the completely alpha or beta alloys. The alpha-beta alloys such as Ti-7A1-4Mo reach a tensile strength of 168,000 psi after solution heat treatment and aging. References should be consulted prior to heat treating these titanium alloys.

It should be noted here that some ferrous metals are also hardened by solution and precipitation heat treatments. Notably among these are the precipitation hardening stainless steels such as 17-4 PH, 17-7 PH, PH 15-7 Mo, and the AM-350 and AM-355 alloys. These alloys when work hardened and aged may approximate the high hardness of the martensitic stainless steels while retaining much more ductility.

Self-Evaluation

1 Match the correct numbers on the phase diagram (Figure 7) with the following:

A. _____Solvus lines
B. _____Mushy areas of liquid and solid solutions
C. _____Area of solid solutions
D. _____Line denoting the temperature of highest solubility of the solid solution *a* or *b*
E. _____Mixture of solid solutions *a* and *b*

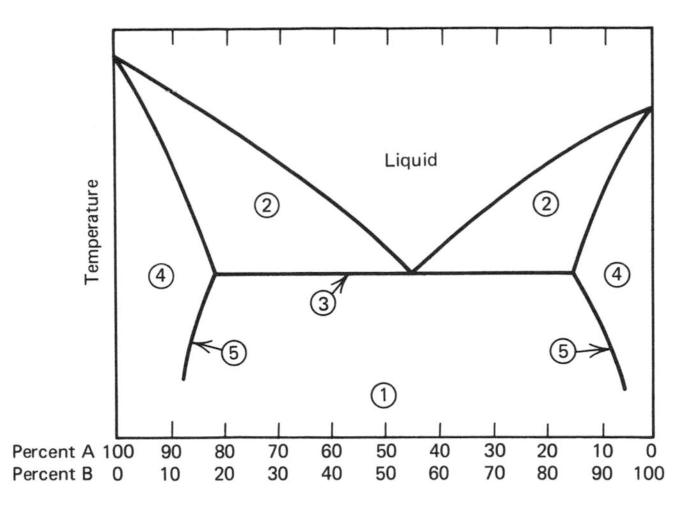

Figure 7.

2 What is the result when a solution of 4 percent copper in aluminum is quenched so that the cooling curve crosses the solvus line (as shown on Figure 2)?

3 What are the two necessary steps needed to harden heat treatable nonferrous metals?

4 What causes hardening in the aluminum-copper alloy?

5 What is the purpose for using clad aluminum?

6 Aluminum that is not heat treatable may be stress relieved for further cold working by what process?

7 How can the grain size of metals that are stress relieved or annealed be controlled?

8 Name two kinds of copper alloys that can be hardened.

9 Nickel and most other metals form an oxide scale on the surface when they are heated in a furnace in the presence of oxygen. Name two methods mentioned in this chapter to control this problem.

10 When titanium is alloyed with other elements, what effect do they have on the crystal structure?

Worksheet

Objective Demonstrate the process of solution heat treatment and aging.

Materials Furnace and two small strips of 18-22 gage 2024 aluminum alloy.

Procedure 1 Heat one strip to 920°F (493°C) and hold at this temperature for 20 minutes.

2 Quench the strip in cold water.

3 Make a hardness test on both the control (not heat treated) strip and the quenched strip. Record the test on a sheet of paper.

4 Note that both strips are easily bent with the fingers at this stage.

5 Make a hardness test each day for four days and record your results on your paper.

6 On the fourth day, make the bending test with your fingers.

Conclusion

1 Did the hardness change on the solution heat treated strip each day? How much?

2 Does it bend as easily as the strip that was not heat treated?

3 Show your results to your instructor.

15 The Effects of Machining on Metals

Most forms of metal working, such as shearing, forging, and all forms of cold working, alter the grain structure of metals in ways that can affect their behavior. Machining practices can have a profound effect on the grain structure, usefulness, and working life of a metal part. This can be the result of distortion, notches that are produced by machining, and by disturbance of the surface while metal cutting is taking place.

Modern machine tools are very powerful and with modern tooling much more production can often take place when the operator uses the correct feeds and speeds. Sometimes an operator will take a very light roughing cut when it is possible to take ten times the amount. Often the cutting speeds are too low to produce good surface finishes and for optimum metal removal. Because of the investment and cost of labor, time is very important and a larger amount of metal being removed can shorten the time to produce a part and often improves its physical properties.

OBJECTIVES

After completing this chapter, you will be able to:

1. Determine how metal cutting affects the surface grain structure of metals.
2. Explain notch sensitivity and stress concentration related to design and operating procedures, and how these can affect the working life of a part.
3. Analyze chip structures, chip breakers, speeds and feeds, and their effect on metals.
4. Describe machining behavior of different metals, the machinability ratings, tool materials, and cutting fluids.

INFORMATION

Effects of Machining on Surface Structures

A machine operator often notices that by using certain tools having certain rake angles at greater speeds he gains a better finish on the surface. The operator may also notice that when speeds are too low, the surface finish becomes rougher. Not only do speeds, feeds, tool shapes, and depth of cut have an effect on finishes, but the surface structure of the metal itself is disturbed and altered by these factors.

Tool materials can be either high speed steel, cast alloy, carbide, or ceramic. Most manufacturing today is done with carbide tools. Greater amounts of materials may be removed and tool life extended considerably when carbides, as compared to high speed steels, are used. Much higher speeds can also be used with carbide tools than with high speed steel tools. A good machine operator can analyze the chips he is removing from the work and gain a fairly good idea of the kind of cutting that is being made. The chip color reveals the temperature of the chip. If it is not colored, the chip is at a low temperature, below 400°F (204°C). A light straw-colored chip is approximately 450°F (232°C), a brown about 475°F (246°C), purple about 525°F (274°C), blue about 600°F (315.5°C), and gray is above 700°F (371°C). Of course, chips should never be allowed to become red hot when using carbide tools.

High speed steel tools should show a chip that is from light straw to purple in color. Cast alloy tools can operate in the chip temperature of purple to blue. Carbide tools should produce chips that are from blue to gray. Ceramic tools should produce chips with temperatures above a blue or gray color.

The various tool rakes from negative to positive have an effect on the chip and on the surface finish. There is a great difference in the surface finish between metals cut with coolant or lubricant and metals that are cut dry. This is because of the cooling effect of the coolant and the lubricating action that reduces friction between the tool and chip. The chip tends to curl away from the tool more quickly and there is a more uniform chip when coolant is used. Also, the chip becomes thinner and pressure welding is reduced. Where metals are cut dry, pressure welding is a definite problem,

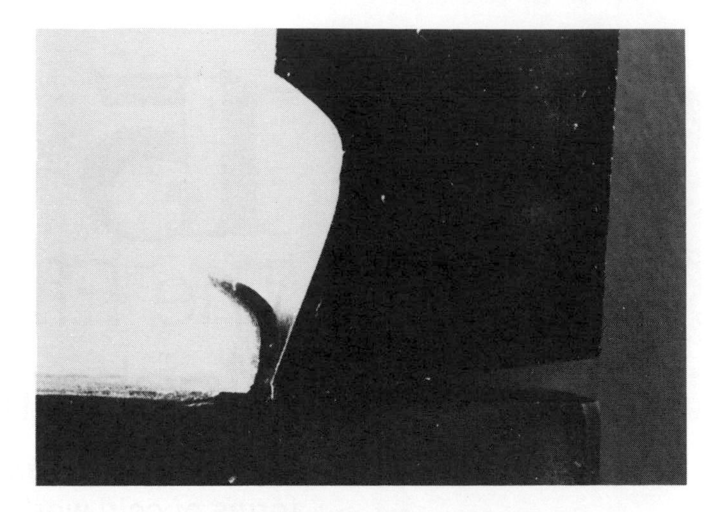

Figure 1*a*. A continuous form chip is beginning to curl away from this positive rake tool.

Figure 1*b*. A thick discontinuous chip being formed at slow speed with a negative rake tool.

Figure 1*c*. A continuous, but thick, chip is being formed with a zero rake tool.

Figure 2. The crater on the cutting edge of this tool was caused by chip wear at high speeds. The crater often helps to cool the chip *(Machine Tools and Machining Practices)*.

especially in the softer metals such as 1100 aluminum and low carbon steels. Pressure welding produces a **built-up edge** that causes a rough finish and a tearing of the surface of the workpiece. Figures 1*a* to 1*c* show chips that are formed with tools having positive, negative, and zero rakes. These chips are all formed at low surface speeds and consequently are thicker than they would have been at higher speeds. There is also more distortion at low speeds. The material was being cut dry without coolant.

At high speeds, cratering begins to form on the top surface of the tool because of the wear of the chip against the tool; this causes the chip to begin to curl (Figure 2). The crater makes an air space between the chip and the tool, which is an ideal condition since it insulates the chip from the tool and allows the tool to remain cooler; the heat goes off with the chip. The crater also allows the coolant to get under the chip, an ideal condition.

Various metals cut in different ways. Softer, more ductile metals produce a thicker chip and harder metals produce a thinner chip. A thin chip indicates a clean cutting action with a better finish. The more rapid the cutting speed, the thinner and more uniform is the chip that is formed.

A common misconception is that the material splits ahead of the tool as wood does when it is being split with an axe. This is not true with metals; the metal is sheared off and does not split ahead of the chip (Figures 3*a* to 3*c*). The metal is forced along in the direction of the cut and the grains are elongated and distorted ahead of the tool and forced along a shear plane as can be seen in the micrographs. The surface is disrupted more with the tool having the negative rake than with the tool with the positive rake; in this case, because it is moved at a slow speed. Negative rake tools require more power than positive rake tools.

Higher speeds give better surface finishes and produce less disturbance of the grain structure. This can be seen in Figures 4*a* and 4*b*. At a lower speed of 100 SFM, the metal is disturbed to a depth of 0.005 to 0.006 in., and the grain flow is shown to be moving in the direction of the cut. The grains are distorted and in some places the surface is torn. This condition can later produce fatigue failures and a shorter working life of the part than would a better surface finish. At 400 surface feet per minute (SFM), the surface is less disrupted and the grain structure is altered only to a depth about 0.001 in. When the cutting speeds are increased to 600 SFM and above, little additional improvement is noted. Zero and negative rake tools are stronger, have a longer working life, and give a good finish at higher cutting speeds. Positive rake tools at lower cutting speeds can produce a good finish when sharpened properly.

Coolant also has a tremendous effect on finishes and surfaces. With carbide tools, it must be applied correctly. Coolant should be flooded over the carbide tool to avoid intermittent cooling and heating that create thermal shock and consequent cracking of the tool. An ideal coolant application, as shown in Figure 5, is made from underneath and from behind the chip across the top of the tool using about 15 to 20 pounds of pressure.

Tool load, speed, feeds, rakes, and types of tools all have an effect on the final product. These effects are similar on other alloys of steel and on most nonferrous metals. See Table 1 for high speed lathe tool angles.

Analysis of Chip Structures

Machining operations performed on various machine tools produce chips of three basic types: **the continuous chip, the continuous chip with built-up edge on the tool, and the discontinuous chip.** The

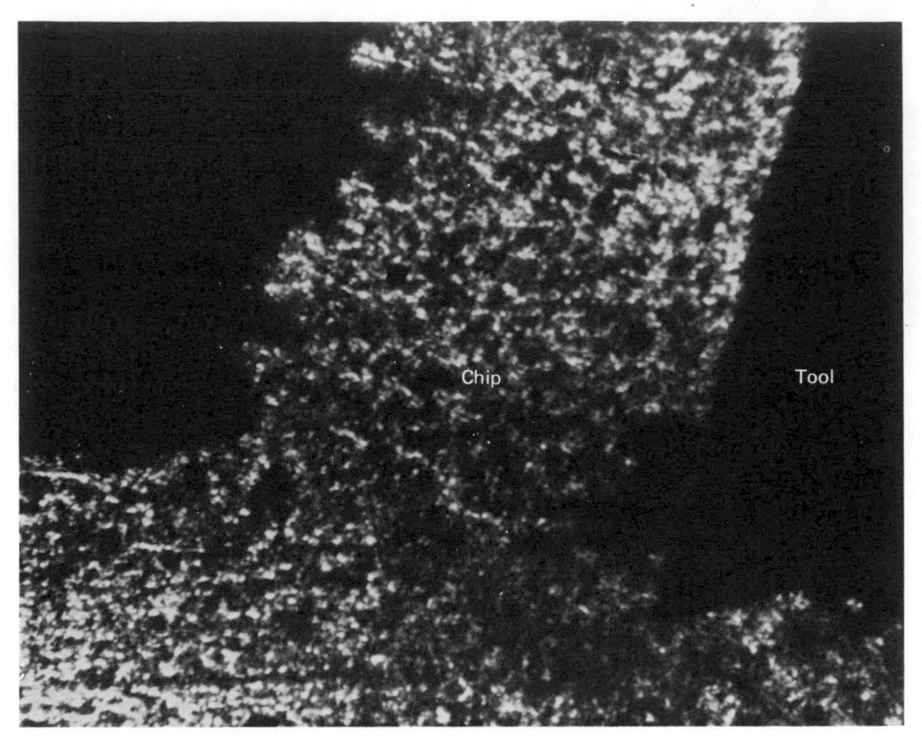

Figure 3*a*. The positive rake chip magnified 100 diameters at the point of the tool. The grain distortion is not as evident as in Figures 3*b* and 3*c*.

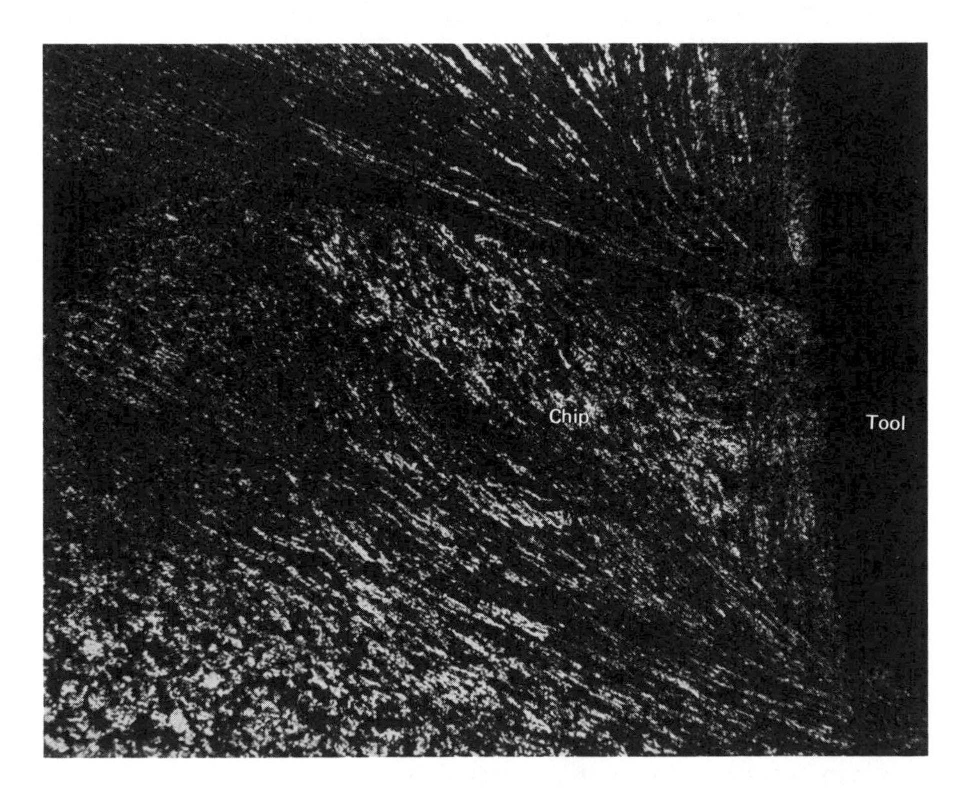

Figure 3*b*. Point of negative rake tool magnified 100 diameters at the point of the tool.

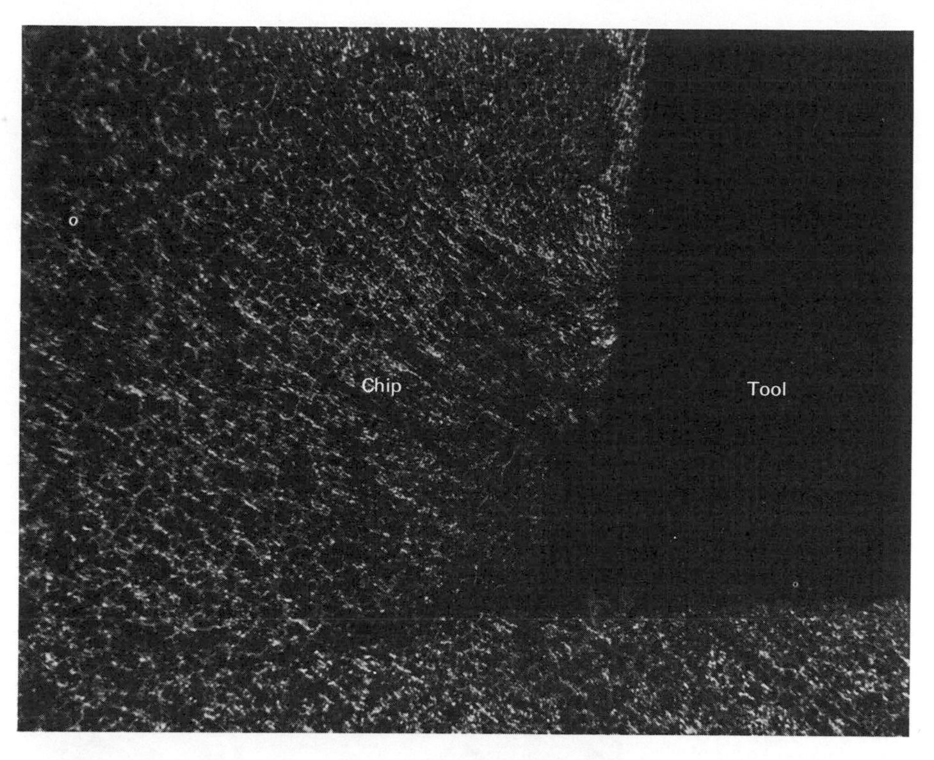

Figure 3*c*. Zero rake tool at 100 diameters shows similar grain flow and distortion to the negative rake tool.

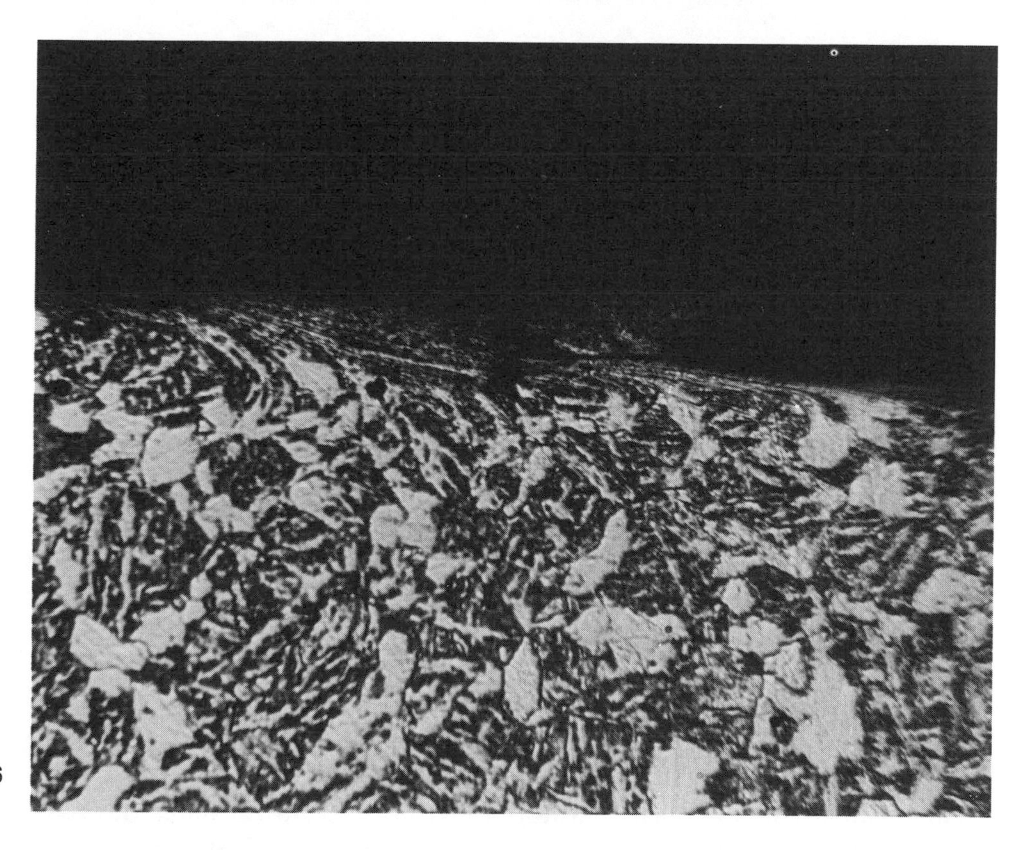

Figure 4*a*. This micrograph shows the surface of the specimen that was turned at 100 SFM. The surface is irregular and torn, and the grains are distorted to a depth of approximately 0.005 to 0.006 inch (250×).

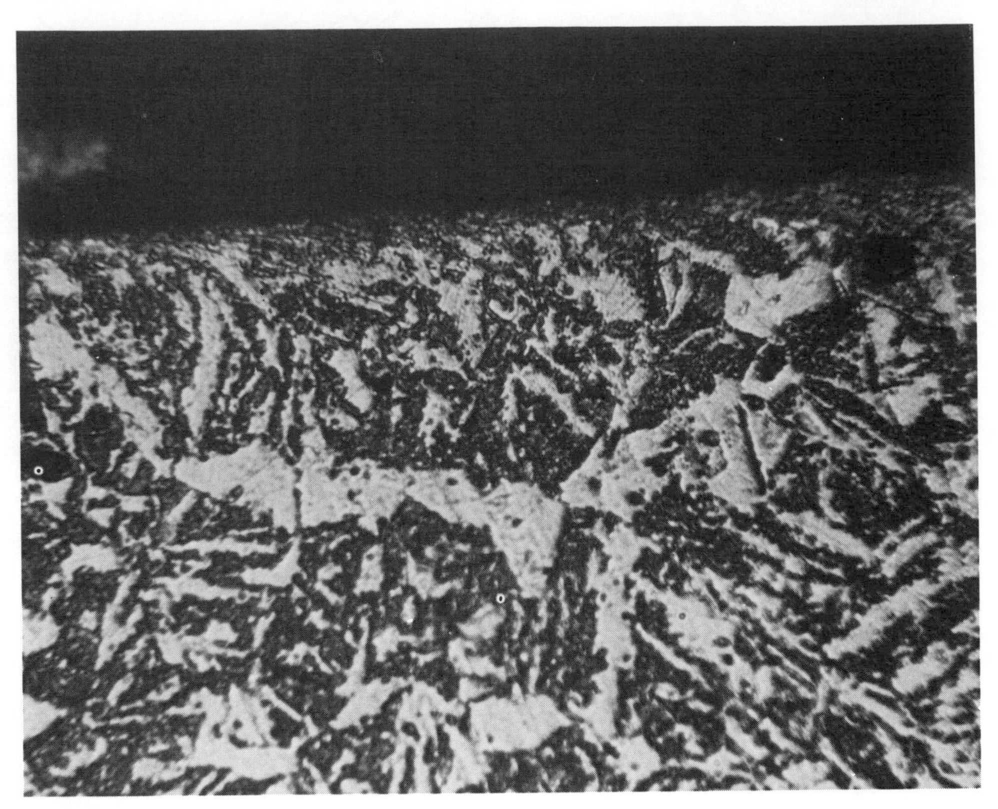

Figure 4*b*. At 400 SFM, this micrograph reveals that the surface is fairly smooth and the grains are only slightly distorted to a depth of 0.001 inch (250×).

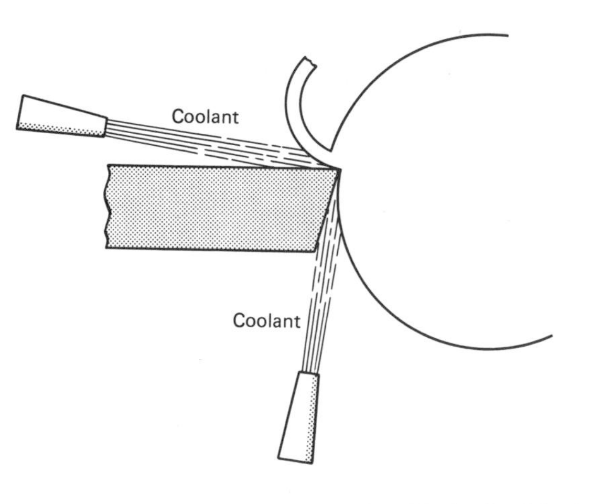

Figure 5. Though not always possible, this is the ideal way of applying coolant to a cutting tool.

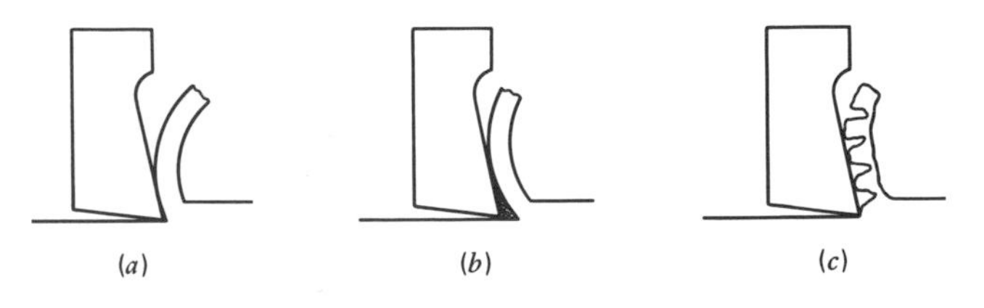

Figure 6. The three types of chip formations. *(a)* Continuous. *(b)* Continuous with built-up edge. *(c)* Discontinuous (segmented).

formation of the three basic types of chips can be seen in Figure 6. Various kinds of chip formations are shown in Figure 7. High cutting speeds produce thin chips and tools with a large positive rake angle favor the formation of the continuous chip. Any circumstances that lead to a reduction of friction between the chip-tool interface, such as the use of coolant, tend to produce a continuous chip. The continuous chip usually produces the best surface finish and has the greatest efficiency in terms of power consumption. Continuous chips create lower temperatures at the cutting edge, but at very high speeds there are higher cutting forces and very high tool pressures. Since there is less strength at the point of positive rake angle tools than in the negative rake tools, tool failure is more likely with large positive rake angles at high cutting speeds or with intermittent cuts.

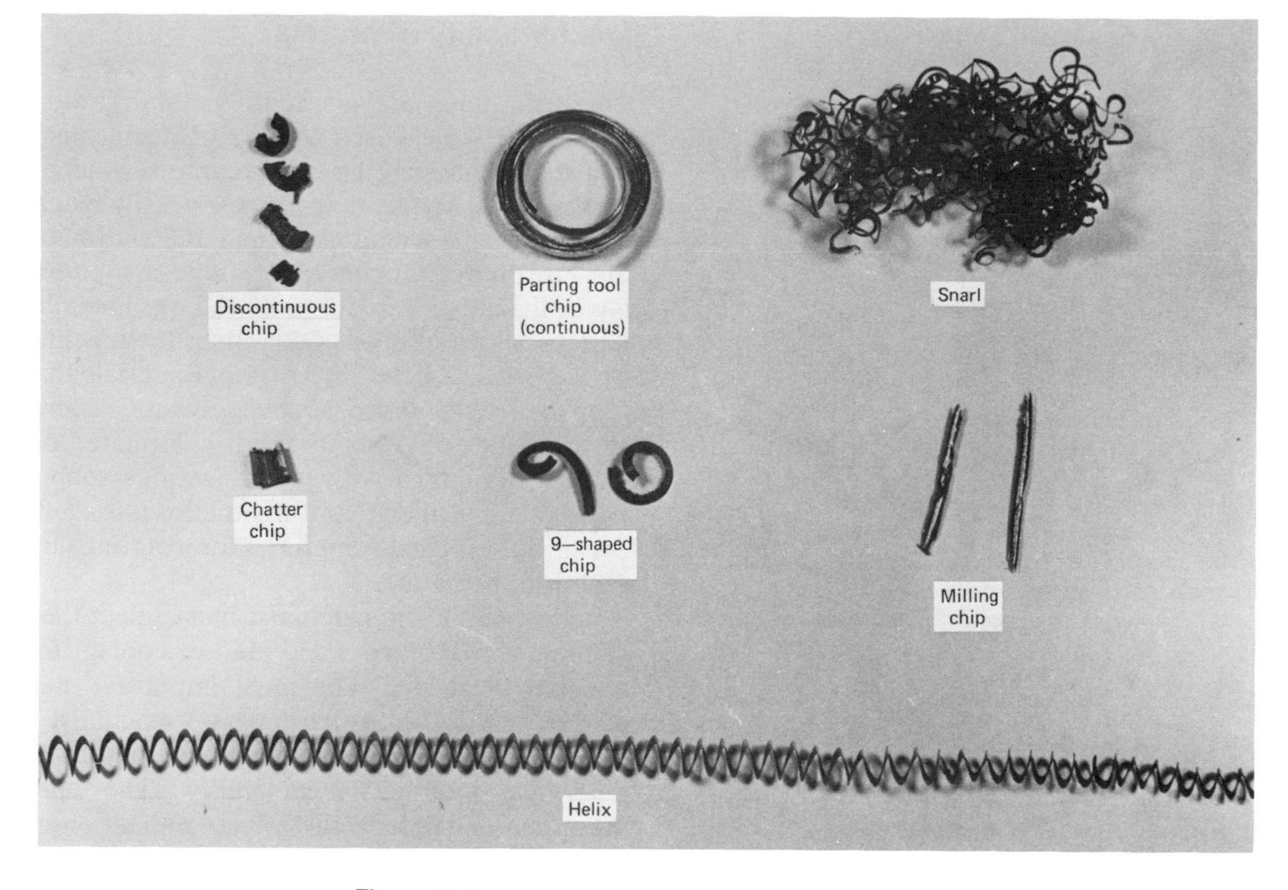

Figure 7. Some of the kinds of chips that are formed in machining operations.

Negative rake tools are most likely to produce a built-up edge with a rough continuous chip and a rough finish on the work, especially at lower cutting speeds and with soft materials. Positive rake angles and the use of coolant plus higher speeds decrease the tendency for a built-up edge on the tool.

The discontinuous or segmented chip is produced when a brittle metal, such as cast iron or hard bronze, is cut. Some ductile metals can form a discontinuous chip when the machine tool is old or loose and a chattering condition is present, or when the tool form is not correct. The discontinuous chip is formed as the cutting tool contacts the metal and compresses it to some extent; the chip begins then to flow along the tool and, when more stress is applied to the brittle metal, it tears loose and a rupture occurs. This causes the chip to separate from the work material. Then a new cycle of compression, tearing away of the chip, and its breaking off begins.

Low cutting speeds and a zero or negative rake angle can produce discontinuous chips. The discontinuous chip is more easily handled on the machine since it falls into the chip pan.

The continuous chip sometimes produces snarls or long strings that are not only inconvenient but dangerous to handle. The optimum kind of chip for operator safety and for producing a good surface finish is the 9-shaped chip that is usually produced with a chip breaker.

Chip breakers take many forms and most carbide tool holders either have an inserted chip breaker or the chip breaker is formed in the insert tool itself. Figure 8 shows how the action of a chip

Table 1 High Speed Lathe Tool Angles in Degrees

	Side Relief	End Relief	Side Rake	Back Rake
Aluminum	12	10	15	35
Brass	10	8	0–5	0
Bronze	10	8	0–5	0

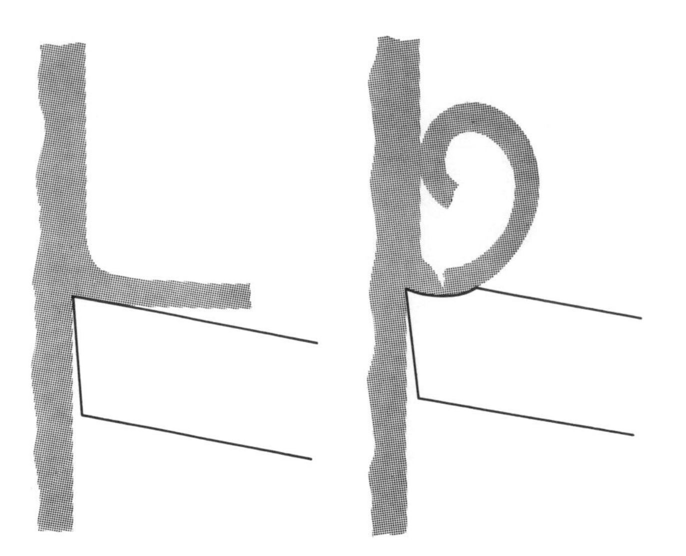

Figure 8. The action of a chip breaker to curl a chip and cause it to break off *(Machine Tools and Machining Practices). (a)* Plain tool. *(b)* Tool with chip breaker.

breaker takes place to curl the chip against the work and then to break it off to produce the proper type of chip.

Operators of the machine usually must form the tool shape themselves on a grinder when using high speed tools and they may or may not grind a chip breaker on the tool. If they do not, a continuous chip is formed that usually makes a wiry tangle, but if they grind a chip breaker in the tool, depending on the feed and speed, they produce a more acceptable type of chip. A high rate of feed will produce a greater curl in the chip, and often even without a chip breaker, a curl can be produced by adjusting the feed of the machine properly. Figure 9 shows how chip breakers may be formed in a high speed tool.

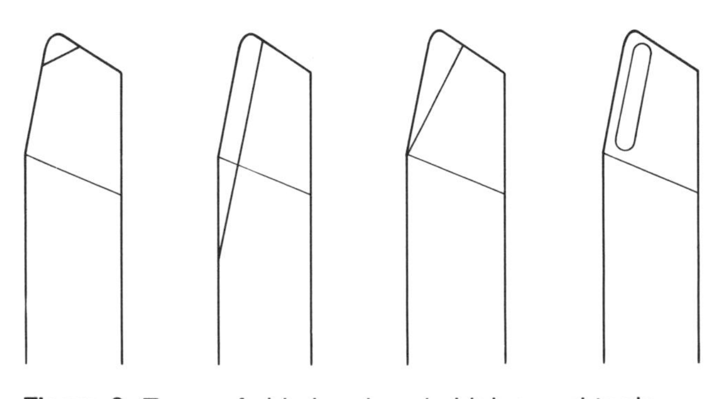

Figure 9. Types of chip breakers in high speed tools *(Machine Tools and Machining Practices).*

Machinability of Metals

Various ratings are given to metals of different types on a scale based on properly annealed carbon steel containing 1 percent carbon, which is 100 on the scale. Metals that are more easily machined will have a higher number than 100 and materials that are more difficult to machine are numbered lower than 100. Table 2 gives the machinability of various alloys of steel. Of course, the machinability is greatly affected by the tool materials that are used and even more so by coolants. There are several kinds of coolants used: chlorinated or sulfurized oils mixed with water, various soap solutions or emulsifiers, and synthetic cutting fluids. An air blast is used sometimes for a cutting fluid for cast iron. See Table 3.

In general, the machinist must select the type of tool, speeds, feeds, and kind of coolant for the material being cut. The most important material property, however, is hardness. A machinist often shop tests the hardness of material with a file to determine relative machinability. The operator must also understand the effects of heat on a normally machinable metal such as C4140. A weld on C4140 will harden the base metal near the weld and make it impossible to cut, even with carbide tools. It must first be annealed.

Most alloy steels can be cut with high speed tools but at relatively low cutting speeds, which produce a poor finish unless a large back rake is used. These steels are probably best machined by using carbide tools at high cutting speeds from 400

TABLE 2 Machinability Ratings of Annealed Steels

AISI Classification	Machinability Rating	Approximate Hardness BHN (Brinell)
B1113	135	200
B1112	100	205
C1118	80	160
C1020	65	150
C1040	60	200
A8620	50	220
A3140	55	200
A5120	50	200
C4140	50	200
Cast iron	40–80	160–220
302 Stainless	25	190

Table 3 Cutting Fluid and Speed Table for Machining Various Metals with High Speed Tools

Material	Cutting Fluid	Speed (ft/min)
Aluminum and alloys	Soluble oil, kerosene, light oil	200–300
Brass	Dry, soluble oil, synthetic solution, light mineral oil	150–300
Bronze, common	Dry, soluble oil, mineral oil, synthetic solution	200–250
Cast iron, soft	Dry or air	100–150
Cast iron, medium	Dry or air	70–120
Cast iron, hard	Dry or air	30–100
Magnesium and alloys	60-second mineral oil	300–600

to 600 SFM, or even higher; in some cases, speeds as high as 1200 SFM are used accompanied by a coolant. Some alloys of steel, however, are more difficult to machine and they tend to work harden. Examples of these are austenitic manganese steels, inconel, stainless steels, and some tool steels.

Machining Stainless Steels

Machining characteristics of stainless steels with high speed tools are given in Table 4. These figures can be multiplied by a factor of 3 or 4 when carbide tools are used. A general rule is to use slower cutting speeds and higher feeds for stainless steel than are used for mild steels. Some stainless steels are free machining types, but others tend to work harden quickly.

Drills should be ground with a 130-degree included angle and the web should be thinned to reduce the dead center to one-sixth of the drill diameter. If the tap is sharp and kept properly aligned, 75 percent threads can be tapped in free machining stainless steel. In many cases, a 60 percent thread will hold just as well and will be much easier to tap. Many tap drill charts provide for a selection of tap drills of various percentages of thread.

Mineral oil and sulfurized oils are used for machining, threading, reaming, and tapping stainless steels. Water soluble oils are used for high speed machining with carbide tools. The mixture is generally heavier to give better lubrication than that used with other metals.

Machining Aluminum Alloys

The aluminum alloys all have good machinability, especially those having the harder tempers. Soft (1100) aluminum tends to build up on the tool, and a rough finish usually results. For cutting aluminum alloys, edges of tools should be sharp and without burrs or wire edges. Honing the cutting edges with a very fine oilstone is recommended after grinding. Back rake (Figure 10) on turning and parting tools should be 30 to 35 degrees. Carbide tools should be used for the high silicon alloys. Machining of aluminum alloys can be done without lubricant or coolant. For most purposes a soluble oil or a soluble synthetic concentrate is satisfactory, but various solutions of kerosene and oil are also used.

Machining Cast Iron

Cast iron is machined dry or with a cooling jet of air. Tapping in cast iron may be done dry, but it is greatly improved by using various commercial cutting fluids designed for cast iron tapping.

Gray cast irons, whether of the ferritic or pearlitic (high strength) types, offer no great problems for machining unless welds are involved. Unless proper preheating and postheating are done

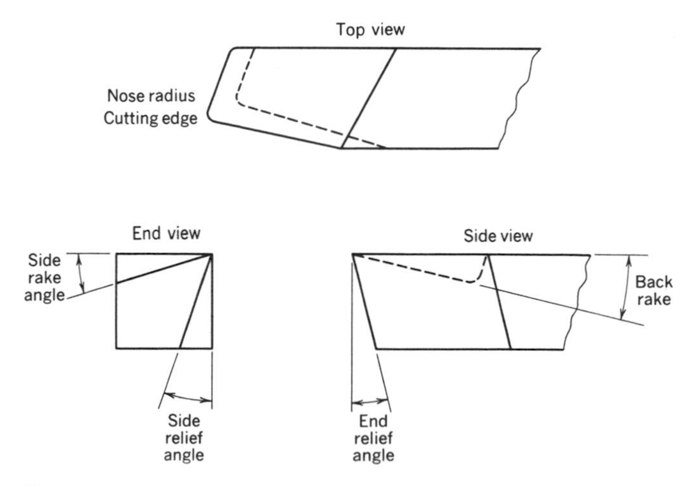

Figure 10. A right-hand turning tool bit used for aluminum *(Machine Tools and Machining Practices).*

when welding cast iron, the heat affected zones of the weld section will contain extremely hard carbides. Even the use of "machinable" ENi electrodes will not entirely eliminate this hard zone under the weld if the base metal is cold when welded.

Malleable (pliable or workable) and nodular cast irons are both easily machined in most cases. White or chilled cast iron is virtually nonmachinable and must usually be ground because of its extreme hardness. Some chilled irons can be machined to some extent with certain grades of carbides and with ceramics. These may be selected from the manufacturer's catalogs for the particular job. White cast iron is not considered to be weldable.

Stress Concentration

The design and machining practices of parts made on various machine tools have a great effect on their working life. Stress concentration takes place wherever there is a sharp shoulder, corner, groove, or roughness. The root of a thread can also create stress concentration that can sometimes initiate fatigue of the metal and cause ultimate failure. In Figure 11, a photoelastic material is being stressed with the sharp edge of a punch. The lines caused by polarizing the light reveal the stresses. The concentration of the stresses can be seen where the lines converge to a point. This is one method of studying the behavior of metals that are in similar situations under stress concentration. This particular plastic model would suddenly fail where the notch shows the stress concentration if very little additional pressure were applied to the punch.

The stress concentration that can take place on a shaft with a shoulder is shown in Figures 12*a* and 12*b*. Figure 12*a* shows the concentration at the sharp corner of a simulated piece of shafting that has been machined to a sharp shoulder. The point of concentration can be seen where the pencil is pointing. In Figure 12*b*, a fillet radius was formed at the shoulder of the simluated photoelastic part and the pencil is pointing to the long curved flow lines that reveal an even distribution of stresses over a larger area and indicate little stress concentration. The type of design in Figure 12*b* will not tend to fail by fatigue at the shoulder, but the type of design in Figure 12*a*, having the sharp shoulder or corner, will tend to fail by fatigue if there are cyclic (reversing) stresses involved. See Chapter 22 "Service Problems" for more information on fatigue failures and stress concentration.

Table 4 Machining Characteristics for Stainless Steels for High Speed Tools (with Coolant)

		Machining Operations			
Stainless Steels	**Brinell Hardness**	**Turning**	**Drilling**	**Milling**	**Characteristics**
		Feed per Revolution		Feed per Tooth	
		0.003-0.008 in.	0.003-0.005 in.	0.003-0.005 in.	
		Cutting Speeds (Surface ft per Min)			
301, 302, 304, 304L	150–250	60–120	30–50	40–60	Nonhardening by heat treatment. Tends to work harden rapidly.
316, 316L	150–240	60–120	30–50	30–50	Machines similar to 302.
303	150–240	70–140	50–80	40–60	Austenitic. Free machining.
430F	170–230	80–150	60–90	50–80	Ferritic. Free machining.
410	180–240	80–130	40–80	40–60	Martensitic. Low chromium.
416	180–240	80–150	60–90	50–80	Martensitic. Free machining. Short, brittle chips.
440A, 440B, 440C, 440F	200–265	40–80	30–50	30–50	Martensitic. Carbide tools suggested because of its abrasive action.

Source. Armco Steel Corporation, *Machining Armco Stainless Steels*, "Cutting Rates for Armco Stainless Steels," 1973. (Condensation of speed and feed tables.)

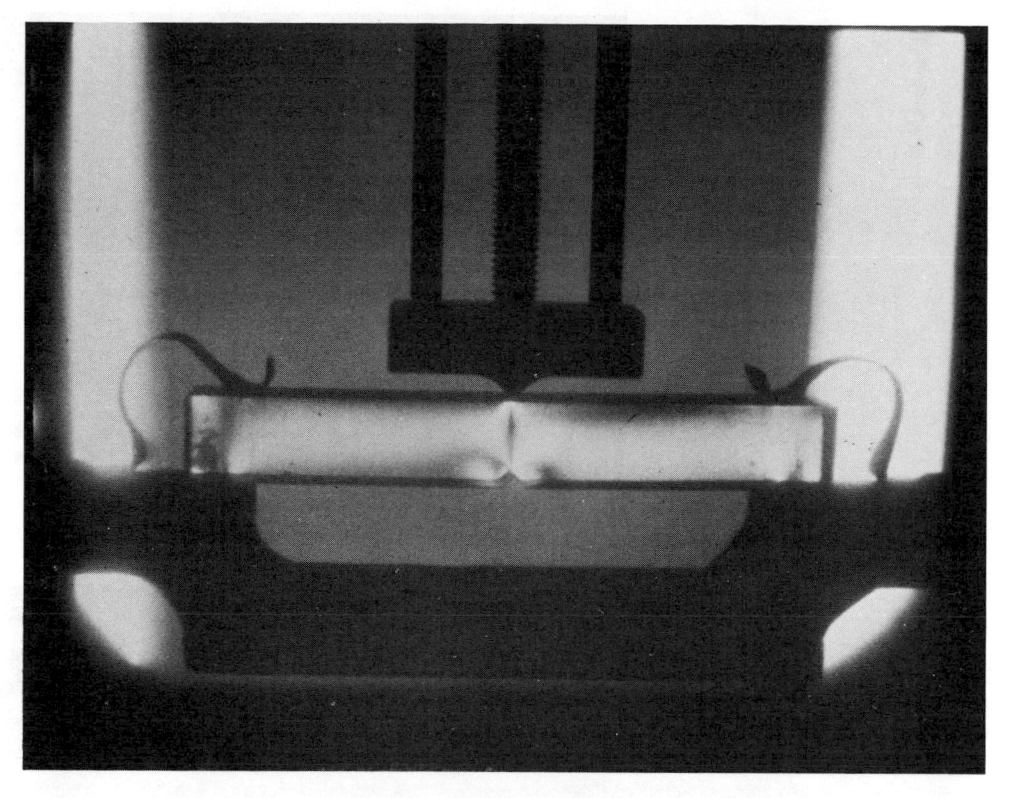

Figure 11. Stress concentration caused by pressure applied at a point on this photoelastic specimen. The stress can be seen in converging lines that point to the origin of stress.

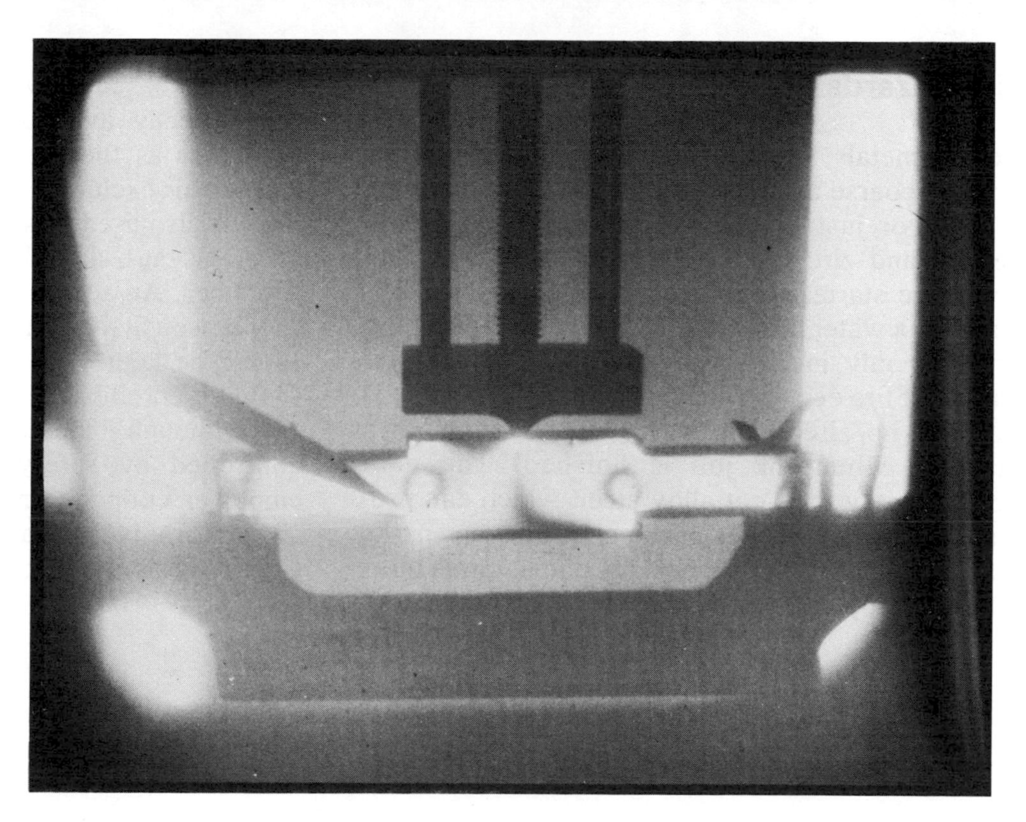

Figure 12*a*. This specimen, having sharp shoulders and being under stress, shows a definite stress concentration at the sharp shoulder.

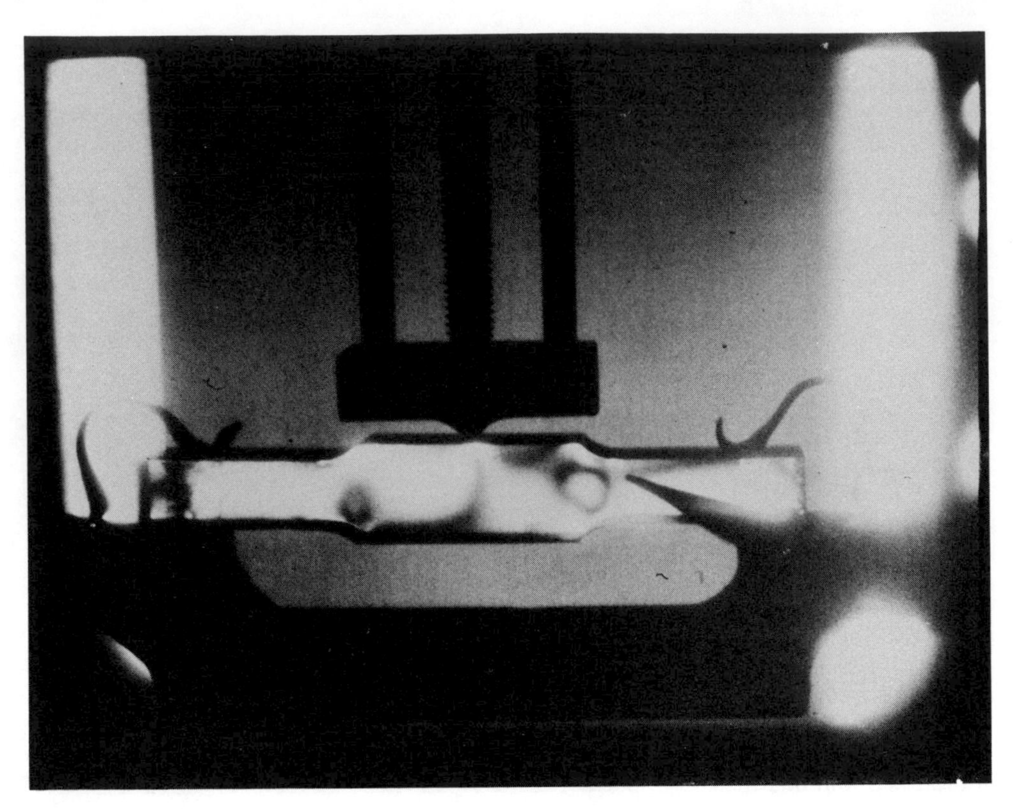

Figure 12*b*. In this specimen the shoulders are rounded and the curved lines indicate an even distribution of the stress.

Fire Hazards

Certain metals, when divided finely as a powder or even as coarse as machining chips, can ignite with a spark or just by the heat of machining. Magnesium and zirconium are two such metals. The fire, once started, is difficult to extinguish, and if water or a water based fire extinguisher is used, the fire will only increase in intensity. The greatest danger of fire occurs when a machine operator fails to clean up zirconium or magnesium chips on a machine when the job is finished. The next operator may then cut alloy steel, which can produce high temperatures in the chip or even sparks that can ignite the magnesium chips. Such fires often destroy the entire machine if not the shop. Chloride based powder fire extinguishers are commercially available. These are effective for such fires as they prevent water absorption and form an air-excluding crust over the burning metal. Sand is also used to smother fires in magnesium.

Oily rags should be kept in a covered steel container. An accumulation of oily rags can ignite spontaneously and cause a fire. Solvents and oils should be kept away from open flames; smoking should be prohibited in their vicinity.

Although industrial hazards exist, they are controlled by safety programs and employee-employer cooperation. Machine shop or welding shop work is a relatively safe occupation, but safety rules must be observed.

Self-Evaluation

1. What does the chip color reveal to the machine operator?
2. In what way do tool rakes, positive and negative, affect surface finish?
3. Soft materials tend to pressure weld on the top of the cutting edge of the tool. What is this condition called and what is its result?
4. Which indicate a greater disruption of the surface material: thin uniform chips or thick segmented chips?
5. In metal cutting does the material split ahead of the tool? What does it do?
6. Which tool form is stronger, negative or positive rake?
7. What effect does cutting speed have on surface finish? On surface disruption of grain structure?
8. How can surface irregularities caused by machining later affect the usefulness of the part?
9. Which property of metals is directly related to machinability? How do machinists usually determine this property?
10. Why should machinists never make a practice of producing sharp corners on shaft shoulders? What does this cause?

Worksheet

Objective Determine the effects of machining on a workpiece by microscopic examination.

Materials Engine lathe, carbide tool with negative rake, a short piece of 4140 or other alloy shafting, and the needed metallurgical equipment.

Procedure

1. Turn a distance of 2 or 3 inches with 0.050 in. depth of cut with the lathe set at a finish feed (about 0.003 in.) at 100 SFM.
2. Turn a distance of 1 or 2 in. at a 0.050 in. depth of cut with a finish feed of 400 SFM.
3. Turn a short distance (about $\frac{1}{2}$ in.) at 600 SFM.
4. Saw off slices of each of the different cutting speeds on a cutoff saw (about $\frac{3}{16}$ in. thick).

5 Using a vertical bandsaw or hacksaw, cut out a segment of each sample as shown in Figure 13.

Figure 13. Two specimens are sawed from a thin slice of a metal bar that was turned at various speeds on the lathe. The segments are then mounted in plastic to facilitate grinding and polishing for microscopic examination of their outer edges.

6 Encapsulate each specimen in plastic and mark each with its cutting speed.

7 Polish and etch each specimen with nital.

8 View the specimens on the curved (outer) edge with the microscope first at 100× and then at higher magnification.

Conclusion

1 Describe the surface condition of each sample.

2 How were the grains disturbed in each sample? How do you think this will affect the usefulness of the part?

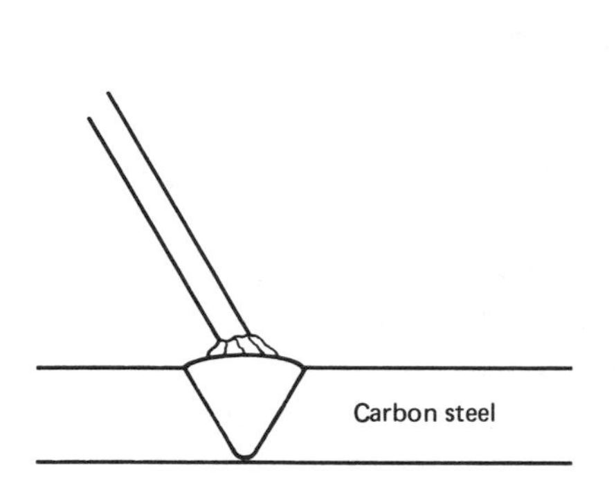

16 Metallurgy of Welds: Carbon Steel

In the first chapters of this book, you learned the basic principles of metallurgy; that is, process metallurgy involving the reduction of ores and refining of metals. You also learned about the alloying, casting, and working or shaping of metals to produce the finished products and about physical metallurgy, which includes mechanical testing, metallography, and heat treatment. Physical and process metallurgy are both involved in welding processes. In fact, the process of welding is very similar on a small scale to the melting of metals in a furnace and to casting them into the ingot forms.

The welding of metals is quite similar to the manufacture of metals because the elements that go into the making of the metal in the furnace and the grain structures that develop as it cools are often the same (Figure 1). There is also a similarity in the peening of welds to the hot and cold working of steels. Peening work hardens cold metal just as cold rolling work hardens metal. There are other comparisons such as thermal cycles that take place in

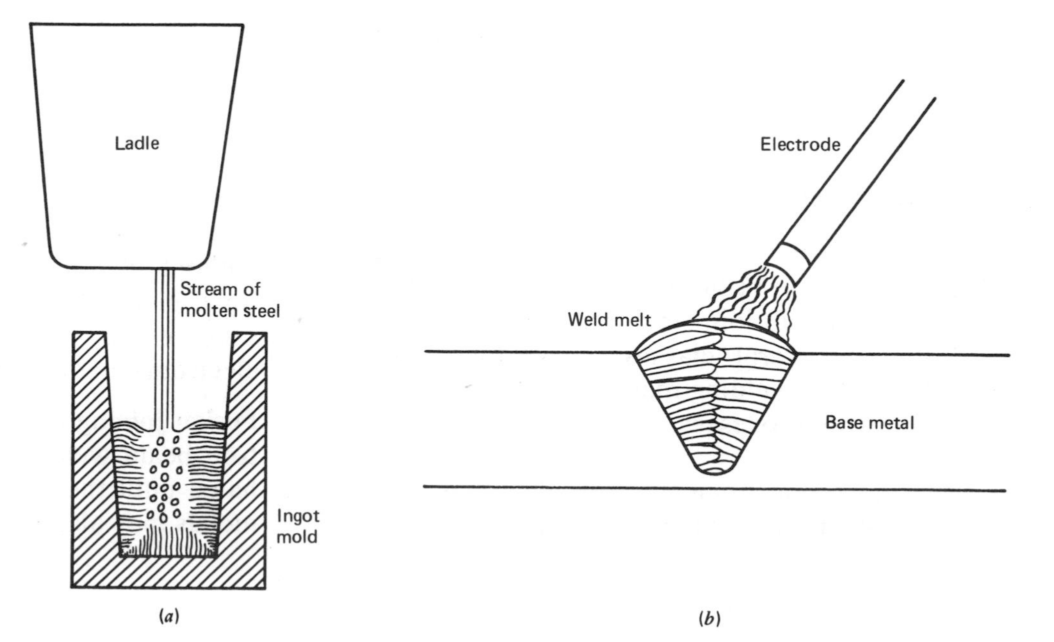

Figure 1. Similarity between *(a)* casting of ingots and *(b)* the welding process.

the melting and solidification of metals that are also related to welding. These cycles cause changes in grain size, strength, and ductility.

There is more included in welding than the selection of the electrode or wire and "burning" the rod to make a weld, although much welding is performed in just this way. If the welder is not aware of the type of material in the base metal on which he is welding, and if he does not understand the metallurgical conditions in the weld areas, he could experience difficulties such as cracking, porosity, and numerous other problems.

Before modern welding processes were developed, welding of iron and steel was done by a blacksmith with a forge and anvil. The ends of two pieces such as the ends of a wagon tire hoop were scarfed (tapered) and heated to a white heat. Silica sand was tossed into the joint for a flux to remove oxides and the pieces to be joined were then brought to the anvil. There they were hammered together to cause the joining of the metal by pressure while excess slag was being forced out of the joint. The resultant weld was only partially bonded over the weld area even though the correct heat and procedure may have been used. Modern methods of welding produce a more complete bonding of the metals, but other problems that were not found in forge welding, such as hard brittle zones, have arisen.

OBJECTIVES

After completing this chapter, you will be able to:

1. Describe the changes in welds and heat affected zones because of the heat of welding and their effects upon the welded structure.
2. Select a correct welding process and filler metal for a carbon steel base metal in order to have the optimum metallurgical condition in the weld.
3. Explain the types of welding that are done in industry.
4. Explain the effects of slags and fluxes in welding.
5. Macroetch weld sections for observation of their columnar structure and heat affected zones.

INFORMATION

A great many welding processes such as gas, arc, induction, electron beam, resistance, and pressure welding are used to join steel, but these can all be listed within several classes of welding. Most methods used in welding may be classified as one of the following processes.

1. The application of heat (from electrical resistance, friction, or flame) and pressure (forge welding) may be classed as **solid phase welding.** The joining is done without filler metals and without melting or changing the base metal. Solid phase welds may also be made by using other energy sources such as ultrasonics or electromagnetic induction.
2. **Fusion joining** (arc, gas, plasma arc, and electron beam welding) requires that the parts be hated until they melt and flow together. Filler metals may also be used.

3 **Liquid-solid phase welding** requires the base metal parts to be heated but not melted. A dissimilar molten metal is used to join the parts together. Brazing and soldering are examples of this kind of welding. Diffusion often occurs, but is not necessary for adherence of the filler metal.

In Chapter 5 you learned of a steel classification system that indicated the carbon and metal alloy content of steels. The SAE-AISI systems are used for steels used in machinery, tools, products, and bar stock.

Welders in steel construction, pipe lines, and pressure vessels typically use The American Society for Testing and Materials (ASTM) standards to determine material specifications, practices, definitions, and methods of testing. See Table 1. These standards for steel all carry the prefix letter A; for example, A27-62 denotes a low to medium strength carbon steel casting, and A7-61T covers steel for buildings and bridges. The American Welding Society (AWS) also has a system of codes, recommended practices, standards, and procedures. The AWS deals with such areas as welding and testing procedures. AWS also has specifications for welding rods and electrodes.

Most welding on steel is performed on low carbon steel having about 0.08 to 0.2 percent carbon. Let us examine the structures pertaining to these welds. You have already learned that steel undergoes certain changes when it solidifies from the molten state and cools to room temperature; that is, it begins to form a lattice structure based on a face-centered cube that is called austenite, and cools through a transformation temperature or critical point that marks the change from a face-centered to body-centered lattice structure called ferrite. The formation of grains takes place during this cooling period from the molten state to the solid, and the size and arrangement of the grain structure in the weld zone and the heat affected zone of the weld are permanently fixed unless the metal is reheated.

There are four basic zones in welding: the **weld zone,** the **fusion zone,** the **adjacent zone,** and the **heat affected zone** (Figure 2). The weld zone (often called nugget) is the weld melt itself after it solidifies. Although the fusion zone (Figure 3) is represented by a fairly narrow area (almost a sharp line), a certain amount of diffusion or base metal

Table 1 Some ASTM Standard Numbers for Steels

ASTM Number	Type of Steel
A1–58T	Open hearth carbon steel rails
A27–62	Low to medium strength carbon steel castings
A7–61T	Steel for bridges and buildings (tentative)
A8–54	Structural nickel steel
A20–56	Boiler and firebox steel
A36–61T	Structural steel (tentative)
A94–54	Structural silicon steel
A120–61T	Black and hot-dipped zinc coated welded and seamless pipe for ordinary uses (tentative)
A120–60	Austenitic manganese—steel castings
A216–60T	Carbon steel castings suitable for fusion welding for high temperature service (tentative)
A240–61T	Corrosion-resisting chromium and chromium-nickel steel plate, sheet, and strip for fusion-welded unfired pressure vessels (tentative)
A415–58T	Hot rolled carbon steel sheets, commercial quality (tentative)
A429–58T	Hot rolled and cold finished corrosion resisting chromium-nickel-manganese steel bars (tentative)

Note. Complete lists and specifications may be found in ASTM Standards reference books.

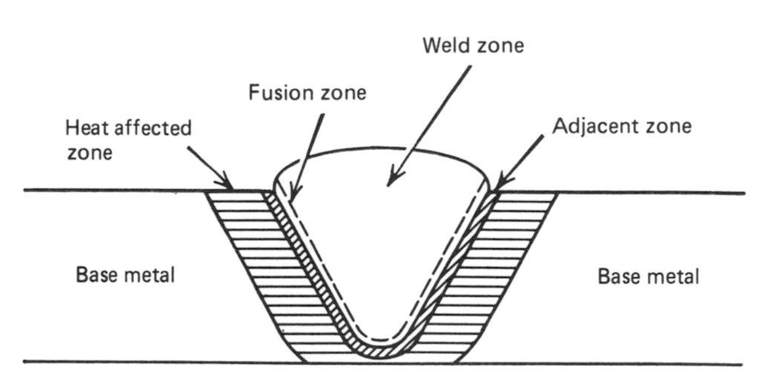

Figure 2. The zones that are found in welding.

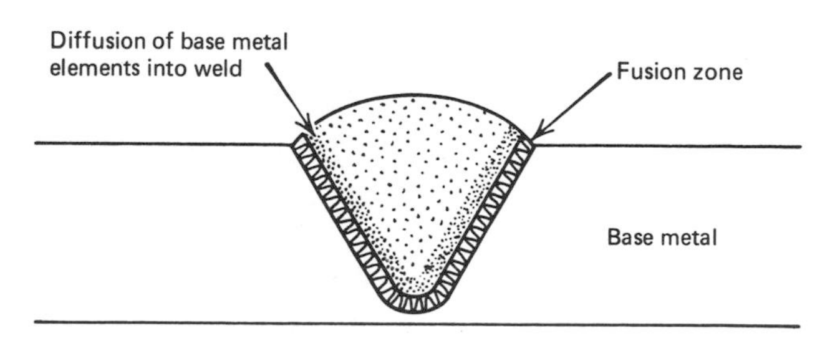

Figure 3. The fusion zone.

pick up is to be found in the weld area near the line of fusion (Figure 4). When the base metal contains considerable carbon, for instance, some of this carbon will diffuse into the melt of the weld zone. This addition of carbon is why the root pass is often more brittle than other passes; its cooling rate is also higher. Any material of high hardenability can promote root cracking. The adjacent zone is near the junction between the base metal and the weld metal. If the cooling rate is rapid, a hard structure will form in the fusion and adjacent zones. The heat affected zone is near the weld in the base metal and it is so called because it is affected by the heat of welding.

While the weld is molten, the atomic structure is not in a lattice arrangement, but is amorphous with a random movement of the iron atoms (Figure 5). When solidification occurs, the weld metal begins to form a lattice structure similar to that in the base metal. There is always a definite, sharp boundary line between the molten weld metal and the base metal.

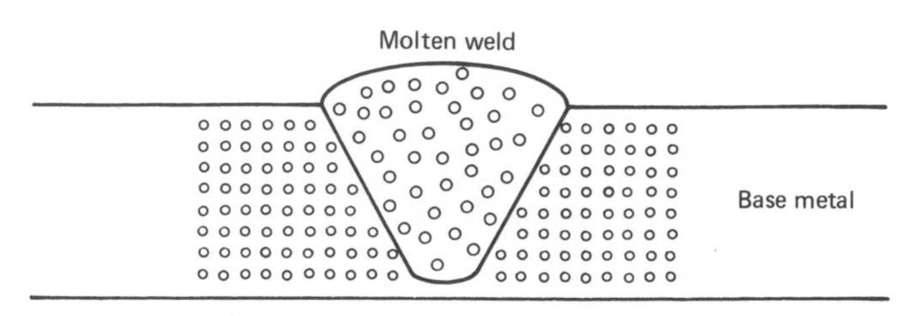

Figure 5. The lattice structure in the base metal remains uniform while the weld melt shows atoms moving at random.

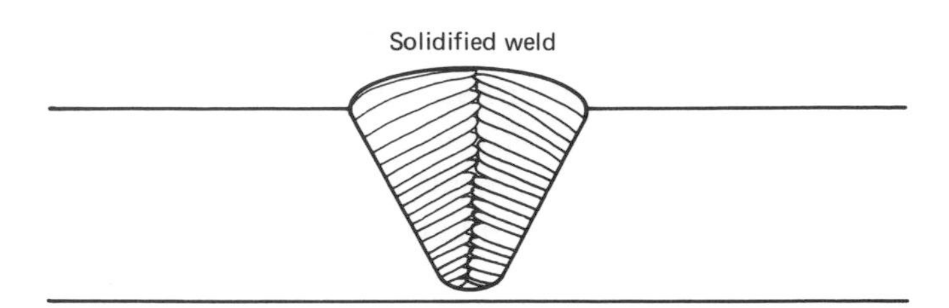

Figure 6. The typical grain formation of welds. Note the sharp boundary line between the molten weld metal and the base metal.

Welds, like ingots, tend to form a columnar structure when they solidify (Figure 6). When steel is slowly cooled from the molten state or heated by welding far above the A_3 line in the austenite phase, the grains begin to grow and join together to form large grains. For example, the casting of metals into ingots generally involves a large mass of material that has a very slow cooling rate. This produces low strength metal with large grains that usually grow in a columnar orientation. These co-

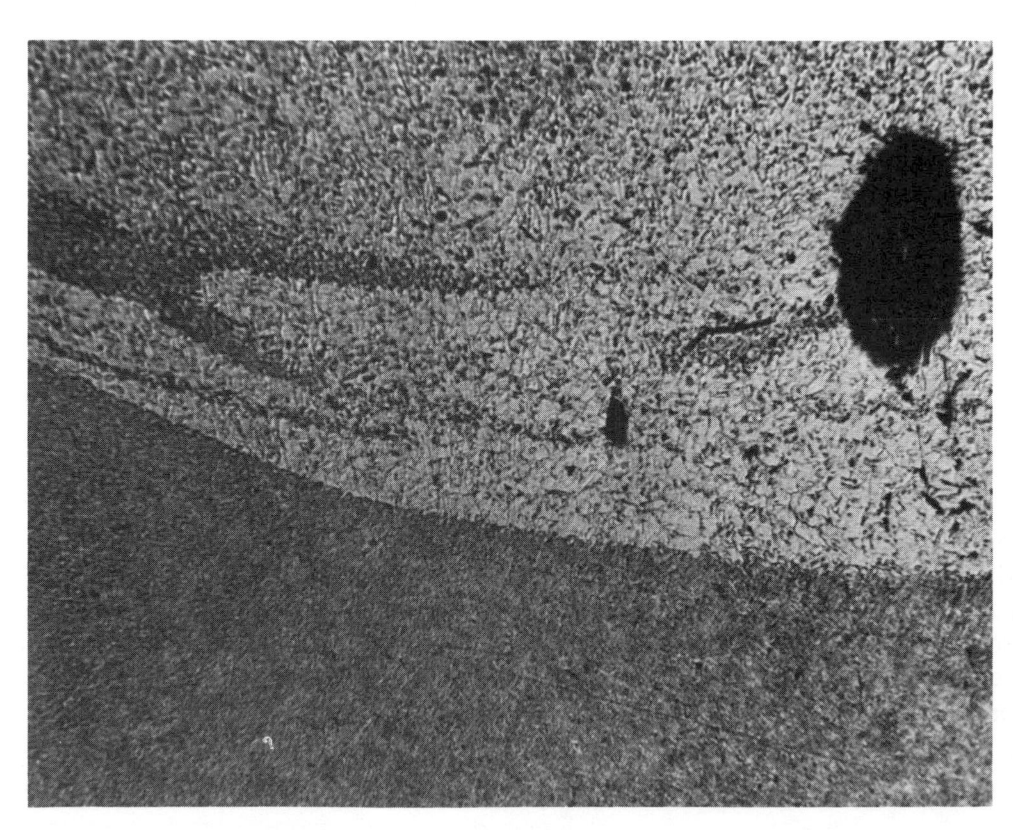

Figure 4. Micrograph showing the diffusion from the base metal into the fusion zone of the weld. Note the gas pockets (250×).

lumnar grains in welds are most evident in those having a large weld nugget, high heat input, and a slow cooling rate.

When the weld metal begins to form these columnar grains, the crystals tend to grow in the cooling molten steel in lines perpendicular to the base metal and in the direction of welding (Figure 7). These lines or grains are called isotherms. The crystals orient themselves toward the direction of welding and outward from the base metal. A single pass weld generally produces a coarse columnar structure (Figures 8*a* and 8*b*) that is somewhat undesirable since it is not as strong as a finer, less oriented grain structure. A two-pass weld will

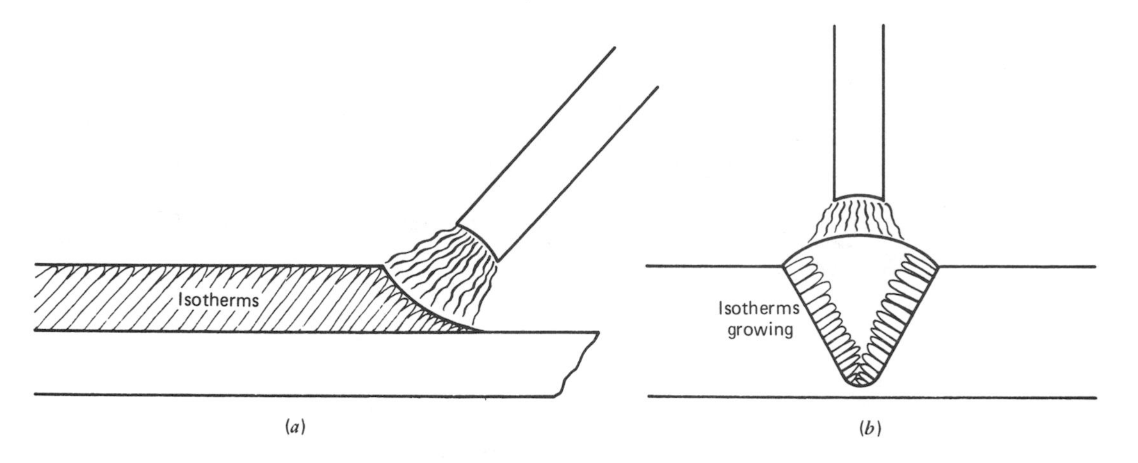

Figure 7. Two views of the weld melt showing the typical formation of isotherms.

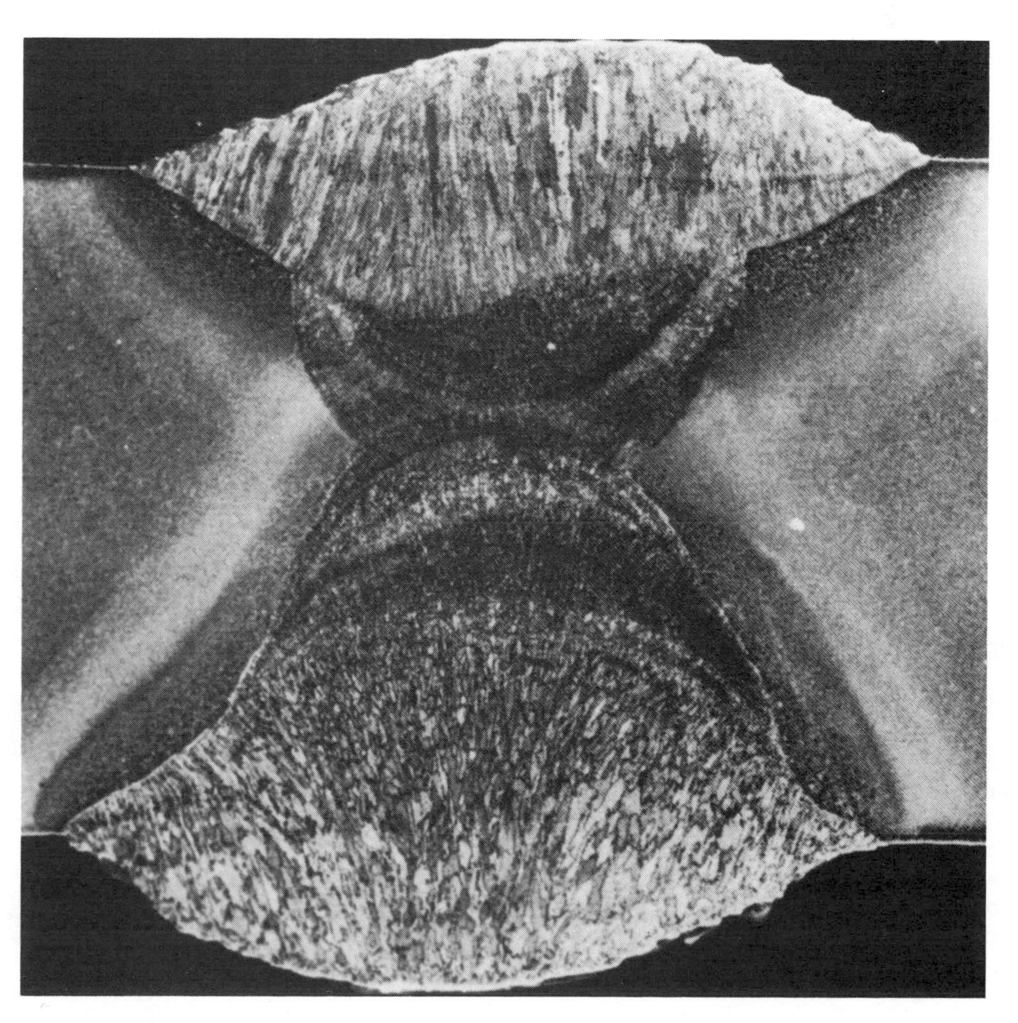

Figure 8*a*. Macrostructure of an arc weld showing the columnar crystallization in the weld metal and heat affected zone in the base metal (4×) (By permission, from *Metals Handbook,* Volume 7, Copyright American Society for Metals, 1972).

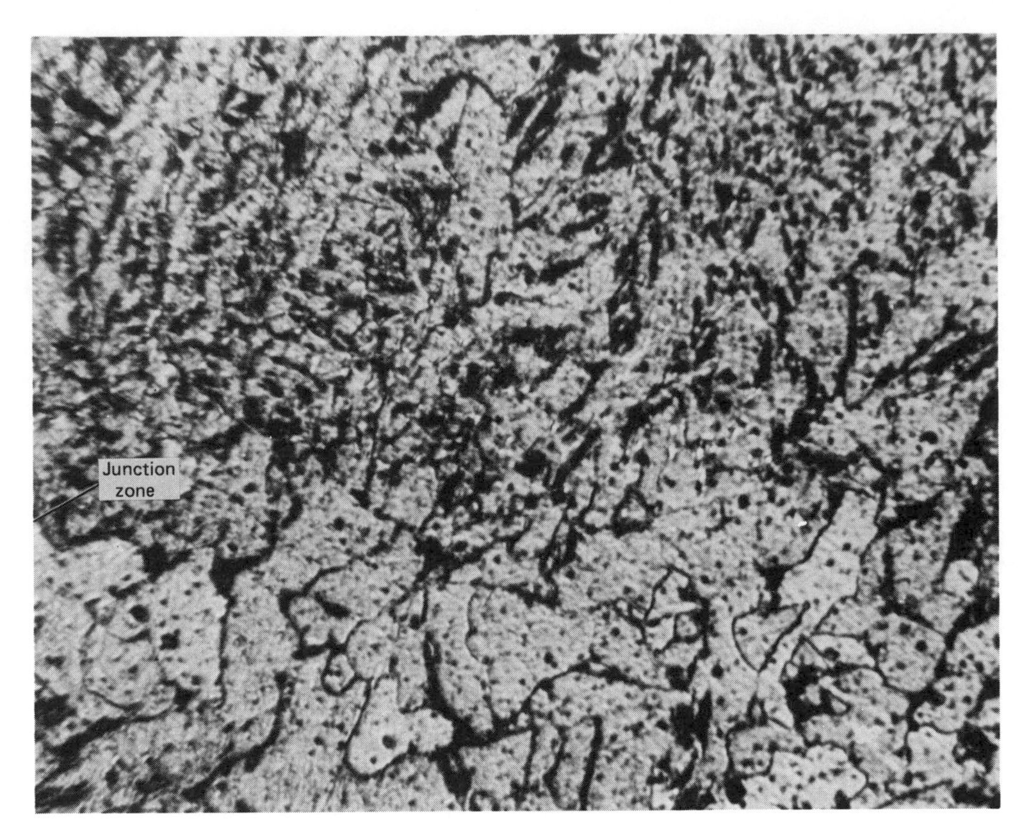

Figure 8*b*. Columnar structure in the weld near the junction zone (100×).

often recrystallize the ferrite grains to some extent, thus stress relieving the first pass weld zones. Nevertheless, the weld zone is usually stronger than the base metal in welded structures (Figure 9).

If the cooling rate is slow, the grains in the weld are irregular and those in the heat affected zone tend to become large (Figures 10*a* and 10*b*). If the cooling rate is rapid, the coarsening effect is minimal. If the weld consists of several passes, the heating effect of the passes will normalize the previously solidified structure, leading to a refinement of the grains since each pass tends to reheat the previous weld pass but at a lower temperature.

The adjacent and heat affected zones near the weld are also those in which the grains begin to grow and coarsen when held at higher temperatures for longer periods, usually having little or no hard structures. In the arc welding process, however, relatively small quantities of metal are molten and consequently often tend to cool very rapidly since the base metal acts as a **heat sink** (heat absorber) that quickly cools the weld nugget.

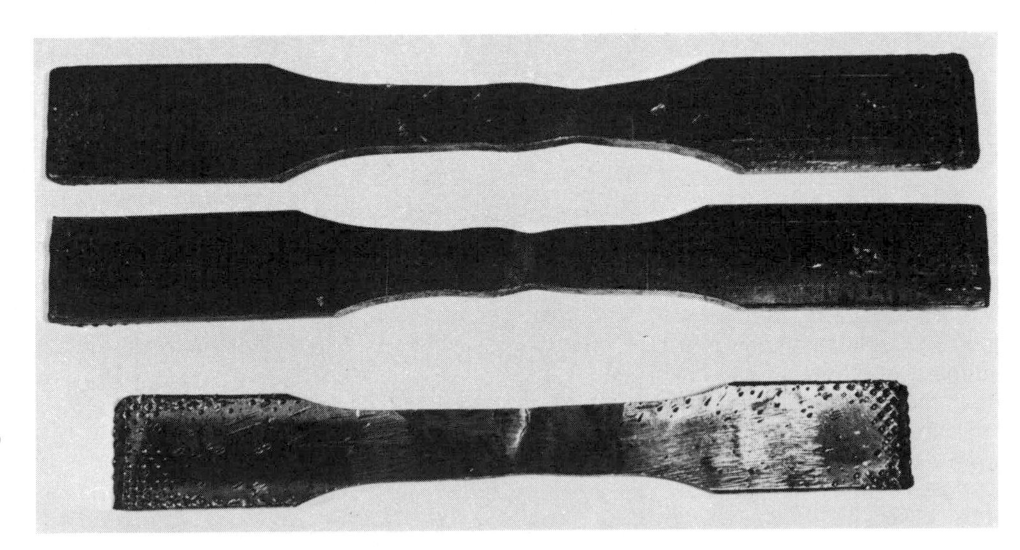

Figure 9. Tensile test samples of the three different welds. The upper two samples begin to fail (neck down) in the heat affected zone near the weld since the larger recrystallized grains in that area are more ductile. The lower test sample began to fail in the weld itself, possibly due to carbon pick up from the base metal and the consequent brittle structure or from hydrogen embrittlement.

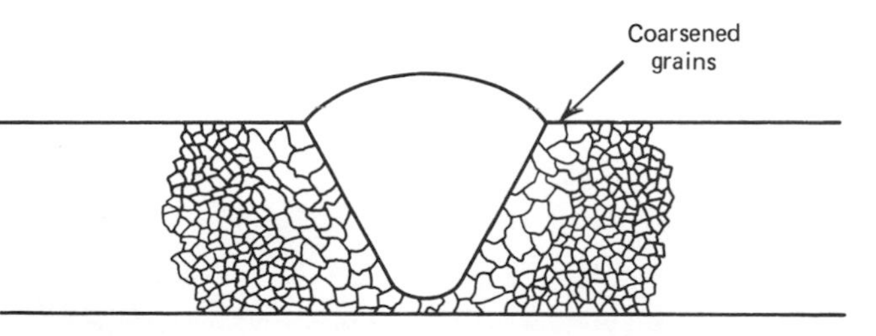

Figure 10*a*. The coarsened grains in the base metal in the heat affected zone are caused by high temperature grain growth.

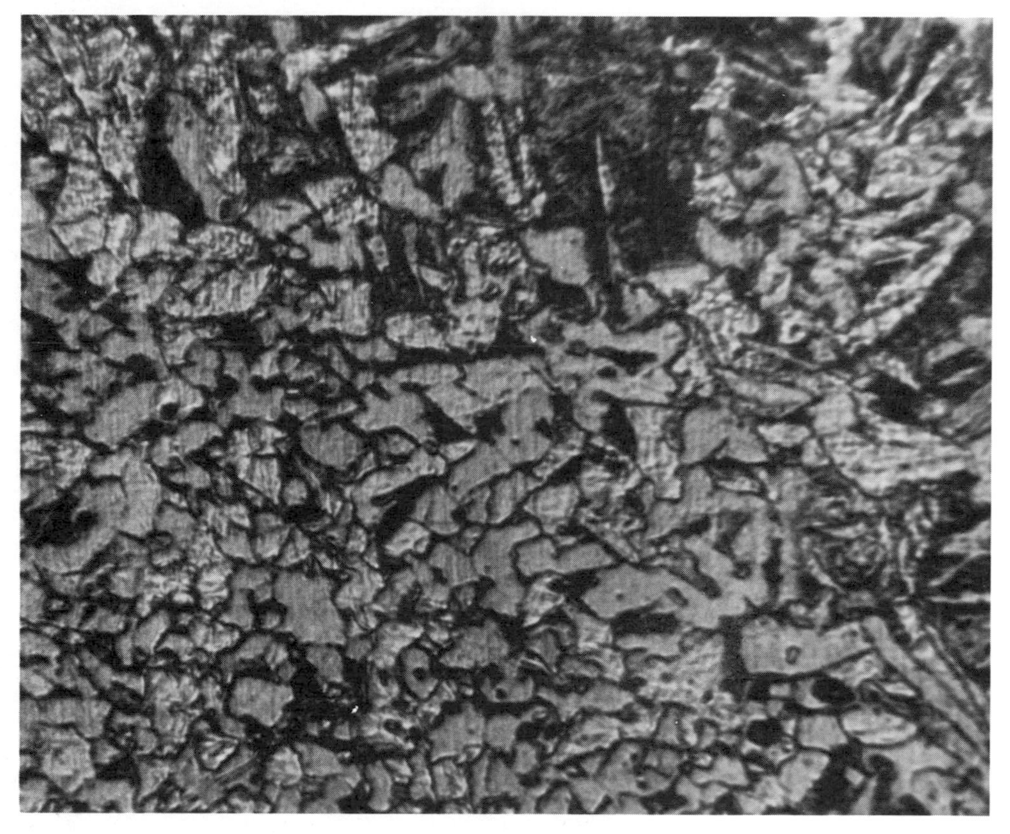

Figure 10*b*. This micrograph shows slow cooling in the weld zone (upper right); slow cooling creates irregular grain structures (180×).

The adjacent zone is the point at which hardening is most likely to take place in steels containing even small amounts of carbon, since the cold base metal quickly quenches the heated adjacent zone as the weld is being made. Multiple passes tend to normalize the previous passes thus reducing the hardness, as shown in Figure 11. Carbides that cause brittleness can often be transformed by postheat treatment or prevented by preheating (Figure 12).

Various metallurgical structures are also formed in the weld zone as cooling or heating is taking place. Low carbon steel welding rods (0.08 to 0.15 percent C) contain ferrite grains with some pearlite and do not harden appreciably unless carbon is picked up from the base metal or from carbon containing contaminants such as oil or grease. When high tensile strength rod that contains a higher percentage of carbon is used, a slow cooling produces pearlite (Figure 13). Other microstructures may also appear. Bainite, often produced while welding (Figure 14), is a dark, acicular (needlelike) microstructure and it is harder and tougher than pearlite depending on the temperature of its formation. Martensite, however, is an extremely hard, brittle structure appearing as a fine, acicular microstructure (Figure 15).

The formation of martensite in the heat affected zone or in the adjacent zone results in many failures in welds because of its brittleness. Hardenability increases with carbon content and promotes the formation of the hard, brittle martensite structure that may crack after welding.

Preheating and **postheating** minimize the formation of the brittle martensitic structure. Preheating to 200 to 400°F (93 to 204°C) prevents the formation of martensite by inducing slower cooling to the base metal. Postheating by tempering or annealing eliminates the brittle martensite that is already formed. Hot cracking can also be reduced

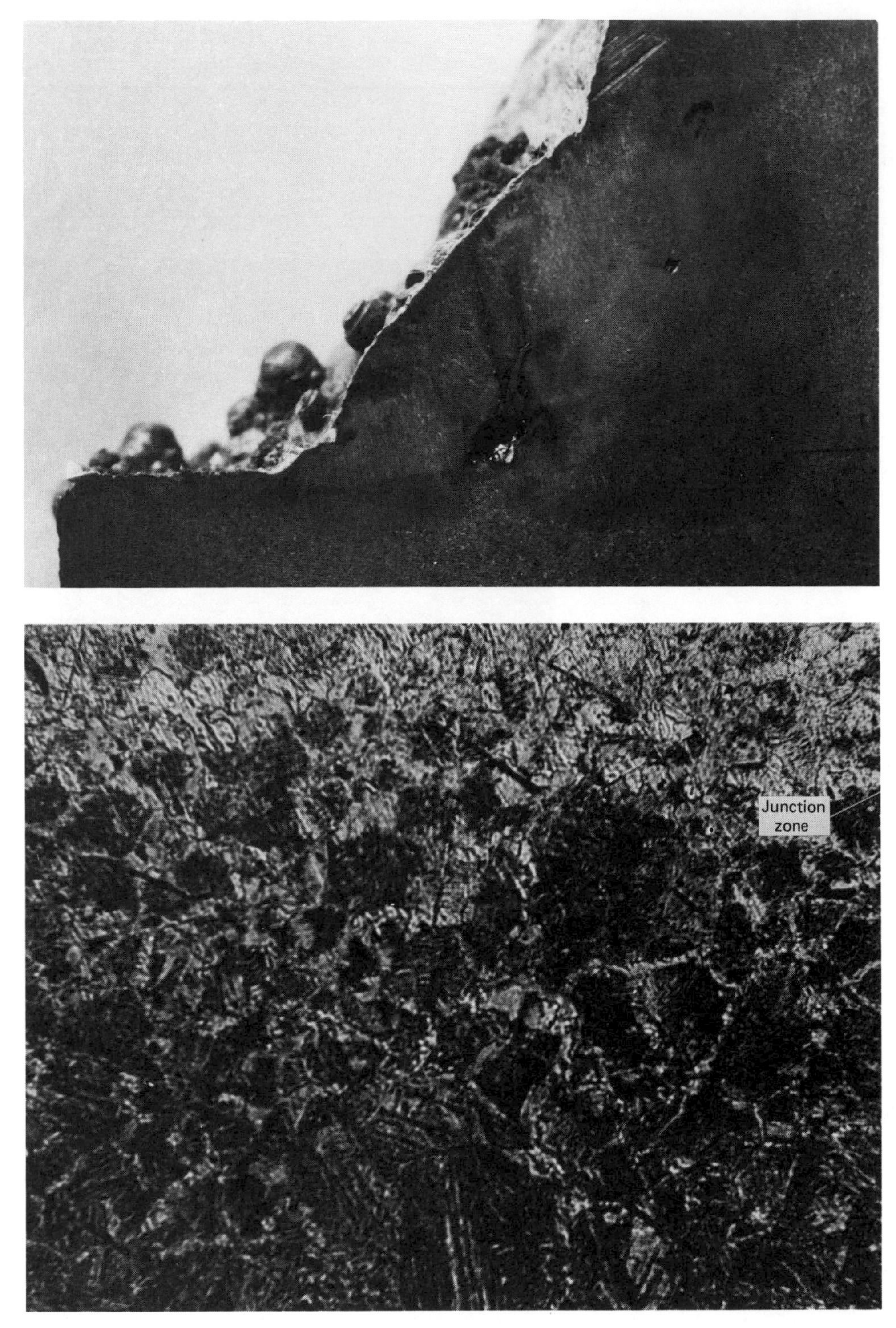

Figure 11. Micrograph of an arc weld made by the "narrow-bead" technique (plate 1 inch thick). The successive passes have partially normalized the heat affected zone giving the narrow, lighter areas.

Figure 12. Micrograph of weld junction zone showing martensite structure in the base metal (500×).

by lowering the cooling rate with preheating. See Chapter 17 for a discussion of weld difficulties.

Ferrite structures in weld areas may require a stress relief if the structure is distorted during welding. The stress relief is done by heating to a temperature between 950 and 1200°F (510 and 649°C) for a short period of time. This will recrystallize the ferrite grains. The heating of the structure lowers its yield strength to a value lower than the residual stress value and allows plastic flow to occur, which relieves the stress. Pearlite grains are not affected by this stress relief (Figures 16*a* and 16*b*). If they are held for several hours at 1200 to 1300°F (649 to 704°C), however, the lamellar or network formation will spheroidize and ductility will increase.

In arc welding, the various zones in the base metal are confined to a very narrow region extending to $\frac{3}{16}$ inch on either side of the weld edge, depending on the thickness of the material (Figure 17). This is true even when the weld is built up from small passes that tend to normalize (refine) the structure of the earlier passes. The successive passes also affect the structures in the heat affected zone to a certain extent so that particularly in thick plates you can find alternating layers of heat affected and partially normalized material.

The tempering effect can be detected by the

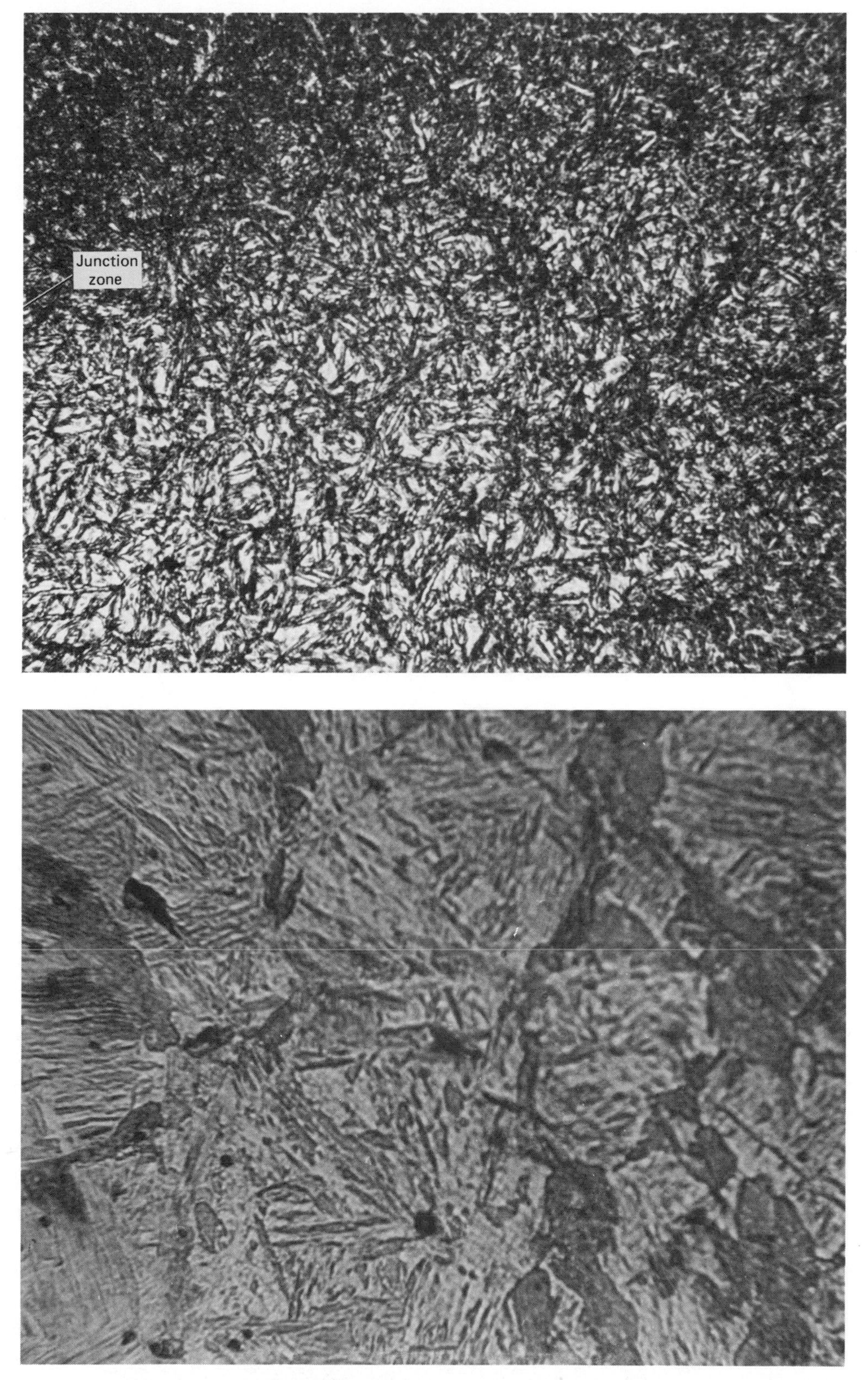

Figure 13. Junction zone of a large weld in SAE 1045 steel that cooled slowly, causing large irregular pearlite grains to form in the base metal. Finer ferrite grains may be seen in the weld zone (top) (250×).

Figure 14. Bainite, such as this microstructure shows, is often found in welds. The bainite appears as dark acicular areas in a matrix of ferrite. Bainite produces a tougher weld than martensite would (500×).

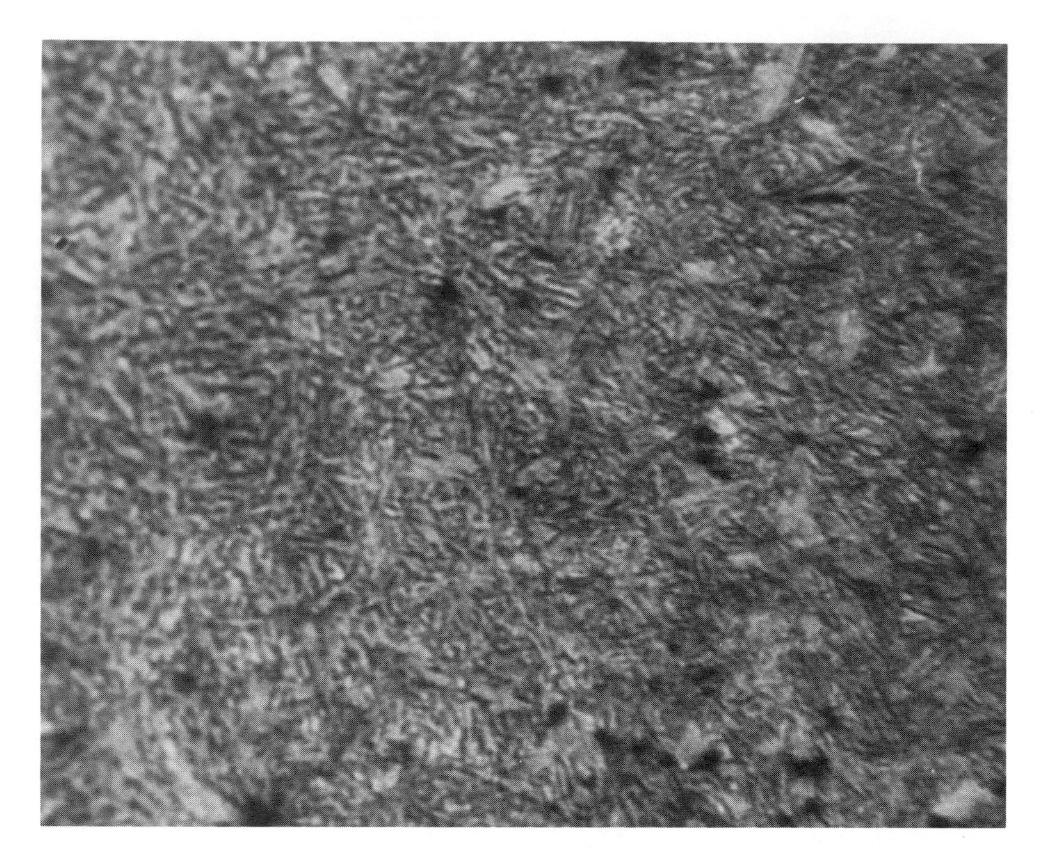

Figure 15. Photomicrograph showing light untempered martensite (500×).

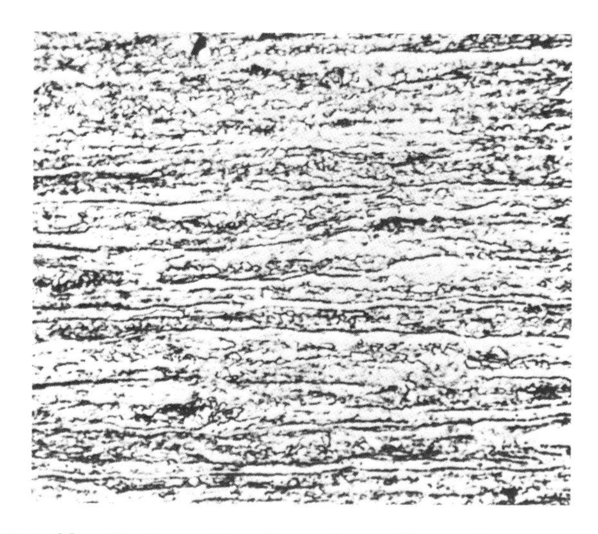

Figure 16*a*. Flattened ferrite and pearlite grains caused by cold working or weld stresses (1000×) (By permission, from *Metals Handbook,* Volume 7, Copyright American Society for Metals, 1972).

Figure 16*b*. The flattened ferrite grains are recrystallized when heated to 1025°F (552°C), but the pearlite grains are not affected by this stress relief (1000×) (By permission, from *Metals Handbook,* Volume 7, Copyright American Society for Metals, 1972).

variations in hardness from point to point in this zone (Figure 18). For example, in area *A* the first pass produces a substantial increase in hardness, depending on the carbon content and rate of cooling; subsequently, the hardness at this point falls because of higher interpass temperatures. This example is only an approximation, since many factors are involved in the hardening and normalizing of the weld zones. The same occurs in the other areas (*B*, *C*, and *D*) in the overheated zone. The mechanical properties and structures are eventually averaged out, except perhaps in the small

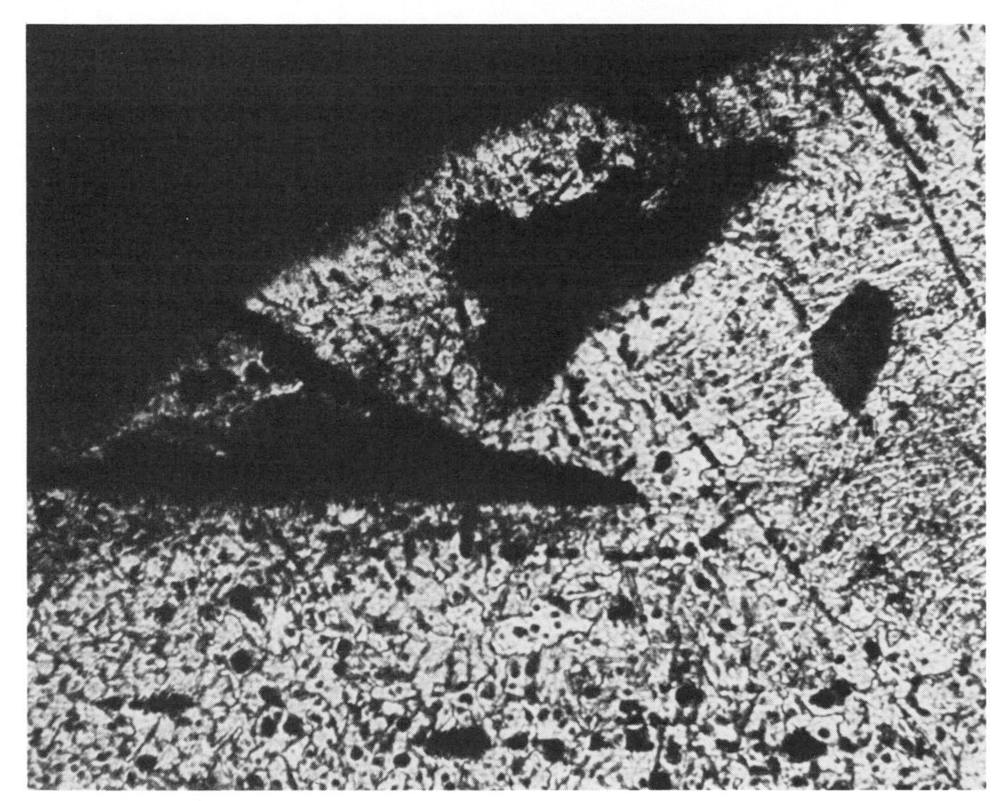

Figure 17. Micrograph showing the heat affected zone of a weld (100×). This is the toe area of the weld showing an overlap and gas pockets.

areas in the fourth pass at the corners that tend to retain some hardness, sometimes causing small toe cracks.

Medium or higher carbon and alloy steels tend to "contaminate" the weld metal with carbon or other elements. Thus the weld zone can also become brittle and develop cracks. Thick mild steel plates contain more carbon than thinner plates of the same ASTM designation. This is because thick plates cool more slowly while being rolled and therefore have a lower tensile strength and larger grain size. Carbon is added to strengthen the plate and bring it up to specifications.

The strength of the final joint does not altogether depend on the weld metal. If cold worked steel is heated to above 950°F (510°C), which is the temperature of recrystallization, the distorted ferrite grains will recrystallize and the steel in the heat affected zone will lose its cold worked strength. When the base metal is hot worked steel (soft and low strength), there will not be much change in the strength or hardness of the heat affected material because it does not recrystallize (Figure 19).

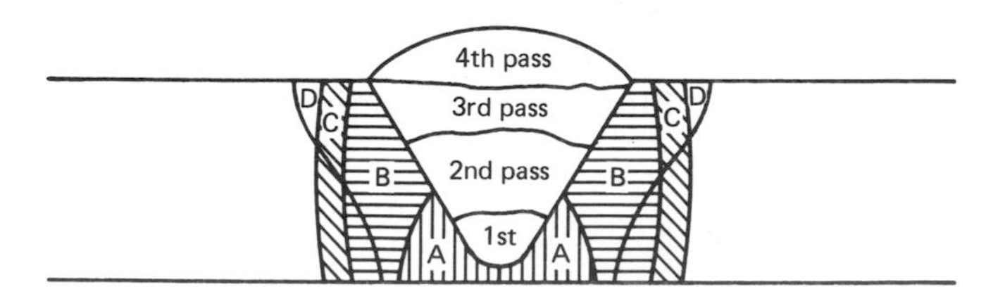

Figure 18. The effect of the deposition of successive passes on transformations in the base metal.

The Effects of Base Metal Mass

The term "heat sink" is often used when referring to the base metal mass. A large mass such as a thick plate will absorb more heat than a small mass. Rapid cooling rates are therefore associated with larger masses; this tends to develop more hardened structures. This is especially true in view of the fact that manufacturers raise the carbon content of thicker plates. Large, heavy plates will not have a heat affected zone as wide as thinner plates with the same weld size (Figures 20*a* and 20*b*). It is important to maintain a temperature for the particular steel (Figure 21). This is done by preheating, which results in a reduction of the cooling rate, thus preventing the formation of martensite.

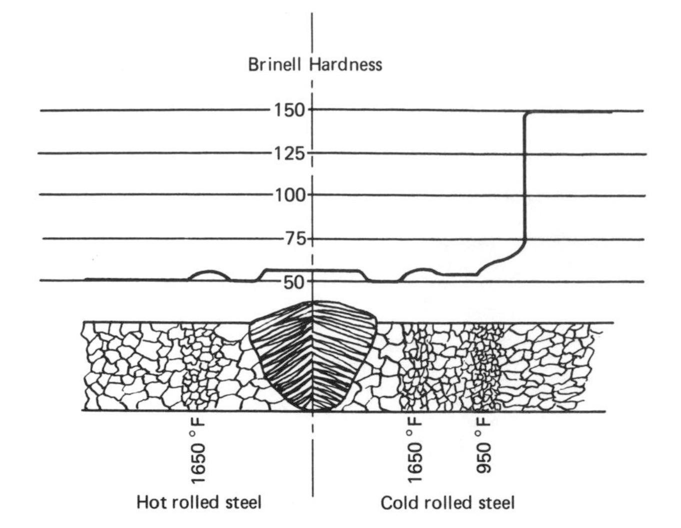

Figure 19. Single pass welds in mild steel. The section in the annealed condition shows grain refinement in the heat affected zone where the temperature reached 1650°F during welding. The graph on the left side shows that welding did not affect hardness. The section on the right shows cold worked (cold rolled) mild steel before welding. Here grain refinement is seen in the vicinity of the recrystallization zones that reached 950° and 1650°F.

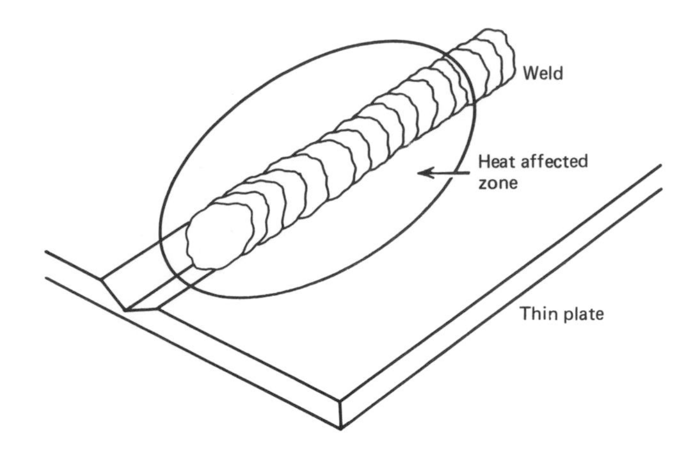

Figure 20*a*. A thin plate, when welded, has a wide heat affected zone (up to $\frac{3}{16}$ inch) because of its lower cooling rate.

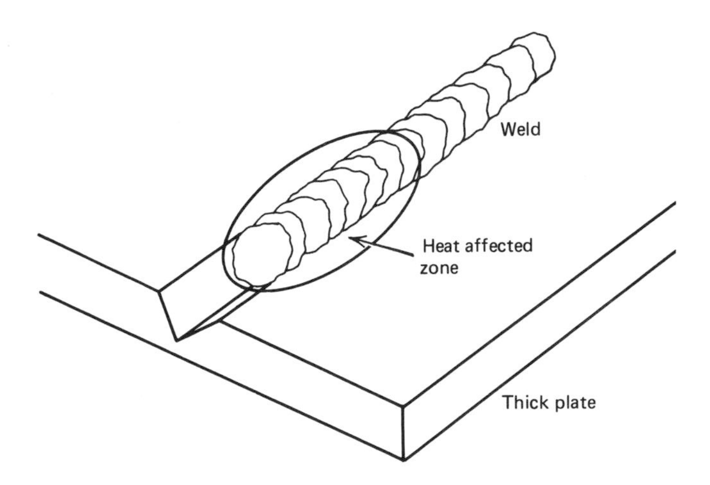

Figure 20*b*. A large heavy plate with its high cooling rate produces a narrow heat affected zone.

Preheating

Preheating requires the raising of the base metal to a specific temperature prior to making the weld. Preheat may be localized or involve the entire part, and the temperatures may range from 100 to 1200°F (38 to 649°C). Several methods used to measure the preheat temperature are chalk or crayons that melt at a specific temperature (Figure 22), a thermometer, or by placing the part in a temperature controlled furnace. The reduced cooling rate prevents the formation of martensite (Figure 23). Even with preheat, it is difficult to effect transformation to the desired pearlite. Preheating is done to prevent cold cracks, reduce distortion and residual stress, and to reduce hardness in heat affected zones. Higher temperature preheat has the disadvantage of widening the heat affected zone and enlarging the grain structure; multipass welds tend to minimize this condition, however.

Postheating

Postheat treatments for welds are used to relieve stresses, remove the effects of cold work, and in-

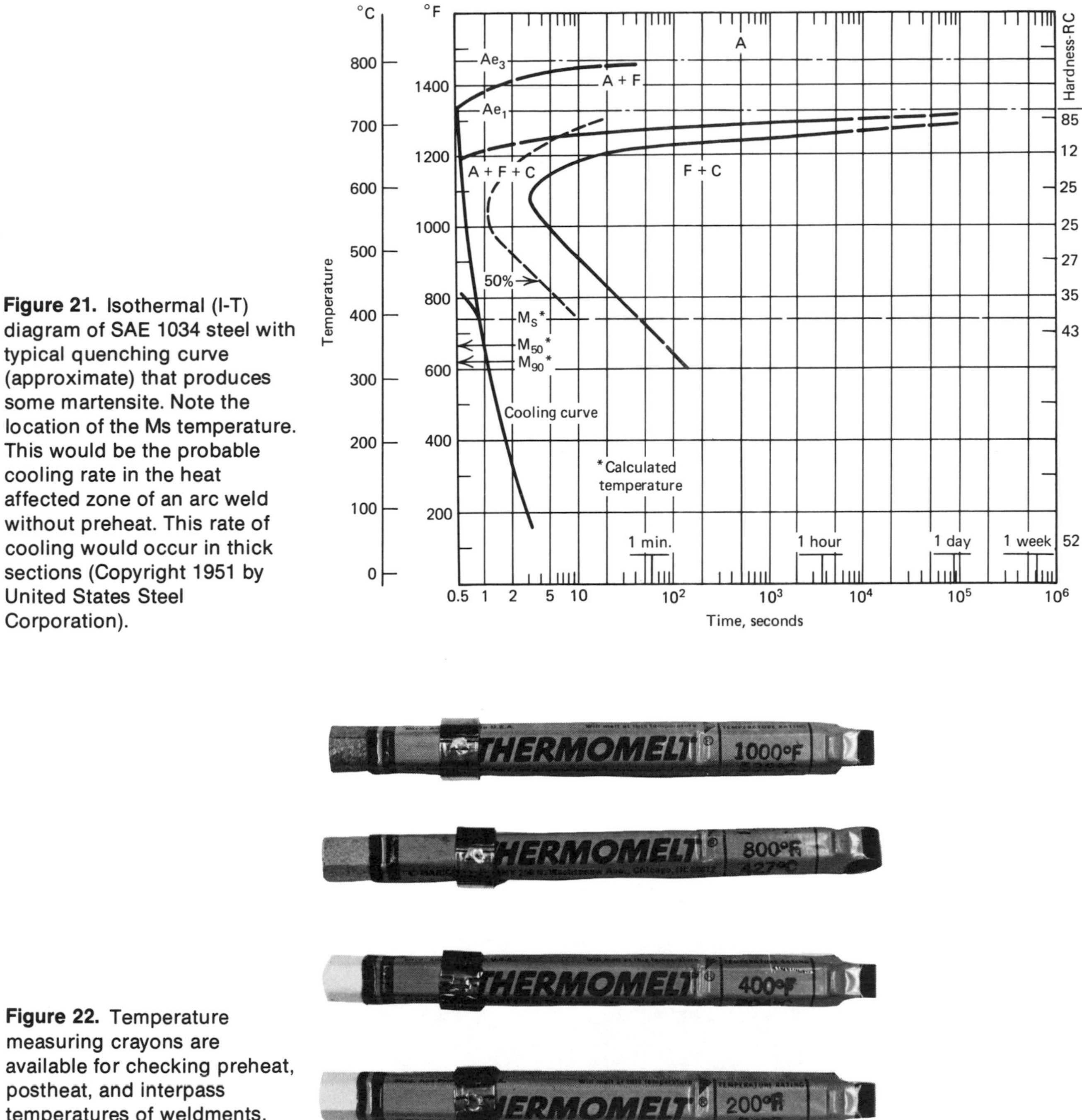

Figure 21. Isothermal (I-T) diagram of SAE 1034 steel with typical quenching curve (approximate) that produces some martensite. Note the location of the Ms temperature. This would be the probable cooling rate in the heat affected zone of an arc weld without preheat. This rate of cooling would occur in thick sections (Copyright 1951 by United States Steel Corporation).

Figure 22. Temperature measuring crayons are available for checking preheat, postheat, and interpass temperatures of weldments.

crease toughness, strength, and corrosion resistance. Some postheat treatments are stress relief anneal, normalizing, full anneal, hardening, and tempering. These can be carried out in a furnace or by localized heating. Thermal insulating blankets are sometimes used on very large weldments such as water or fuel tanks and heat is then applied to the inside.

Hardened structures may be altered and softened by the spheroidizing process. This process requires that the metal be held at the temperature of 1300°F (704°C) for a period of time. The spheroidized structures (Figure 24) will then be much softer having a lower tensile strength than the original steel. Pearlite can also be changed by this process from alternating plates to spheroids of cementite in a matrix of ferrite. An anneal above the transformation or critical temperature may also be used for completely removing hardened structures such as acicular martensite or carbides.

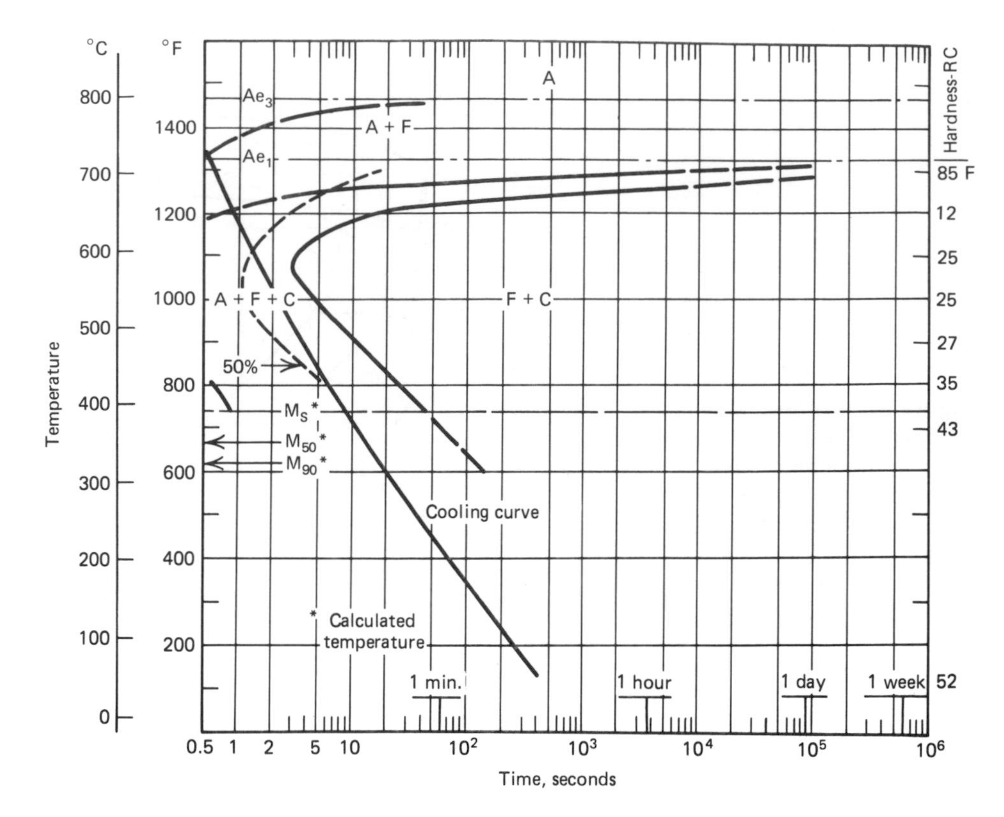

Figure 23. I-T diagram with cooling curve. Same as Figure 21 except that a cooling curve has been added, having a probable cooling rate like that found in most arc welds that have been preheated. Only about 50 percent martensite is produced (Copyright 1951 by United States Steel Corporation).

Figure 24. Spheroidized adjacent zone of a weld after stress relief at 1300°F (704°C) (500×).

Peening of Welds

Welds are often peened when either hot or cold. This can be compared to hot and cold rolling of metals in the steel mill. Peening on welds when they are cold introduces compressive stresses in the weld and base metal, which helps to relieve tensile stresses caused by welding. This strengthens the weld metal but lowers its ductility. A normal low carbon steel microstructure has regular polygonal grains and the same microstructure, after cold working, has elongated, flattened grains. These stresses may be removed by stress relief. A practice of peening while the weld is hot will minimize the formation of stressed grain structures if weld ductility is a requirement. Peening is either done by hand with a hammer, with air tools, or by shot peening.

Oxy-Acetylene Welding

In oxy-acetylene welding, a fuel gas and oxygen mixture is burned to produce an extremely hot flame. Flame temperatures with oxy-acetylene range from 5800 to 6300°F (3204 to 3482°C) When natural gas is used with oxygen, the flame temperature is a little over 5000°F (2760°C), but when acetylene is mixed with compressed air, the flame temperature is 3400°F (1871°C), too low for welding steel by fusion.

Gas welds are made on steel, brass, aluminum, and copper. Braze welds are usually made by gas welding.

In brazing and braze welding, a metal or alloy of lower melting point than either of the base metals to be joined is used. Some common brazing alloys are bronze, copper, and silver alloys. Brazing is done by heating the base metal to a temperature above the melting point of the brazing rod metal but not high enough to melt the base metal. A flux is usually applied to clean the surfaces and the melted alloy flows over the surface or between the parts by surface or capillary action. The molten metal must "wet" the alloy by diffusing into the surface to some extent to create a bond that makes a strong joint when the metal solidifies. Figure 25*a* shows a brazed overlay on carbon steel that failed to bond because it was not properly "wetted." Brazed welds can also become contaminated by base metal pick up (Figure 25*b*) if the base metal is overheated.

Gas welding with steel filler rods is done with a neutral flame (neither excess oxygen nor acetylene) with a low carbon steel filler rod. Welding rods for gas welding are similar to arc welding electrodes of the E-60XX series and also come under AWS specifications.

In the weld zone of gas welds on steel using steel filler rods, the metal is maintained in the molten state for a relatively long time with this welding process, depending on the volume of the molten pool and on the thickness of the base metal. Consequently, the grain size increases in the heat affected zone to a greater distance from the weld zone, depending on the material thickness. The structure of the weld zone in gas welds is very

Figure 25*a*. This bronze overlay on a steel shaft shows a lack of bond in some places. The steel base metal was not adequately preheated to ensure "wetting" of the steel with bronze and sufficient flux was not used.

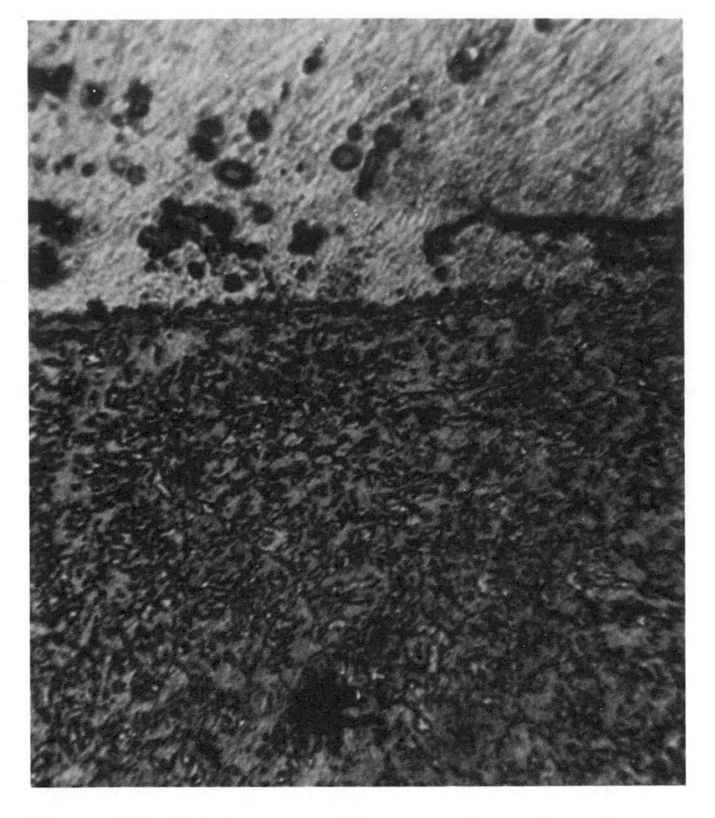

Figure 25*b*. This bronze weld was overheated causing incipient melting of the steel surface and subsequent diffusion of the displaced base metal into the molten bronze weld (100×).

coarse and irregular (Figure 26). In the junction zone, which consists of a mixture of weld metal and base metal, the structure is still coarse, but with the slow cooling rate that is typical of gas welding, hard structures such as martensite are not likely to form.

The very coarse grains formed in the base metal, when it is overheated, tend to produce a brittle structure with cleavage planes. This is sometimes called the Widmanstatten structure by metallurgists. Failure by cleavage is usually transgranular (through the grain) as contrasted to shear failure, which is often along grain boundaries (intergranular) (Figures 27*a* and 27*b*). Cleavage is always a brittle failure occuring suddenly, but shear failures show some ductility. An example of this problem is sometimes seen when low or medium carbon tubing is overheated when welded or brazed. The failure occurs, not in the weld, but beside it, sometimes as much as $\frac{1}{2}$ inch away (Figure 28). The thicker the base metal, the more likely the problem will occur. Alloying elements such as manganese, chromium, and molybdenum tend to promote this cleavage pattern in the grains.

Figure 29 shows the difference in the distance

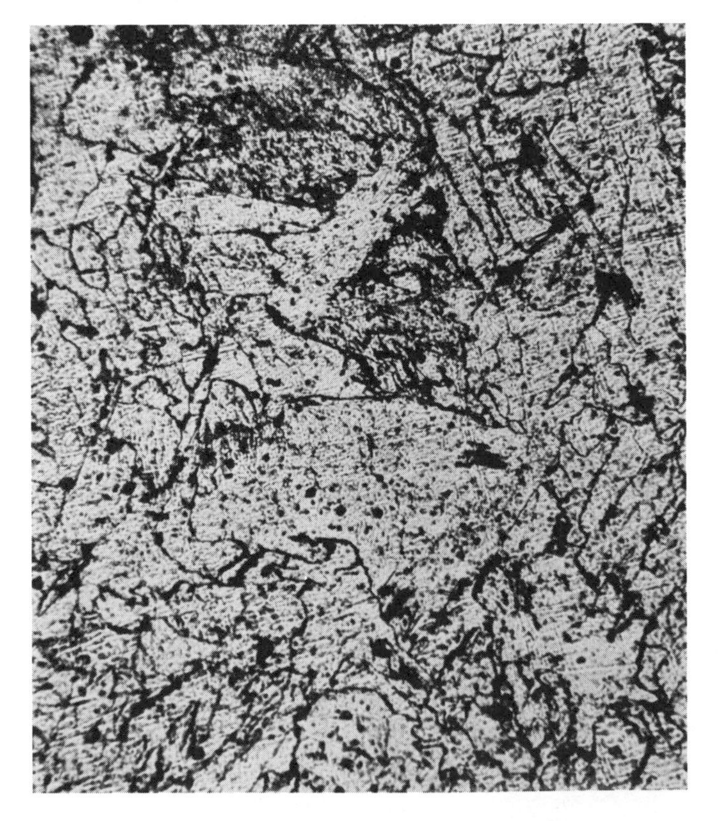

Figure 26. Structure of the weld metal in an oxy-acetylene weld showing a coarse irregular structure (500×).

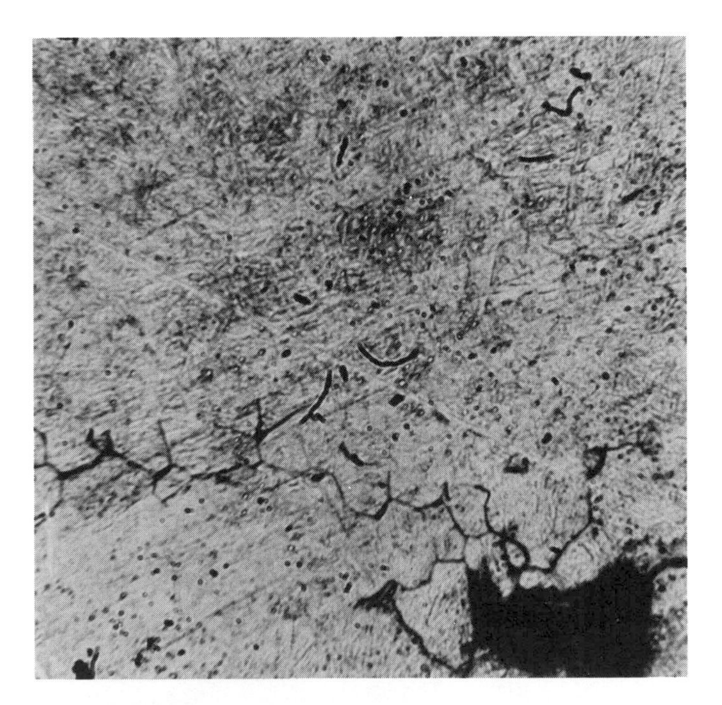

Figure 27. An intergranular weld failure, possibly developed at high temperature while the weld was solidifying. The black area at the bottom is a gas pocket probably from hydrogen entrapment (500×).

from the weld axis through the heat affected grains from the arc welding and that of the oxy-acetylene welding process. This grain growth is taking place in the base material that was in the normalized condition prior to welding.

Plasma-Arc Welding and Cutting

In arc welding, the electric current flows between two electrodes through an ionized column of gas called a plasma at a temperature about 6500°F (3593°C). A new method of welding called **plasma-arc,** or plasma torch, welding is sometimes used in place of the gas tungsten-arc (TIG) process (Figure 30). The heat in plasma-arc welding originates in an arc but not in the same way as in an ordinary arc where it is diffused (spread out). Instead, it is forced through a constricted orifice with a plasma gas that is supplemented with an ordinary shielding gas.

There are two systems: the **transferred-arc** and the **nontransferred system.** The workpiece is part of the circuit in the transferred-arc system as it is in the ordinary arc welding system (Figure 31). The constricting nozzle surrounding the electrode in the nontransferred system acts as an electrical terminal with the arc forming between it and the electrode end. The plasma gas carries the heat to the workpiece. Gas temperatures are from 10,000 to 60,000°F (5538 to 33,315°C) and theoretical temperatures of 200,000°F (111,093°C) are possible.

The advantages of plasma-arc welding over the gas tungsten-arc process are higher welding speeds, narrower welds since there is greater energy concentration, and improved arc stability. The extremely high temperatures generated make possible the welding of high temperature metals and the melting of refractory materials. Plasma-arc heating and welding are most used for the more exotic metals and manufacturing processes.

Plasma torch cutting has the advantage of making smooth cuts, relatively free from contamination with only a shallow melted zone and less

Figure 28. This tubular motorcycle frame failed in the heat affected zone of an oxy-acetylene welded part due to extreme grain growth away from the weld.

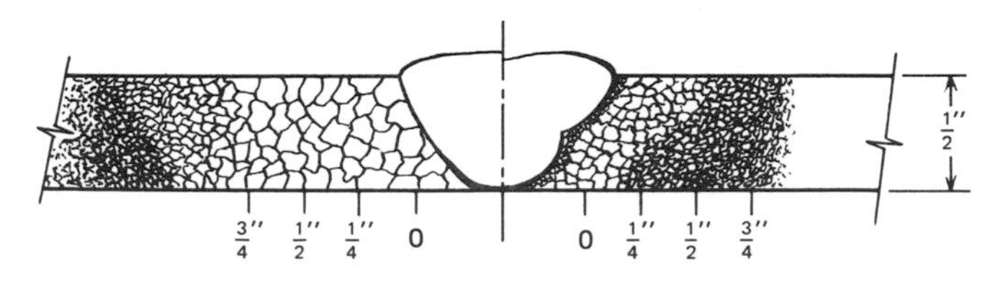

Figure 29. Comparison of oxy-acetylene welding (left) and arc welding (right).

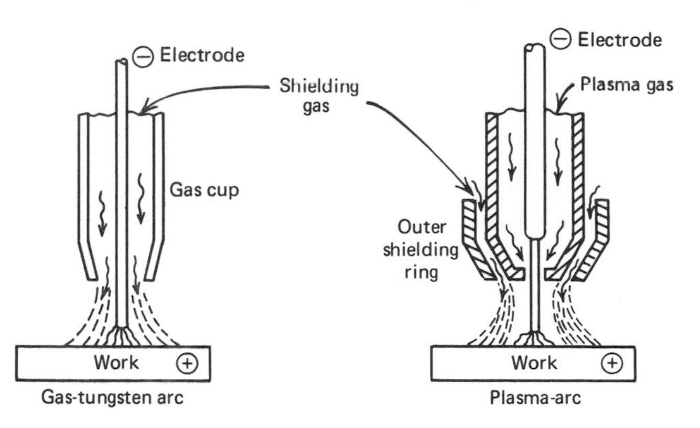

Figure 30. Comparison of plasma-arc and gas tungsten arc welding torches.

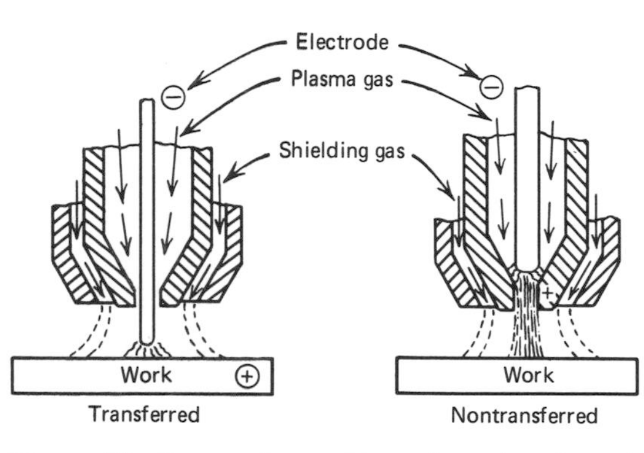

Figure 31. Comparison of transferred and nontransferred plasma-arc torches.

metallurgical effects than with oxy-acetylene cutting. Unlike oxy-acetylene cutting, a plasma torch can easily cut nonferrous and high temperature melting alloys. Aluminum, for example, may be cut with a plasma arc in thicknesses ranging from $\frac{1}{8}$ to 5 inches. Stainless steel cannot be cut with the oxy-acetylene torch unless a ferrous powder is fed into the cut, but it is easily cut with the plasma-arc torch.

Filler Metals

Many of the welding processes involve a deposition of filler metal. In some cases, a consumable electrode is used. The electrode is covered with a flux. In other cases, a wire is used with an inert gas or a flux core that protects or shields the weld from the surrounding atmosphere. Other types of welding make use of a rod or wire that is melted in the joint by a heat source such as an acetylene torch or by tungsten inert gas (TIG). Brazing and soldering, for instance, make use of the filler metals in this way.

Both submerged arc welding and electroslag welding make use of a flux under which the bare rod or bare wire makes the weld. Different types of wires or electrodes used for welding are classified in the AWS-ASTM system of classification for filler metals. In this classification system, the initial letters designate the basic process of deposition. The letter E stands for electrode, R for welding rod, and B for brazing filler metal. The combinations of ER and RB indicate suitability for either of the processes that are used. In all of the types of welding processes, the following is a list of the filler metals that could be used.

1. Steel with little or no carbon in the composition (0.08 to 0.15 percent carbon)
2. Carbon steel
3. Low alloy steel
4. High alloy or stainless steels and manganese steels
5. Nickel and nickel base alloys
6. Copper base and copper alloys, tin base and tin solders, cobalt base alloys

Some electrodes are tubular and contain a granular metal or compound. An example is the electrode used for depositing tungsten carbide. Table 2 gives the designation and composition of some covered electrodes used in carbon and low alloy steel. Table 3 shows the mechanical properties of some electrodes.

Fluxes and Slags

When molten iron is exposed to oxygen, the oxygen atoms dissolve in the surface and penetrate by diffusion to some degree throughout the melt. Iron oxide (FeO) begins to form at the surface and in time the entire melt will oxidize. If the melt is suddenly cooled, the solid metal state retains the oxygen, which causes porosity. Nitrogen and hydrogen also create porosity and cracking in welds by becoming entrapped in the metal as it solidifies.

In the early days of metal arc welding, bare filler wires were used for electrodes, but these produced a poor weld because the arc passed through the atmosphere and entrapped oxygen and nitrogen in the weld, thus producing a very porous and brittle weld. To prevent this contamination from the atmosphere, some form of shielding is necessary. Some of the many methods of accomplishing this are:

1. Fluxes
2. Slags
3. Gases (controlled atmospheres)
4. Vacuum

Fluxes are primarily used in welding processes to prevent oxidation of the metal by floating as a liquid on the surface of the metal. The flux chemically or physically combines with the surface oxide to remove it.

Shielding slags have a slightly different purpose as they prevent oxidation of the molten metal and often combine with the melt to produce desired alloys or compounds in the weld metal. Some slags contain deoxidizers such as aluminum to remove trapped oxygen in the melt.

The use of gases for shielding or a vacuum in welding require no slag to protect the weld surface. Shielding gases such as argon or helium are inert and are made to surround the arc and the molten pool to protect it from the atmosphere. Some electrodes have a covering that shields the arc. As the covering burns off, it produces a gas that protects the melt from the atmosphere, and a viscous slag forms and solidifies to protect the weld nugget

Table 2 AWS A5.1-69 and A5.5-69 Designations for Manual Electrodes

a. The prefix "E" designates arc-welding electrode.

b. The first two digits of four-digit numbers and the first three digits of five-digit numbers indicate minimum tensile strength:
- E60XX .... 60,000 psi minimum tensile strength
- E70XX .... 70,000 psi minimum tensile strength
- E110XX ....110,000 psi minimum tensile strength

c. The next-to-last digit indicates position:
- EXX1X ....All positions
- EXX2X ....Flat position and horizontal fillets

d. The suffix (Example: EXXXX-A1) indicates the approximate alloy in the weld deposit:
- — A1 ....0.5% Mo
- — B1 ....0.5% Cr, 0.5% Mo
- — B2 ....1.25% Cr, 0.5% Mo
- — B3 ....2.25% Cr, 1% Mo
- — B4 ....2% Cr, 0.5% Mo
- — B5 ....0.5% Cr, 1% Mo
- — C1 ....2.5% Ni
- — C2 ....3.25% Ni
- — C3 ....1% Ni, 0.35% Mo, 0.15% Cr
- — D1 and D2 ....0.25-0.45% Mo, 1.75% Mn
- — G ....0.5% min. Ni, 0.3% min. Cr, 0.2% min. Mo, 0.1% min. V, 1% min. Mn (only one element required)

Source. The Procedure Handbook of Arc Welding, Twelfth Edition, The Lincoln Electric Company, 1973.

Table 3 Typical Mechanical Properties of Mild-Steel Deposited Weld Metal

	Condition							
	As-Welded				Stress-Relieved at 1150°F			
Electrode Classification	Tensile Strength (psi)	Yield Strength (psi)	Elong. in 2 in. (%)	Impact* (ft-lb)	Tensile Strength (psi)	Yield Strength (psi)	Elong. in 2 in. (%)	Impact* (ft-lb)
E6010	69,000	60,000	26	55(1)	65,000	51,000	32	75
E6011	70,000	63,000	25	50(1)	65,000	51,000	30	90
E6012	72,000	64,000	21	43	71,000	62,000	23	47
E6013	74,000	62,000	24	55	74,000	58,000	28	
E6020	67,000	57,000	27	50				
E6027	66,000	58,000	28	40(1)	66,000	57,000	30	80
E7014	73,000	67,000	24	55	73,000	65,000	26	48
E7015	75,000	68,000	27	90				
E7016	75,000	68,000	27	90	71,000	60,000	32	120
E7018	74,000	65,000	29	80(1)	72,000	58,000	31	120
E7024	86,000	78,000	23	38	80,000	73,000	27	38
E7028	85,000	78,000	26	26(2)	81,000	73,000	26	85

*Charpy V-notch at 70 °F. except where noted.

(1) Charpy V-notch at –20 °F.

(2) Charpy V-notch at 0 °F.

Source. The Procedure Handbook of Arc Welding, Twelfth Edition, The Lincoln Electric Company, 1973.

from contamination. Covered electrodes for just about every type of welding are available for the shielded metal arc process. Many different elements or materials are used in various coverings for specific purposes. Cellulose, limestone, asbestos, iron powder, sodium silicate, and many others are used for a rod covering. See Table 4 for typical composition of rod coverings.

By contrast, separate fluxes are sometimes used for gas welding. These fluxes are either applied to the joint being welded or applied to the rod by the operator while the rod is in a heated condition. In the gas metal arc, also termed MIG (Metal Inert Gas), welding process, a wire is used for the electrode and it is fed continuously while the welding is being performed. Some types use a flux in the core of the wire to help the welding process (in some cases with a gas), while other types use bare solid wire and an inert gas such as argon or helium and, to a large degree, carbon dioxide is used. Carbon dioxide is not an inert gas, and it produces a small amount of slag.

Shrinkage and Distortion

Whenever heat is applied to steel, it will expand; when cooled, it contracts (Figure 32). Uneven heating and cooling, or heating and cooling in a localized area, can produce distortions in the base

Table 4 Typical Functions and Composition Ranges of Constituents of Coverings on Mild Steel Arc Welding Electrodes

	Function of constituent	
Constituent of Covering	**Primary**	**Secondary**
Cellulose	Shielding gas	—
Calcium carbonate	Shielding gas	Fluxing agent
Fluorspar	Slag former	Fluxing agent
Dolomite	Shielding gas	Fluxing agent
Titanium dioxide (rutile)	Slag former	Arc stabilizer
Potassium titanate	Arc stabilizer	Slag former
Feldspar	Slag former	Stabilizer
Mica	Extrusion	Stabilizer
Clay	Extrusion	Slag former
Silica	Slag former	—
Asbestos	Slag former	Extrusion
Manganese oxide	Slag former	Alloying
Iron oxide	Slag former	—
Iron powder	Deposition rate	Contact welding
Ferrosilicon	Deoxidizer	—
Ferromanganese	Alloying	Deoxidizer
Sodium silicate	Binder	Fluxing agent
Potassium silicate	Arc stabilizer	Binder

(a) Used (in place of constituent on line above) in E6011 and E6013 electrodes to permit welding with alternating current

Source. By permission, from *Metals Handbook,* Volume 6, *Welding and Brazing,* Eighth Edition, Copyright American Society for Metals, 1971.

Table 4 (continued)

	Composition Range, %, in Covering on Electrode of Class:								
Constituent of Covering	**E6010, E6011**	**E6012, E6013**	**E6020**	**E6027**	**E7014**	**E7016**	**E7018**	**E7024**	**E7028**
Cellulose	25 to 40	2 to 12	1 to 5	0 to 5	2 to 6	—	—	1 to 5	—
Calcium carbonate	—	0 to 5	0 to 5	0 to 5	0 to 5	15 to 30	15 to 30	0 to 5	0 to 5
Fluorspar	—	—	—	—	—	15 to 30	15 to 30	—	5 to 10
Dolomite	—	—	—	—	—	—	—	—	5 to 10
Titanium dioxide (rutile)	10 to 20	30 to 55	0 to 5	0 to 5	20 to 35	15 to 30	0 to 5	20 to 35	10 to 20
Potassium titanate	(a)	(a)	—	—	—	—	0 to 5	—	0 to 5
Feldspar	—	0 to 20	5 to 20	0 to 5	0 to 5	0 to 5	0 to 5	—	0 to 5
Mica	—	0 to 15	0 to 10	—	0 to 5	—	—	0 to 5	—
Clay	—	0 to 10	0 to 5	0 to 5	0 to 5	—	—	—	—
Silica	—	—	5 to 20	—	—	—	—	—	—
Asbestos	10 to 20	—	—	—	—	—	—	—	—
Manganese oxide	—	—	0 to 20	0 to 15	—	—	—	—	—
Iron oxide	—	—	15 to 45	5 to 20	—	—	—	—	—
Iron powder	—	—	—	40 to 55	25 to 40	—	25 to 40	40 to 55	40 to 55
Ferrosilicon	—	—	0 to 5	0 to 10	0 to 5	5 to 10	5 to 10	0 to 5	2 to 6
Ferromanganese	5 to 10	5 to 10	5 to 20	5 to 15	5 to 10	2 to 6	2 to 6	5 to 10	2 to 6
Sodium silicate	20 to 30	5 to 10	5 to 15	5 to 10	0 to 10	0 to 5	0 to 5	0 to 10	0 to 5
Potassium silicate	(a)	5 to 15a	0 to 5	0 to 5	5 to 10	5 to 10	5 to 10	0 to 10	0 to 5

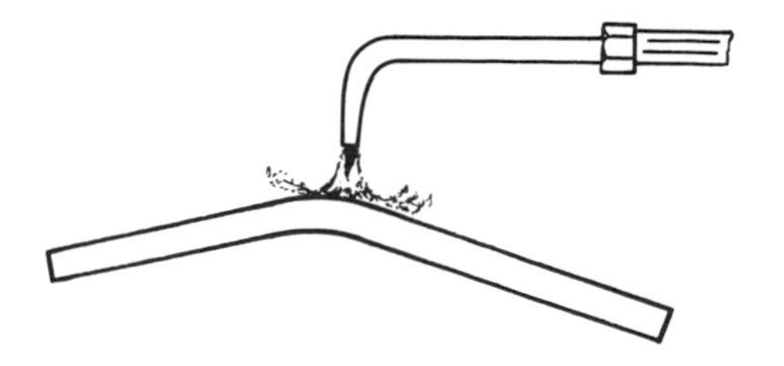

Figure 32. When heat is applied to one side of a bar of metal, it will bend as shown and return to its original position when cooled.

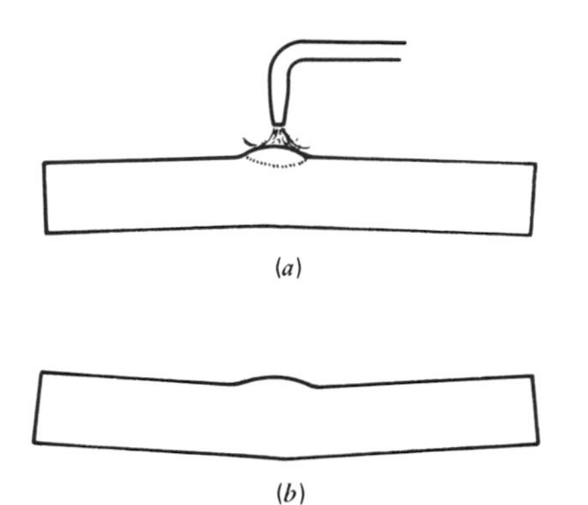

Figure 33. When heat is applied to a small spot so that it becomes red hot on one side of a bar *(a)*, the heated spot will upset because of the restraint of the colder metal surrounding it. When it has cooled, it will be bent in the opposite direction as shown at *(b)*.

metal (Figure 33). Steel expands and contracts at the rate of approximately 0.0000065 inch per degree Fahrenheit change in temperature per unit length. When steel cools to normal temperatures from a molten condition, this amount of contraction is sufficient to cause a great strain in rigidly held members and in the weld zone, but if they are free to move, considerable distortion can take place in the base metal.

It is also true that iron undergoes shrinkage or a reduction in volume of approximately 1.5 percent when it solidifies. This solidification in the weld zone itself induces a considerable amount of distortion upon the weldment as well as the contraction following solidification while cooling.

Welds on one side of a butt-joint, for example, such as is seen in Figure 34, will cause the members to be drawn to one side and tilt toward the direction of welding. If both sides are welded alternately, the stress is equalized so that both members remain in their correct position.

If there is an undercut or notch at the edge of the weld, a stress concentration is produced that increases the stress level at that point since it is an abrupt change in section. A notch such as this, which is called a stress raiser, can initiate failure through cracking or fatigue (Figure 35).

Welding that is done on one side of a plate, for example, even if it is relatively light welding, will cause a warping or dishing of the plate as shown in Figure 36. In both cases, restraints can be put upon the base metal parts so that they will not distort; but when this is done, considerable internal strain is placed upon the weld and the heat affected zone of the weld. Sometimes a result of this high stress is failure by cracking. This is a common problem

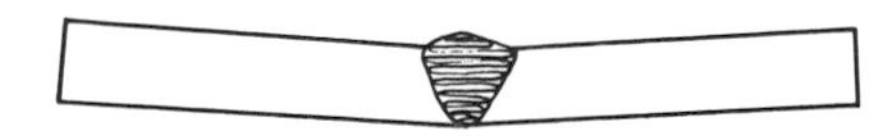

Figure 34. Welds, as they shrink, tend to move the welded parts out of square. This tendency of welds to shrink and distort weldments should always be kept in mind when making welds.

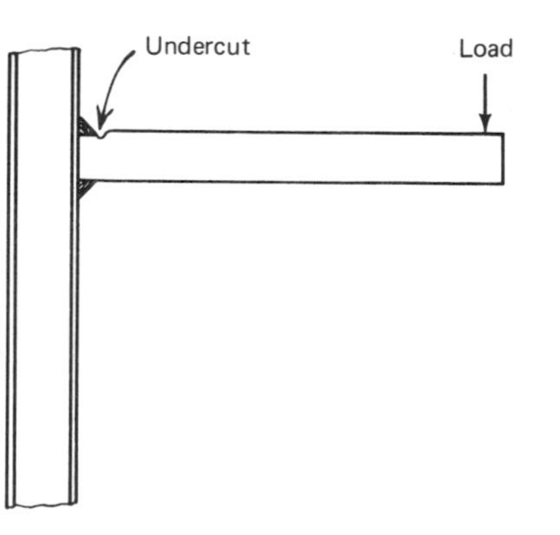

Figure 35. Weld undercuts, such as this one, create a stress concentration in the notch. Fatigue failures are often initiated by these stress raisers.

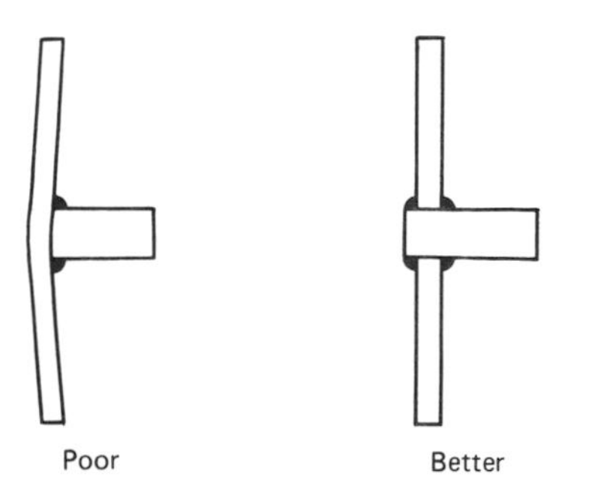

Figure 36. A piece of solid round stock is welded to a steel plate. The weld causes the steel plate to dish. If the round stock were extended through the part and welded on both sides, the dishing problem will be minimized.

and it is often overcome by the use of shot peening or peening with air hammers or slag hammers. This is done preferably when the weld is hot so that it is a form of hot working of the steel, although cold peening is more effective in relieving the primary stress. Of course, stress relief heat treatment would be even more effective.

Distortion of weldments through weld contractions can be offset to some extent by prepositioning the weldment on a bias that takes into consideration its probable movement or distortion so that when it assumes its final position after being drawn by the weld, it will be in the desired alignment (Figure 37). Another example of this is when two plates are butt welded together and the weld is started from one end, they will be drawn together when the weld is finally completed (Figure 38). This method eliminates any restraint on the base metal and allows the weld to contract normally with fewer internal stresses. However, it is not always easy to predict the outcome of the final position of the weldment when using this method.

Rolled metals possess a fibrous quality, called **anistropy,** in the direction of rolling. Compared to hot rolled steels, cold rolled or drawn steels are even stronger in the direction of elongation or rolling than they are in the traverse (crosswise) axis. Since metals have a tendency to split in the direction of rolling, the welder should be aware of the direction of weld stresses when making a weld so that weld stresses will be lengthwise and not crosswise (Figure 39).

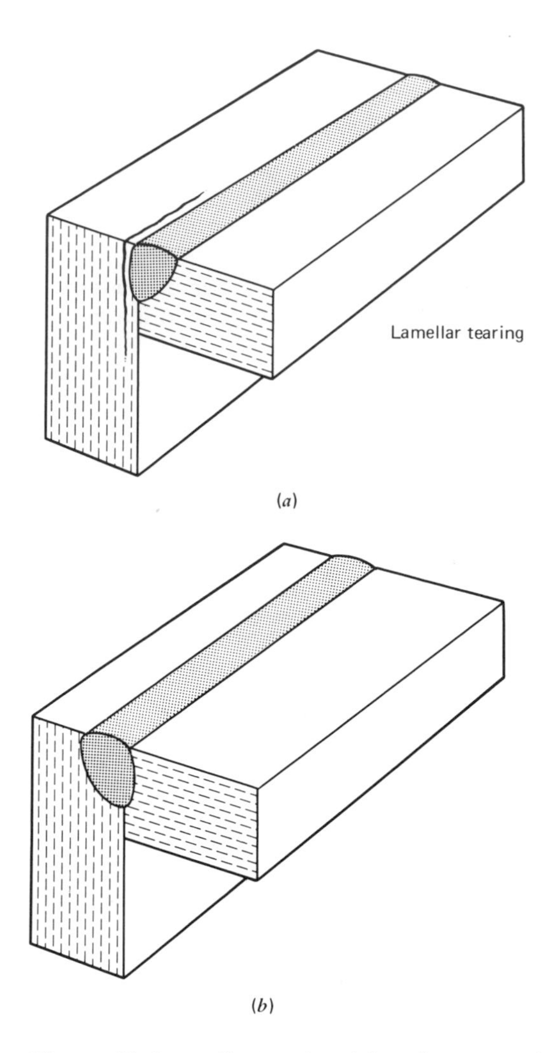

Figure 39. Lamellar tearing *(a)* and a suggested solution *(b)* (The Lincoln Electric Company).

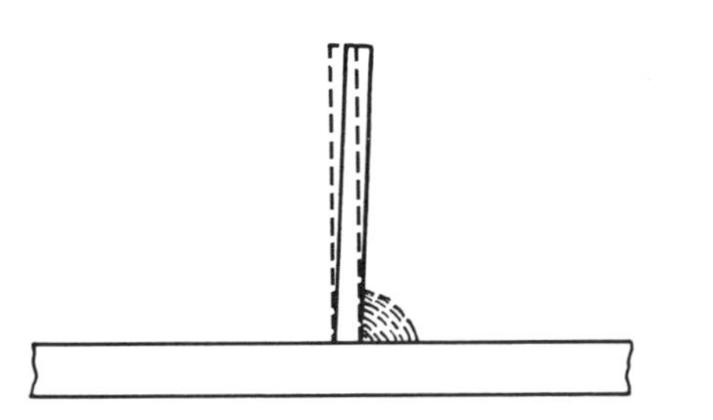

Figure 37. Biasing a part to be welded to allow for weld shrinkage will help to keep the weldment aligned.

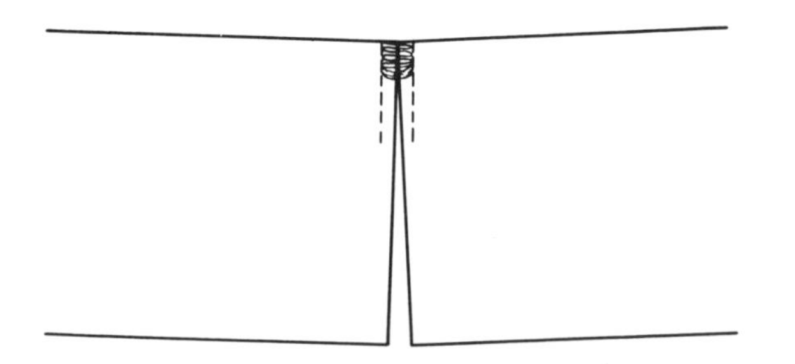

Figure 38. Plates can sometimes be separated slightly on one end to allow for weld shrinkage and thus avoid restraint.

Testing of Welds

The quality assurance of good welds can be promoted by a weld inspection program and by a welder qualification program. The first program is used to assure the company that good welds are being made on their products. The second program is to qualify the welders for specific assignments or projects. Either program involves the following:

1 Visual inspection of the weld.

2 Nondestructive testing such as radiographic,

magnetic particle, ultrasonic, and liquid penetrant.

3 Standard mechanical tests such as bending or tensile tests. These are usually destructive tests.

4 Laboratory tests such as hardness, micrographic, and macrographic.

5 Field tests. Simple methods of determining weldability of base metals with particular electrodes in field conditions where no other equipment is available for testing.

Visual Inspection This requires the good judgment of the welder to interpret the condition of the weld skillfully. Some of the defects he or she might detect in fillet joints and butt joints are an undersize weld, surface porosity, an undercut, and cracks (Figure 40).

Nondestructive Testing Internal porosity can only be detected by radiographic tests or destructive tests such as breaking the part in a bend test. Lack of penetration in a weld can be detected by destructive, ultrasonic, or radiographic testing. Slag inclusions may also be detected by the same methods of testing.

Surface cracks may be detected by magnetic particle, dye penetrant, or a fluorescent dye penetrant type of testing procedures. Metallurgical tests such as macroetching and micrograph or microscopic inspection will be discussed in Chapter 17 "Metallurgy of Welds: Alloy Steels."

Standardized Mechanical Tests Standardized tests and tensile specimens may be found in publications such as *AWS Standard Qualifications Procedure, ASME Boiler and Pressure Vessel Code,* and *AWS Welding Handbook.* Some of the basic tests that are used are tensile tests to determine proportional limit, yield points, modulus of elasticity, elongation, and reduction of area. Hardness tests are often made on the cross section of the weld (Figure 41). Bend tests elongate the outer fibers of the weld to reveal the occurrence of cracks (Figure 42). Notch bar, impact tests, or fatigue tests are used to determine endurance limits or the fatigue strength of the weld or base metal. Torsion tests are used to determine yield strength and ultimate shear stress, and creep tests are used on weldments under stress to determine the amount of permanent set in a given time for a given temperature.

Details of the AWS Procedure Qualification Tests may be found in Figures 43 to 45. Complete specifications for welding qualifications may be found in welding data handbooks.

Laboratory Tests Hardness testing and microscopic or macroscopic evaluation of welds are often performed on weld sections in the laboratory. Samples of the weld must be taken for these tests, making them destructive types of tests. These samples are prepared for microscopic inspection and detection of carbon content, inclusions, porosity, and cracking can then be made.

Figure 40*a*. Weld having porosity and cracks plus slag inclusions. This is a very poor weld.

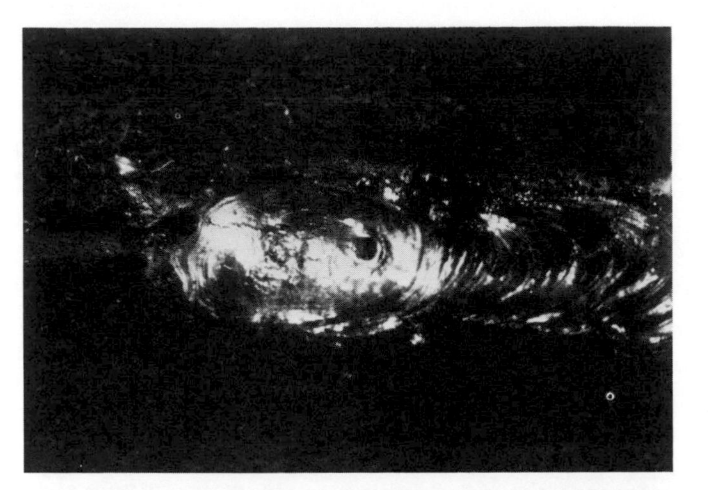

Figure 40*b*. Crater crack at end of pass is a hot metal crack caused by removing the electrode too quickly.

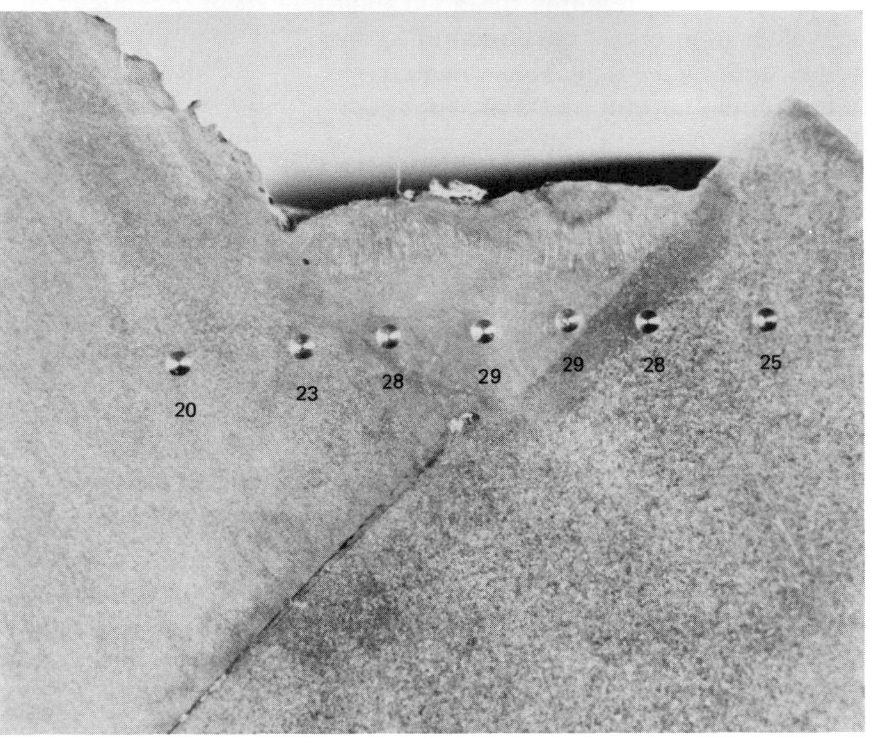

Figure 41. The hardness of various areas on the weld on Rockwell C scale numbers.

Figure 42. In this bend test the weld (bottom) failed along the junction zone of the weld. The top weld began to fail in the center of the weld.

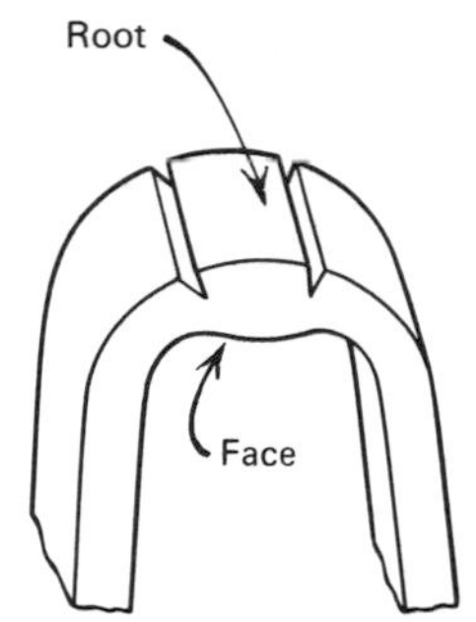

Figure 43. If weld reinforcement is not removed, stretching is concentrated in two places and failure results (The Lincoln Electric Company).

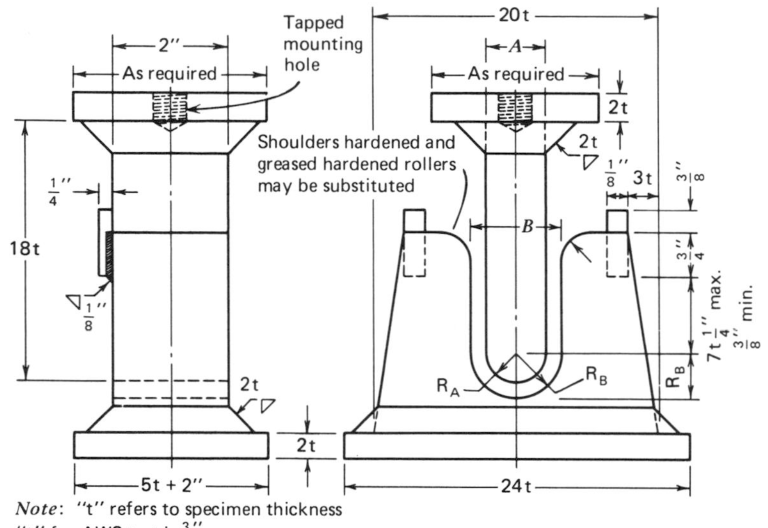

Note: "t" refers to specimen thickness
"t" for AWS test is $\frac{3}{8}$"
"t" for API Std. 1104 is tabulated wall thickness of pipe

(*a*)

Jig Dimensions	AWS TEST For Mild Steel Mimimum Yield Strength–psi			API Std. 1104 For All Pipe Grades
	50,000 and under	55-90,000	90,000 and over	
Radius of plunger R_A	$\frac{3}{4}$	1	$1\frac{1}{4}$	$1\frac{3}{4}$
Radius of die R_B	$1\frac{3}{16}$	$1\frac{7}{16}$	$1\frac{11}{16}$	$2\frac{5}{16}$

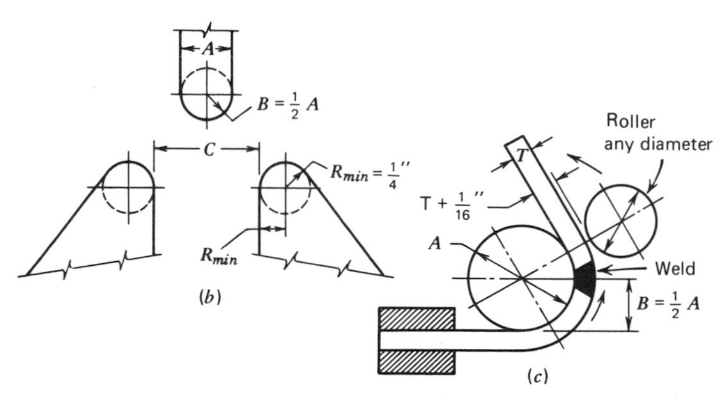

A	B	C	Yield Point–psi*	A	B
$1\frac{1}{2}$	$\frac{3}{4}$	$2\frac{3}{8}$	50,000 and under	$1\frac{1}{2}$	$\frac{3}{4}$
2	1	$2\frac{7}{8}$	55,000 to 90,000	2	
$2\frac{1}{2}$	$1\frac{1}{4}$	$3\frac{3}{8}$	90,000 and over	$2\frac{1}{2}$	$1\frac{1}{4}$

*Minimum specified

Figure 44. *(a)* Jig for guided bend test used to qualify operators for work done under AWS and API specifications. *(b)* Alternate roller-equipped test jig for bottom ejection. *(c)* Alternate wrap-around test jig (The Lincoln Electric Company).

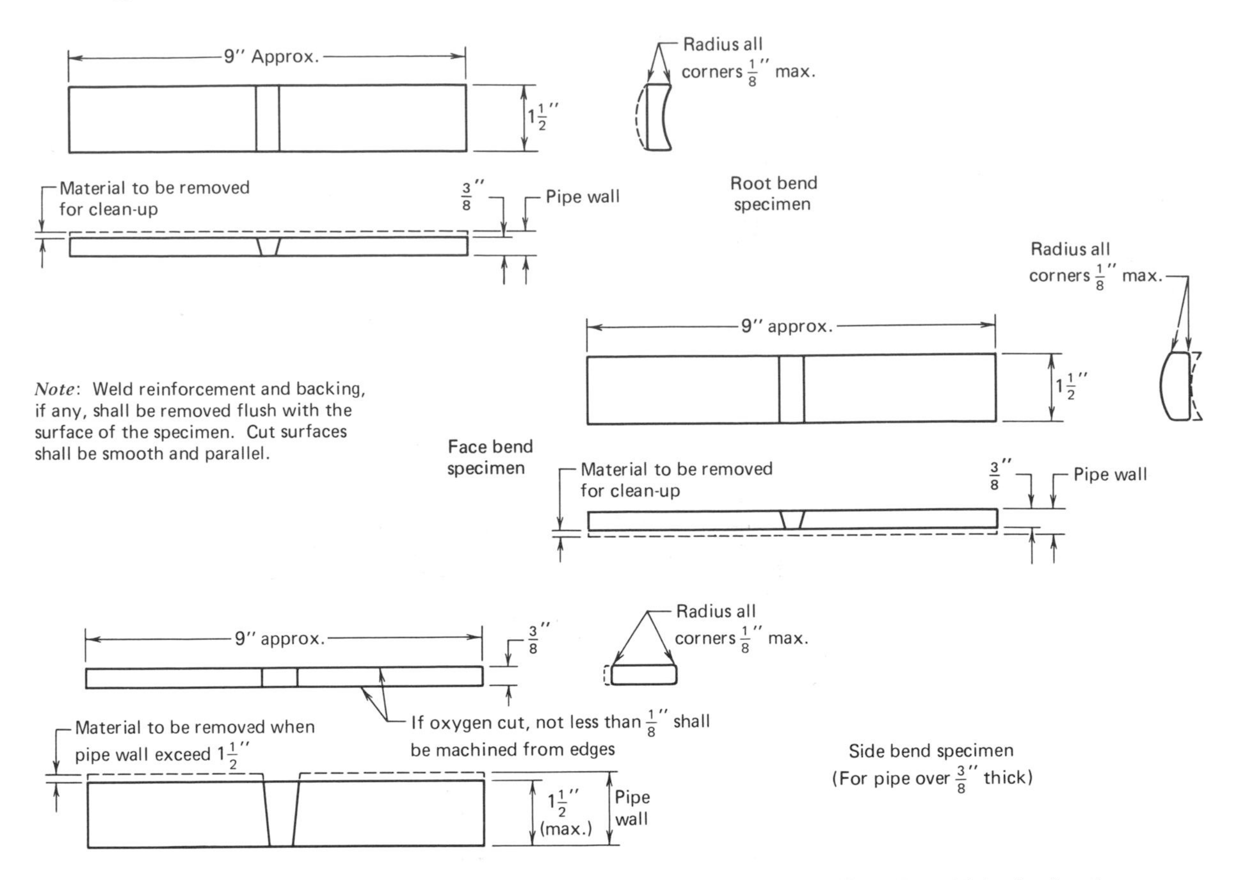

Figure 45. Method of preparing test specimen from 5 and 8 inch pipe for procedure qualification tests (The Lincoln Electric Company).

Field Tests A simple field test, for example, could be the use of spark testing techniques to determine the carbon content of a base metal since the carbon would greatly affect the weldability of the part, especially if preheating and postheating methods were not practicable. Hardness may be determined to some degree with a file test or a portable hardness tester.

Another very simple field test that welders sometimes use is a fracture test (clip test) in which a small piece of steel is welded at a right angle to the base metal with a fillet weld and then deliberately broken off using a hammer (Figures 46*a* through 46*d*). If the weld itself fractured, it would indicate that a good bond was maintained on the base metal, but if the weld remained intact and a section of the base metal were torn out, it would indicate that the base metal had too much carbon or alloy content, causing it to harden sufficiently to break in a brittle fracture. While this very simple test might be useful for some field conditions, it is

Figure 46*a*. Clip test for determining weldability (Lane Community College).

Figure 46*b*. Clip being broken off with hammer (Lane Community College).

Figure 46*c*. If the base metal is torn out and the weld remains intact, the base material is not weldable without preheat and postheat operations (Lane Community College).

Figure 46*d*. If the weld breaks, but the bond is good in the base metal, it can be considered weldable without preheat or postheat (Lane Community College).

not satisfactory where many stringent requirements are placed upon welds such as in building construction and steel pipeline operations. Ultrasonic and radiographic testing are methods often used in construction and pipeline welds.

Hot Metal Safety

Oxy-acetylene torches are often used for cutting shapes, circles, and plates in welding and machine shops. Safety when using them requires proper clothing, gloves, and eye protection. It is also very important that any metal that has been heated by burning or welding be plainly marked, especially if it is left unattended. The common practice is to write the word HOT with soapstone on such items. Wherever arc welding is performed in a shop, the arc flash should be shielded from the other workers. **Never** look toward the arc because if the arc light enters your eye, even from the side, the eye can be burned.

When handling and pouring molten metals such as babbitt, aluminum, or bronze, wear a face shield and gloves. Do not pour molten metals where there is a concrete floor unless it is covered with sand.

Hazardous Fumes

Some metals such as zinc give off toxic fumes when heated above their boiling point. Some of these fumes when inhaled cause temporary illness, but other fumes can be severe or even fatal. The fumes of mercury and lead are especially dangerous, as their effect is cumulative in your body and can cause irreversible damage. Cadmium and beryllium compounds are also very poisonous. Therefore, when welding, burning, or heat treating these metals, adequate ventilation is an absolute necessity.

Uranium salts are toxic and all radioactive materials are extremely dangerous. When working with any metals with which you are not familiar it is best to check on toxicity and proper handling by consulting an appropriate reference book or safety representative.

Self-Evaluation

1 In what ways are welding and welds similar to the manufacture of metals?

2 Name three classes of welding.

3 If the base metal contains a medium to high carbon percentage (0.40 to 1 percent, for example), what happens in the fusion zone?

4 What are the four basic zones in welds?

5 In which zone is columnar grain structure likely to be found?

6 Is arc welding or oxy-acetylene welding more likely to cause grain growth and large grains in the heat affected zone?

7 A very rapid cooling rate in a carbon steel base metal can cause the formation of what microstructure?

8 What effect does multiple passes have on the fusion and adjacent zones?

9 Name three causes of failures in welds.

10 How can preheating help to prevent hard zones in carbon steel?

11 How can postheating eliminate hard zones in welds?

12 Explain why the heat affected zone in cold rolled steel is not as strong as the unheated base metal.

13 What affect does the mass of the base metal have on cooling rates?

14 What is the purpose for the covering on electrodes or the shielding gases used?

15 Distortion due to shrinkage is mostly caused by (a) the weld or (b) the base metal expansion and contraction?

Worksheet

Objectives

1 Macroetch the cross section of a carbon steel weld on low carbon steel using E6013 electrodes to determine the weld zones.

2 Macroetch the cross section of SAE1040 carbon steel and weld using E11018 electrodes to determine the weld zones.

3 Hardness test a specimen across the various weld zones.

Note Extreme care must be exercised when using acids to avoid spilling or splashing on your skin. Face masks and rubber gloves should be worn. If spilling on skin does occur, rinse with **cold** water immediately or apply a bicarbonate of soda solution.

Materials A 10 percent solution of hydrochloric acid in water, nital (5 percent solution of nitric acid in methyl alcohol), pyrex bowls or beakers, and stainless steel tongs.

Note Carburized or decarburized surface, structure of welds, or depth of hardening may be brought out with nital. Porosity, cracks, and segregation may be seen by using hydrochloric acid. Heating is not necessary for these reagents although higher temperatures accelerate the rate of etching.

Procedure 1

1. Make a butt weld as shown in Figure 47 with mild steel plate and 6013 rod and a similar one on 1040 plate with 11018 rod.

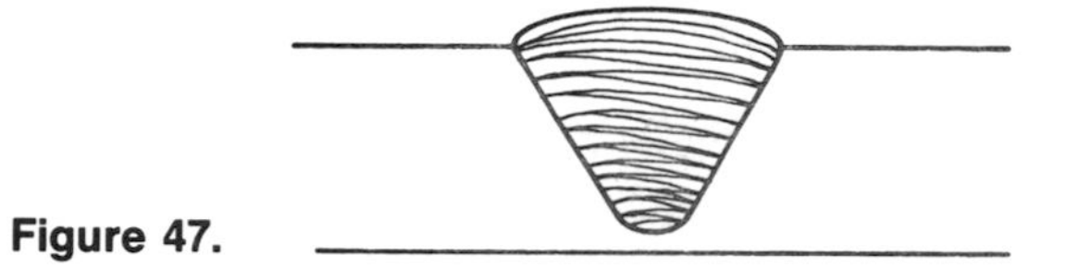

Figure 47.

2. Saw a section out of the center of the welded piece.
3. Polish the section on a belt sander.
4. Pour a small amount of nital in a flat pyrex dish. Pour a small amount of hydrochloric acid (10 percent solution) in another pyrex dish.
5. Using tongs, place the specimen in the nital for about 5 min., then remove and rinse in water.
6. Place the specimen in the hydrochloric acid for about one second. Rinse in water.
7. Grains and outline of the weld should now be evident. If not, etch the part again for a longer time.

Conclusion After close visual inspection of the macroetched specimen, can you see the various zones? Describe what you see by writing a short paragraph and make a sketch of your weld section, detailing the parts or zones of the weld.

Procedure 2

1. Check for hardness with a hardness tester across the weld as shown in Figure 41.
2. Check at $\frac{1}{16}$ inch increments and record the hardness of each test on a sketch of your weld.

Conclusion What did you find out about the hardness of various zones of the weld? Write a paragraph explaining the reason for the variation in hardness.

Alloy steel

17 Metallurgy of Welds: Alloy Steel

An alloy steel has properties that result from some element other than carbon. An alloy steel must contain at least a small percentage of manganese, silicon, and copper. However, all steels contain a small amount of manganese and other trace elements. Tool steels are a special category of hardenable steels. This chapter deals with the weldability and metallurgy of many of the alloy and tool steels.

OBJECTIVES

After completing this chapter, you will be able to:

1. Describe the effects of welds on the microstructure and properties of several alloy steels.
2. Prepare specimens of cross sections of welds on alloy steels for microscopic study.
3. Describe and analyze the microstructure in a given specimen.

INFORMATION

Among the hundreds of alloy steels, some can be heat treated such as the quenched and tempered alloy steels. Others are high strength low alloy steels, stainless steels, abrasion resisting alloy steels, heat resisting high alloy steels, and high temperature service alloy steels. Alloying elements affect hardenability and other properties of steels, all of which greatly affect the weldability of the metal and the condition of the weld.

The purposes for alloying steel with other elements are as follows:

To increase strength, either at low or high temperatures.

To increase hardenability and improve toughness.

To increase wear resistance.

To improve magnetic properties.

To increase corrosion resistance.

The effect of alloying elements on the hardenability of carbon steels was discussed in Chapter 11. There it was noted that the hardenability of carbon steels is increased by most alloying elements and that the element chromium is a ferrite former; that is, chromium dissolves primarily in ferrite, thus reducing the austenite range. Other alloying elements such as molybdenum, silicon, and titanium also tend to reduce the austenitic region. These elements tend to make steels transform into ferrite when cooling.

Carbide Formers

Elements that are carbide formers (form hard compounds with carbon) are chromium, tungsten, molybdenum, titanium, and vanadium. These also dissolve to some degree in ferrite and readily combine with any carbon present. All carbides found in steel are hard and brittle, but the carbides formed by chromium, vanadium, and tungsten are extremely hard and have high wear resistance.

Effect of Alloying Elements on Hardenability

The diagram (Figure 1) shows how carbon steel is affected by the presence of alloying elements to change the critical range, and how the position of the eutectoid is moved to the left. Although the carbon content is reduced at the shifted eutectoid point, the steel will be about as hard as 0.83 percent carbon steel when quenched. It can be seen then that the hardenability is increased in carbon steel by the addition of these alloying elements. Nickel and manganese tend to lower the temperature of the A_1-$A_{3,1}$ line. Rapid cooling rates also tend to lower the A_1-$A_{3,1}$ line. When the temperature is lowered sufficiently, austenite cannot transform to ferrite because in the cooler temperature the movement of atoms in the lattice is too sluggish so that the alloy steel remains austenitic at low temperatures. Nickel and manganese also lower the upper critical temperature on heating, while molybdenum, aluminum, silicon, tungsten, and vanadium tend to raise it.

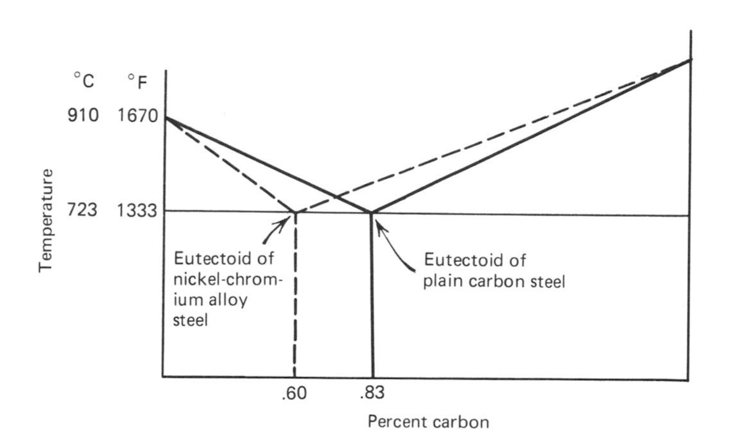

Figure 1. Iron carbon diagram showing how eutectoid can move to the left. Most alloying elements tend to shift the eutectoid point to the left, thus increasing the hardenability in the medium carbon range. The nickel-chromium alloy will get as hard as the plain carbon steel.

Tempering Effects of Alloying Elements

Another important factor in welding is that the complex carbide forming elements such as chromium, tungsten, molybdenum, and vanadium affect the tempering process in that they raise the tempering temperature for a particular hardness and, in some cases, even increase their hardness when the tempering temperature (or postheat temperature) is raised. This is because of the delayed transformation (decomposition) of retained austenite; that is, not all of the austenite is transformed into martensite when it is quenched. Instead, the resultant precipitation of the remaining austenite into fine carbides takes place during the tempering process. This is known as secondary hardness that would, of course, increase the brittleness of the weld. Some **high alloy steels** should not be postheated for this reason. These factors are not only important in heat treating alloy steels, but they are extremely important to the final condition of the metal when welding it.

Specifications for Steels

Standard ASTM specifications are used to designate carbon and alloy structural steels. Shapes such as wide flange beams, angles, plates, and bars are used for the construction of bridges, buildings, pressure vessels, and for other structural purposes. A36, for example, includes carbon steels and structural grades of plate, bars, and shapes. Full ASTM specifications may be found in welding data handbooks.

The high strength, low alloy (HSLA) steels have improved properties over carbon steel such as higher strength, abrasion resistance, and corrosion resistance. These steels are usually ferritic in structure and are used in the as-rolled condition. Carbon content is usually between 0.16 and 0.22 percent and manganese content is from 0.85 to 1.60 percent. Alloy content includes nickel, chromium, phosphorus, silicon, vanadium, columbium, and copper (Figure 2). ASTM specifications for these steels include A242, A441, A572, and A588 SAE steels.

C	Mn	P	S	Si	Ni	Cr	Mo	V	Cu	B
0.15	0.92	0.014	0.020	0.26	0.88	0.50	0.46	0.06	0.32	0.0031

Figure 2. A typical composition of USS "T-1" steel (United States Steel Corporation).

Some high yield strength HSLA constructional steels are quenched and tempered at the steel mill for desired properties. For this reason, these construction steels should be fabricated according to the manufacturer's specifications. Even though the carbon content is usually low, they can have high yield strengths of 170,000 psi. Some trade names for these steels are SSS-100 (Armco), T1 (United States Steel), RQ 100 (Bethlehem Steel Corporation), and Jalloy-S-100 (Jones and Laughlin Steel Corporation).

The constructional alloy T1 Type B has a high yield strength as shown in Figures 3*a* and 3*b*. Some wear or abrasion resistant alloys are similar but contain higher percentages of chromium and carbon and work harden to 400 BHN or more. A fairly high preheat is used on T1 steels for welding or gas cutting to reduce hard zones. Those containing a higher carbon content require a higher temperature preheat. Preheat temperatures vary with these steels. Figures 4 and 5 show a field test for determining approximate preheating temperatures. Low hydrogen electrodes E90XX or E100XX grades are suggested for welding these steels. Complete data on properties, applications, and procedures are available for these steels from the manufacturers.

Nickel Steels

The nickel steels contain from 1 to 5 percent Ni, which increases strength, hardenability, and toughness at low temperatures. Nickel (in lower percentages) also reduces the carbon content of the eutectoid and lowers the brittle transition temperatures of steel. When large percentages are added, the steel becomes austenitic. Since higher percentages (10 to 12 percent) of nickel (or manganese) tend to lower the critical temperatures of

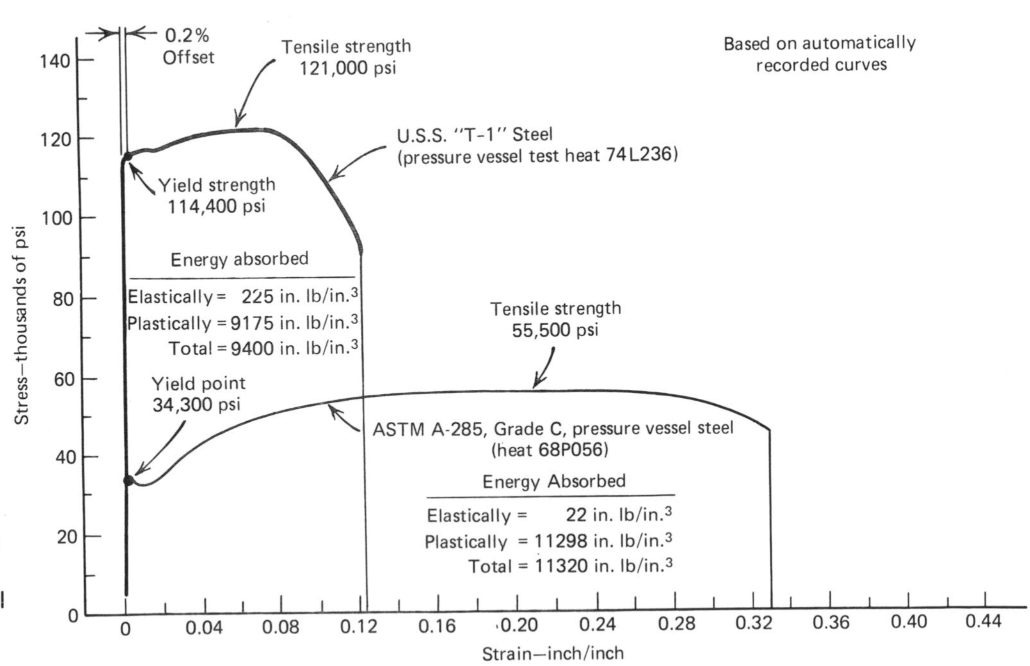

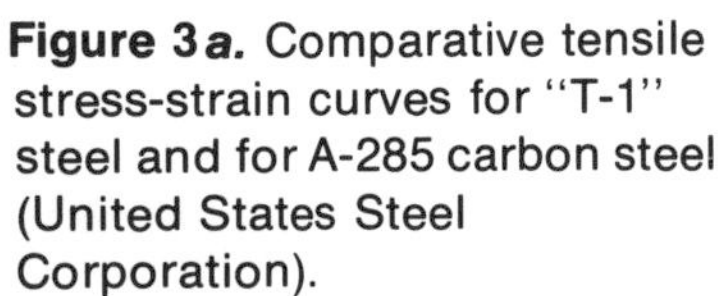

Figure 3*a*. Comparative tensile stress-strain curves for "T-1" steel and for A-285 carbon steel (United States Steel Corporation).

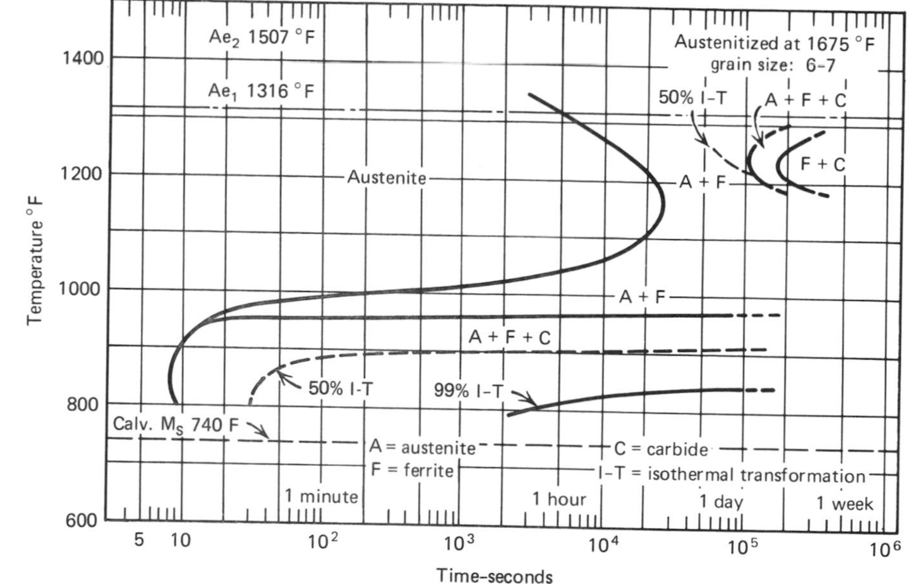

Figure 3*b*. In this USS "T-1" steel isothermal transformation diagram, a considerable period of time elapses before transformation starts in the pearlite transformation temperature range of 1100 to 1300°F (593 to 704°C). A relatively short period of time is required for transformation in the bainite range to begin (United States Steel Corporation).

Figure 4. This clip test shows that this base metal will not produce satisfactory welds without preheat and postheat. A prior weld broke out in the base metal instead of the weld, revealing the embrittled heat affected zone. With preheat and postheat the second weld broke through the weld metal, indicating good ductility in the heat affected zone.

Figure 5. This weld is being postheated. The preheat was made to 400°F (204°C) using temperature crayons. The clip will again be broken off to determine weld strength and weldability at these preheat and postheat temperatures.

low carbon steel retarding the decomposition of austenite, rapid cooling rates anneal or soften the steel instead of hardening it, especially if it is fully austenitic. These austenitic steels should not be preheated since complex carbides can form if the steel is held in the 1200 to 1600°F (649 to 871°C) range for any length of time. Carbides embrittle and weaken the steel. Examples are austenitic stainless steels and high manganese steels.

Chromium Steels

Chromium tends to form carbides, which increase the toughness and strength of ferrite since it goes into solution to some extent in ferrite. Chromium is also used as an alloying element for steels that are used for magnetic purposes. Chromium steels containing more than 0.30 percent carbon often require solution heat treatment when they are welded in order to inhibit the formation of carbides.

Nickel-chromium steels of the 3XXX series are among the heat treatable alloy steels and are deep hardening steels. These steels are used for drive shafts, gears, pins, and axles. Preheat is advised for welding nickel-chromium steels in the carbon ranges above 0.35 percent. No preheat is required in these steels below 0.25 percent C. Low hydrogen rod of similar carbon content should be used for welding these alloys. Choice of filler metal is critical.

Manganese Steels

Manganese is principally used in all steels for a deoxidizer and to control sulfur content, but the steel is classed as an alloy steel when above 1 percent manganese is alloyed with steel. Manganese alloyed with steel promotes deep hardening, strengthens steel, and is a carbide former to some extent. Like nickel, manganese lowers the critical temperature range and decreases the carbon content of the eutectoid.

Austenitic manganese steel containing 12 percent or more manganese has high wear resistance and is used in power shovel teeth and buckets, grinding machinery, rock crushing machinery, and railway tracks. This alloy has the peculiar property of hardening into large brittle carbides surrounding austenite grains when it is cooled slowly from 1700°F (927°C). On the other hand, if the same alloy, after giving the carbides time to dissolve, is quenched from 1850°F (1010°C), the structure will be fully austenitic with much higher ductility and strength but softer than before. When this alloy is placed in service and subjected to repeated impact, the hardness increases to approximately 550 BHN. This is because the manganese steels have the property of work hardening to a great extent. Reheating, as in welding, and cooling slowly will embrittle the metal. Low interpass temperatures are therefore needed when welding this metal. Water quench is sometimes used. Peening should not be used. Preheat or postheat is not desirable. An underlayer of austenitic stainless steel is sometimes applied to provide ductility between the work hardened manganese steel and the base metal. This tends to prevent spalling (pieces of weld coming loose on the surface). A disadvantage of this method is that subsequent oxy-acetylene cutting through the stainless steel weld would be difficult.

Molybdenum, Tungsten, Vanadium, and Silicon

Molybdenum, like chromium, has a strong effect in promoting hardenability and increases the high temperature hardness of steels. Molybdenum is often alloyed along with nickel and chromium to steel.

The effect of tungsten in steel is similar to that of molybdenum in that it has a great effect on hardenability, is a strong carbide former, and causes steel to require a higher tempering temperature because it tends to retard the softening of martensite. Tungsten is mostly used in tool steels.

Vanadium is also used for tool steels, springs, and bearings where high hardness and wear resistance are required. It is a deoxidizer and a strong carbide former. Vanadium also has the unique capacity to retard grain growth at higher temperatures in steels.

Silicon is used as a deoxidizer also and it is much less expensive than vanadium. Above 0.06 percent silicon, it is considered an alloy of steel (silicon steel). Silicon increases strength and toughness. With 3 percent silicon in steel, the steel has useful magnetic properties and is used in electrical machinery.

Stainless Steels

Stainless steels, as you have learned, are alloys of chromium, chromium-nickel, or chromium-nickel-manganese. They are classified in three types: the 400 series martensitic, the 400 series ferritic, and the 300 series austenitic. As you remember, martensite types are hardenable having sufficient carbon content to harden by quenching; and others, such as ferritic, have no appreciable carbon. The austenitic type, being work hardening, also contains very little carbon. Each of these types present unique welding characteristics and problems. Proper procedures must be observed and the right filler metals used when welding these metals.

Ferritic chromium grades contain very little carbon so that any brittleness is not caused by hard carbides. Since grain growth is very pronounced in the heat affected zone in these steels causing them to be brittle, they should not be used where vibration is a factor. Welded ferritic stainless steel joints possess a high shock resistance at elevated temperatures and may be used for this purpose. Annealing may be done at 1200 to 1550°F (649 to 843°C), but types 430, 430F, 442, and 446 cannot be normalized.

Martensitic steels contain carbon to 1 percent and may be hardened by heat treatment. Because some types are air hardening, the air cooling of weld zones lowers ductility and raises the hardness considerably. Martensitic steels can be welded, however, in the annealed, hardened and tempered, or hardened condition. Welding will produce a "martensitic" zone adjacent to the weld. Preheating and control of interpass temperatures are the most effective means of avoiding cracking. Postweld heat treatment is required.

Tables 1 and 2 show that full anneal and process anneal for these welds are similar to that of plain carbon steels. However, in Table 2 the process is done at lower temperatures. Table 3 shows annealing procedures for ferritic stainless steel.

Austenitic grades cannot be hardened by heat treatment and contain only small amounts of carbon, but can be hardened by cold working. Heating (welding) of unstabilized grades to temperatures from 800 to 1600°F (427 to 871°C) will cause chromium carbide to form and precipitate out along the grain boundaries (Figure 6), reducing corrosion resistance. (Rust will form parallel to the weld on the heat affected zone.) The chromium carbides can be dissolved by solution heat treatment of the welded part. This is done by heating at 1850 to 2100°F (1010 to 1149°C) and quenching in the correct medium for the grade. Stabilized grades contain columbium or titanium (Types 321, 347, and 348) and are designed to limit the formation of harmful chromium carbides. The small amount of carbon instead becomes titanium or columbium carbides as these metals have a greater affinity for carbon than does chromium, and the carbides disperse more uniformly in the grains.

Grades having a very low carbon content (304L and 316L) can be heated for short periods without carbide precipitation. Table 4 reveals the quite different behavior of austenitic metals in annealing procedures. After heating to the correct temperature, the metal is quenched in water or oil instead of slow cooled to soften it. This restores a ductile grain structure and disperses any chromium carbides or prevents their formation at the grain boundaries. The austenitic stainless steels (except free machining grades) are more weldable than ferritic and martensitic grades. Weld joints are tough.

Stainless steel may be welded by several processes, such as oxy-acetylene, TIG, GMAW (MIG), or metal arc welding. When arc welding, flux coated electrodes of the correct alloy content

Table 1 Full Annealing Procedure

Stainless Steel	Temperature °F (°C)	Time	Cooling[a]	Hardness (Typical) Brinell	Rockwell
Type 410 Armco 410 Cb Type 403 Armco 416	1550–1650 (843–899)	1–3 hours	Slow cool	137–159	B75–83
Type 420 Type 420F	1600–1650 (871–899)	1–2 hours	Slow cool	170–201	B86–93
Type 440A	1625–1675 (885–913)	1–2 hours	Slow cool	207–229	B94–98
Type 440B	1625–1675 (885–913)	1–2 hours	Slow cool	217–241	B95–C20
Type 440C Type 440F	1625–1675 (885–913)	1–2 hours	Slow cool	229–255	B98–C23

[a]25–50°F (14–28°C) per hour to 1100°F (593°C).

Source. Armco Steel Corporation, Middletown, Ohio, *Heat Treating Armco Stainless Steels,* 1966.

Table 2 Process Annealing Procedure

Stainless Steel	Temperature °F (°C)	Time	Cooling Method	Hardness (Typical) Brinell	Rockwell
Type 410 Armco 410 Cb Type 403 Armco 416	1350–1450 (732–788)	1–3 hours	Any	170–197	B86–92
Type 414	1200–1300 (649–704)	2–6 hours	Any	241–255	B99–C24
Type 431	1375–1425 (746–774) 1150–1225 (621–663)	2 hours 4–8 hours	Air Any	241–255	B99–C30
Type 420 Type 420F	1350–1450 (732–788)	1–4 hours	Any	207–223	B94–97
Type 440A	1350–1450 (732–788)	1–4 hours	Any	229–248	B97–C22
Type 440B	1350–1450 (732–788)	1–4 hours	Any	235–255	B98–C23
Type 440C Type 440F	1350–1450 (732–788)	1–4 hours	Any	255–277	B100–C26

Source. Armco Steel Corporation, Middletown, Ohio, *Heat Treating Armco Stainless Steels,* 1966.

Table 3 Annealing Procedures for Ferritic Stainless Steels

Stainless Steel	Annealing Temperature Range, °F (°C)	Time[a]	Quench	Typical Hardness (as annealed) Brinell	Rockwell
Type 405	1200–1500 (649–816)	1 to 2 hours	Air or water	140–163	B77-85
Type 430	1400–1525 (760–829)	1 to 2 hours	Air or water	140–163	B77–85
Type 430F	1200–1450 (649–788)	1 to 2 hours	Air or water	163–192	B85–91
Type 442	1400–1525 (760–829)	1 to 2 hours	Air or water	149–174	B80-88
Type 446	1400–1525 (760–829)	1 to 2 hours	Air or water	159–183	B84-90
Armco 400	1500–1700 (816–927)	3 min per 0.100 in. thickness	Air	121–137	B69-75
Armco 409	1500–1700 (816–927)	3 min per 0.100 in. thickness	Air	130–150	B72-80
Armco 18 SR	1500–1700 (816–927)	3 min per 0.100 in. thickness	Air	163–210	B85-95

[a] 1 to 2 hours recommended for heavy sections. Hold at temperature 3 minutes for every 0.100 inch (2.54 mm) of thickness.
Source. Armco Steel Corporation, Middletown, Ohio, *Heat Treating Armco Stainless Steels,* 1966.

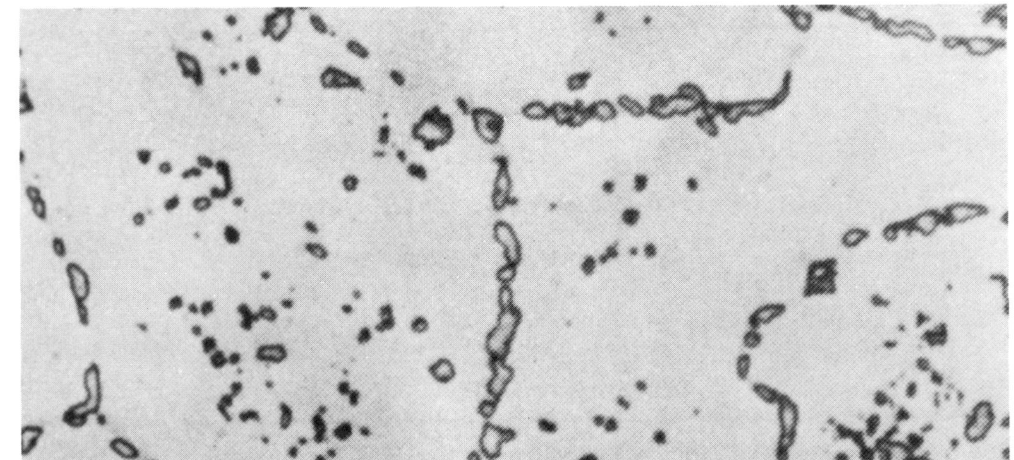

Figure 6. Type 310 stainless steel hot rolled plate annealed at 1950°F (1066°C), water quenched in less than three minutes, exposed for 27 months at 1400°F (760°C), slowly air cooled. Chromium carbides are precipitated at the austenite grain boundaries (250×) (By permision from *Metals Handbook,* Volume 7, Copyright American Society for Metals, 1972).

Table 4 Solution Annealing Temperatures for Austenitic Stainless Steels

Stainless Steel	Solution Annealing Temperature Range, °F (°C)
Type 301	1900–2050 (1038–1121)
Type 302	1900–2050 (1038–1121)
Type 302B	1900–2050 (1038–1121)
Armco 303	1900–2050 (1038–1121)
Type 303 Se	1900–2050 (1038–1121)
Type 304	1850–2050 (1010–1121)
Type 304L	1750–1950 (954–1066)
Type 308	1850–2050 (1010–1121)
Type 309	1900–2050 (1038–1121)
Type 310	1900–2050 (1038–1121)
Type 316	1900–2050 (1038–1121)
Type 316L	1750–1950 (954–1066)
Type 317	1950–2050 (1066–1121)
Type 317L	1750–1950 (954–1066)
Type 321	1750–1950 (954–1066)
Type 347	1800–1950 (982–1066)
Type 348	1800–1950 (982–1066)
Armco 18-9 LW	1900–2050 (1038–1121)
Armco NITRONIC 32	1850–1950 (1010–1066)
Armco NITRONIC 33	1850–1950 (1010–1066)
Armco NITRONIC 40	1950–2050 (1066–1121)
Armco NITRONIC 50	1950–2050 (1066–1121)
Armco NITRONIC 60	1950 (1066)

Source. Armco Steel Corporation, Middletown, Ohio, *Heat Treating Armco Stainless Steels,* 1966.

should be used. Flux coated electrodes also provide a fast, economical weld deposit with good penetration.

Welding Tool Steels

Tool steels are similar to alloy steels in that they often contain many of these alloying elements in varying amounts. Tool steels may be welded if the proper electrode is used and correct preheat and postheat procedures are used. A filler metal must be used that closely approximates the base metal in carbon and alloy content. Preheat is usually necessary to avoid cracking since the HAZ (heat affected zone) is typically embrittled by welding. Arc welding procedures are normally used. In order to restore the tool or part to its original hardness, a hardening and tempering procedure that is correct for that type of tool steel must be followed in most cases. An anneal may be necessary to allow remachining of the welded part prior to hardening. In some cases, a hardened and tempered tool steel may be welded without anneal by using special electrodes and procedures. For example, austenitic stainless steel electrodes are often used where a ductile joint is required, such as when welding dissimilar metals or broken tools. A buttering technique is used when welds are large, followed by a fill-in of low hydrogen rod.

Welds on **high speed** tools may be done either with the tungsten base (T) type tool steel electrode or with the molybdenum high speed steel electrode (M). The tungsten type should be used where red hardness is needed and the molybdenum type where full hardness must be maintained as high as 1000 °F (538 °C). Expensive dies and cutting tools are often repaired with these electrodes, but correct procedures must be used for making a successful weld. In most cases, only a preheat is used; the electrode deposit being hard as-welded for use. Heat treating procedures may be used if desired, but distortion or warping may occur in the part and subsequent grinding would then be required.

Preheating temperature for high speed steel should be at 1000 °F (538 °C) for a length of time that will insure uniform and thorough heating of the part (this will not affect the hardness of high speed steel). A short arc should be maintained with the proper size electrode to avoid the formation of undercuts and craters. Use a skip-weld procedure with immediate hot peening of the weld to relieve stresses. Do not peen the weld when cold. When the weld has been completed, allow it to cool slowly to room temperature, then reheat to 1000 °F (538 °C) to temper and stress relieve the part. Allow it to cool slowly to room temperature.

High speed steel tools such as milling machine cutters can be repaired by welding (Figure 7). Since the weld is usually sufficiently hard as-welded, no hardening heat treatments are needed on cutting edges.

Cracking in the Weld Metal

Cracking down the center of the weld or microcracking of the weld metal usually takes place at high temperatures (hot cracking) during solidification and is caused by stresses of expansion and contraction in the weld. This is sometimes the result of trying to fill in a gap that is too wide with the

Figure 7. This expensive high speed milling cutter was cracked and has been repaired by welding. It has been reground to give many more hours of service.

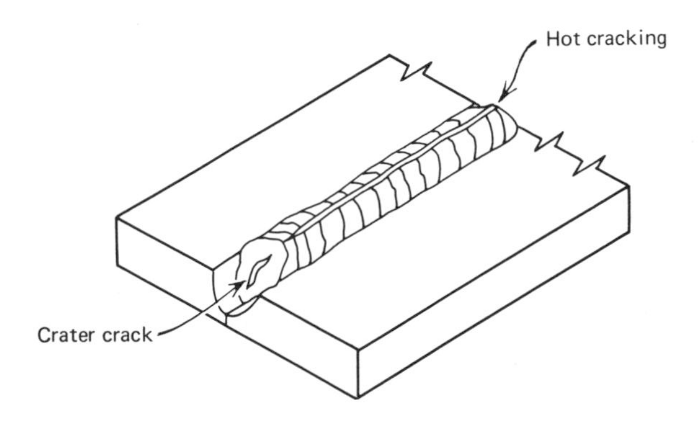

Figure 8. Hot cracking in the weld is often seen as a separation down the center of the weld.

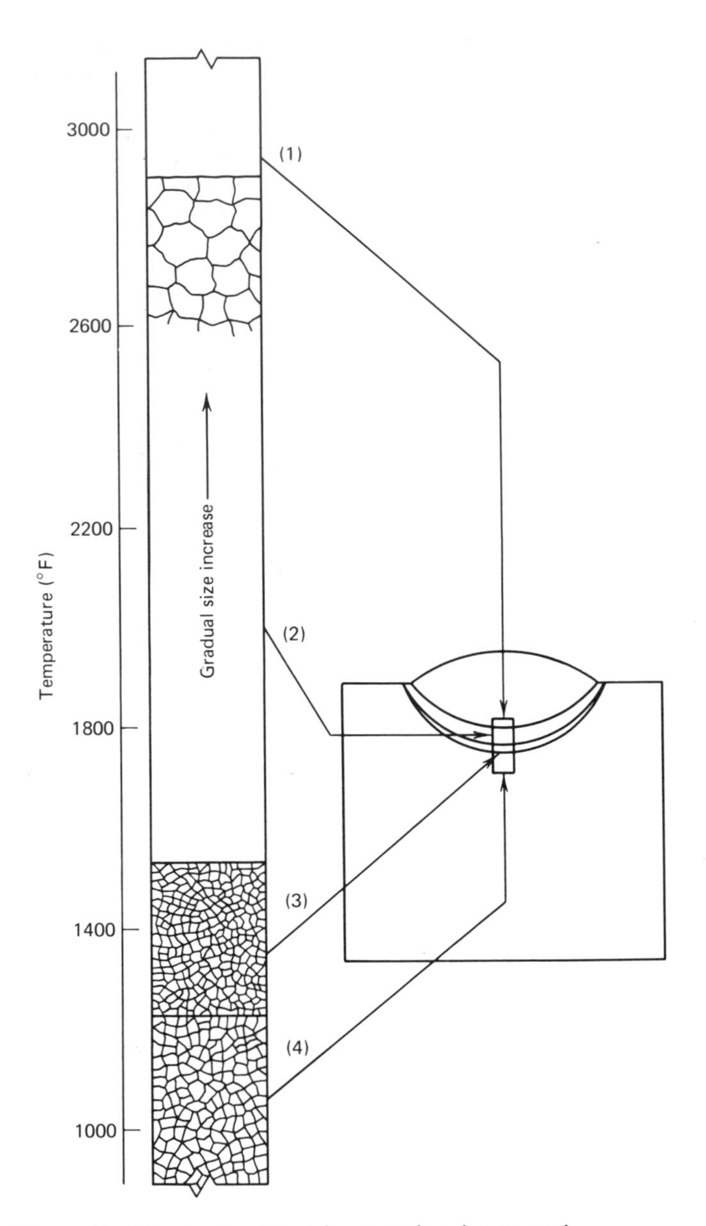

Figure 9. Effect of welding heat on hardness and microstructure of an arc-welded 0.25 percent carbon steel plate. The schematic diagram represents a strip cut vertically through the weld shown. Significance of the four numbered zones are: (1) Metal that has been melted and resolidified. Grain structure is coarse. (2) Metal that has been heated above the upper critical temperature 1525°F (829°C) for 0.25 percent carbon steel but has not been melted. This area of large grain growth is where underbead cracking can occur. (3) Metal that has been heated slightly above the lower critical temperature 1333°F (723°C) but not to the upper critical temperature. Grain refinement has taken place. (4) Metal that has been heated and cooled, but not to a high enough temperature for a structural change to occur (The Lincoln Electric Company).

weld. High cooling rates tend to increase hot cracking in the weld metal. Slowing down the cooling rate will help to prevent hot cracking (Figure 8). The end or terminal crater crack is one example of hot cracking. This can often be eliminated by back filling of the crater before breaking the arc.

Preheating is also used to slow down the cooling rate in low and high carbon alloy steels. Tables in welding manuals provide correct preheat temperatures for a specific steel and electrode. Some alloy and carbon steels develop coarsened grains if the preheat temperature is too high and the interpass temperatures are consequently above the specified degree (Figure 9). If this occurs, a postheat normalizing procedure is necessary to refine the coarse brittle grain in the heat affected zone (HAZ).

Weld metal cracks may be caused by the following.

1 Wrong choice of filler metal or electrode. For example, a high tensile filler metal on low carbon steel base metal is the wrong application.
2 Welding conditions; too much gap between base metal parts or pick up of embrittling elements such as carbon, nitrogen, or hydrogen.
3 Defects such as porosity, slag or oxide inclusions and hydrogen gas pockets (hydrogen embrittlement).
4 Too rapid cooling rates (no preheat).
5 Constraint of base metal parts. For example, two heavy plates clamped rigidly so that no movement can take place while welding: all the stresses involved are concentrated in the weld metal.

Cracking in the parent or base metal is a serious problem when welding high carbon or tool steels. These cracks form in the fusion zone or in the heat affected zone adjacent to the weld bead (Figure 10). This is called underbead or hard cracking and cannot normally be detected on the surface of the base metal. These cracks are difficult to detect even with magnetic particle inspection or X-ray. Any welding method performed on medium to high carbon steels will result in a hard zone adjacent to the weld if the cooling rate is too fast. The underbead crack follows the contour of the weld bead.

Radial cracks are sometimes a result of welding stress in hard or brittle zones. Hydrogen entrapped in the base metal and weld metal often initiates such underbead cracking. Hydrogen is very soluble in molten steel, but it is forced out of the metal by diffusion if the cooling rate is slow. Hydrogen may be picked up by the use of damp electrodes or by those having a cellulose type of coating. Oil, grease, paint, or water on the base metal will also cause hydrogen embrittlement. When low hydrogen rod that has not picked up any moisture is used on higher carbon steels and preheat is used, underbead cracking is not likely to be a problem.

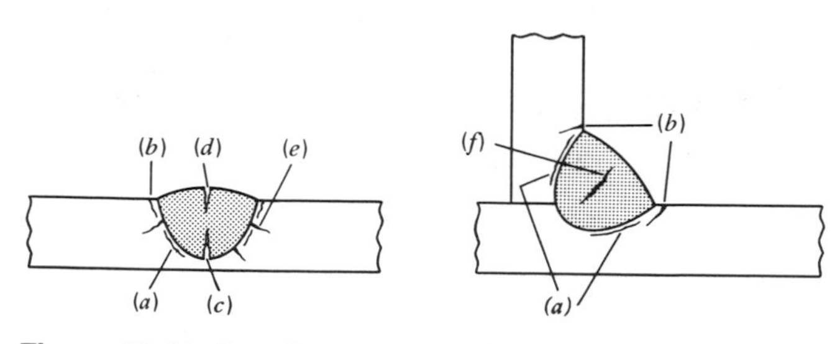

Figure 10. Various forms of cracking in welds: *(a)* Underbead cracking, *(b)* toe cracks, *(c)* root cracks, *(d)* hot cracking, *(e)* radial cracking and *(f)* internal cracks.

Toe cracks (Figure 10) are caused by stress at the toe of the weld and across the heat affected zone. Preheating reduces the problem that is usually found in metals with low ductility. A root crack (Figure 10) originates in the root bead. This is caused by a rapid cooling rate of the small weld bead, carbon pick up from oily metal intensifies this problem. Preheat should be used.

Hard Surfacing

Machine part surfaces subject to wear and abrasion such as power shovel teeth and rock crushing machinery are sometimes built up with wear resistant rod materials. There are scores of hard facing alloys from which to choose. When this selection is made, the composition of the base metal must be known if a successful weld is to be made. The proper selection of rod should also be made to have the correct application to the wear problem (impact resistance, abrasive wear, heat, pressure, or corrosion).

A tough weld deposit should never be put over a harder, more brittle hard surfacing weld. These welds will come loose (spall). A hard weld should always be on top of a softer one. Stainless steel is often used as an underlay weld for hard surfacing, as mentioned earlier.

Wear resistance of a hard surfacing weld will decrease when base metal dilution is increased. The oxy-acetylene process produces the least amount of dilution (5 percent), while arc deposits vary from 20 to 40 percent (Figure 11). Small surface cross checks (cracks) or hairline cracks are desirable in the harder surfacing alloys since they relieve stresses and minimize distortion.

Welding temperatures are very important. A preheat will often eliminate underbead cracking in alloy or carbon steel base metal, while high man-

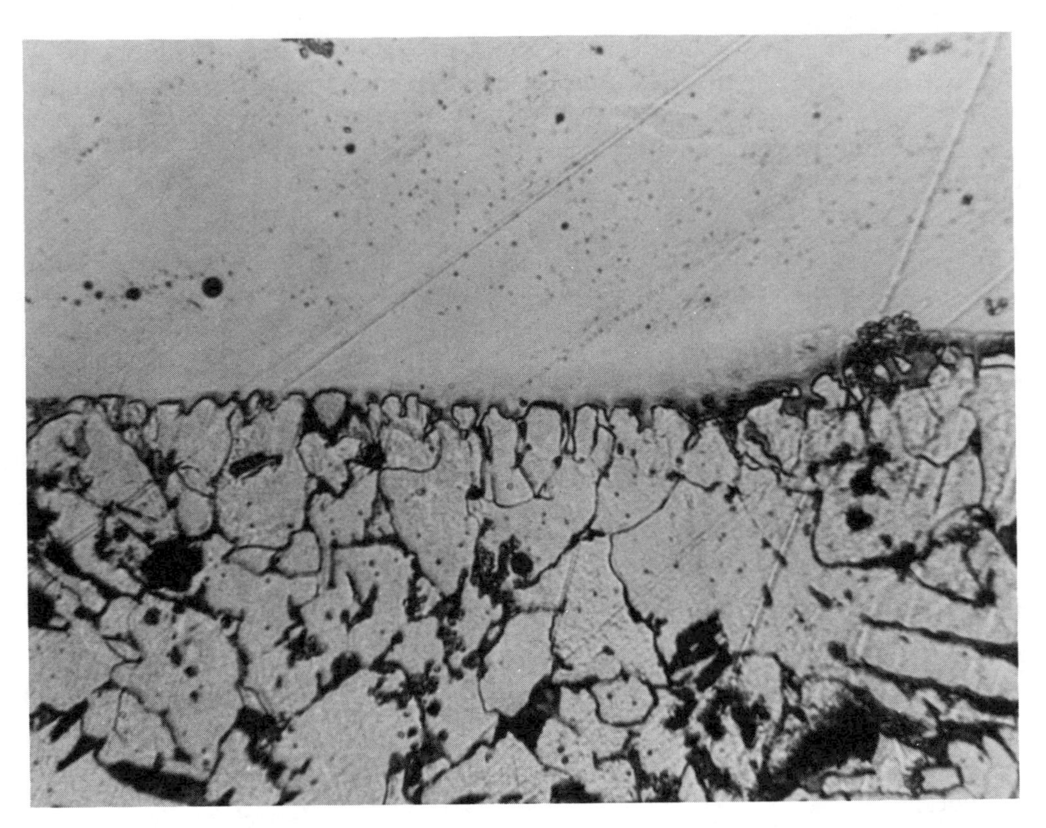

Figure 11. Hard surface arc weld deposited on low carbon steel. Note that there is very little dilution from the base metal.

ganese (austenitic) steel should not be preheated at all since temperatures above 600°F (315.5°C) may remove the tough properties of high manganese steel. High manganese steel derives its wear resistance from its ability to work harden rapidly. As welded, its hardness is about Rc 15 and after being subjected to working or impact, it can increase to over Rc 50. Arc welding only is recommended for manganese steel since oxy-acetylene welding tends to heat the base metal more, thus reducing the cooling rate that can cause weld embrittlement.

Admixture (Pick Up) in Welds

Admixture is the combining of the base metal and the molten weld. This natural mixing is of little significance when welding low carbon steels with mild steel electrodes. Alloy and high carbon steels can greatly affect the weld metal through admixture, however. Some combinations of base metal and weld cause segregations or concentrations that weaken the weld. Weld ductility is reduced by carbon pick up and cracking can result. Using welding electrodes that closely approximate the carbon and alloy content of the base metal reduces base metal and weld admixture and consequently the difficulties it can cause. In some cases admixture is an advantage as it tends to strengthen the weld metal to some extent.

Microscopic Examination of Welds

Many of the conditions of weld metal and base metal zones that have been described in this chapter can be identified by using microscopic examination techniques. A method of learning to identify the condition of welds is to purposely make welds with a known electrode on a known base metal

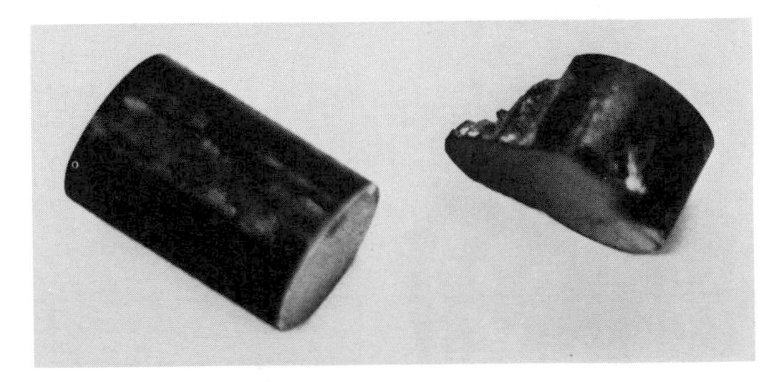

Figure 12. Weld passes made on a SAE 1040 steel bar. The piece at left is a section of bar prior to making the welds. The section at right is polished and etched for microscopic examination.

(Figure 12). Different preheat and postheat temperatures may be used and recorded. The specimen is sectioned using a metallurgical cutoff abrasive saw and polishing and etching techniques are followed as explained in Chapter 7.

Self-Evaluation

1. What affect do alloying elements have on hardenability of steels (welds)?
2. Alloying elements such as nickel or rapid cooling rates sometimes inhibit the decomposition of austenite into martensite. What affect does postheating have on this untransformed austenite?
3. Why should high manganese steels (such as used for hard facing) not be slow cooled or preheated?
4. Which alloying element has the unique ability to retard grain growth at high temperatures?
5. In what way is the brittleness of martensitic stainless steels in the HAZ different from the brittleness in ferritic stainless steels?
6. Name three things that should be remembered when welding on tool steels.
7. SAE-AISI specifications are used for carbon steel and alloy bar stock. What standard specification system is used for carbon and alloy structural steels such as beams, angles, and plates?
8. Describe the location of a hot crack and give two causes of this problem.
9. Welds should never be made where there is oil, grease, paint, or water. Why is this so?
10. Why should a tough but ductile weld not be placed over a brittle, hard surface weld?

Worksheet

Objective By microscopic examination identify weld grain structure, admixture, inclusions, carbon content, and condition of the juncture and heat affected zones.

Material Three small pieces of SAE 4140 bar or round stock; E60B, E70XX, and E100XX welding rod; metallurgical polishing equipment and microscope.

Procedure

1 Make a pile of weld metal on each of the three pieces of SAE 4140 steel. Allow to cool.

2 Make a section through the center of each weld with the metallurgical saw or a hacksaw. Do not overheat the weld when cutting it off.

3 The piece should not be more than one inch wide to facilitate polishing.

4 Polish and etch the specimen as explained in Chapter 7.

5 Examine the weld metal and heat affected zones of each sample with the microscope.

6 Make a drawing on a sheet of paper of the welded piece surrounded by several circles in which you may sketch a micrograph representing what you see in the microscope. Draw a line from each circle to that part of the weld it represents. See Figure 13.

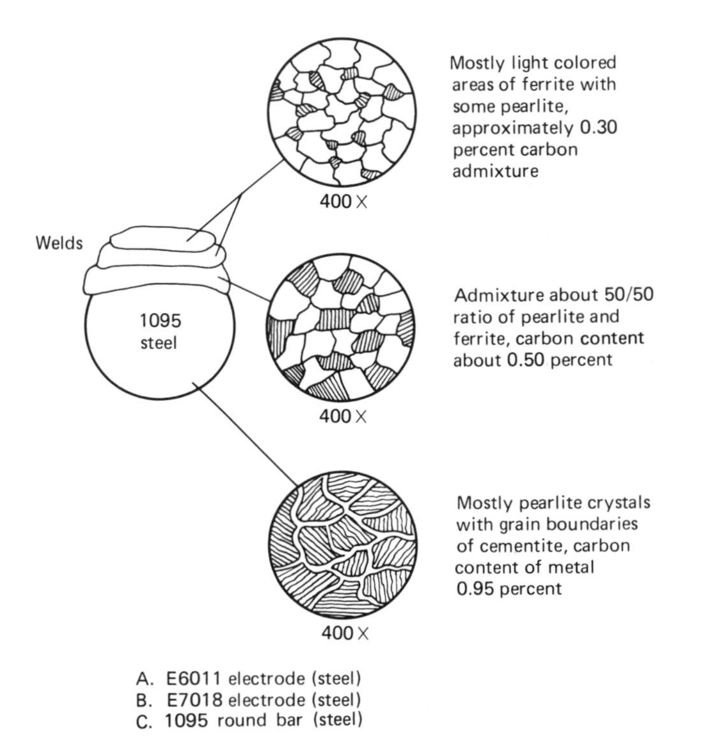

Figure 13. Example of sketch and conclusions for SAE 1095 steel.

7 Write a conclusion under each of the three representative specimens on your paper describing the metallurgical condition of the weld.

8 Turn in your specimens and paper to your instructor.

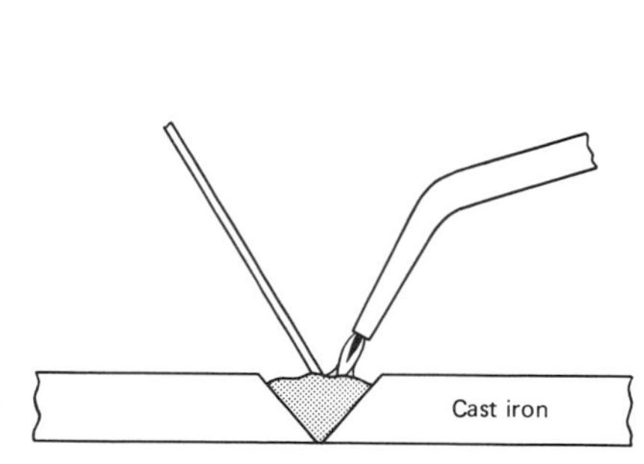

18 Metallurgy of Welds: Cast Iron

A great many products used today are made of cast iron and possess varying characteristics. They range from lower quality gray cast irons used for manhole covers or furnace grates to the more complex high strength alloy irons such as those used in automobile crankshafts and automotive engine blocks. Ductile and malleable irons are used where impact strength is a requirement. This chapter deals with the techniques and problems in the welding of various cast irons.

OBJECTIVES

After completing this chapter, you will be able to:

1. List procedures for making welds on several cast irons.
2. Recognize by microscopic examination weld structures of three cast iron welds.

INFORMATION

Steel castings and steel weldments are making substantial inroads in the replacement of cast irons for the manufacture of machinery. The advantages of steel castings are their higher strength. Weldments can often be made at lower cost than steel or cast iron castings. Nevertheless, large quantities of iron castings continue to be made and these sometimes require welding. The greater amount of welding is done on gray cast iron and very little on malleable iron; some ductile is welded. The welder should always remember, however, that cast iron is considerably less weldable than low carbon steels. The large amounts of carbon and silicon in cast irons result in a complicated, less ductile, metallurgical weld structure; and, under no circumstances should cast iron be welded in areas that will be subjected to high stresses.

Most welding applications on cast iron are found in the repair of casting defects in foundries. Other applications are in production of welded cast iron assemblies and for the repair of castings that have become damaged or worn in service (Figures 1*a* and 1*b*).

In every type of cast iron welding, a filler metal is required in order to accomplish a strong joint. Almost all the common welding processes can be used for welding cast iron. Cast irons are difficult to weld simply because they contain a large amount of carbon and silicon. Gray cast iron can contain anywhere from 2 to 4 percent carbon and 0.5 to 2 percent silicon. The weldability of cast iron depends almost entirely on the type and the distribution of the carbon and the kind of microstructure that develops on cooling. Remember, cast iron can have, in addition to the expected con-

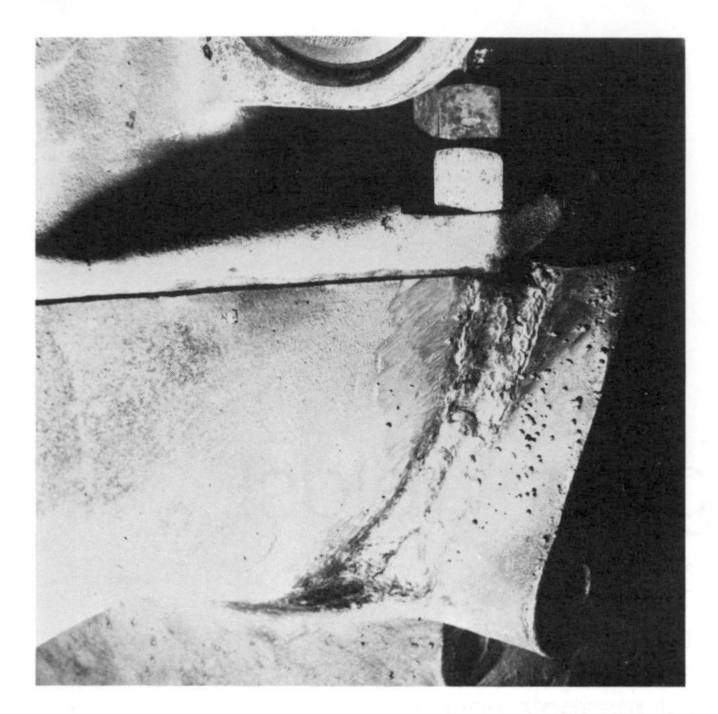

Figure 1*a*. Weld on a cast iron engine housing.

Figure 1*b*. Cracks between cylinders in an engine block are welded and remachined before the sleeves are put back in place.

stituents (ferrite, pearlite, and graphitic carbon), all the structures associated with heat treatable steels; that is, martensite, bainite, free carbides, and so forth.

All cast irons may be welded with the exception of white cast iron on which welding is not recommended. It is rather difficult to make strong welds on gray cast iron because of its tendency to form carbides (Figure 2) near the weld and because of the lack of ductility in the base metal. The contraction of the weld metal (which is usually ductile and has a higher tensile strength than the base metal) when solidifying causes stress in the base metal, which is usually under restraint. Since the base metal is very brittle and cannot move, cracking often occurs near the weld. Class 20 (ferritic) is easier to weld than Class 50 (pearlitic) iron.

The junction zones near the weld tend to form extremely hard and brittle carbides. The weld metal also tends to pick up carbon from the base metal, especially if the filler metal is low carbon steel. This can cause the weld metal to become brittle also. Since gray cast iron has no yield strength and its tensile strength may be only half its compressive strength, the concentrated stress caused by the weld can cause microcracking in the junction zone and massive cracking in the casting itself.

The method of welding can greatly influence the kind of microstructure that is formed. Brazing or gas welding, for instance, causes less weld stress and weld cracking than arc welding. It is almost always necessary to use some kind of preheating and postheating techniques with the welding process for cast iron.

The Keep-It-Cool Method

If preheating is not practical, the welding procedure often used is to keep the casting as cool as possible. Usually only shielded metal arc welds are made when using this procedure and nickel alloy or steel electrodes with special coatings are used. This method is used on a high percentage of repair jobs on large heavy machinery. ENi (austenitic nickel electrodes) or ESt (steel electrodes) are used with this method. Welding can be performed for only a few minutes or seconds at a time, after which the casting must be allowed to cool. Multibead welds are often used on heavy repairs. A second pass of ENi rod will help to soften the fusion zone of the original pass. Cracks in large castings that cannot be preheated or postheated are welded for a short distance, allowed to cool, and then followed by another short weld. The object is to avoid thermal stresses in the casting by reducing the heat input. Spot preheating is not

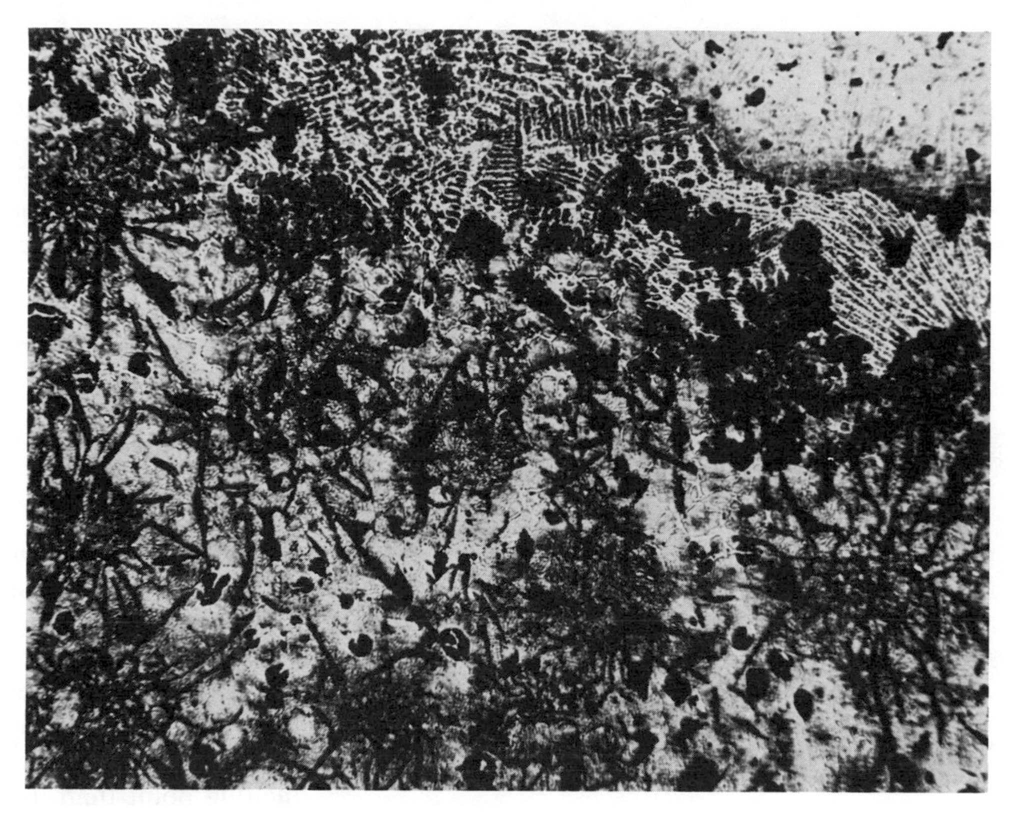

Figure 2. Microstructure of weld junction zone showing the formation of carbides. The upper right area is the nickel base deposited weld. The center area is a band of brittle dendritic white iron interspersed with small particles of graphite. The lower area is a transition from iron carbide to a typical gray cast iron microstructure (500×).

practical in cast irons as it tends to cause cracking and should not be used. Any localized heating can cause cracking or extended cracking in the part that is being welded (Figure 3).

One of the difficulties with the keep-it-cool (reduced heating) method is that the porous structure of cast iron may contain foreign materials such as oil, water, and other liquids. These impurities tend to volatilize on heating and produce a weak, porous weld. When preheat is used, many of these foreign materials are driven out prior to welding and consequently the weld is more homogenous without porosity.

The keep-it-cool method has the advantage of providing the narrowest possible heat affected zone, which can be as small as 0.00015 to 0.00020 inch; but one objection to a very narrow heat affected zone is that it is more concentrated and contains extremely hard substances that produce tiny microcracks from which larger cracks can later radiate (Figure 4). Because of the microcracks, the bond between the weld and the base metal is much weaker than a weld that has been preheated. Because of this, when correct procedures are not followed, welds in gray cast iron sometimes fail. If it is at all possible, when using this technique, the use of even a low temperature preheat from 400 to 800°F (204 to 427°C) can make a much stronger weld. The casting after welding should be allowed to cool very slowly and be protected against cold drafts of air.

Figure 3. Transverse cracking in arc weld on cast iron caused by too high heat input on cold base metal.

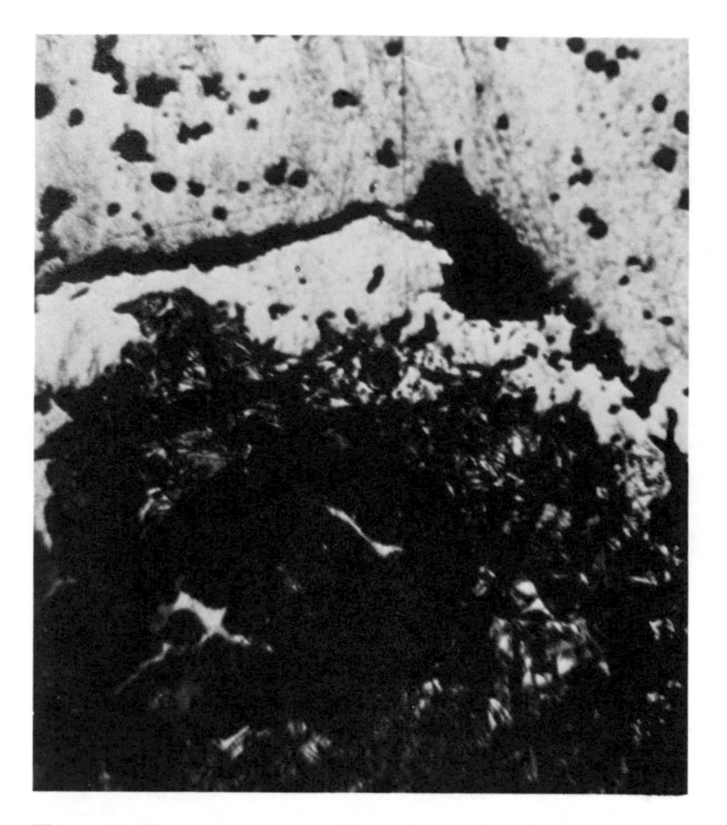

Figure 4. The narrow heat affected zone in this micrograph of a cast iron weld junction zone shows a band of martensite adjacent to an underbead crack, which is commonly found in cast iron welds made by the "keep-it-cool" method (500×).

Preheat and Postheat

The best way to eliminate hardening in the fusion zone in a cast iron weld is to both preheat and postheat the casting. As with steel, cast iron structures are greatly affected by the rate of cooling of the welds. Large weldments or sections will cause a fast cooling rate in the weld, which will transform the adjacent zone into martensite. On small sections or when preheat is used, the slower cooling rate will prevent the formation of martensite.

If at all possible when making any weld in gray cast iron, it is recommended to make a preheat up to 800°F (427°C). The preheat should be held for a period of time before welding (to an hour) to allow the contaminates to vaporize.

Postweld heat treatments may either consist of full annealing or stress relieving. The greatest softening affect may be achieved by annealing from 1650°F (899°C). Annealing lowers the as-cast tensile strength of most irons. Stress relief at 1150°F (621°C) followed by furnace cooling to 700°F (371°C) usually produces the best results.

Preheat and postheat on large, heavy castings should be done very slowly so that localized stresses will not develop and cause cracking in the casting. Also, after preheating or postheating has been done, a slow cooling period should be allowed by burying the casting in an insulating material or by placing a thermal blanket around the casting to keep it warm for a longer period of time (Figure 5).

Joint Preparation

For welding cast iron with most methods, the weld groove must be prepared by machining, chipping, or sawing. Grinding should be done only to a limited extent because of its tendency to smear the graphite flakes on the surface to be welded. When this happens, the weld often will not "take" or fuse; that is, it will tend to form into balls and roll off rather than make a bond. Graphite has a higher melting point than iron and will not flux out with most cast iron welding procedures. Steel cast iron electrodes (ESt) may be used where large graphite flakes have been formed from repeated high temperature heating cycles as in exhaust manifolds and engine castings (Figure 6).

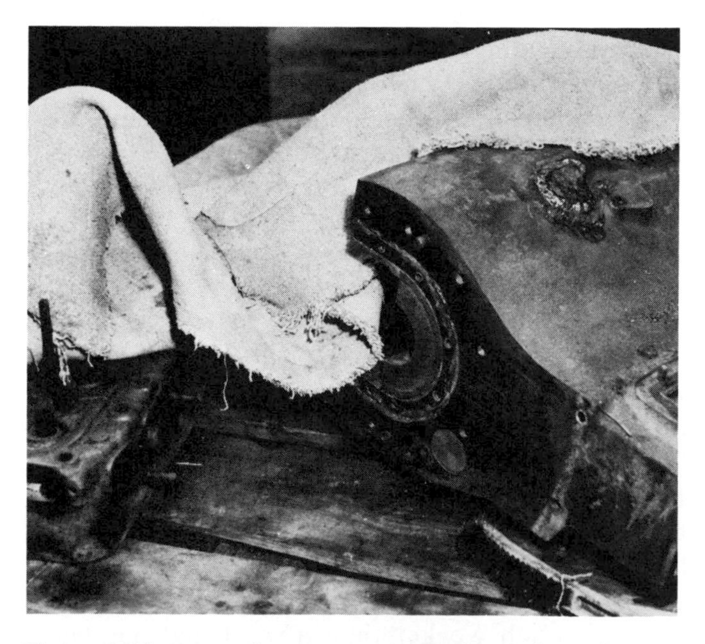

Figure 5. These castings were wrapped in a thermal (insulated) blanket to prevent them from cooling too rapidly. The blanket has been removed to reveal the castings.

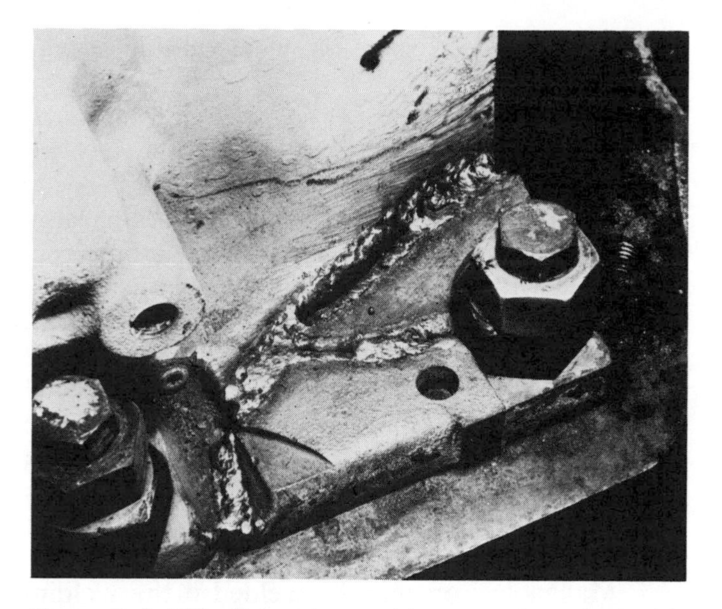

Figure 6. A difficult cast iron weld was made easier in this casting by using a buttering technique on the sides of the weld surface using ESt rod to flux out the large graphite flakes. This was followed by ENi rod to fill in the weld giving the weld more ductility.

If grinding is used for preparation of the V grooves for welding, one way to remove the graphite that is smeared is to heat the surface to be welded with a torch having an oxidizing flame and then allowing it to cool. This causes the surface graphite to form into small, loose spheres that can be removed with a wire brush.

For metal arc welds, the groove can be V shaped with a 60-degree included angle. For brazing, the groove should be wider having a 45-degree or more included angle. This provides a wider bond area that is needed for brazed joints. A hole should be drilled at the ends of the cracks to be welded in a casting in order to eliminate more cracking from the welding stresses.

Welding Rods and Methods Used in Welding Cast Iron

Steel type cast iron welding rods (ESt) are very useful when arc welding gray cast irons. These ESt rods have a coating that fluxes out large graphite inclusions making for easier welding; but these welds are usually too hard to machine. Ordinary steel welding electrodes will not work very well for welding cast iron since they do not have the correct coating. When welding cast iron to steel, the ESt type rod is used, but the weld will not be machinable. These electrodes are also lower in cost than the ENi rods and can be used on small parts where no machining will be needed. A preheat of 400 to 600°F (204 to 315.5°C) is recommended even on these small parts.

The other type, nickel rod (ENi) is used where machinability is a factor. While the ESt rod does not produce a very machinable weld, the ENi rod does, except in the heat affected zone in some cases. ENi (nickel or nickel-iron) rods are widely used for cast iron welding and are most successful.

Castings can also be gas welded using one of the RCi types or they can be metal arc welded using an ECi type rod. Since both rod types are made of cast iron, the weld will approach the characteristics of the base metal and will be machinable if the casting has cooled slowly.

Studding One method of reinforcing a weld that is sometimes used is called studding. This method is used on highly stressed parts such as fly wheels that are worn or broken and need repair. Studding is also used where narrow thin sections cannot be adequately welded for strength and the reinforcing helps to give the part a longer service life, and is often used to restore gear teeth (Figure 7). Where preheat is impractical, such as on large castings and where maximum strength is needed, studding provides the added strength that is often needed.

Buildup Nickel rods are often used to build up worn surfaces that are to be remachined. Sometimes a difficulty with build up repair on cast iron is that the worn surface is only a few thousandths of an inch deep and the fusion zone of the weld is at the place where the final finish cut will be taken. A hard structure sometimes develops at or under the fusion zone in the base metal when the cooling rate is rapid. Consequently, machinability is sometimes

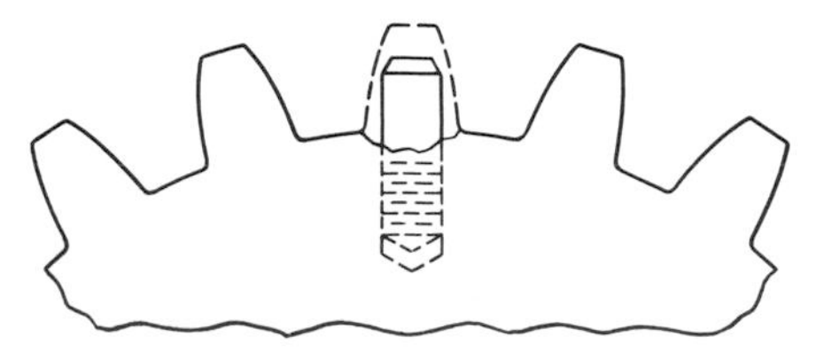

Figure 7. Reinforcing a cast iron weld can be done by studding.

poor even with the so-called machinable ENi rods when it is necessary to make cuts at or under the junction zone. When a situation like this exists, 20 to 30 thousandths of an inch of the surface should be removed by machining prior to welding so that the final finish cut will be taken in the weld zone, which is soft (Figure 8).

Gas Welding If cast iron filler rods are used with gas welding techniques, for example, the postheat should be to a dull red and held for a period of time. But if the arc welding techniques are used, then a postheat just under a dull red is all that is needed. There should be a soaking period, however, to remove stresses.

Brazing Braze welding cast iron makes a very strong bond and is one of the best ways of welding cast iron, but it does not match the original metal in color and is not acceptable in some cases. The welder must be careful not to overheat the base metal when braze welding. The prepared surfaces must be wider than that type of V groove used for arc welding and the welder should make certain that the surface is properly wetted with the bronze before he fills in the weld. If the part is to be used in high temperature operations, such as in an exhaust manifold or for stove parts, braze welding should not be used since the weld will fail at high temperatures.

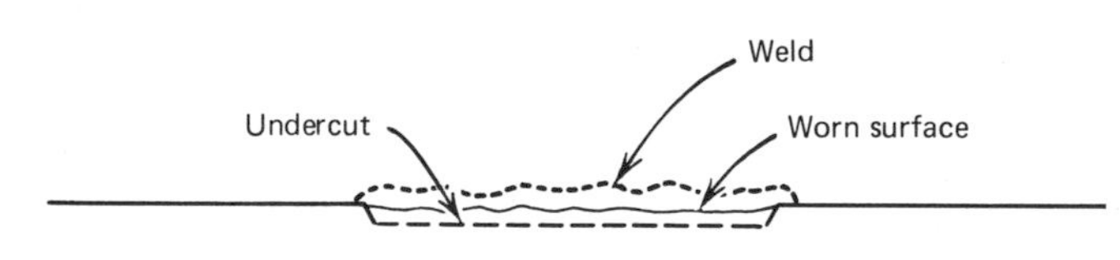

Figure 8. A worn cast iron surface built up with machinable rod should be prepared as shown.

Special Welding Processes

A method of making heavy welded joints in large castings at less expense is in the use of a buttering technique with an ESt or ENi rod followed by a lower cost rod such as E7018 to fill in the weld. This helps avoid the stresses usually caused by welding. The sides of the groove are first surfaced with ESt rod and then the center is filled with the lower cost metal. The buttered halves also can be heat treated before the final welding is done in order to remove stresses in the adjacent and heat affected zones.

Malleable iron may be welded if the welding preheat temperature is kept fairly low. The recommended welding process is by brazing because welding temperatures are low enough to prevent the malleable iron from reverting to white cast iron. Malleable iron will revert to white iron if it is heated above the transformation temperature.

Metal arc welding on malleable cast iron produces brittle structures just as in other cast irons and, if preheating is used, the brittle area will be very wide. If cast iron filler rod and gas techniques are used, the malleable iron must go through a postheat treatment that will restore the malleable characteristics of the material. This heat treatment consists of a prolonged heating, annealing, and cooling process.

Nodular cast iron may be welded with nickel ENi rod or buttered and welded with low hydrogen rod. A preheat should be used to 800°F (427°C), which limits the formation of martensite and helps to eliminate under bead cracking. Postheat (annealing) also should be used to 1200°F (649°C).

Self-Evaluation

1 Is cast iron normally fusion welded without the use of filler rod? Which cast irons can be welded?

2 What is the most common application of cast iron welding?

3 State the major reason why cast irons are somewhat difficult to weld.

4 Which kinds of welding methods may be used for welding cast iron?

5 Why does the lack of ductility of the gray cast iron base metal tend to cause cracking near the weld and not usually in the weld?

6 Name one advantage and one disadvantage of the "keep-it-cool" method of welding gray cast iron.

7 The fusion zone of "machinable" cast iron welds is sometimes hard. What can be done to eliminate or avoid this problem?

8 Why does the weld metal sometimes roll up into balls without fusing to the base metal? How can this be avoided?

9 How can a cast iron weld be reenforced for greater strength?

10 Why must malleable iron go through a prolonged postheat treatment if cast iron rod and gas welding techniques are used?

Worksheet

Objective Determine the condition of three cast iron welds by the use of microscopic examination techniques.

Materials Pieces of gray cast iron prepared to make three different butt welds. ENi and ESt electrodes, ECi oxy-acetylene cast iron rod, metallurgical cutting and polishing equipment, and metallurgical microscope.

Procedure

1 Make three welds with no preheat using three different rod materials: ENi and ESt by the arc process and the ECi rod by gas weld. Allow them to cool to room temperature.

2 Cut a section of each weld and mark the samples for identification purposes.

3 Polish each specimen, but do not etch.

4 Observe the various zones of the weld with the microscope. Draw a sketch of the specimen and make circles around it. Make micrograph sketches in the circles that approximate what you see. Write a conclusion beneath the circles.

5 Now etch each specimen to bring out other microstructures and repeat the process of making sketches as in step 4. Write a conclusion stating the probable condition of the welds, such as hardness, brittleness, admixture, inclusions, microcracking, and bonding.

19
Metallurgy of Welds: Nonferrous

In the early days of welding, procedures for joining nonferrous metals such as copper and aluminum were either nonexistent or very difficult to use and consequently very little welding of these metals was done. These difficulties have long since been overcome and nonferrous metals are now easily welded, soldered, or brazed by means of modern technological methods. This chapter deals with welding techniques and methods for such nonferrous metals as aluminum, copper and copper alloys, and monel metals. The emphasis is on the metallurgical condition of the welds and weldments.

OBJECTIVES

After completing this chapter, you will be able to:

1. List some of the metallurgical conditions and problems related to welding nonferrous metals.
2. Macroetch weld specimens in nonferrous metals to observe and identify the weld structure.
3. Prepare nonferrous welds for microscopic examination to determine the metallurgical condition of the weld.

INFORMATION

While nonferrous metals are for most purposes hardened by cold working, many of them can also be hardened by precipitation heat treatment. Welding on precipitation hardening alloys presents a somewhat different welding problem than welding on nonhardenable, nonferrous metals. Probably one of the major reasons for past difficulties in welding nonferrous metals has been their great affinity for oxygen, which causes the formation of an oxide film. Many nonferrous metals also have a very high thermal conductivity, requiring a high heat input in the weld area in order to maintain a molten puddle. In some ways, the welding of nonferrous metals is similar to the welding of ferrous metals. Grain growth is a problem in both kinds of welding. Porosity, inclusions, cracking, and embrittlement are all common to both types of welding. As you study this chapter, you should refer to Chapter 14 ''Heat Treating of Nonferrous Metals'' to help clarify some of the metallurgical concepts that are seen in this chapter.

Joining Processes Used for Aluminum

Of all nonferrous metals aluminum is probably the most easily welded metal; it can be readily arc welded, brazed and soldered. All three of these joining processes are used extensively in industrial manufacturing. Aluminum also may be resistance welded, induction welded, ultrasonic welded, gas welded, pressure welded, and flow welded. These major groups of welding may also be broken down into subgroups that include many other kinds of welding processes.

Probably the most common kind of arc welding done on aluminum in small shops is that of TIG (tungsten inert gas), where an inert gas is used to shield the aluminum surface from the atmosphere. Metal arc welding with specially coated electrodes is also used but less often.

There are three characteristics of aluminum that make it behave differently from steel. These are its great affinity for oxygen, its high thermal conductivity, and the fact that it does not change color before reaching its melting point. This persistence of color often causes the operator to have some difficulty in detecting welding temperatures. Aluminum is very active chemically and has such an affinity for oxygen that on clean metal an oxide film forms immediately. This oxide film, which is thin, transparent, but very hard, prevents the weld metal from joining with the base metal. If any welding is to be done, this film must be removed. It could be removed mechanically by scraping, by using a wire brush, by sanding, or it can be dissolved chemically with a flux. A chemical reaction can replace the oxide with another material that does not hinder bonding. Also the heat of welding can sometimes remove the oxide and float it off into the slag. Aluminum welds are very subject to contamination, even the filler metal itself can be contaminated, causing a poor weld. Aluminum sheet or plate to be welded should be carefully cleaned, and preferably the plate should be sheared, sawed, or machined just prior to the welding operation. All lubricants, oils, oxides, and greases should be removed (Figures 1*a* and 1*b*).

Aluminum does not change color before it becomes molten and it only begins to glow a dull red a few hundred degrees above the melting point. This fact causes some difficulties when gas or TIG welding. One way of determining temperatures of the base metal is by marking it with temperature indicating crayons that melt at a specific temperature. Another method, and the most common one, used to determine the proper welding temperature is to watch the adjacent area near the weld for a "wet" appearance of the surface. This indicates that there is melting at the surface. Because aluminum conducts heat so rapidly, the whole area around the weld tends to become weakened because of the high temperature it develops, especially when gas welding. The base metal, if it is overheated in the vicinity of the melting point, can fail, causing the weldment to collapse near the weld. With arc welding, this problem does not occur since the weld is made so rapidly that there is not sufficient time for the base metal to overheat. In either case, however, mechanical support should be provided for the weldment because it loses strength at higher temperatures.

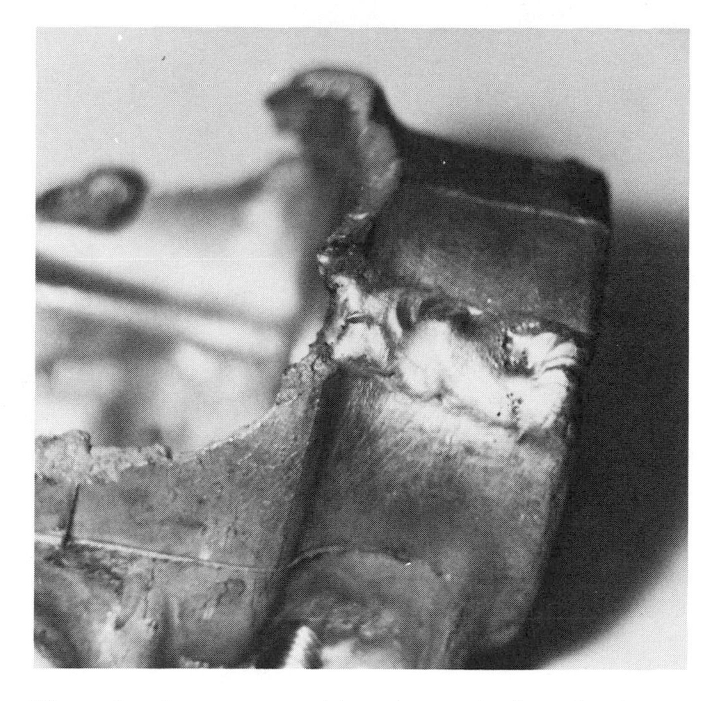

Figure 1*a*. A very poor weld made on a broken aluminum casting, showing surface porosity.

Weldability of Aluminum Alloys

Since most aluminum is strain hardened and most of the aluminum used is in the form of an alloy, the base metal is almost always stronger than the weld metal, which is cast and not strain hardened. The base metal near the weld is softened by recrystallization of the grains; this gives the heat affected zone more ductility but a lower tensile strength than the rest of the base metal. It is extremely

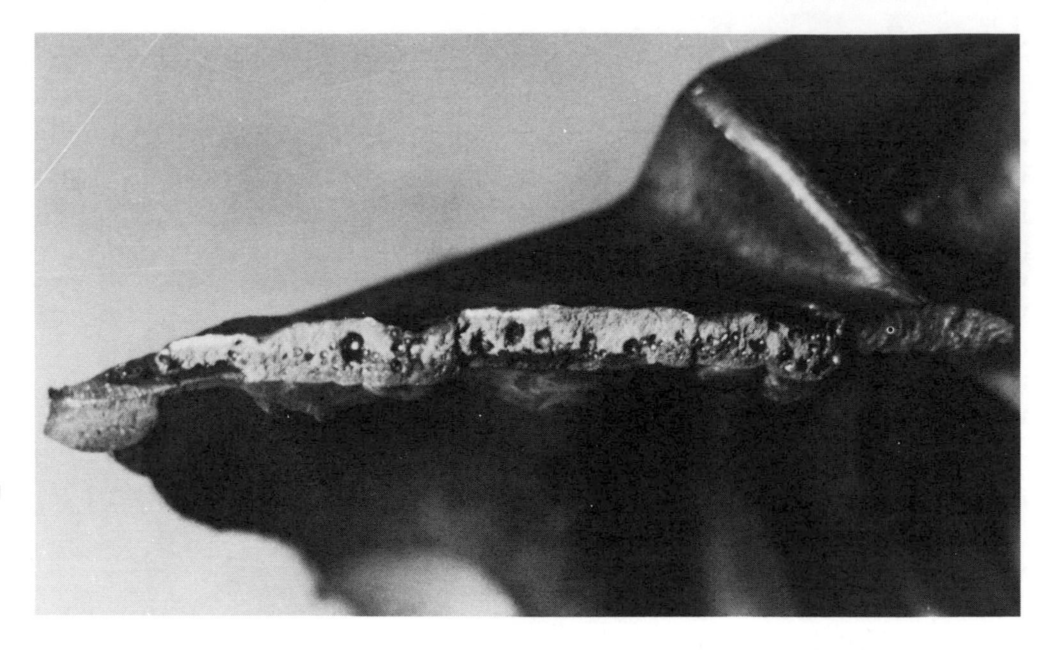

Figure 1*b*. Broken section of a weld made on the same casting shown in Figure 1*a*. The great amount of visible porosity is due to contaminants on the base metal.

important for the welder to know the kind of aluminum alloy on which the weld is being made. The four digit system of designation for aluminum alloy may be found in Chapter 4 "Identification of Nonferrous Metals." Of the aluminum alloys, the 1000, 3000, and 5000 series are not heat treatable and they can be hardened only by cold working. These and other alloys such as the heat treatable 6000 series can be gas welded or arc welded with no difficulty. The 2000, 4000, and 7000 series present the greatest difficulty in welding. Arc welding and resistance welding are preferred when welding is done under controlled conditions. Gas welding is not recommended. See Table 1. The 2000, 4000, and 7000 series of aluminum alloys are heat treatable since they contain copper, magnesium, manganese, chromium, zinc, or silicon alloying additions; postweld heat treatments should be used.

The heat treatable alloys such as the 2014, 2017, and 2024 types can only be welded by using special techniques, fixtures, and heat treatments (if necessary).

The Effects of Welding Heat on Aluminum Alloys

The weld zone of almost all aluminum alloys is adversely affected by high temperatures. The weld area becomes weaker than the unaffected base metal because of grain growth. Even on annealed aluminum the cast structure of the weld itself has a lower strength than the base metal. Sometimes a filler alloy will increase the weld metal strength. When a heat treatable alloy such as 2024 (which is solution heat treated and aged) is used in a structural application and welded, the heat affected zone can be adversely affected because copper aluminide precipitates to the grain boundary. However, there are methods used for welding such alloys.

The heat treatable alloys of the 2000, 4000, and 7000 series should not be welded unless very stringent controls are used to produce the original heat treated condition in the weld zones; exceptions are the use of one of the more weldable alloys. Some alloys that can be heat treated by postweld solution heat treatment permit partial or complete restoration of their strength and avoid grain boundary precipitation after welding. Welding of 7075, 7079, and 7178 is not recommended, but 7005 and 7039 were developed for welding.

Aluminum alloys have a tendency to be hot short; that is, they possess low strength at solidification temperatures and as the weld melt cools, the strain that is placed on the high temperature weld metal sometimes causes hot cracking (or tearing) that is usually evident in the center of a weld. The factors that cause or influence hot cracking are the rate of cooling, the weld size, restraint of the joint, and the solidification temperature of the alloy. For example, pure aluminum melts at about 1200°F (649°C) while some aluminum alloys melt at lower temperatures, some as low as 900°F (482°C).

Table 1 Relative Weldability of Heat Treatable Wrought Aluminum Alloys and Cast Aluminum Alloys That Are Not Heat Treatable

Welding Process	Wrought Aluminum ASA Alloy Designation																	Cast Aluminum ASA Alloy Designation														
	2014	2017	2024	2218	2219	2618	6061	6063	6070	6101	6201	6951	7005	7039	7075	7079	7118	443.0	A444	208.0	213.0	514.0	B514.0	F514.0	C712.0	A712.0	D712.0	413.0	A514.0	518.0	360.0	380.0
Resistance	A	A	A	A	A	A	A	A	A	A	A	A	A	A	A	A	A	A	C	B	C	B	B	B	A	C	C	C	B	C	C	C
Pressure	C	C	C	C	C	C	B	B	C	A	A	A	B	B	C	C	C	O	O	O	O	O	O	O	O	O	O	O	O	O	O	O
Metal arc	C	C	C	C	C	C	A	A	B	A	A	A	O	O	O	O	O	A	C	B	C	C	C	C	A	C	C	B	C	O	O	O
TIG	C	C	C	C	C	A	A	A	A	A	A	A	A	A	C	C	C	A	B	B	B	A	B	A	A	B	B	B	A	B	B	B
Brazing	O	O	O	O	O	O	A	A	C	A	A	A	B	C	O	O	O	C	A	O	O	O	O	O	B	A	A	O	O	O	O	O
Soldering	C	C	C	C	C	C	B	B	B	A	B	A	B	B	C	C	C	C	B	C	C	O	O	O	C	B	B	O	O	O	O	C
Gas	O	O	O	O	O	O	A	A	C	A	A	A	O	O	O	O	O	A	C	C	C	C	C	C	A	C	C	B	C	O	O	O

A. Readily weldable.

B. Weldable in most applications; special techniques may be necessary.

C. Limited weldability.

O. Not recommended for that joining method.

Note. Ratings are based on the most weldable temper.

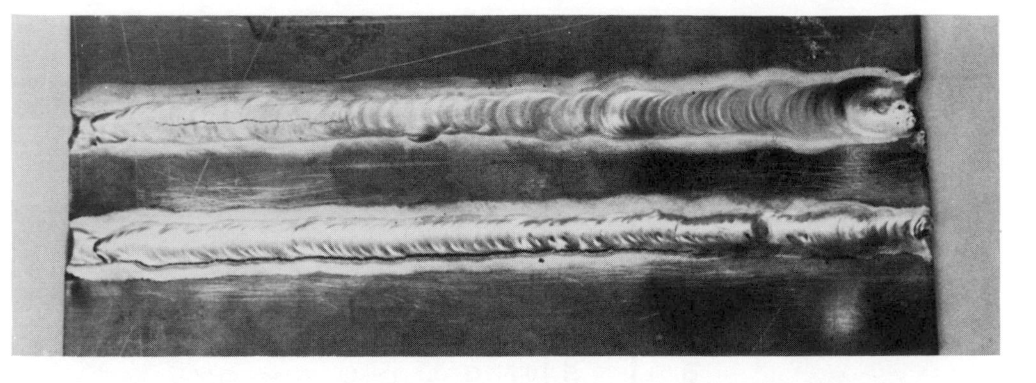

Figure 2. (Top) Hot cracking in the center of the weld. (Bottom) Toe cracking in aluminum alloy weld caused by overheating.

The use of preheat helps to minimize the effects of hot cracking by reducing the rate of cooling, but preheating should not be used if the joint is restrained. Also the mechanical properties of the base metal can be decreased by preheating, especially with the heat treatable alloys (Figure 2). Hot short cracking can also be controlled by selection of the proper filler metal.

Welding Copper

Pure copper presents some welding difficulties because it has an affinity for oxygen and because it contains a small amount of oxygen in the form of oxides that tend to migrate to the grain boundaries, thus weakening the copper. The gas metal arc process is recommended for welding deoxidized copper, and the weld should be made rapidly to prevent the formation of oxides. Gas welding is not recommended because of its slow rate of deposition and its high heat input. Gas metal arc welding may be used to join electrolytic and oxygen-free copper, but the welds are generally of poor quality. Shielded metal arc welding with the proper electrodes is used on some brasses, bronzes, and deoxidized copper.

All of the numerous copper alloys are weldable with various types of welding processes. The AWS classification for welding rods and electrodes for copper and copper alloys may be found in welding handbooks.

In copper and copper alloy welding the same problems and defects occur that are found in other types of welding and, usually, the same inspection methods can be used. Visual inspection will usually detect porosity and high temperature cracking. Other defects could be found by ultrasonic, radiograph, and other methods of testing. If copper must be welded by oxy-acytelene methods, then deoxidized copper should be used. This is copper that contains a small percentage of silicon, phosphorus, or other deoxidizers. Where any weld strength is required, oxygen-free copper should be used. Many of the same problems are associated with the welding of coppers that are found in aluminum welding, such as its high thermal conductivity, which rapidly removes the heat of welding, distributing it throughout the material. Also its high affinity for oxygen and other gases such as hydrogen and carbon monoxide creates problems. Copper has a high coefficient of expansion that tends to promote cracking in the weld during cooling, especially if there is a restraint in the base metal.

All of the hundreds of copper alloys, brasses, bronzes, and beryllium alloys are weldable. The copper alloys containing lead are the most difficult to weld since the lead is volatized (becomes a gas) before the copper melts and the alloy becomes very weak at welding temperatures.

Some aluminum bronzes and beryllium copper are heat treatable alloys. A suitable postheat treatment should be used when welding these alloys to restore the original strength and hardness.

Welding Dissimilar Metals

The nickel alloy electrodes are often used for joining dissimilar metals. Monel electrodes are used to join nickel to carbon steel by metal arc welding, or for welding copper to carbon steel. Austenitic stainless steel or inconel electrodes are also often used for joining dissimilar metals such as nickel to stainless steel.

Welding Nickel Alloys

Monel, nickel, and inconel are all readily welded. They can be welded with filler metal containing the same analysis as the base metal; the heat affect of welding does not alter the properties of the base metal to any great extent. In nickel, the age hardening alloys of the K-monel, Z-nickel, and Hastalloy group are hardenable and may be welded by several methods. Shielded metal arc is usually preferred, but gas welding may be used, though excessive grain growth may take place in the heat affected zone. These alloys are hardenable simply by heating between 1100 and 1600°F (593 and 871°C) and then cooling slowly. A quenching step is not needed as in aluminum alloys. Therefore, a simple annealing procedure is all that is required after welding to give the weld and the heat affected zone uniform hardness and higher corrosion resistance.

Welding Titanium Alloys

In Chapter 14 ''Heat Treating of Nonferrous Metals'' you learned about titanium in its various allotropic conditions such as the alpha and beta transformations resulting from varying alloying elements. For example, the addition of aluminum raises the transformation point of the alpha phase and chromium, vanadium, iron, or zirconium slow the transformation from beta to alpha when cooling. The least weldable group of titanium alloys are those in the mix or alpha-beta alloys; although some of these have been fusion welded under special conditions. The most important aspect of welding titanium alloys is to avoid contamination by the formation of oxides. This is true of all the reactive metals such as zirconium, magnesium, and tantalum. TIG welding is often used for joining titanium where an inert gas is used to prevent contamination. Weld areas should be extremely clean; a very small amount of oil, even from a fingerprint, can cause embrittlement. Titanium is usually welded in the annealed state to prevent embrittlement (Figure 3).

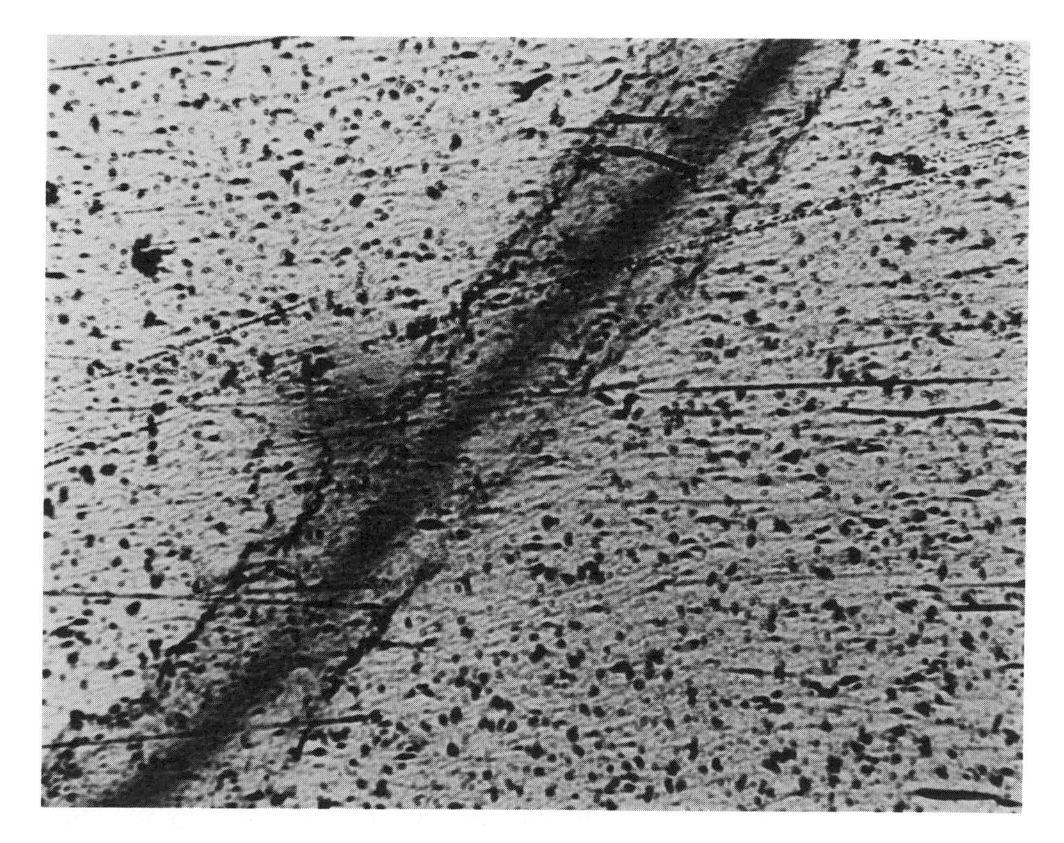

Figure 3. Junction zone of a weld on titanium alloy done with TIG process using argon gas for shielding.

Self-Evaluation

1 What is the single most important reason why nonferrous metals were difficult to weld in the past?

2 Why do some nonferrous metals require a high heat input when they are welded?

3 Why should aluminum be carefully cleaned for welding by brushing or scraping?

4 What takes place in the heat affected zone of welds in strain (work) hardened aluminum? How does this affect strength?

5 Many aluminum alloys that are readily welded of the heat treatable alloys such as the 2000, 4000, and 7000 series present some difficulties in making strong welds. What is the reason for these difficulties?

6 Oxygen may be present in copper or may be introduced into the weld by the welding process. How can it severely weaken the final weld?

7 When welding heat treatable nickel alloys, what procedure is needed to restore the original strength?

8 Which kind of filler rods are most useful for welding dissimilar metals?

9 Reactive nonferrous metals (magnesium, tantalum, titanium, and zirconium) are not welded with coated electrodes in the presence of the atmosphere. Why is this so? What forms of welding can be used?

10 How can a welder avoid embrittlement in titanium welds?

Worksheet 1

Objective By macroetching, illustrate changes due to recrystallization in aluminum welds in the heat affected zone that are accelerated by prior cold working.

Material Two pieces of $\frac{1}{4}$ in. thick aluminum, 1000 or 3000 series H0 to H5 (soft) prepared for welding, two more pieces of the same size aluminum H34 to H36 (hard), TIG welder and appropriate aluminum welding rod, a 5 percent solution of sodium hydroxide in methanol, a petri dish, and tongs.

Procedure

1 Preheat to 300 or 400°F (149 or 204°C) and weld each of the specimens together.

2 When cool, section each weld by sawing across them. Make an identifying mark on each specimen.

3 Grind them flat on a belt grinder and finish on 400 grit paper.

4 Pour the sodium hydroxide into the petri dish and, using tongs, place each specimen in the petri dish face up so that you can observe the progress. There must be enough solution to cover the specimens.

5 When the grain structure is visible, remove the specimens and hold them under running water.

Conclusion Which specimen shows the greatest amount of grain growth in the heat affected zone? Which weld do you think is stronger? If you wish, you may prepare these two specimens for microscopic examination to further study the various weld zones.

Worksheet 2

Objective By microscopic examination, determine the affects of welding heat on heat treatable aluminum alloys.

Material Two $\frac{1}{4}$ in. thick specimens of 2024 aluminum alloy preferably in a solution heat treated, aged condition, a 5 percent sodium hydroxide and methanol enchant, polishing equipment, and a metallurgical microscope.

Procedure

1. Prepare the welded specimen for polishing and etching.
2. Begin with a light etch. Etch again, if necessary, to bring out the grain structure.
3. Mount in the microscope and set at 400× or 500×.
4. Observe the normal grain in the base metal well away from the weld, then compare them with the grains in the heat affected zone. You should see some evidence of copper aluminide precipitation at the grain boundaries.
5. Search the adjacent and weld zones for unusual grain structures.

Conclusion Do you think the heat affected zone has been weakened by welding? How could this be corrected by postheat treatment? Draw a circle and make a sketch of the grain structure that you see in the microscope; identify important areas. Write a conclusion to your paper and submit it to your instructor.

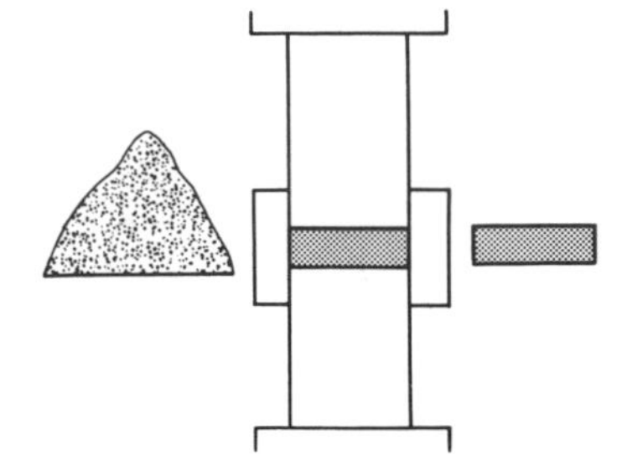

20 Powder Metallurgy

Powder metallurgy, commonly referred to as P/M, is one of four major methods of forming metals. The other three are by casting of molten metal, by the plastic deformation of hot or cold metal, and by machining. P/M is the process of producing useful metal shapes from metallic powders. Metals in powder form has had limited use for many centuries. Iron tools were made by the Egyptians by the powder metallurgy principle 3000 years ago. Other metals such as gold, silver, tin, and copper have been made into ornaments from ancient times to the present through powder metallurgy. Thomas Edison, in his search for a suitable filament for the incandescent electric lamp, could not use tungsten although it was known to be an ideal material for a filament. Metallurgical techniques had not yet been developed to produce tungsten wire. The P/M method of forming tungsten wire was finally developed in 1909, making possible the manufacture of modern electric light bulbs. In this chapter, you will learn how P/M parts are made and how they are used.

OBJECTIVES

After completing this chapter, you will be able to:

1. Describe the methods used to manufacture P/M parts and some of their characteristics.
2. Determine the density of P/M parts by microscopic examination.

INFORMATION

Modern powder metallurgy was actually an early development of the automotive industry. During World War II and since that time, new methods have been developed and many new products have been produced by this method. With P/M it is possible to make very complex, unique shapes that are not practical with other metal forming processes. Very good surface finishes can be obtained, and controlled porosity or permeability for filtration is possible. Powder metallurgy systems are suited to high volume production of small parts. Almost any combination of alloys can be used to produce high temperature components and very hard or tough products such as tungsten carbide tools. Close dimensional tolerances can be maintained. Machining is eliminated or reduced. Scrap losses as found in conventional methods of manufacture are eliminated by the powder metallurgy method.

Some of the products that are made by P/M are cutting tools such as tungsten carbide inserts and cermets (ceramic tools). Precision parts such as cams, gears, and links are also made by P/M. P/M techniques make possible the manufacture of antifriction materials such as self-lubricated porous bearings, filters such as those used in gasoline lines in automobiles, and high strength magnets such as the well-known alnico magnet. High strength metal parts used at high temperatures, such as turbine blades in jet engines, which cannot easily be fabricated by conventional methods, are made possible by P/M (Figure 1).

There are three basic steps in the manufacture of P/M parts:

1 *Blending*. Metal powders are mixed together with alloy additions and lubricants until they are thoroughly blended.

2 *Compacting*. The blended metal powders are fed into a precision die and pressed, in most cases, at room temperature with pressures from 10 to 60 tons per square inch. The part, called a green compact or briquette, is ejected from the die.

3 *Sintering*. The green compact is heated in a controlled atmosphere furnace to just under the melting point of the metal to bond the compressed powders into a strong homogenous structure.

Metal Powders

Metal powders used for powder metallurgy range in size from microscopic, less than 0.0001 inch, to about 0.002 inch in diameter. Powders are available for almost any of the metals and alloys. Metals and other elements such as graphite, which are insoluble in the solid state, are easily combined as mixed powders and compacted to produce a strong metal part. Powders are classified according to particle size and shape in addition to other considerations such as the presence of impurities and the metallurgical condition of the grains.

Metal powders are produced by several methods, some of which are listed here:

1 *Mechanical*. Milling, using crushers, rollers, and ball mills (Figure 2) where metals are disintegrated to produce the size of powder required. The final grinding is usually done in a rotating ball mill that has many steel balls that impinge upon the powder and grind it to the needed size.

2 *Chemical*. Reduction, which converts metals from the oxides (ores) directly to metal powders at a temperature below their melting point. For example, iron powder is produced by direct reduction from the iron ore or from the mill scale.

3 *Shotting*. The method of making powders (usually rather coarse) by passing molten metal through a sieve and dropping the particles into water.

Figure 1. Some of the hundreds of metal parts made by powder metallurgy.

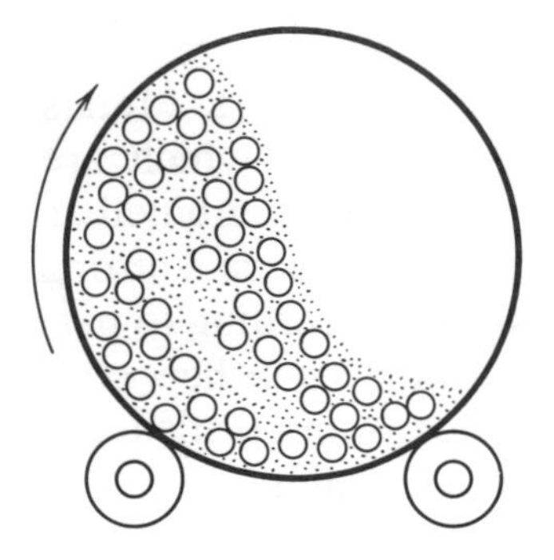

Figure 2. The action of the ball mill is shown as a continuous grinding as the drum rotates.

4 *Atomization.* The spraying of molten metal to produce powders. This process is generally only used with low temperature metals such as tin, lead, and zinc.

When the powders are produced by any of these methods, they must be classified and blended to be ready for use. Vibratory screens or sieves are used to separate or classify the granules according to particle size. Other vibrating mills aid in the blending and mixing of various components such as would be found in sintered tungsten carbide that has a cobalt matrix powder plus the tungsten carbide granules. Two or more metallic powders may thus be blended together to impart special properties to the final product. The mixed and blended powders are pressed into a die in a hydraulic or crank type press to form the correct shape. The powder is fed through a chute into an opened die cavity; space in the die is allowed so that when the punch exerts the pressure on the powder, cold welding the particles together, the part is formed into what is known as a green briquette. The part is then ejected from the die and it is hard enough at this stage to be handled, but it can be damaged if it is dropped or mishandled (Figure 3).

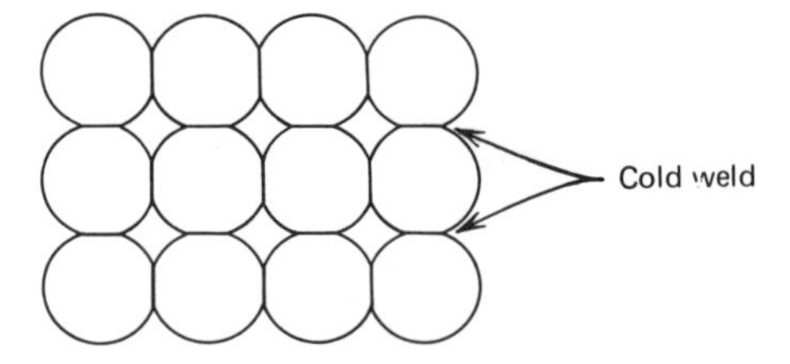

Figure 3. The compressed particles in the briquette are cold welded together at this stage with very weak bonding.

The depth and shape of the die cavity, the amount of powder, and the amount of pressure all determine the density that is required. The briquette is then moved to a furnace where it is sintered; that is, held at a predetermined temperature for a given length of time. Here the powders begin to bond together (Figure 4). The steps needed to produce a powder metal part are as shown in Figure 5.

Secondary Operations

Unless more precision is needed, the parts are now ready for use. For greater precision the parts can be sized in a coining die that brings them to the needed dimension. Other secondary operations include impregnating the parts with antifriction material such as graphite or oil to make self-lubricating bearings. The density and strength of the part can be increased by infiltrating the pores with a lower melting point alloy. Sintered products can also be modified by machining, plating, and heat treatment.

Sintering

In **solid phase sintering** (the most common method), the green compact part, or briquette, must be sintered in a carefully controlled atmosphere furnace from $\frac{1}{2}$ to 2 hours at 60 to 80 percent of the melting point of the lowest melting constituent. **Liquid phase sintering** is carried out above the melting point of one of the alloy constituents or above the melting point of an alloy formed during sintering. Sintering, a solid state process, develops metallurgical bonds. The important changes that take place during this process are:

1 *Diffusion.* This takes place especially on the surface of the particles as the temperature rises.

2 *Densification.* Particle contact point areas increase considerably. When this happens, there is a loss in porosity that reduces voids. Also, there is an overall decrease in the size and some distortion of the part during the sintering process. A coarse, green briquette must be made larger to allow for this shrinkage.

3 *Recrystallization and grain growth.* This occurs between particles in a contact area, causing the

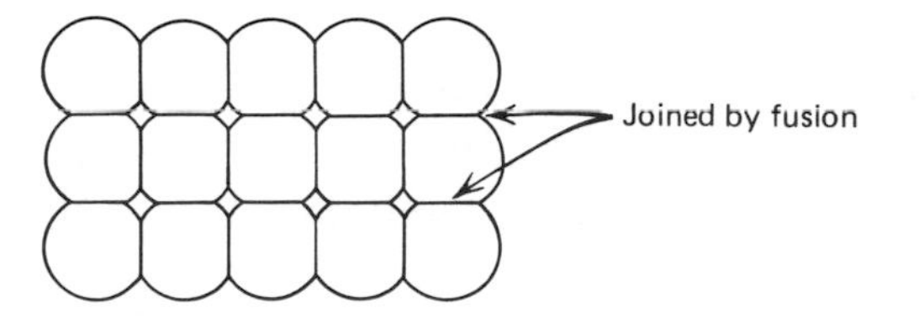

Figure 4. After sintering, the particles in the briquette become fused together.

grain and lattice structure such as those in a solid metal to join but still leaving voids or holes between the particles, depending on the amount of compression in the green briquette and the time of sintering.

Density of Pressings

The density of a pressed article depends on many factors such as particle size, pressure, and the time of sintering. **Hot pressing,** in which the article is pressed while at a high temperature, produces a density approaching that of solid metal and has a higher ductility than solid phase sintered parts. In contrast, large articles are sometimes "loose sintered" in a mold and then forged or cold worked. Loose sintering is also used to produce porous filters. Another means of producing very porous articles is by combining the metal powder with a combustible or volatile substance like sawdust that is removed by the heat of sintering. Density can vary in the pressing depending on the shape of the part. Special methods and molds are needed for some products (Figure 6).

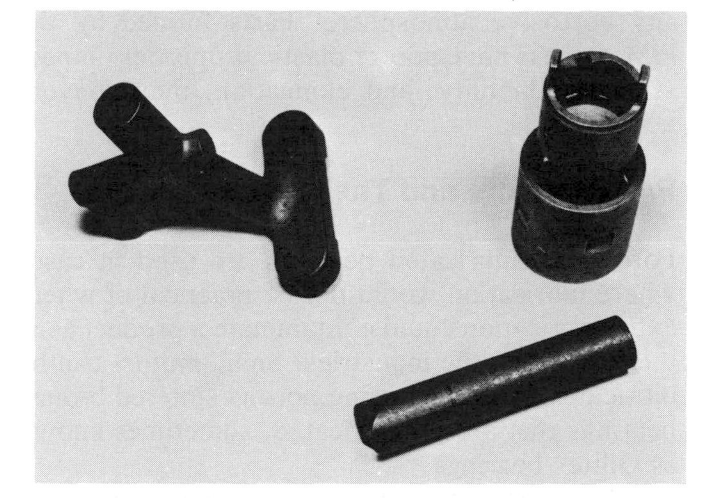

Figure 6. Complex shapes such as these require special molds and techniques to avoid incomplete filling and variations in density.

Advantages of P/M

Production of small intricate parts can be produced at a rate that is as much as three times greater than by conventional means. Tooling costs are relatively low. Certain unique products can be made with a great variety of combinations of metals and nonmetals. P/M parts may be heat treated, machined, plated, or impregnated with lubricants and other materials.

Disadvantages of P/M

P/M parts have a somewhat lower resistance to corrosion than solid metals because they are porous and they present a larger internal surface to

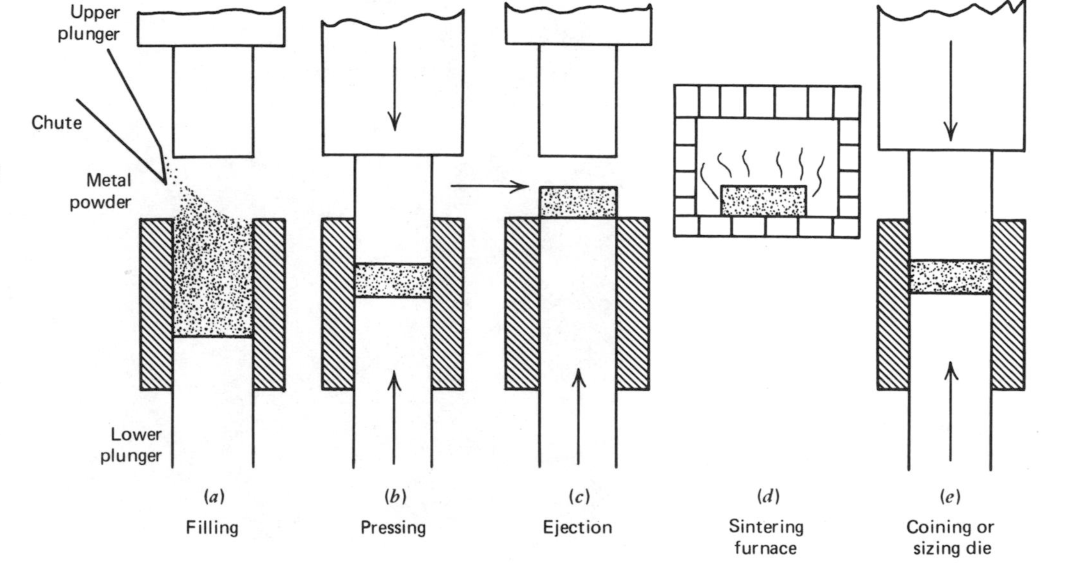

Figure 5. The depth of die cavity and length of plunger stroke is determined according to the density required. *(a)* A measured amount of metal powder is placed in the die cavity through the chute. *(b)* Pressure is applied. *(c)* The briquette is ejected from the die cavity. *(d)* The parts are sintered at a specified temperature for a given length of time. The parts are now ready for use. *(e)* If more precision is needed, they can be sized in a coining die.

any corrosive atmosphere. Parts formed by the P/M process have poorer plastic properties (impact strength, ductility, and elongation) than conventional metals.

P/M Products and Their Uses

Porous, prelubricated bearings are used in cases where lubrication would not be practical or where extra lubrication could contaminate a product as in the food or textile industries. Small motors can be lubricated for life by using porous sintered bronze bearings that are prelubricated, sometimes known as Oilite® bearings.*

Machining of wear surfaces in porous bearings is not recommended because the voids become partially closed during cutting operations such as reaming. This seals the lubricant in the bearing so that it cannot circulate on the wear surface. It is the capillary action of the bearing that allows the impregnated oil to flow onto the bearing area, especially when the temperature rises (Figure 7).

Porous iron-graphite bearings are extensively used in machines as they have a higher wear resistance than babbitt or bronze and are economical to manufacture. Aluminum base antifriction materials are also made by the P/M process. These also contain graphite, iron, and copper in small percentages. Copper-nickel powders are often formed as a layer on a steel strip and sintered. The sintered strip is impregnated with babbitt metal and is then formed into insert bearings for automobile and aircraft engines.

Creep resistant alloys are made from tungsten, molybdenum, tantalum, and other powders such as carbides, nitrides, borides, and silicides. These articles are used in high temperature applications such as jet engine impeller blades, heating elements, and electronic equipment.

Very hard alloys for cutting tools are made by the P/M process. Tungsten carbide, tantalum carbide, and titanium carbide are all used to give various properties to the "carbide" family of cutting tools. These powders are usually bonded in a matrix of powdered cobalt. Ceramic cutting tools are also made by the P/M process (Figure 8).

Friction materials such as brake shoes for automobiles are often made by P/M. P/M friction materials are usually in the form of bimetal powder materials bonded to a steel base. Strips, discs, and other forms are made this way.

Mechanical parts are extensively made by the P/M process. Gears, cams, levers, and link mechanisms are but a few of the many types of small precision P/M parts produced in this country (Figures 9*a* and 9*b*).

*Oilite® is a trademark of the Chrysler Corporation.

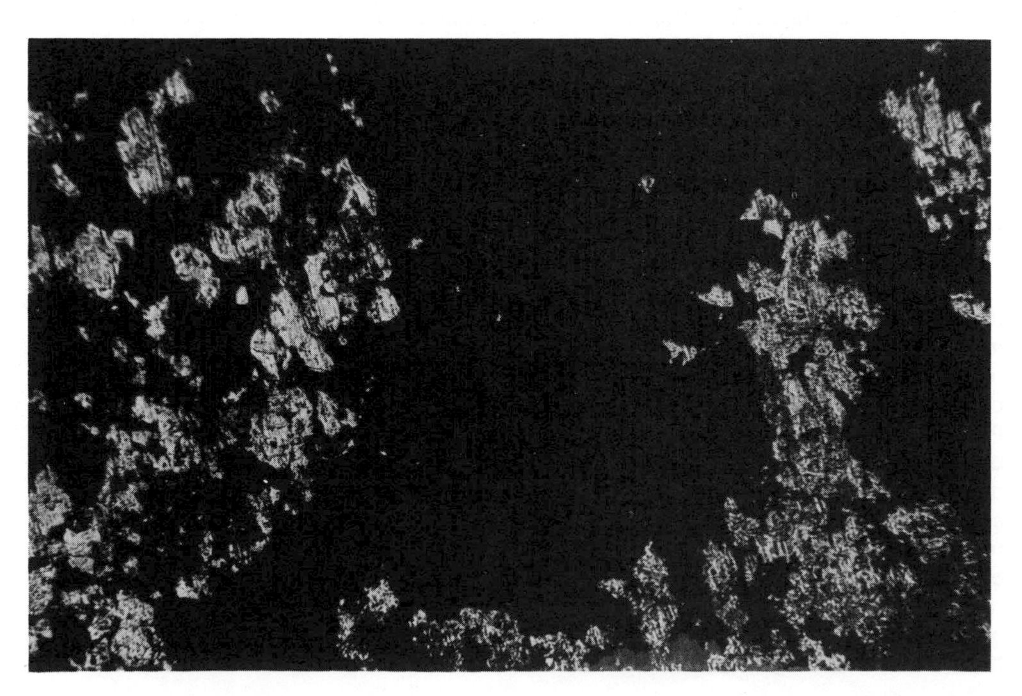

Figure 7. Micrograph of a porous bronze sintered bearing (500×) showing large voids (black areas) that are impregnated with oil.

Figure 8. Ceramic cutting tools such as this insert are produced from aluminum oxide powders by the P/M process.

Figure 9*b*. These gear pump parts made by the P/M process are examples of the very high precision attainable with this method.

Figure 9*a*. Small gears such as these, when made by the P/M process, must be forced through a resizing die.

P/M is widely used with many more applications besides those presented in this chapter. A continuing development in this field is to be expected, which has and will continue to have a great economic importance for the metals industry.

Self-Evaluation

1. Although P/M was used in ancient times to some extent, modern powder metallurgy as we know it was an outgrowth of what industry?
2. Name some kinds of unique parts that can be made by P/M.
3. State the three usual steps for making P/M parts, assuming that the mixed and blended powder is already prepared.
4. What holds the green briquette together before it is sintered and what bonds it together after it is sintered?
5. Name some advantages and disadvantages of P/M.

This chapter has no Post test.

Worksheet

Objective Determine the relative porosity of several sintered parts by microscopic examination.

Material Several discarded sintered metal parts with widely diverse uses, such as a porous bearing or gas line filter, and parts with more density, such as gears or levers. Specimen mounting and polishing equipment. A metallurgical microscope and magnifying glass.

Procedure

1. Fracture each specimen in a vise with a hammer. Cover it with a shop cloth before striking it and wear safety glasses.
2. Observe the fractures with a magnifying glass.
3. Encapsulate the parts for polishing. Put an identifying mark on them.
4. Polish the samples with successively finer grits of paper and finish with 600 grit. Do not polish on a motorized polishing wheel and do not etch. This process must be done with extreme care since the surface metal tends to slide over and cover the pores if excessive pressure is applied while polishing.
5. The specimen is now ready for microscopic study. Use 100× power at first. Compare the various specimens at this magnification.

Conclusion How does the structure of these samples compare to conventional metals? How did each sample compare with the others? Make a sketch of each sample as you see it in the microscope. Write your conclusion and give it to your instructor.

21 Nondestructive Testing

Nondestructive testing methods are among the most useful tools of modern industry. Inspection and testing of each part are sometimes necessary for critical aircraft parts. This is regularly done with the testing systems explained in this chapter.

OBJECTIVES

After completing this chapter, you will be able to:

1. Name the several methods of nondestructive testing and explain the specific uses and operation of each.
2. Use testing equipment for inspecting test pieces.

INFORMATION

Nondestructive testing, as the name implies, in no way impairs for further use the specimen that it tests. Usually these tests do not directly measure mechanical properties, such as tensile strength or hardness, but are intended to locate defects or flaws. When machines or parts have large safety factors built into them, there is little need for nondestructive testing. However, many products used in aircraft, space technology, and other industries require a high level of reliability. This is achieved by inspection at the time of manufacture and, in some cases, continued testing during the service life of the part. The most common types of nondestructive tests are **magnetic-particle inspection, fluorescent-penetrant inspection, ultrasonic inspection, radiography (X ray and gamma ray), eddy current inspection, and visual inspection.**

Magnetic-Particle Inspection

There are several methods of magnetic-particle inspection (Magnaflux) that are used to detect various kinds of flaws in ferromagnetic metals such as iron and steel (Figure 1). A magnetized workpiece is sprinkled with dry iron powder or submerged in a liquid in which the particles are suspended. The Magnaflux Corporation has developed a method called Magnaglo in which fluorescent-magnetic particles are suspended in solution. The solution is flowed over the magnetized workpiece, which is then viewed under a black (fluorescent) light. Cracks that are only a few millionths of an inch wide can be found by this method (Figure 2).

The wet method is useful for inspection in the manufacture of parts and for maintenance inspection. The dry method is generally used for inspec-

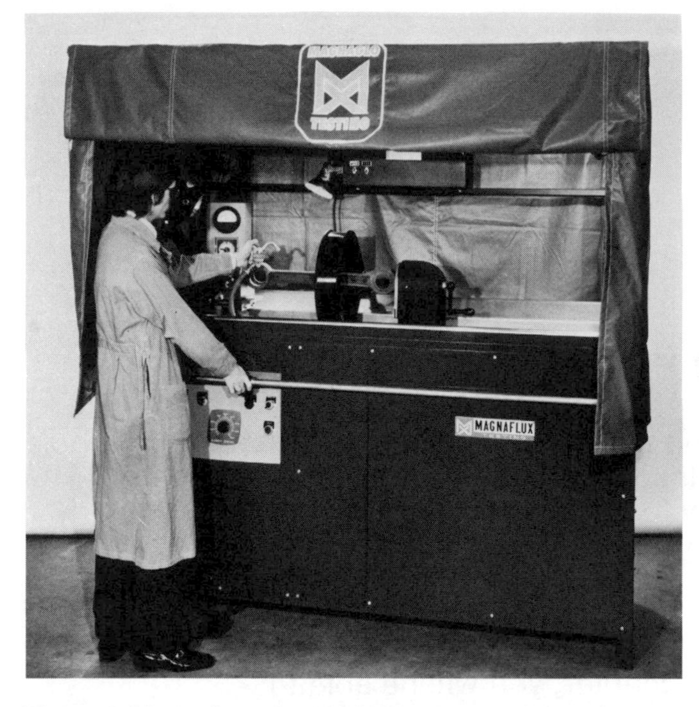

Figure 1. Magnaflux type "H-600" series machine. A versatile shop unit that will handle the requirements of all magnetic particle testing up to the capacity of the machine (Courtesy of Magnaflux Corporation).

tion of welds, large forgings, castings, and other parts having rough surfaces. The basic principle of magnetic testing is shown in Figure 3. A magnetic pole is formed at a crack, which causes the magnetic powder to concentrate at that point. When a part is magnetized lengthwise, as shown in Figure 4, transverse (crosswise) cracks can be detected. This is done by energizing a coil around the bar. If an electric current is passed through the bar, however, a circular magnetic field results and the defects can be found that are lengthwise to the part (Figure 5). Because the part must often be demagnetized after testing, this feature is built into the machine. This system of inspection is limited to the magnetic metals such as iron and steel.

Fluorescent-Penetrant Inspection

Invisible cracks, porosity, and other defects are also found by this method in metals such as iron, steel, aluminum, bronze, tungsten carbide, and nonmetals such as glass, plastics, and ceramics. Fluorescent-penetrant inspection is widely used for testing and inspection on these materials. Zyglo, the Magnaflux Corporation copyrighted name for this method, is similar to Magnaglo in the use of black light to make the defects glow with fluorescence (Figure 6). Once applied, Zyglo penetrant is drawn into every defect no matter how fine or deep, but it must be given the time to do so. The length of time depends on the type of defect. The surface film is rinsed off with water, developer is applied to draw out the penetrant, and the part is inspected for defects under black light where cracks and other flaws will fluoresce brilliantly. Zyglo systems are available in many sizes ranging from hand portable test kits to huge automated systems.

Dye Penetrants

A similar method of nondestructive testing using dye penetrants is visual inspection, but without the black light and fluorescent penetrant. As with Zyglo, it may be used on almost any dense mate-

Figure 2. Front axle king pin for a truck as it appeared (left) under visual inspection, apparently safe for service; (center) with dangerous cracks revealed by Magnaflux; and (right) with the same cracks revealed by fluorescent Magnaglo, under black light (Courtesy of Magnaflux Corporation).

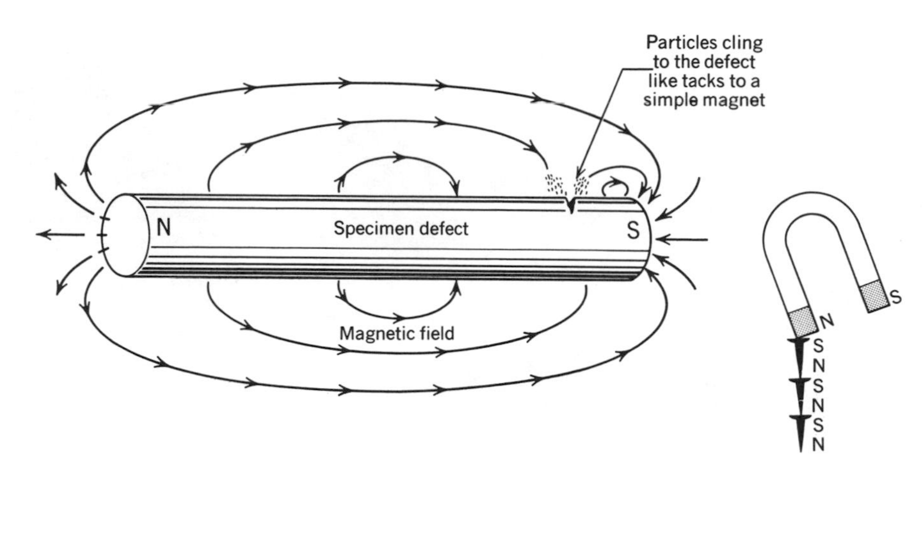

Figure 3. By inducing a magnetic field within the part to be tested, and by applying a coating of magnetic particles, surface cracks are made visible; in effect the cracks form new magnetic poles. Particles cling to the defect like tacks to a simple magnet (Courtesy of Magnaflux Corporation).

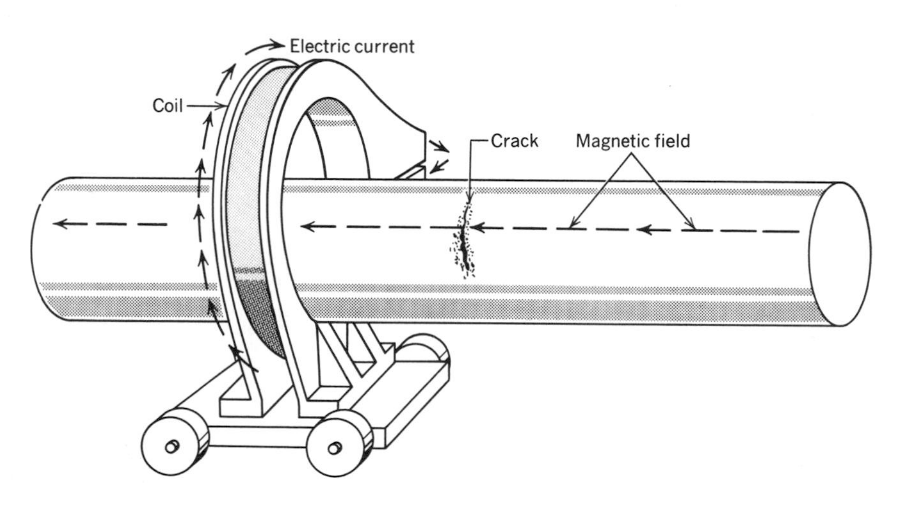

Figure 4. Longitudinal method of magnetization (Courtesy of Magnaflux Corporation).

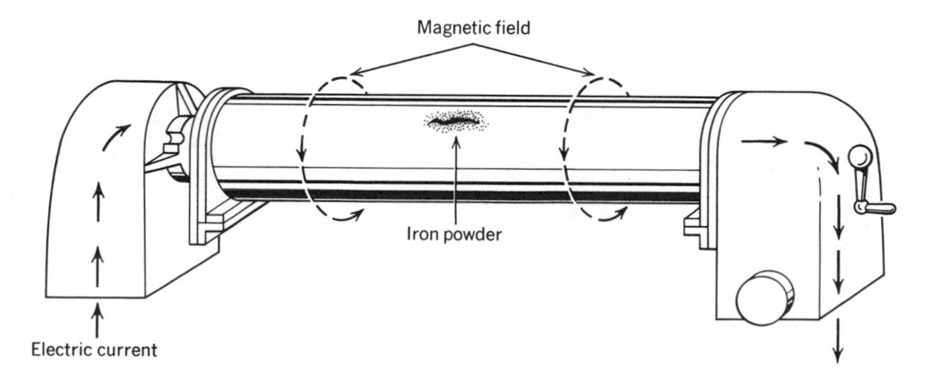

Figure 5. Circular method of magnetization (Courtesy of Magnaflux Corporation).

Figure 6. Freshly reground carbide tipped tools as they appear; (left) to normal visual inspection, perfectly good; (center) test results with Zyglo penetrant; and (right) with supersensitive Zyglo Pentrex. All cracks, however small, are found (Courtesy of Magnaflux Corporation).

rial. This method, called Spotcheck by the Magnaflux Corporation, works as follows:

1 Surfaces must be clean and dry prior to soaking.
2 Dye penetrant is applied to the defect and allowed to soak. Soaking time should be sufficient to get the penetrant into fine cracks.
3 Remove excess, but do not rinse out cracks.
4 The developer is applied. Allow enough time for the developer to find very small cracks.
5 Inspection shows a bright colored indication marking the defect.

Spotcheck (Figure 7) is available in sealed pressure spray cans or for brush or spray gun application. Its advantages are portability for remote uses (Figure 8) or for rapid inspection of small sections in the shop, low cost, and ease of application.

Figure 7. Spotcheck is used to locate fatigue crack in punch press frame (Courtesy of Magnaflux Corporation).

Figure 8. Spotcheck is used to find a dangerous crack in an aircraft wheel (Courtesy of Magnaflux Corporation).

Ultrasonic Inspection

The pulse-echo system and the through-transmission system are two methods of ultrasonic inspection (Figure 9) used to check for flaws in metal parts. Both systems use ultrasonic sound waves (millions of cycles per second) in their testing.

In the pulse-echo system, a pulse generator produces short bursts of sound that activate a transducer (a device that converts mechanical energy into electrical energy) fastened to the metal being tested. The signal pulse is seen as a pattern or ''pip'' on an oscilloscope screen when the sound wave enters the test piece (Figure 10). Part of the sound wave is reflected back when it reaches the other side of the material and shows on the oscilloscope as a second ''pip.'' If there is a flaw in the specimen, a smaller ''pip'' will be seen between the other two. Since the distance between the ''pips'' on the oscilloscope screen represent elapsed time of the reflected pulse, the distance to a flaw can be accurately measured. A trained operator is required to interpret the results.

The through-transmission method uses a transducer on each side or end of the test piece (Figure 11). The signal pulse enters the material at one transducer, travels through the material to be received by the other transducer, and is translated into another signal shown on an oscilloscope. If no flaws are present, a clear, strong signal will be seen on the oscilloscope, but if the material contains any flaws, a weaker or distorted signal will be seen.

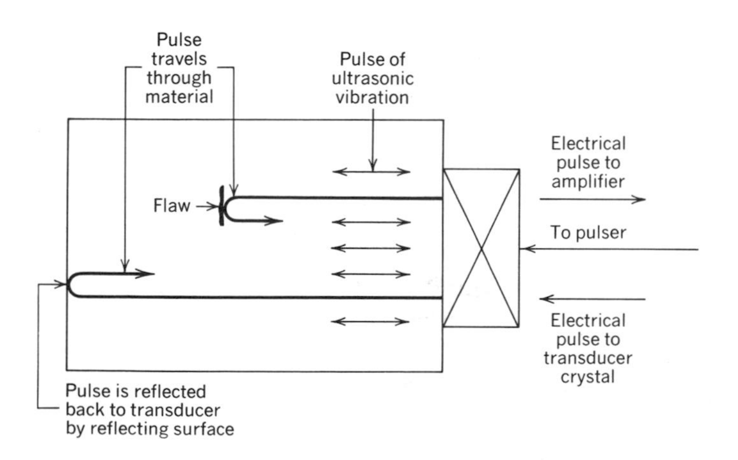

Figure 9. How ultrasonic sound waves are used to locate flaws in material *(Machine Tools and Machining Practices)*.

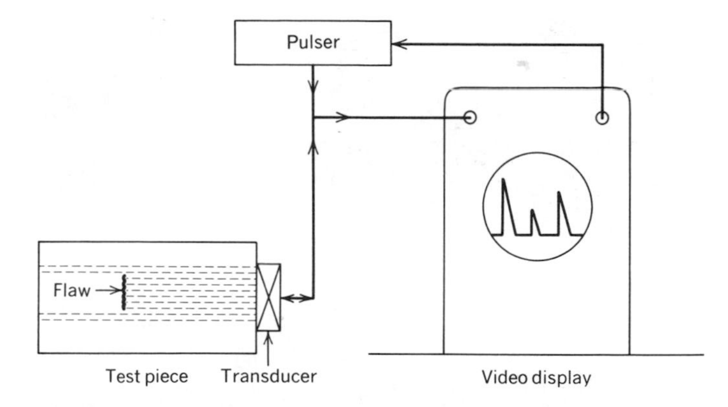

Figure 10. Ultrasonic pulse-echo system *(Machine Tools and Machining Practices)*.

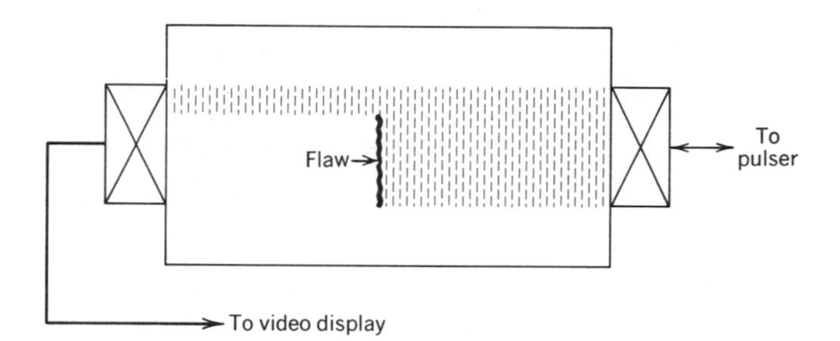

Figure 11. Through-ultrasonic transmission *(Machine Tools and Machining Practices)*.

The reduced signal can be set up to actuate a bell or light to alert the operator when similar parts are being checked.

Ultrasonic inspection is used in production for inspection of manufactured parts and for testing structures such as pipelines and bridges. Internal defects, cracks, porosity, laminations, thickness, and weld bonds are tested in metal or nonporous nonmetallic structures (Figure 12). Some testing equipment is relatively light and portable (Figure 13), and some require the part to be immersed in a liquid. Test results are immediate and accurate with this method. Tests can be made through long bars of steel or through thin sections that require testing on one side only. The reflecting surface must be parallel to the testing surface, however.

Radiographic Testing

Radiographic testing utilizes the ability of X rays (Figures 14 and 15) or gamma rays (Figures 16 and 17) emitted from radioactive materials, such as

Figure 12. Portable ultrasonic machine instruments can be used for routine manual inspection under field conditions. This is very useful for structural weld testing and corrosion surveys (Courtesy of Magnaflux Corporation).

Figure 13. The ultrasonic method results in swift inspection of critical weldments with dependability (Courtesy of Magnaflux Corporation).

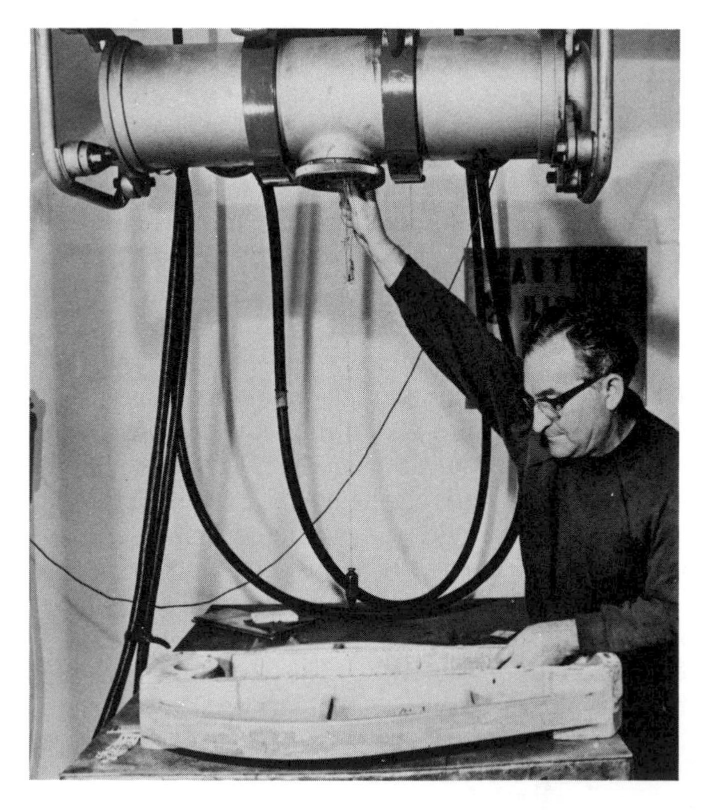

Figure 14. X ray method of testing. A specialist makes a test of pilot run parts to evaluate manufacturing techniques and procedures (Courtesy of Magnaflux Corporation).

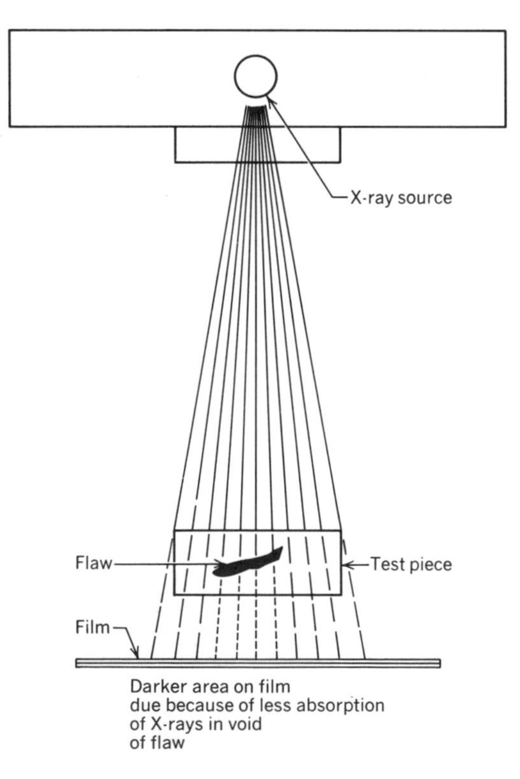

Figure 15. How X ray inspection works *(Machine Tools and Machining Practices)*.

Figure 16. An experienced technician is preparing to make gamma ray inspection of a section of pipe (Courtesy of Magnaflux Corporation).

radioactive radium or cobalt 60, to pass through solids. The test results are determined from a radiograph, which is a film exposed to radiation that has gone through the test materials. Shadows on the radiograph reveal defects since the radiation will pass through a void, crack, or area of lower density at a greater intensity and will appear darker on the negative. This testing method is widely used for forgings, castings, welded vessels including welds on pipelines, and corrosion analysis on pipelines and structures. Portable equipment is used for field inspection. Since there is a radiation hazard, only trained technicians should use the equipment. The test is quite sensitive and provides a permanent record.

A radiation detector may be used to determine material thickness since the radiation that passes through the test material decreases as the thickness increases. A moving, continuous strip of material can then be constantly monitored for thickness without any physical contact to the radiation source or detector.

Eddy Current Testing

Eddy current inspection can only be used to test electrically conducting materials. An alternating current in a coil produces a corresponding magnetic field. Eddy currents, which flow opposite to the main current, are produced in the test material if the coil is placed in, near, or around it. The eddy currents then produce a change in impedance in a magnetic field, which is converted into voltage that is read on a meter or oscilloscope (Figure 18). This method is used for detecting seams or variations in thickness and for sorting alloys in various compositions and heat treatments. Physical differences in mass, dimension, and shape can also be detected.

Eddy current testing has the advantage of being very rapid; it need not touch the specimen and can be used for automatic inspection. One limitation of this inspection method is because eddy currents are present only at the surface of metallic specimens, defects at or near the center of parts or rods will not be detected.

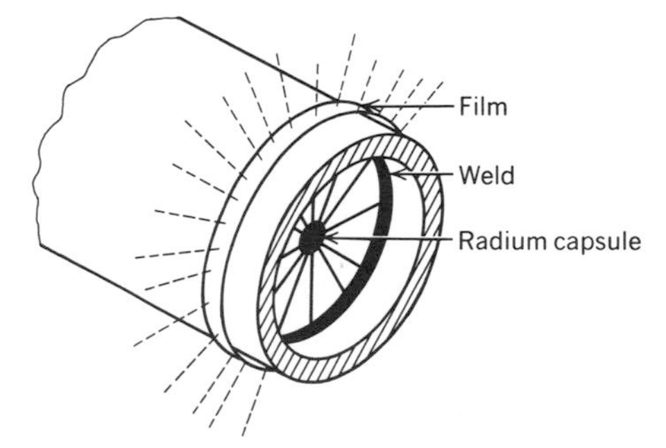

Figure 17. How gamma ray inspection works *(Machine Tools and Machining Practices)*.

Figure 18. New Magnatest ED-100 for nondestructive comparison of magnetic materials. Used to separate metallurgical properties as alloy variations, heat treat condition, hardness differences, internal stress, and tensile strength (Courtesy of Magnaflux Corporation).

Self-Evaluation

1 What is the purpose of nondestructive testing?

2 Name a limiting factor of magnetic-particle inspection.

3 How must the part be magnetized to form new magnetic poles on a crosswise crack? On a lengthwise crack?

4 Describe how fluorescent penetrant works to reveal defects. On what materials can this method be used?

5 What are some of the advantages of using a dye penetrant method (without black light)?

6 Name the two basic systems of ultrasonic testing. Briefly explain their differences.

7 What can be determined in a specimen by the use of ultrasonic inspection systems?

8 How can X rays help us to detect flaws in solid steel or other metals?

9 In what way does gamma ray inspection differ from X ray inspection?

10 Explain some of the uses of eddy current inspection. To what metals is it limited?

Worksheet

Objective Inspect for cracks or other flaws by using any nondestructive testing equipment or method available.

Materials Three different mechanical parts that have been in rough service, but appear sound otherwise.

Procedure Check to see if your shop has a fluorescent penetrant, dye penetrant set, magnetic inspection equipment, or any other type of nondestructive testing equipment. Choose the type of inspection available to you and obtain the operation instruction manual. Follow the step-by-step instructions for testing the three specimens.

Conclusion List your results.

	Specimen 1	Specimen 2	Specimen 3
Type of material			
Kind of test			
Type of flaw			
Location of flaw			
Service condition (Safe or unsafe for use)			

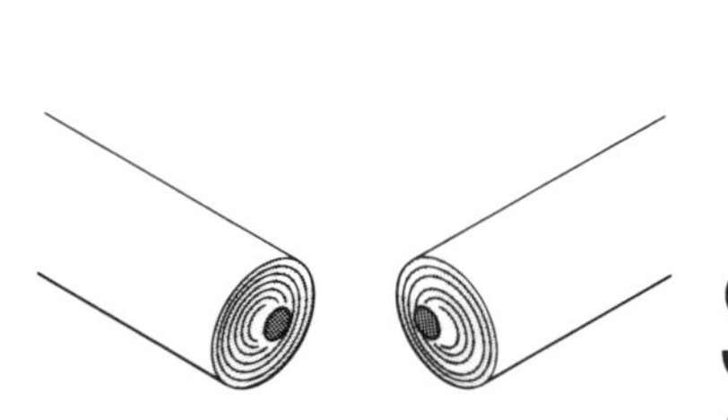

22 Service Problems

One of the most common problems found in industrial equipment and other machinery is mechanical breakdown due to wear, overload, or fatigue. Operational failures sometimes cause a loss of production, loss of work time for employees and, occasionally, hazardous conditions. Many service problems can be prevented by proper maintenance, selection of materials, and design. This chapter will help you recognize some industrial problems and help you to determine the needed corrective measures.

OBJECTIVES

After completing this chapter, you will be able to:

1. Explain the causes of several industrial problems that lead to failures and list corrective measures for them.
2. Identify and describe the causes of failure in 10 metal parts.
3. Make recommendations for change in material design or in heat treatment for each of the 10 metal parts to correct their service failures.

INFORMATION

Metal failure occurs only in an extremely low percentage of the millions of tons of steel fabricated every year. Those that do occur fall into the following groups.

Operational failures
- Overload
- Wear
- Corrosion and stress corrosion
- Metal fatigue
- Brittle failures

Failures caused by thermal treatments
- Forging
- Hardening and tempering
- Welding
- Grinding

Overload

Overload failures are usually attributed to faulty design, to extra loads applied, or to an unforeseen machine movement. Shock loads or loads applied above the design limit are quite often the cause of the breakdown of machinery.

Although mechanical engineers always plan for a high safety factor in designs (for instance, the 10 to 1 safety factor above the yield strength is used in fasteners), the operators of machinery often tend to use machines above their design limit. Of course, this kind of overstress is due to operator error. Inadequate design can sometimes play a part in overload failures. Improper material selection in the design of the part or improper heat treatment can cause some failures when overload is a factor.

Often a machinist or welder will select a metal bar or piece for a job based upon its ultimate tensile strength rather than its yield point. In effect this is a design error and can ultimately cause overload and breakdown (Figure 1). The strength of any material selected for a job should be based on its yield strength plus an adequate safety factor. When a part of a mechanism or structure is stressed beyond its yield point, it has been damaged or ruined.

Wear

Excessive wear can also be caused by continuous overload, but wear is ordinarily a slow process that is related to the friction between two surfaces (Figure 2). **Rapid wear** can often be attributed to lack of lubrication or the improper selection of material for the wear surface. Some wear is to be expected, however, and could be called **normal wear.**

Wear is one of the most frequent causes of failure. We find normal wear in machine tooling, such as carbide, and high speed tools that wear and have to be replaced or resharpened. Parts of automobiles ultimately wear until an overhaul is required (Figure 3). Machines are regularly inspected for worn parts, which when found are replaced; this is called preventative maintenance. Often normal wear cannot be prevented; it is simply accepted, but it can be kept to a minimum by the proper use of lubricants. Rapid wear can occur if the load distribution is concentrated in a small area because of the part design or shape. This can be altered by redesign to offer more wear surface. Speeds that are too high can increase friction considerably and cause rapid wear.

Metallic wear is a surface phenomenon and it is caused by the displacement and detachment of surface particles. All surfaces subjected to either rolling or sliding contact show some wear. In some severe cases, the wear surface can be cold welded to the other surface. In fact, some metals are pressure welded together in machines, taking advantage of their tendency to be cold welded. This happens when tiny projections of metal make a direct contact on the other surface and produce friction and heat, causing them to be welded to the opposite surface if the material is soft. Metal is torn off if the material is brittle. Insufficient lubrication is usually the cause of this problem (Figure 4).

High pressure lubricants are often used while pressing two parts together in order to prevent this sort of welding. Two steel parts such as a steel shaft and a steel bore in a gear or sprocket, if pressed together dry, will virtually always seize or weld and cause the two parts to be ruined for further use (Figure 5). In general, soft metals, when forced together, have a greater tendency to cold weld than harder metals. Two extremely hard metals even when dry will have very little tendency to weld together. For this reason, hardened steel bushings and hardened pins are often used in earth moving machinery to avoid wear. Some soft metals when used together for bearing surfaces (aluminum to aluminum) have a very great tendency to weld or seize. Among these metals are aluminum, copper, and austenitic stainless steels.

Cast iron, when sliding on cast iron as is found in machine tools on the ways of lathes or milling machine tables, has less tendency than most met-

Figure 1. This CR shaft was stressed beyond its yield point by an overload. This could have been prevented by using a tougher alloy shafting.

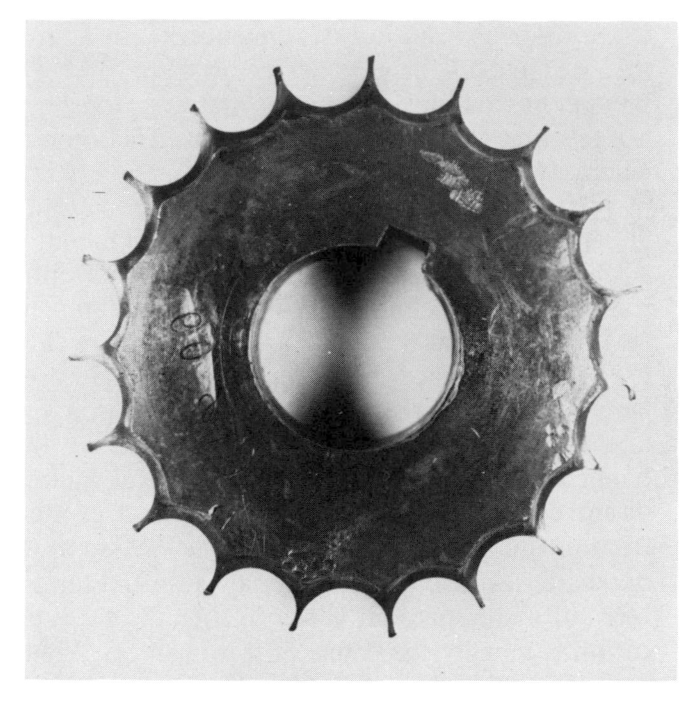

Figure 2. The teeth on this mild steel roller chain sprocket have worn almost to the point of complete failure. Wear life would be increased considerably if an alloy steel were used for the sprocket.

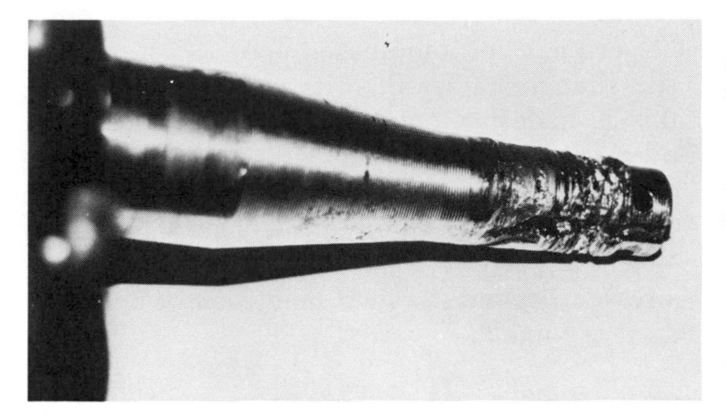

Figure 3. This type of failure can be hazardous. The spindle bearing seized up due to lack of lubrication and began to twist the spindle. Had the operator of the automobile not stopped very quickly, the wheel would have come off.

Figure 4. The babbitted surface of this tractor engine bearing insert has partially melted and torn off. This failure was not due to normal wear but to lack of lubrication.

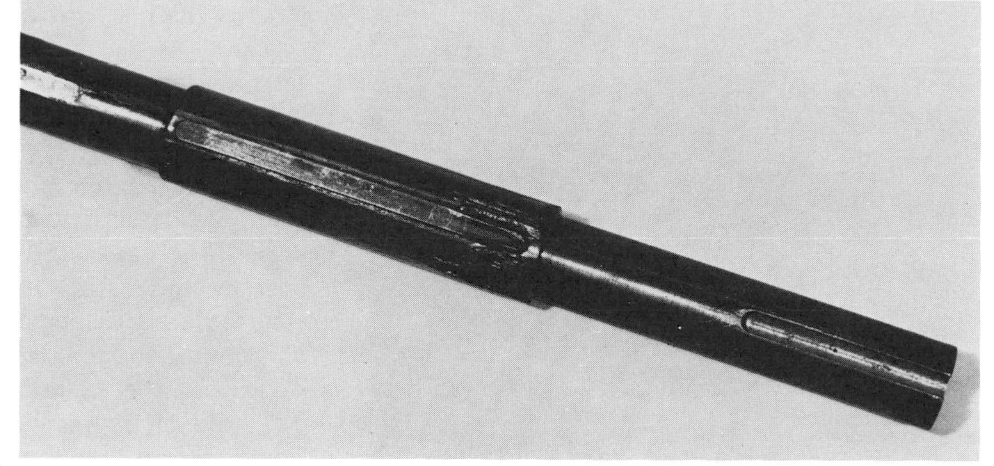

Figure 5. This shaft had just been made by a machinist and was forced into an interference fit bore for a press fit. No lubrication was used, and it immediately seized and welded to the bore, which was also ruined.

als to seize because the metal contains graphite flakes that provide some lubrication, although additional lubrication is still necessary.

As a general rule, however, it is not good practice to use the same metal for two bearing surfaces that are in contact. However, if a soft steel pin is used in a soft steel link or arm, it should have a sufficiently loose fit to avoid seizing. In this application it is better practice to use a bronze bushing or other bearing material in the hole and a steel pin because the steel pin is harder than the bronze and when a heavy load is applied, the small projections of bronze are flattened instead of torn out. Also, the bronze will wear more than the steel and usually only the bushing will need replacing when a repair is needed. (Figure 6).

Some metals have a tendency to work harden and, although they would gall or seize in their soft condition as they begin sliding together, they begin to harden on the surface and minimize the tendency to cold weld. An example of this is in the austenitic manganese steels used in rock crusher machinery. When these have work hardened sufficiently, they do not tend to cold weld to their own surfaces.

In **abrasive wear** small particles are torn off the surface of the metal creating friction. Friction involving abrasive wear is sometimes used or even required in a mechanism such as on the brakes of an automobile. The materials are designed to minimize wear with the greatest amount of friction in this case. Where friction is not desired, a lubricant is normally used to provide a barrier between the two surfaces. This can be done by heavy lubricating films or lighter boundary lubrication in which there is a residual film. Lubrication will be further discussed in Chapter 25.

Erosive wear is often found in areas that are subjected to a flow of particles or gases that impinge on the metal at high velocities. Sand blasting, which is sometimes used to clean parts, utilize this principle.

Corrosive wear takes place as a result of an acid, caustic, or other corrosive medium in contact with metal parts. When lubricants become contaminated with corrosive materials, pitting can occur in such areas as machine bearings.

Surface fatigue is often found on roll or ball bearing races or sleeve bearings where excessive

Figure 6. The bronze bushing in this arm has seen severe use, is badly worn, and will be replaced, while the steel shaft that turns in the arm shows relatively little wear.

Figure 7*a*. These aircraft engine cylinders must withstand high temperatures and wear. The inside of the cylinder wall is porous chromium plated.

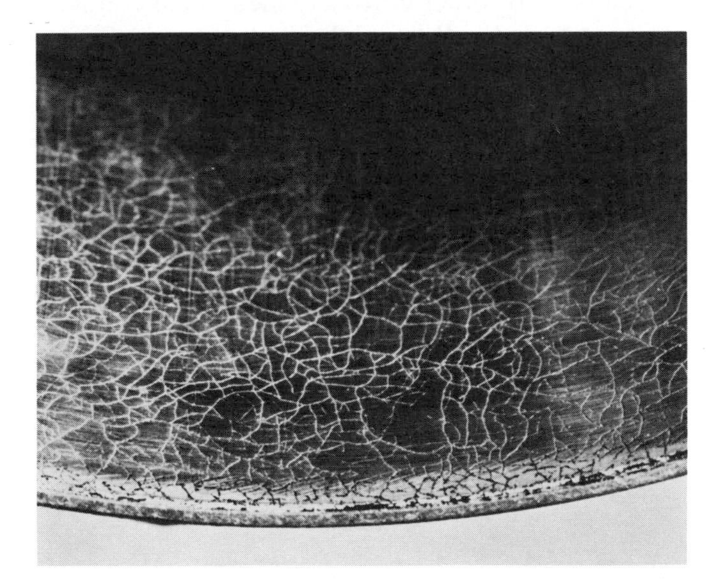

Figure 7*b*. Close up of porous chromium plate. The many grooves or channels will hold the lubricant. This plate is applied in small droplets which, when ground off, produce this effect.

side thrust has been applied to the bearing. It is seen as fine checks (cracks) or spalling (small pieces falling out on the surface).

Protection Against Wear

Various methods are used to limit the amount of wear in the part. One of the most commonly used methods is simply to harden the part. Also the part can be surface hardened by **diffusion** of a material, such as carbon or chrome, into the surface of the part. Parts can also be **metallized, hard faced,** or **heat treated.** Other methods of limiting wear are electroplating (especially the use of hard industrial chromium) and anodizing of aluminum. Chromium plate can either be hard or porous. The porous type can hold oil to provide a better lubrication film. Some internal combustion engines are chromium plated in the cylinders and piston rings (Figure 7*a* and 7*b*). Some nickel plate is used and also rhodium, which is very hard and has high heat resistance.

The oxide coating that is formed by anodizing on certain metals such as magnesium, zinc, aluminum, and their alloys is very hard and wear resistant. These oxides are porous enough to form a base for paint or stain to give it further resistance to corrosion.

Some of the types of diffusion are **carburizing, carbo-nitriding, cyaniding, nitriding, chromizing,** and **siliconizing.** Chromizing consists of the introduction of chromium into the surface layers of the base metal. This is sometimes done by the use of chromium powder and lead baths in which the part is immersed at a relatively high temperature. This, of course, produces a stainless steel on the surface of low carbon steel or an iron base metal, but it may also be applied to nonferrous materials such as tungsten, molybdenum, cobalt, or nickel to improve corrosion and wear resistance.

The fusion of silicon, which is called **ihrigizing,** consists of impregnating an iron base material with silicon. This also greatly increases wear resistance. **Hard facing** is put on a metal by the use of several types of welding operations, and it is simply a hard type of metal alloy such as alloying cobalt and tungsten or tungsten carbide that produces an extremely hard surface that is very wear resistant. Metal spraying is used for the purpose of making hard wear resistant surfaces and for repairing worn surfaces (Figures 8*a* to 8*c*). Metalizing is usually done by either feeding a metal powder or a metal wire at a controlled rate through

Figure 8*a*. The undercut that is made ranges from 0.015 to 0.020 inch deep where the build-up is required. Grooves are often made to insure bonding.

Figure 8*b*. Metal being sprayed on prepared surface of shaft. Often a light undercoat of molybdenum is applied first to create a bond that has a physio-chemical nature. Steel spray has only an adhesive bond, not a metallic bond as in welding.

a tool that provides a heat source; the molten particles of metal are forced onto the surface of the base metal at a high velocity. In this process which is not the same as welding, the liquid metal particles simply flatten out on the base metal and make a mechanical bond with the base metal and the previously deposited material instead of a metallurgical bond since the cooling is rapid and an oxide film forms over the particle, preventing fusion of the metal particles. Thus there is only a loose metallic or oxide bond between the particles. This determines to a great extent how strong or how porous the deposited material becomes. Welding is also used to build up surfaces for repair, but the stress concentration at the edge of the weld is often a cause of fatigue failure. Metallizing does not produce stress concentration but undercutting is necessary, which reduces the effective stress area of the part and weakens it to some extent.

Corrosion

Corrosion and stress corrosion are related to the atmosphere surrounding the part and to its resistance to corrosion. This subject will be further discussed in Chapter 23 "Corrosion in Metals."

Figure 8*c*. When sufficient material has been applied, the surface may be machined to the correct diameter after it has cooled.

Figure 9. Shafts sometimes fail by fatigue as shown here. The cracks were initiated from a badly worn keyway and slowly moved toward the center until the load was greater than the tensile strength of the remaining area.

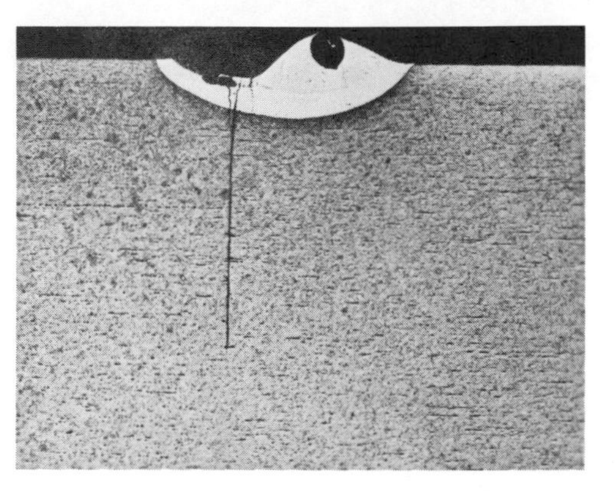

Figure 10. Fatigue failure in a leaf spring nucleated by a weld spatter. The smooth "oyster shell" character is typical of fatigue cracks. Note the nature of brittle failure of the spring in service. The polished section through another weld spatter clearly illustrates the formation of fatigue failure (10 ×) (Courtesy Republic Steel Corporation.)

Fatigue

Failures caused by fatigue are found in many of the materials of industry. Some plastics and most metals are subject to fatigue in varying degrees. Fatigue is caused by a crack that is initiated by a notch, bend, or scratch that continues to grow gradually as a result of stress reversals on the part. The crack growth continues until the cross sectional area of the part is reduced sufficiently to weaken the part to the point of brittle failure (Figure 9). Even welding spatter on a sensitive surface such as a steel spring can initiate fatigue failure (Figure 10). Fatigue is greatly influenced by the kind of material, grain structure (isotropy), and the kind of loading. Some metals are more sensitive to sharp changes in section (notch sensitive) than others.

Parts should be designed and fabricated to utilize directionality of grain flow. Figure 11 is an example of a forged shaft that shows slag inclusions running the length of the material. Wrought

Figure 11. Wrought iron forging, showing inclusions that run the length of the material.

iron and resulfurized drawn shafting contain inclusions that give the material high transverse (crosswise) fatigue strength. These metals have a lowered resistance to high torque values, however, tending to split along the stringers of manganese sulfide (Figure 12).

Fatigue is caused by a concentration of tensile stress that can often be corrected by a change in design (Figures 13*a* and 13*b*). Parts that are subject to stress reversals (cyclic stresses) can have their fatigue life extended considerably by raising compressive stresses on the surface (Figure 14). This can be done in several ways: by carburizing, nitriding, or surface hardening by induction or flame. To increase fatigue life or the endurance limit, parts are sometimes subjected to shot peening to produce residual compressive stresses on the surface.

A study of individual fatigue problems based on the service conditions and by direct observation of the failure can often lead to a conclusion that explains the cause or causes and suggests some corrective measures. The loading of the part can be high or low for its size; it can have high or low speed or stress reversals caused by misalignment. Vibration is often a cause of fatigue, the frequency and intensity being a factor. Occasional overloads are also instrumental in initiating failure.

Fatigue failures are usually characterized by three distinct surfaces (Figure 15).

1 A smooth surface with wave marks such as seen on clam shells. This area represents a slow progression of the initial crack.

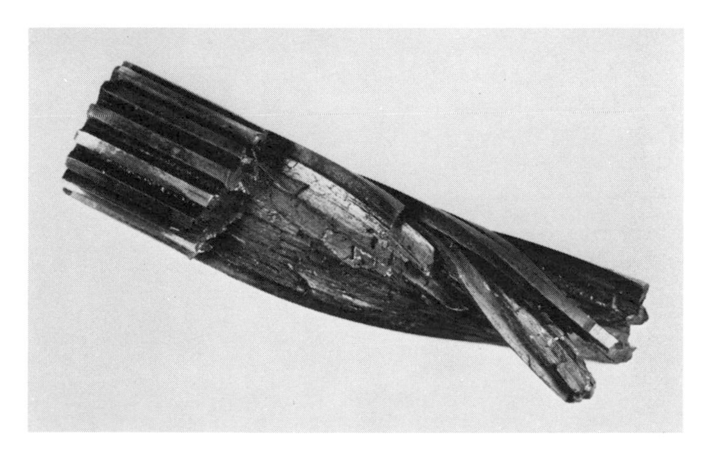

Figure 12. Splined shaft that has split along its lengthwise fibers caused by overload.

2 A similar but rougher surface showing coarse wave marks progressing toward the center.

3 A crystalline area showing the final brittle and sudden failure of the part. It is this last portion that prompts the erroneous conjecture that the part has "crystallized."

The symbols and nomenclature related to fatigue failures are illustrated in Figure 16.

Types of Fatigue Failures

Some of the more common fatigue fracture appearances associated with various bending conditions are illustrated in Figure 17. The following cases are examples.

CASE 1 One-way Bending Load

No Stress Concentration A small elliptically shaped fatigue crack usually starts as a surface flaw such as a scratch or tool mark. The crack tends to flatten out as it grows. It is caused by the stress at the base of the crack being lower because of the decrease in distance from the edge of the crack to the neutral axis. The degree of overstress in the part is indicated by the amounts of smooth-textured area compared to the crystalline-textured area of the fracture. A large crystalline area indicates high overstress (Figure 17, 1-*b*). A smaller crystalline area indicates a lower overstress, which would require a greater number of cycles necessary to produce failure (Figure 17, 1-*a*).

Mild Stress Concentration If a distinct stress raiser such as a notch is present, the stress at the base of the crack would be high, causing the crack to progress rapidly near the surface, and the crack tends to flatten out sooner. The degrees of overstress by the relative areas of smooth and crystalline textures on the fracture surface are shown in Figure 17, 1-*c* and 1-*d*.

High Stress Concentration The smooth fracture growth can change from concave (as in 1-*b*) to convex as the rate of crack growth circumferentially at the surface exceeds the radial crack growth. In high overstress, the convex texture can occur extremely early in crack formation.

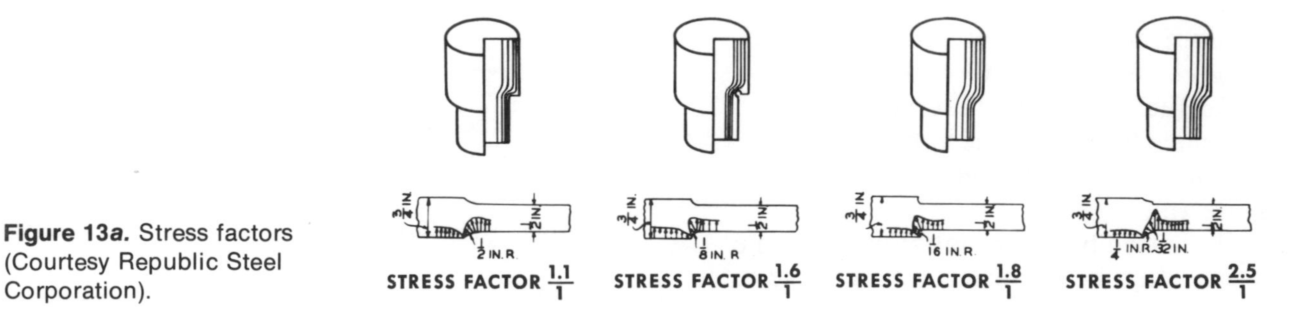

Figure 13*a*. Stress factors (Courtesy Republic Steel Corporation).

Design Considerations

CORRECT PRACTICE

INCORRECT PRACTICE

SMOOTH SURFACE

ROUGH SURFACE

LARGE DIAMETER FOR THREADING

SMALL DIAMETER FOR THREADING

GENEROUS FILLET

NO FILLET

RELIEF

PAD

NO PAD

Figure 13*b*. Concentration of stress is a function of design (Courtesy Republic Steel Corporation).

Intermittant loading

Compressive stresses induced by shot peening or carburizing tend to cancel out tensile stresses

Tensile stresses are induced here and tend to start fatigue cracks

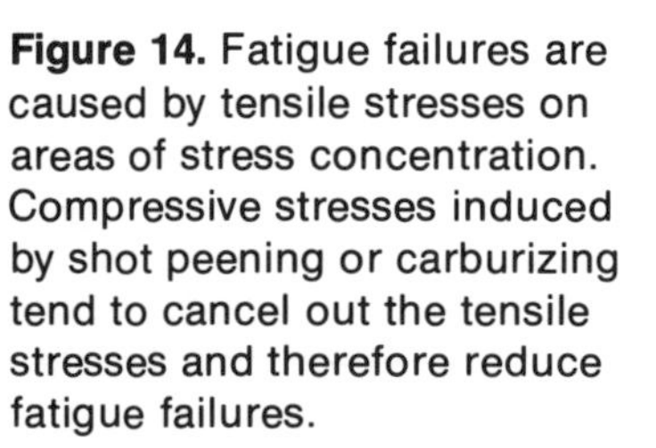

Figure 14. Fatigue failures are caused by tensile stresses on areas of stress concentration. Compressive stresses induced by shot peening or carburizing tend to cancel out the tensile stresses and therefore reduce fatigue failures.

Figure 15. Classic example of fatigue in a shaft. The three distinct areas can be seen here.

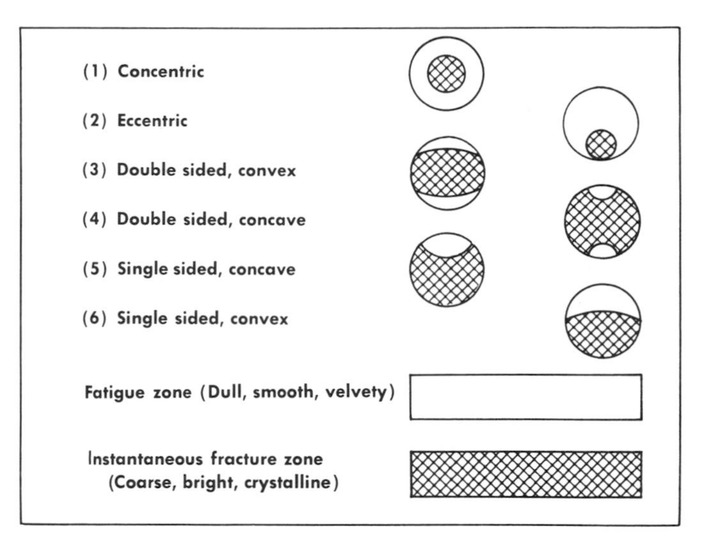

Figure 16. Symbols and nomenclature established by Bacon are useful for the designation of fracture appearances (Courtesy Republic Steel Corporation).

CASE 2 Two-Way Bending Load

No Stress Concentration Cracks start almost simultaneously at opposite surfaces when the surfaces are equally stressed. The cracks proceed toward the center at similar rates and result in a fracture that is rather symmetrical (2-*b* of Figure 17). In low overstress conditions (2-*a*), cracks may not begin at the same time; consequently the fracture is less symmetrical.

Mild Stress Concentration Higher stress concentration and the increased rate of circumferential crack growth cause the fracture to flatten out more quickly. Rapid radial crack growth tends to promote a concave zone with a relatively small radius of curvature. As the relative rates of circumferential and radial crack growth tend to become equalized, the radius of curvature tends to increase.

High Stress Concentration In this case, the circumferential crack rate increases rapidly, quickly exceeding the radial crack rate and the radius of curvature changes from concave to convex. The relative areas of smooth and crystalline textures discussed earlier also apply here.

CASE 3 Reversed Bending and Rotational Load

No Stress Concentration As in Case 2, stress occurs at two extreme surface fibers. Usually the weaker area will fail first. The fracture tends to progress and flatten out from the initial small, concave crack. Eventually the fracture tends to become a straight line. The crack tends to propagate **against** the direction of rotation. With low overstress the crack growth can proceed well beyond the center and promote circumferential growth prior to complete failure. With high overstress, the crack does not proceed as far as shown in 3-*b* of Figure 17.

Mild Stress Concentration With high overstress, the notch causes early crack formation and rapid crack growth around the periphery, and the crystalline zone is centrally located as shown in 3-*d*. Low overstress tends to start the crack at the weakest point and the crystalline zone is moved away from the point of crack initiation (3-*c*). Extremely tough material will respond to these conditions with a fracture similar to 3-*a*.

High Stress Concentration A combination of severe stress concentration and high overstress, such as a groove machined about the entire periphery with a sharp notch radius, causes cracks to start all around the circumference at the same time. The resultant crystalline failure is somewhat circular in appearance and centrally located (3-*f*). Lower overstress tends to move the point of failure away from the central location (3-*c* of Figure 17).

When a shaft is subjected to a torsional load, the maximum shear stress is equal to the maximum tensile stress. However, the corresponding two strengths in steel are not equal, the shear strength being approximately one-half the tensile strength. The shear stress, therefore, will reach the shear strength before the tensile stress will reach the tensile strength; therefore, a shear type of failure will result (Figure 18). One reason that transverse (crosswise) cracks are more prevalent than longitudinal cracks is that grinding or machining marks, which accentuate the probability of failure, are oriented in the transverse direction. The quality of surface finish is therefore very important.

Splined shafts almost always produce a characteristic compound fracture. Fatigue cracks originate almost simultaneously at all the spline roots and grow until the shaft ruptures. The use of fibrous type of steels tends to increase this problem.

Some Examples of Service Failures

The following illustrations are cases of industrial problems resulting from poor practices, error in design or material selection, or simple overload.

1 *Fatigue Failure in Helical Gear* (Figure 19). This pinion gear is part of a heavy reduction drive that was operating on a relatively light load. The smooth clam shell surface extends completely through the gear because the irregular break and the meshing of the gear teeth on both sides of the break prevented any brittle fracture. It was noted from the wear surface on the teeth that there was a slight axial misalignment of the gears and mismating of the teeth because of either an involute error or center distance error. This caused a high stress tooth contact that resulted in fatigue cracking.

2 *Brittle Failure Caused by Weld* (Figure 20). Such brittle failures can be initiated by a stress concentration when there is high overstress. This three inch diameter SAE 1040 bar was loaded as a cantilever beam and used as a special fork on a lift truck. A gusset was welded at the top to strengthen it. The short weld across the end of the gusset produced sufficient stress concentration to cause the sudden brittle failure.

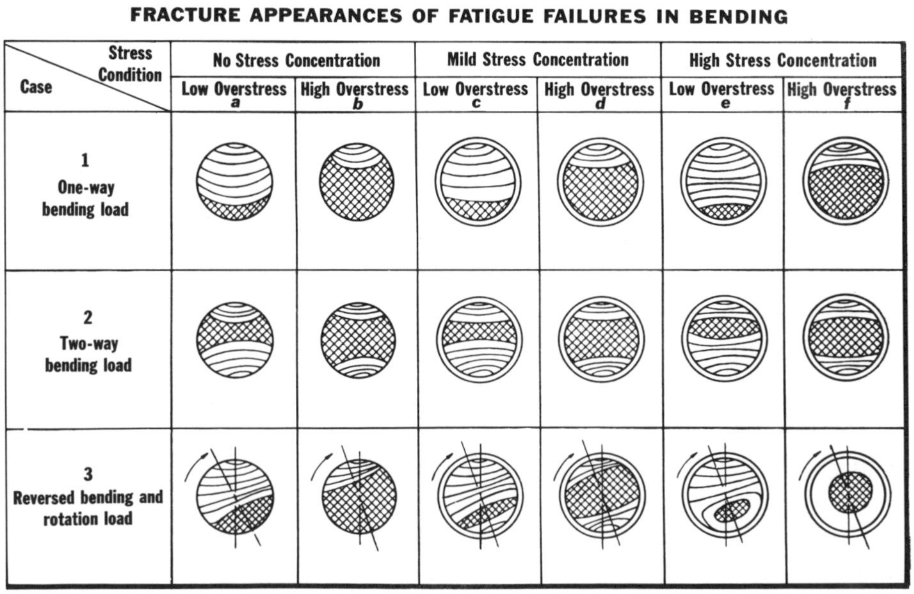

FRACTURE APPEARANCES OF FATIGUE FAILURES IN BENDING

Case / Stress Condition	No Stress Concentration		Mild Stress Concentration		High Stress Concentration	
	Low Overstress a	High Overstress b	Low Overstress c	High Overstress d	Low Overstress e	High Overstress f
1 One-way bending load						
2 Two-way bending load						
3 Reversed bending and rotation load						

Figure 17. Fracture appearances of fatigue failures in bending (Courtesy Republic Steel Corporation).

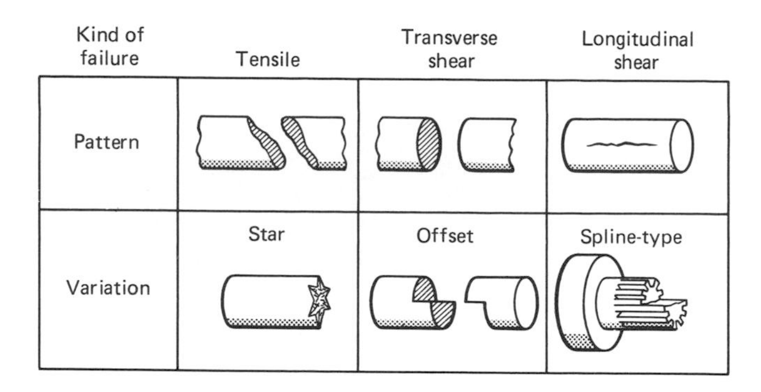

Figure 18. Basic types of torsional failure fractures.

Figure 20. Brittle failure caused by weld.

3 *Weld on Overloaded Shaft in a Sawmill* (Figure 21*a* and 21*b*). Millwrights sometimes find it necessary to use any means to keep the machines running even though they are aware that they are using poor practices. Figure 21*a* shows an extreme overload on the shaft and coupling. The drive key has been twisted out of position. Figure 21*b* reveals the valiant effort that was made by the millwrights to keep this shaft turning. Welding around a shaft is bad practice since it creates a stress concentration that will rapidly initiate fatigue cracking especially with high overloads. Here weld has been piled upon weld until the final failure ended the process.

Figure 19. Fatigue failure in helical gear.

Figure 21*a*. Extremely overloaded shaft caused this failure, twisting the key out of position.

4 *Fatigue in a Torque Wrench* (Figure 22). The square end that fits into a socket on this torque handle was probably overloaded many times, causing small cracks to form on the surface.

Figure 21*b*. Same shaft and coupling as in Figure 21*a*, showing repeated welding on the shaft in an effort to keep it in operation.

Figure 22. Fatigue in a torque wrench.

Figure 23*a*. A failure in a tractor engine. A connecting rod came loose from the crankshaft and was pushed out through the side of the block.

Figure 23*b*. Close-up showing the hole and connecting rod.

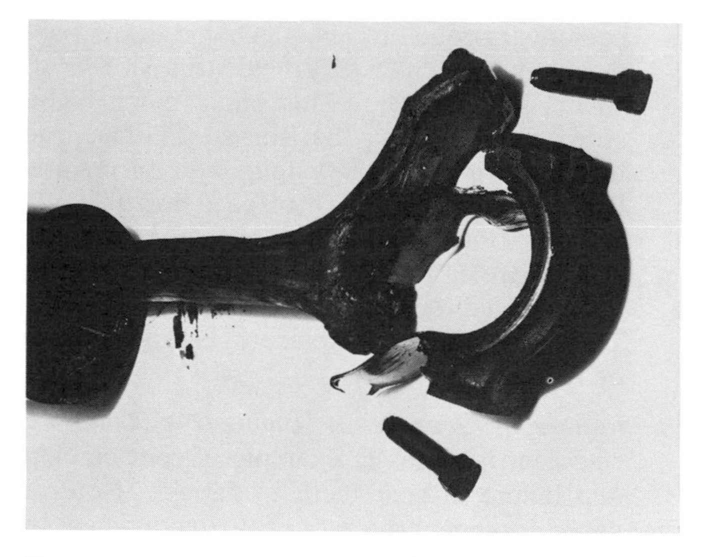

Figure 23*c*. Bent connecting rod and cap. Note the necked down screws indicating that they were overtorqued.

Figure 23*d*. End of one of the cap screws that held the cap on the connecting rod. A typical fatigue pattern is evident.

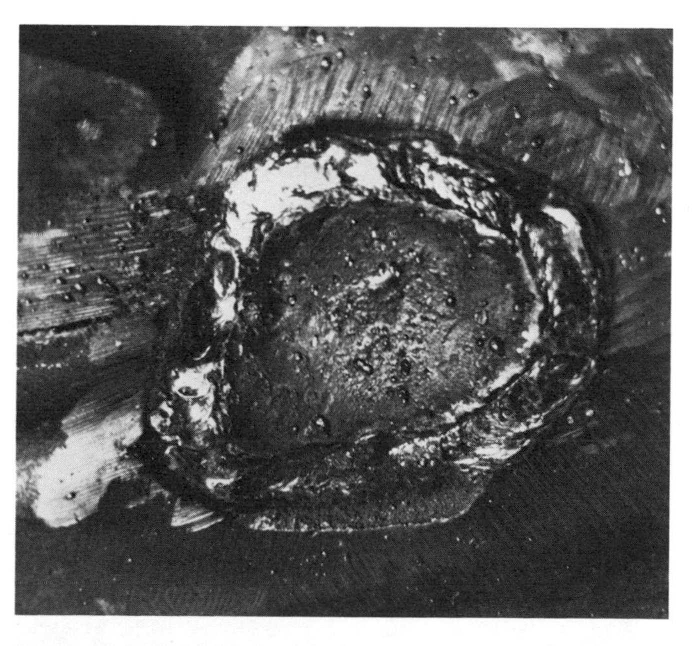

Figure 23*e*. The hole in the side of the engine block has been welded with ENi rod. The entire block was cleaned in a dip tank and preheated a few hundred degrees.

Continued use and many stress reversals continued the propagation of the crack until only a small section was holding.

5 *Failure in a Tractor Engine* (Figures 23*a* to 23*e*). The bolts in the bearing caps suddenly failed and the crankshaft pushed the connecting rod through the side of the engine block (Figures 23*a* and 23*b*). The capscrews that fasten

the connecting rod bearing cap had evidently been overtorqued by the mechanic to the point that they had "necked down" to a smaller diameter (Figure 23*c*). This put a much higher stress on the screws that started a fatigue fracture (Figure 23*d*). The engine will be repaired by welding in the piece of cast iron that was broken out and either by replacing the crankshaft or by metal spraying and grinding the damaged rod bearing surface on the old one. The connecting rod and insert bearings will also be replaced.

6 *Failure of Gear Teeth* (Figure 24). This large ring gear is a classic example of root cracking and failure of gear teeth by fatigue. Note the sharp corners at the root of the teeth where the crack initiated. Redesign of the gear with a rounded root corrected the problem.

Brittle Failures

Small quantities of hydrogen have a great effect on the ductility of some metals. Hydrogen can get into steels by heating them in an atmosphere or a material containing hydrogen, such as pickling or cleaning operations, electroplating, cold working or welding in the presence of hydrogen bearing compounds, or the steel making process itself. There is a noticeable embrittling effect in steels containing hydrogen. This can be detected in tensile tests and can be seen in the plastic region of the stress-strain diagram showing a loss in ductility.

Electroplating of many parts is required because of their service environment to prevent corrosion failure. Steel may be contaminated by electroplating materials that are commonly used for cleaning or pickling operations. These materials can cause hydrogen embrittlement by charging the material with hydrogen. Monatomic hydrogen is produced by most pickling or plating operations at the metal-liquid interface, and it seems that single hydrogen atoms can readily diffuse into the metal. Preventative measures can be taken to reduce this accumulation of hydrogen gas on the surface of the metal.

A frequent source of hydrogen embrittlement is found in the welding process. Welding operations, in which hydrogen bearing compounds such as oil, grease, paint, or water are present, are capable of infusing hydrogen into the molten metal, thus embrittling the weld zone. Special shielding methods are often used that help to reduce the amount of hydrogen aborption.

One effective method of removing hydrogen is a "baking" treatment in which the part, or in some cases the welding rod, is heated for long periods of

Figure 24. Failure of gear teeth.

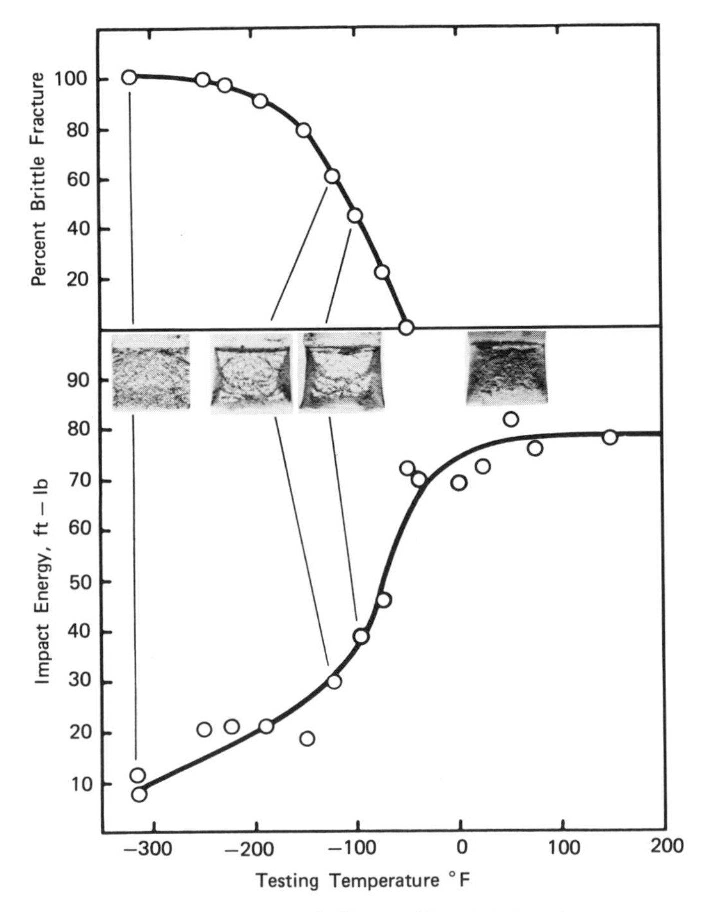

Figure 25. Appearance of Charpy V-notch fractures obtained at a series of tempered martensite of hardness around 30 RC (Courtesy Republic Steel Corporation).

time at temperatures of 250 to 400°F (121 to 204°C). This treatment promotes the escape of hydrogen from the metal and restores the ductility.

Brittle Failures at Low Temperatures

When body-centered cubic metals are subjected to dynamically applied loads at low temperatures, such as those involved in the impact test, the result is a lowering of ductility with a sharp increase of brittleness. This ductile-to-brittle transition phenomenon is commonly referred to as the transition temperature, transition zone, or brittle range (Figure 25). This phenomenon is often exhibited in instantaneous failures of pipelines, storage tanks, ships breaking in half at sea, bridges, and other metal structures subjected to low temperatures.

In Chapter 7 "The Mechanical and Physical Properties of Metals," methods of prevention of brittle transition were listed showing that certain alloying elements in steel, such as nickel, tended to lower the transition temperatures of steels. Some steels containing these elements will not fail by embrittlement even at temperatures below −150°F (−101°C).

Apparently plain carbon steels have higher transition temperatures than similar alloy steels. Also, coarser grained steels go into the brittle range at higher temperatures. Quenched and tempered fine-grained steels can have even higher ductile-to-brittle transition, sometimes at room temperature. When carbides or ferrite are outlining grain boundaries, the transition temperature is raised, which greatly reduces the energy required to produce failure. A mixed microstructure caused by improper quenching (slack quenching, between the A_3 and A_1 temperatures) also produces brittleness in steels. Of all of the microstructures, ferrite produces the highest transition temperature (Figure 26) and pearlite, bainite, and tempered martensite have a slightly lower transition temperature.

Temper Embrittlement

The tempering of steel and some resulting problems were explained in Chapters 12 and 13. It was noted that steels that are tempered for long periods of time in the range of 600 to 1100°F (315.5 to 593°C) tend to become brittle. This brittleness is due to an increase in the ductile-to-brittle transition temperature. When very slow cooling rates from the tempering temperature are used, the same effect is often seen.

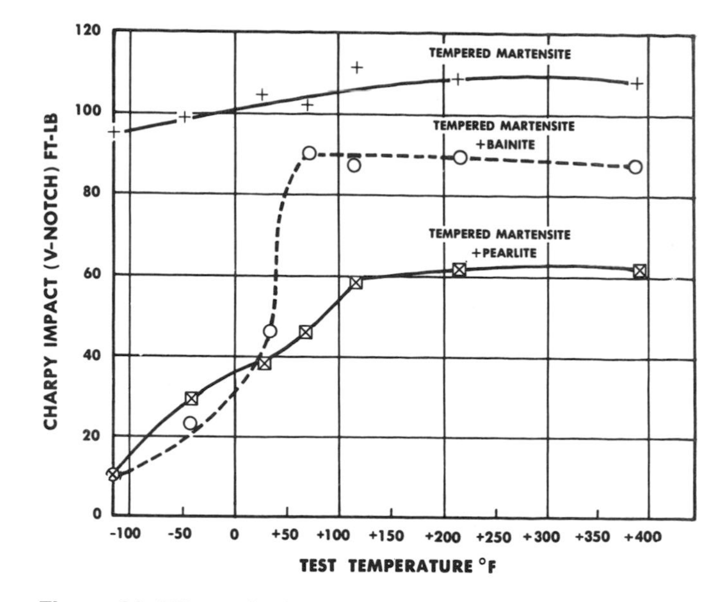

Figure 26. Effect of mixed microstructures, such as may occur in slack quenching, is to reduce impact resistance and to have a pronounced effect on the ductile to brittle transition (Courtesy Republic Steel Corporation).

Failures Caused by Thermal Treatments

Forging is a form of hot working in which metal is heated above the transformation temperature to the forging heat. This temperature must always be well under the solidification temperature to prevent "burning" of the metal (incipient melting of the grain boundary areas) (Figure 27). If this burning occurs, the piece must be discarded. Some metals are hot short (or red short) and cannot be hot forged since they tend to break or split. Many of the copper alloys such as bronze are hot short. Iron and steel containing an excess of sulfur without sufficient manganese to prevent iron sulfide from forming are hot short. Manganese has a greater affinity for sulfur than does iron. Iron sulfide melts at a lower temperature than man-

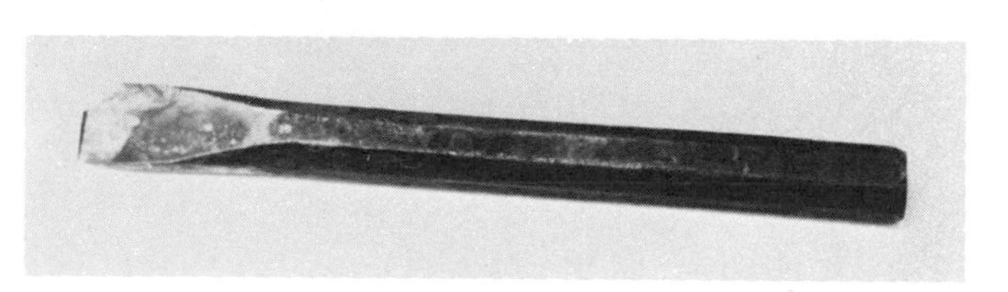

Figure 27. This chisel was burned while being forged and failure occurred as a result. No heat treatment will correct this condition.

ganese sulfide. The melting of the iron sulfide inclusions (which are usually drawn out in long fibrous stringers in rolled steel) causes a separation and splitting of the metal when being forged. Steels in higher carbon ranges must be forged in a narrow temperature range as they are hot short when forged above this range.

Although forging produces a desirable grain flow in metals, it sometimes creates weaknesses such as holes (called flakes or pipes) and laps. When the forging pressure is applied incorrectly, a condition known as piping takes place usually in the center of the piece (Figure 28). Seamless tubing is made from a solid billet using this principle. The forming rolls are set at a slope or angle deliberately to form a pipe that then flows over a plug.

Hardening and tempering problems were discussed in Chapter 13. Quench cracking is probably the greatest single problem related to hardening and tempering. Some of the most common causes for quench cracks are:

1 Overheating during the austenitizing cycle so that normally fine grained steel tend to coarsen. Coarse grained steels increase the depth of hardening and are inherently more prone to quench cracking than fine grained steels.
2 Improper selection of quenchant; that is, quenching too rapidly as with the use of water, brine, or caustic when oil is the proper quenchant for the specific part and type of steel.
3 Improper selection of steel for the design of the part.
4 Time delays between quenching and tempering.
5 Improper design of keyways, holes, sharp changes in section, mass distribution, and other stress raisers.
6 Improper entry of the work into the quenchant with respect to the shape of the part, which results in nonuniform or eccentric cooling.

Figure 28. End of forging showing grain flow and piping at center.

Since welding can alter hardness, grain structure, carbon and alloy content, and corrosion resistance of the base metal, many problems (discussed in Chapters 16 through 19) can arise as a result of welding. Each problem should be analyzed as a metallurgical condition using methods of investigation such as macroscopic, microscopic, destructive, and nondestructive testing. An interesting welding problem is shown in Figures 29*a* to 29*e*. A cold formed flange was bent to 90 degrees near the weld area prior to welding. A cored wire weld was made in one pass adjacent to the flange (which is now broken off). The material is SAE 1020 HR mild steel. It was discovered that the flange was bent too far and an attempt was made to bend it back slightly after the weld area cooled down to room temperature. The result was a brittle failure that is definitely not a characteristic of hot rolled mild steel.

The solution to this problem would be: since recrystallization and grain growth are promoted and accelerated by cold working, a larger more brittle grain structure was formed near the weld at the cold bend. Had the weld been several small, cooler passes instead of one large hot pass, the temperature may have been raised only to the ferrite recrystallization temperature (950 to 1200°F or 510 to 649°C) and small soft grains would have been formed. In both the original HR steels as shown in Figures 29*c* and 29*d* and the overheated stressed area, the grain boundaries show carbide precipitation (migration of carbides to the grain boundaries). This condition promotes brittleness. Figure 29*e* shows the grain formation of a normal hot rolled steel.

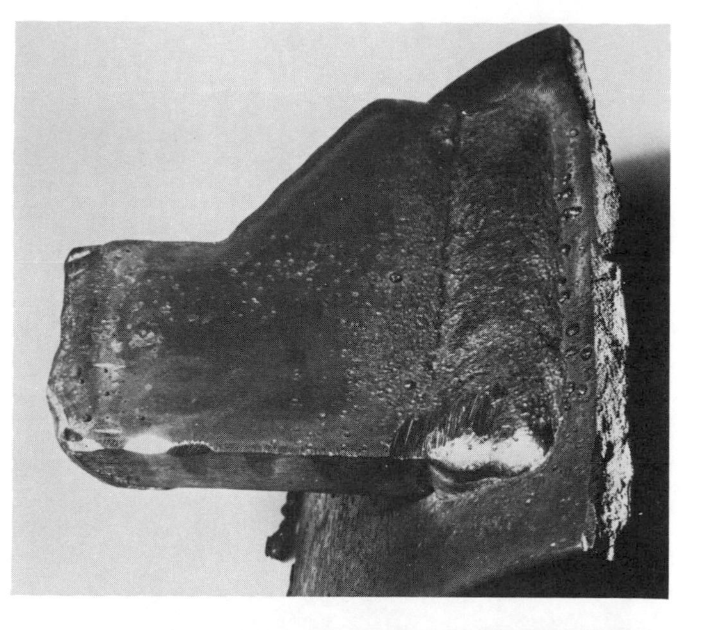

Figure 29*a*. Curved plate that had a cold formed flange shows fracture where the flange broke off in a sudden brittle failure.

Figure 29*b*. Close-up of the fracture showing a brittle crystalline cross section.

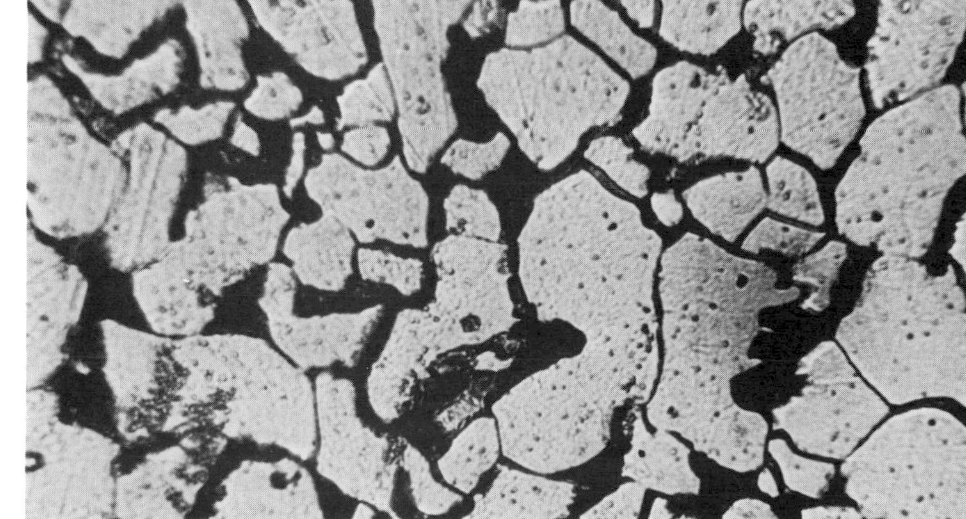

Figure 29*c*. Original (unheated) HR steel showing some carbide precipitation in the grain boundaries (500 ×).

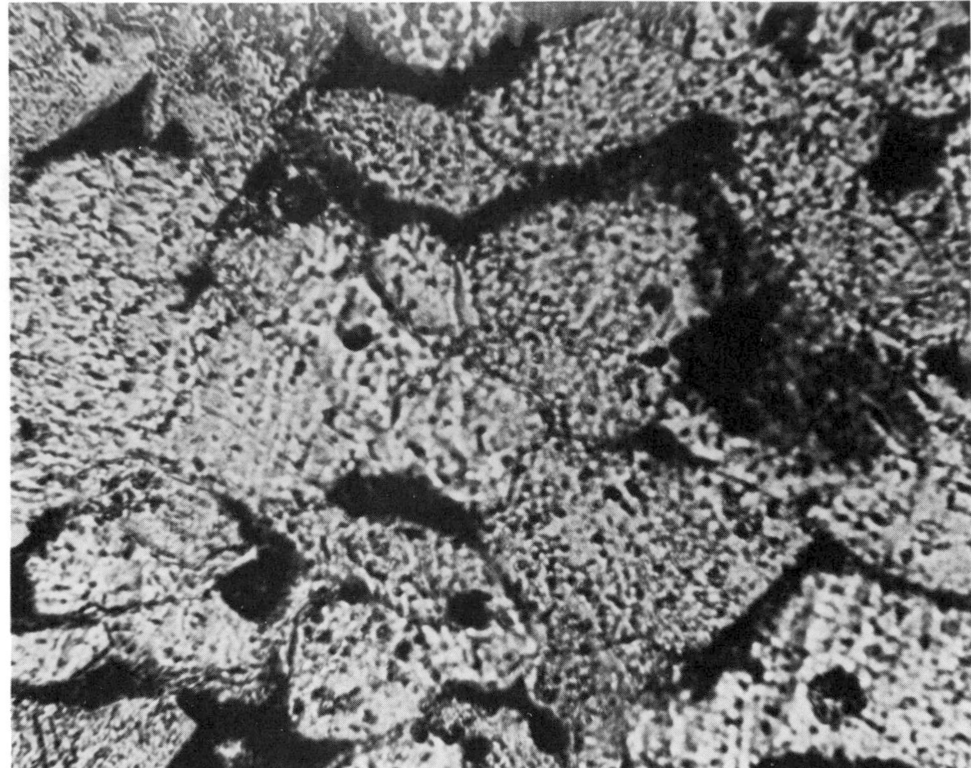

Figure 29*d*. This specimen of the same HR steel was taken from the highly stressed area of the cold formed flange that was overheated by the weld. Considerable carbide precipitation is evident as well as a cleavage pattern, indicating a very weak brittle structure with enlarged grains (500 ×).

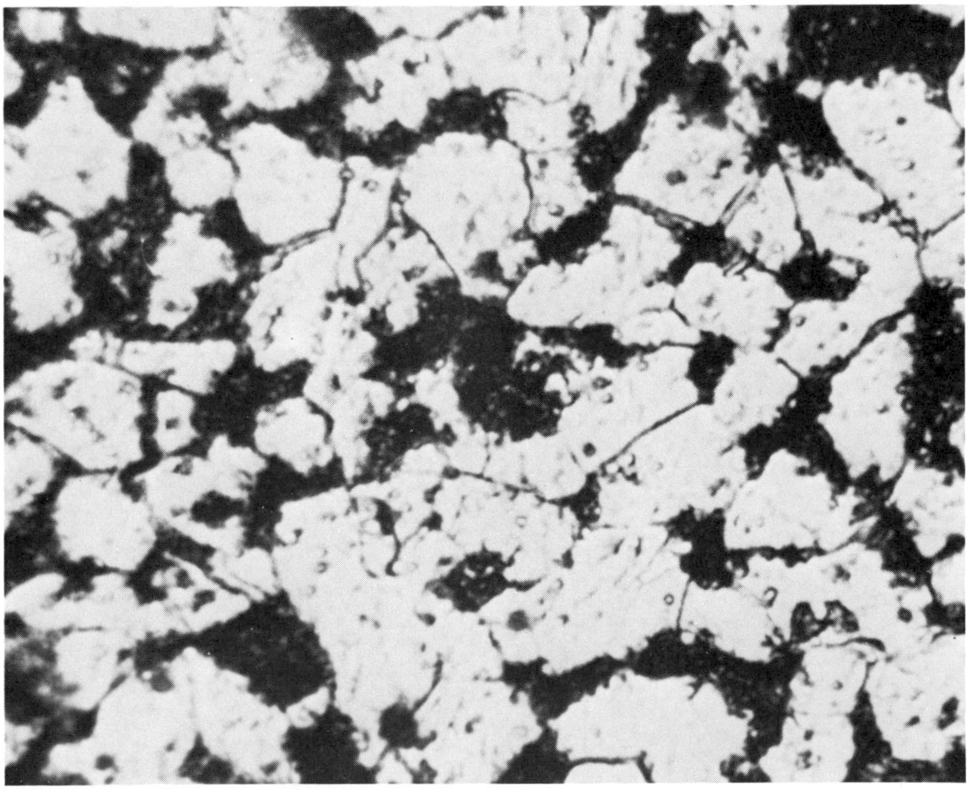

Figure 29*e*. Normal microstructure of a HR steel with the same carbon content (500 ×).

There is also evidence of a preferred orientation of cleavage planes (Figure 29*d*) developing (the Widmanstatten structure) as ferrite needles growing from prior austenite grain boundaries. This is an exceedingly weak and brittle structure that is the result of overheating carbon steel and holding to that temperature for a period of time. Oxy-acetylene welding of steel also tends to promote this structure because of the large heat input that is held for a longer period of time than most arc welding operations. A recommendation would be to bend the flange hot (at forging temperature) prior to welding, or to use smaller passes and cooler welding temperatures if the cold bending must be done.

Welding on highly stressed parts often initiates fatigue in the part (see Figure 9). A weld, due to contraction during cooling, produces a tensile stress in the base metal. This stress concentration can start a crack that continues to grow until the part fails. Shafting, since it often produces stress reversals because of rotation, is particularly sensitive to stress concentrations such as those produced by a weld. It is therefore poor practice to weld anything on the side of the shafting; even a tack weld to hold a key in place can cause a failure (Figure 30). Welding an arm or hub to a shaft is very likely to cause an ultimate failure of the shaft by fatigue.

Figure 30. Stress concentrations caused by welds on this shaft precipitated the fatigue failure.

Grinding Problems

There is a definite relationship between grinding and heat treating. Surface temperatures ranging from 2000°F (1093°C) to 3000°F (1649°C) are generated during grinding. This can cause two undesirable effects on hardened tool steels: development of high internal stresses causing surface cracks to be formed, and changes in the hardness and metallurgical structure of the surface area.

One of the most common effects of grinding on hardened and tempered tool steels is to reduce the hardness of the surface by gradual tempering where the hardness is lowest at the extreme surface but increases with distance below the surface. The depth of this tempering varies with the amount or depth of cut, the use of coolants, and the type of grinding wheel. If high temperatures are produced locally by the grinding wheel and the surface is immediately quenched by the coolant, a martensite having a Rockwell hardness of C65 to 70 can be formed. This gradient hardness, being much greater than that beneath the surface of the tempered part, sometimes causes very high stresses that contribute greatly to grinding cracks. Sometimes grinding cracks are visible in oblique or angling light, but they can be easily detected when present by the use of magnetic particle or fluorescent particle testing.

When a part is hardened but not tempered before it is ground, it is extremely liable to stress cracking. Faulty grinding procedures can also cause grinding cracks. Improper grinding operations can cause tools that have been properly hardened to fail.

Self-Evaluation

Worksheet

Objective Identify and describe 10 kinds of industrial service failures and make recommendations that will help to correct the problem.

	Kind of Failure (Description)	Recommendation to Correct the Problem
1. Figure 31.		
2. Figure 32.		

	Kind of Failure (Description)	Recommendation to Correct the Problem

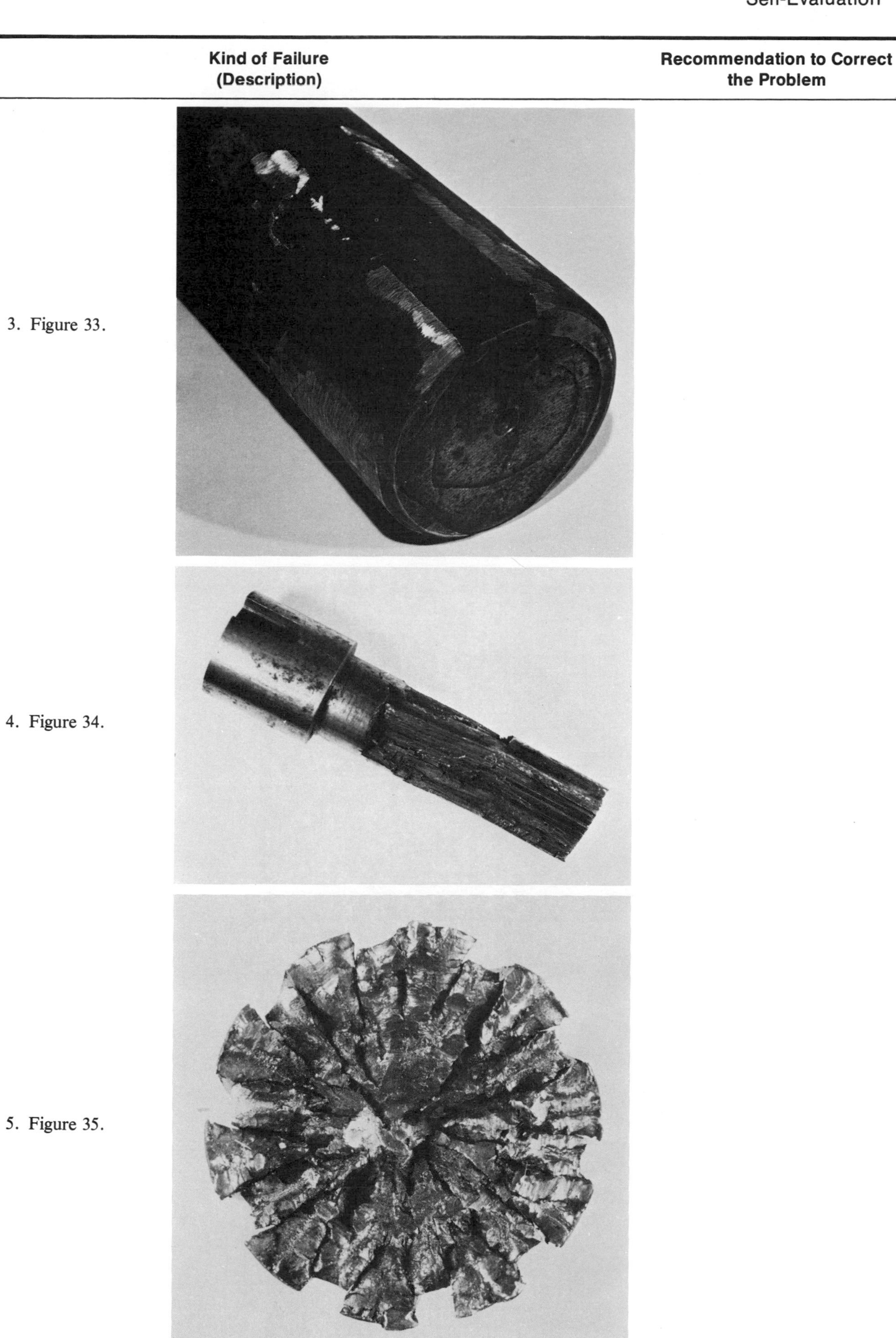

3. Figure 33.

4. Figure 34.

5. Figure 35.

Kind of Failure (Description)	Recommendation to Correct the Problem

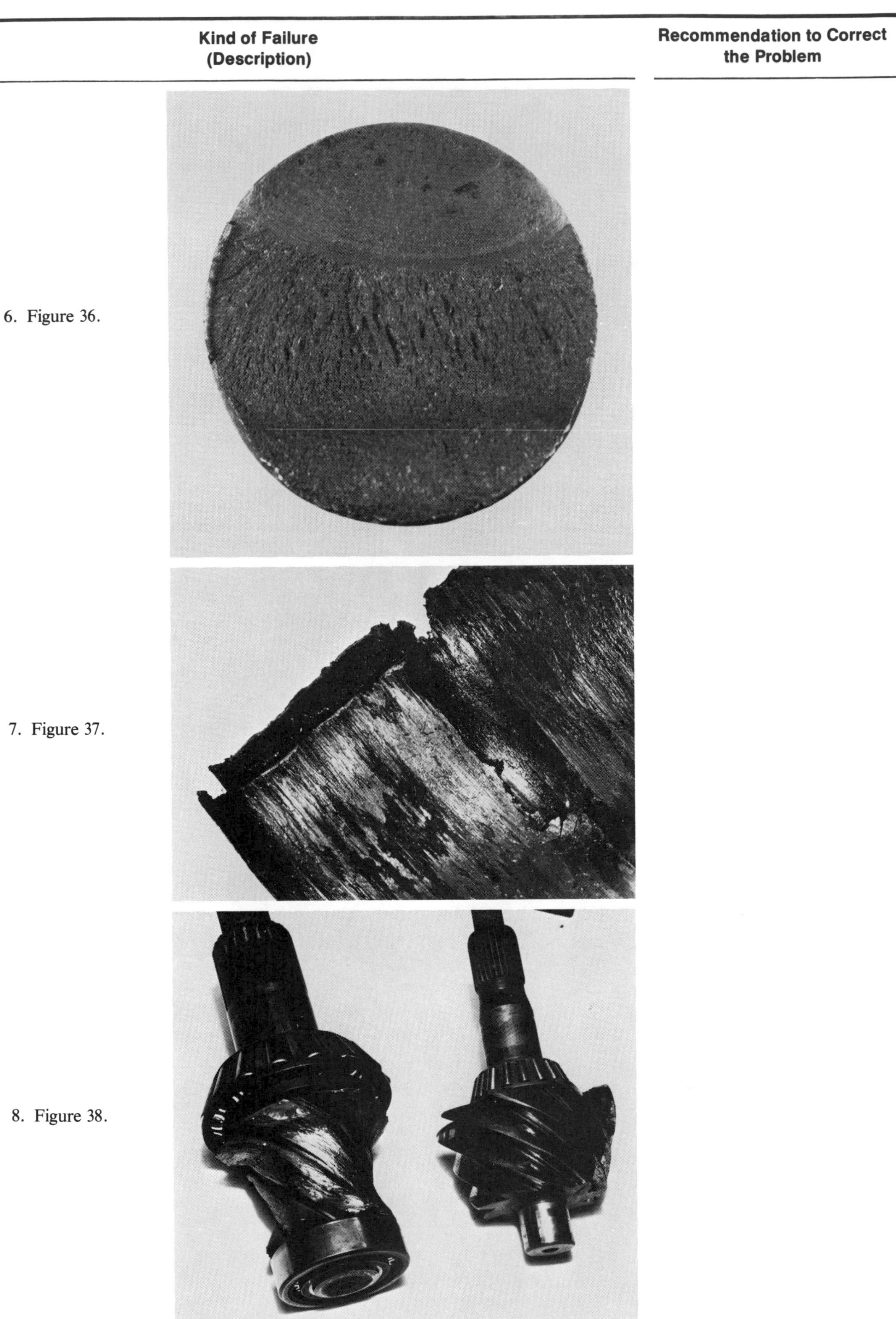

6. Figure 36.

7. Figure 37.

8. Figure 38.

	Kind of Failure (Description)	Recommendation to Correct the Problem
9. Figure 39.		
10. Figure 40.		

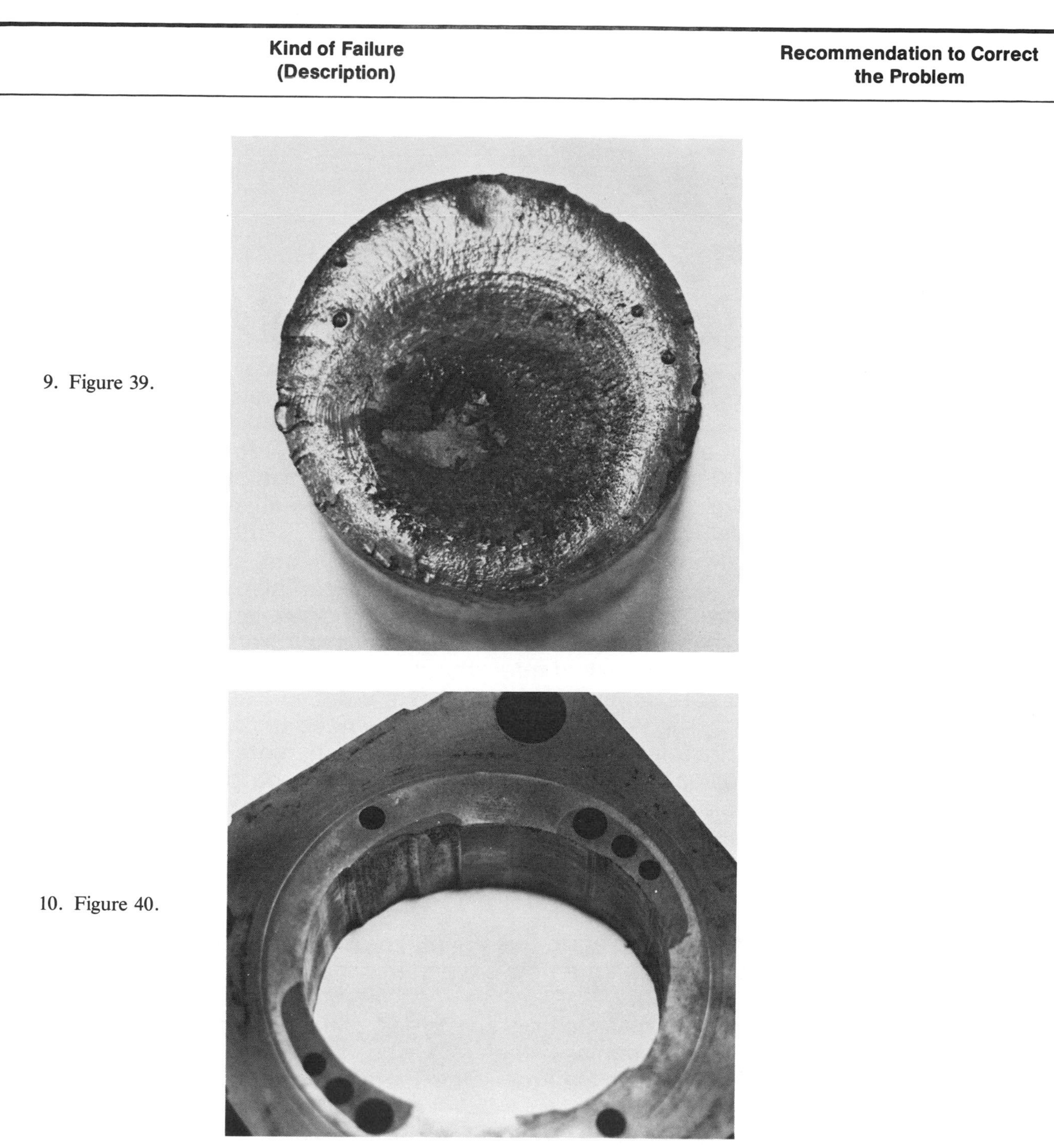

Copy this table format on a sheet of paper, fill in the answers as required, and turn it in to your instructor for grading.

This chapter has no Posttest.

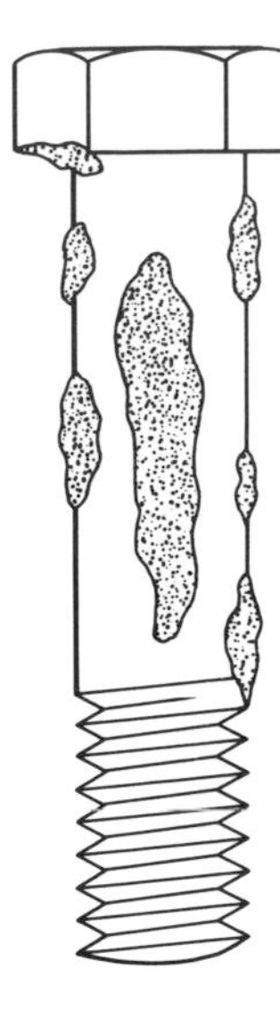

23 Corrosion in Metals

With the exception of gold, silver, and some other noble metals all metals are subject to the deterioration caused by corrosion. Iron, for example, tends to revert back to its natural state of iron oxide. Other metals revert to sulfides and oxides or carbonates. Buildings, ships, machines, automobiles are all subject to attack by the environment. The corrosion that results often renders them useless and they have to be scrapped. Billions of dollars a year are lost as a result of corrosion. Corrosion can also cause dangerous conditions to prevail such as on bridges, where the supporting structures have been eaten away, or aircraft in which an insidious corrosion, called intergranular corrosion, can weaken the structural members of the aircraft and cause a sudden failure.

Those who work with metals need to have a knowledge of the principles involved in metal corrosion in order to be able to use the correct preventative measures. It is the purpose of this chapter to help you understand these principles of corrosion and the methods by which corrosion can be offset.

OBJECTIVES

After completing this chapter, you will be able to:

1. Demonstrate how oxygen in water affects the rate of corrosion of iron.
2. Demonstrate how an oxygen concentration cell works to corrode stainless steel.
3. Demonstrate how active metals can displace ions of a less active metal.

INFORMATION

The most common form of corrosion is a deterioration of metals by an electrochemical action. It is generally a slow and continuous action. High temperature scaling and the formation of oxides on metals is oxidation corrosion. The oxide of iron formed at high temperatures is black, often called mill scale. Corrosion in metals is the result of their desire to unite with oxygen in the atmosphere or in other environments to return to a more stable compound, usually called ore. Iron ore, for example, is in some cases simply iron rust. Corrosion may be classified by the two different processes by which it can take place: **direct oxidation corrosion,** which usually happens at high temperatures, and **galvanic corrosion,** which takes place at normal temperatures in the presence of moisture or an electrolyte.

Direct Oxidation Corrosion

Oxidation at high temperatures is often seen in the scaling that takes place when a piece of metal is left in a furnace for a length of time. The black scale is actually a form of iron oxide, called magnetite (Fe_3O_4) (Figure 1). This oxide coating is also called mill scale because it is formed on heated ingots or slabs that are rolled in steel mills. The red hot steel is constantly scaling since it is in contact with the oxygen in the atmosphere.

Galvanic Corrosion

Galvanic corrosion is essentially an electrochemical process that causes a deterioration of metals by a very slow but persistent action. In this process, part or all of the metal becomes transformed from the metallic state to the ionic state and often forms a chemical compound in the electrolyte. On the surface of some metals such as copper or aluminum the corrosion product sometimes exists as a thin film that resists further corrosion. In other metals such as iron, the film of oxide that forms is so porous that it does not resist further corrosive action, and corrosion continues until the whole piece has been converted to the oxide (Figure 2).

Figure 1. This block of steel was heated to a high temperature in a furnace in the presence of air. It is covered with a loose black scale, which is magnetite-iron oxide (Fe_3O_4). Magnetite, as the name implies, is magnetic.

Positive and Negative Ions

Certain elements have common properties and are arranged into groups in a Periodic Table. See Appendix 1, Table 5. Eight groups in vertical columns are able to form compounds and one group (0) on the right is inactive (inert). The horizontal rows are periods of elements and are arranged in steps in increasing or decreasing atomic numbers. The periodic law states that the properties of elements are periodic functions of their atomic numbers. The atomic number is equal to the number of protons within the nucleus of the atom or the sum of all orbiting electrons.

Valence is the combining power of an atom and refers to the bonding force of atoms. Ions (atoms having a positive or negative charge) of some elements always have the same charge such as +1 or +2 (that is, they are missing electrons in the outer shell); others such as iron vary in valence. When iron has a valence of +2, it is called ferrous; when it has a +3, it is called ferric. Metallic atoms generally carry a positive charge and nonmetals a negative one. Atoms having opposite charge can often combine to form compounds with various bonding arrangements. The charge of the ion is written with a positive or negative sign at the upper right of the symbol; F^{++} for iron, Zn^{++} for zinc, and Cl^{-} for chloride are examples. When a group of atoms in a chemical relation have an electrical charge, they are called radicals. An example is the hydroxyl ion composed of one oxygen atom and one hydrogen atom with a charge of −1 which would be written OH^{-}.

Electrolytes

An electrolyte is any solution that conducts electric current and contains negative or positive ions. Corrosion requires the presence of an electrolyte to allow metal ions to go into solution. The electrolyte may be fresh or salt water, acid or alkaline solutions of any concentration. Even a fingerprint on clean metal can form an electrolyte and produce corrosion. There must be a completed electric circuit and a flow of direct current before any galvanic action can take place. There also must be two electrodes, an anode and a cathode, and they must be electrically connected. (See note at end of chapter.) The anode and cathode may be of two different kinds of metals or they may be located on two different areas of the same piece of metal. The

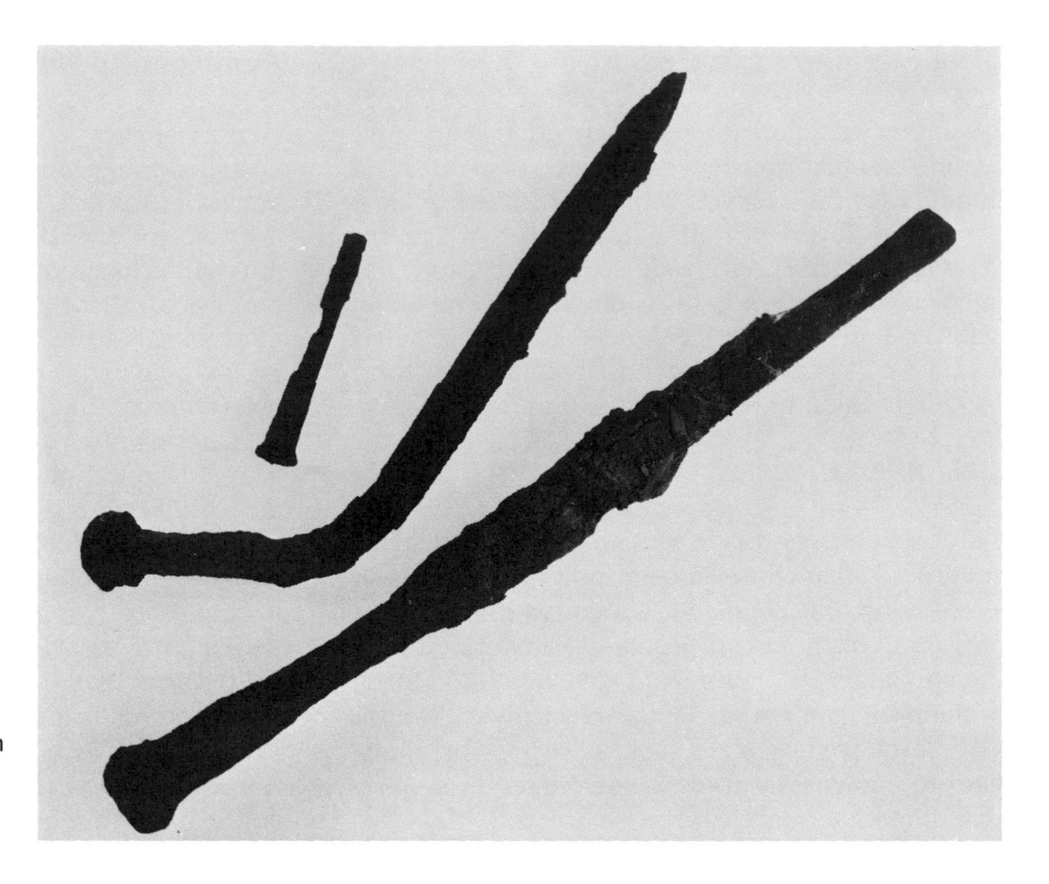

Figure 2. These large iron spikes were found on an ocean beach. They are almost completely changed into iron oxide rust.

connection between the anode and the cathode may be made by the metal itself or by a metallic connection such as a bolt or rivet.

If a piece of metal is immersed in hydrochloric acid, hydrogen bubbles will collect rapidly and be released. Some metals corrode very quickly when placed in acids. In this case, there are anodes and cathodes and the deterioration of the metal occurs at the anodes. Very tiny, well defined cathode and anode areas are formed all over the piece of metal. However they may shift, causing a very uniform corrosion to take place.

How Corrosion Takes Place

When corrosion occurs, positively charged atoms of metal are released or detached from the solid surface and enter into solution as metallic ions while the corresponding negative charges in the form of electrons are left behind in the metal. The detached positive ions bear one or more positive charges. In the corrosion of iron, each iron atom releases two electrons and then becomes an iron ion carrying two positive charges. Two electrons must then pass through a conductor to the cathode area (Figure 3). (Without this electron flow, no metal ions can be detached at all from the anode.) The electrons reach the surface of the cathode material and neutralize positively charged hydrogen ions that have become attached to the cathode surface. Two of these ions will now become neutral atoms and are released generally in the form of hydrogen gas. This release of the positively charged hydrogen ions leaves an accumulation and a concentration of OH negative ions that increases the alkalinity at the cathode. When this process is taking place, it can be observed that hydrogen bubbles are forming at the cathode only. When cathodes and anodes are formed on a single piece of metal, their particular locations are determined by, for example, the lack of homogeneity in the metal, surface imperfections, stresses, inclusions in the metal, or anything that can form a crevice such as a washer, a pile of sand, or a lapping of the material.

Cathodic Polarization

In more neutral electrolytes such as pure water or even a sodium chloride solution, the release of hydrogen gas at the cathode is very slow; thus very

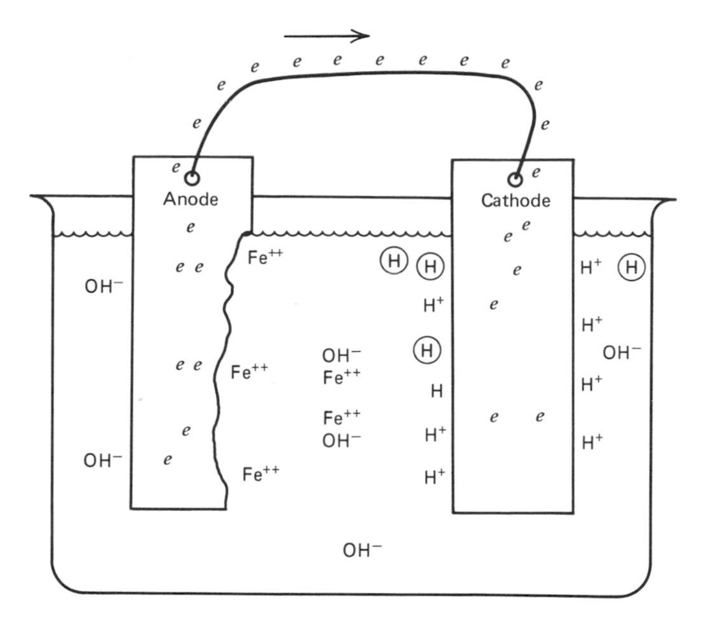

Figure 3. When iron ions (Fe^{++}) become detached from the anode, electrons flow to the cathode and neutralize positively charged hydrogen ions, releasing them from the cathode surface.

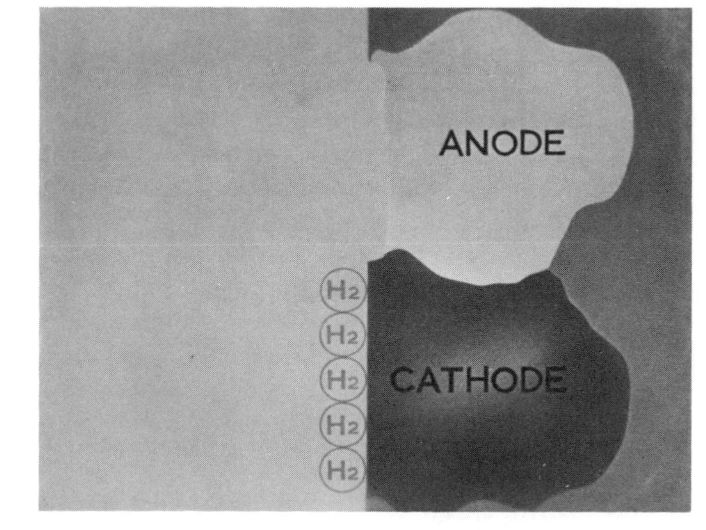

Figure 5. Polarization of the local cathode by a film of hydrogen (Courtesy The International Nickel Company, Inc.).

little corrosion takes place at the anode. This layer of hydrogen on the cathode surface slows down the reaction. The process is called cathodic polarization. If, however, oxygen is dissolved in the electrolyte, a reaction can take place. Oxygen reacts with the accumulated hydrogen to form OH^- ions or water (H_2O). This process permits corrosion to proceed, oxygen acting as a cathodic depolarizer. Often a further combining of the corrosion products and the hydroxyl (OH^-) ions from the cathodic reaction takes place; when ferrous ions, for example, are released from the anode, they combine with the hydroxyl ions at the cathode to form ferrous hydroxide. This then becomes oxidized by oxygen to form ferric hydroxide ($Fe(OH)_2$), which precipitates as ordinary iron rust (Figures 4 to 6).

The action of oxygen in accelerating corrosion in water or in an electrolyte can be demonstrated

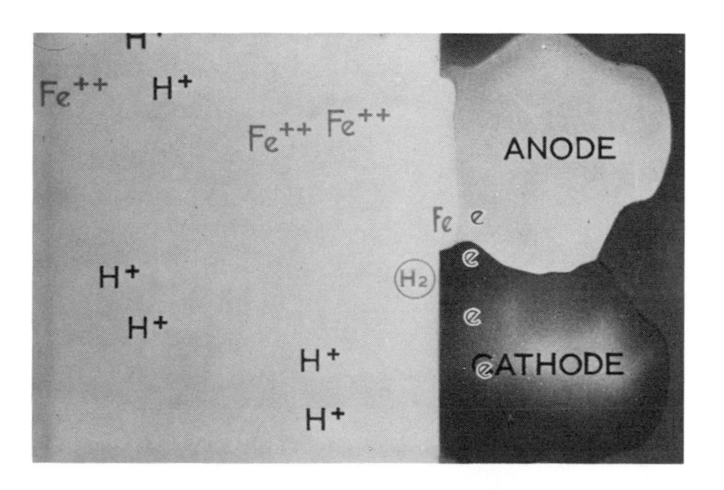

Figure 4. Formation of ions at the anode and hydrogen at the cathode. This is the process of rusting on iron (Courtesy The International Nickel Company, Inc.).

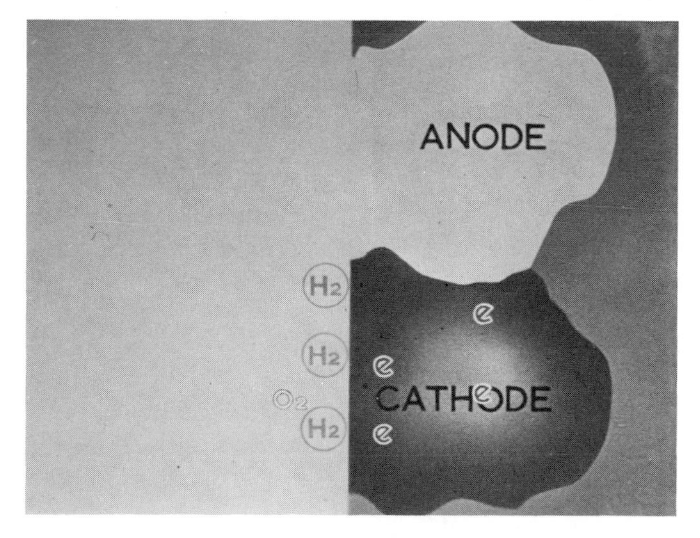

Figure 6. Removal of hydrogen or depolarization of the cathode caused by oxygen in water (Courtesy The International Nickel Company, Inc.).

by placing iron turnings into glass containers of water. Oxygen can be bubbled through one container for a period of time. After several hours, it can be seen that in the container saturated with oxygen, the metal has begun to rust and the one without oxygen still has not corroded. If iron is placed in still water like a stagnant pond, it will not corrode for years because it is oxygen free. In contrast, iron placed in running water containing oxygen often rusts very quickly. Also when water spray or rain drops impinge on the surface of iron, the iron is constantly in the presence of oxygen from the atmosphere, which accelerates corrosion.

The Galvanic Series

Anodes and cathodes can form at various places on the surface of a metal, but on the surface of dissimilar metals the corrosion rates produced are often greater. Some metals have a greater tendency to corrode than others because they are more active chemically and tend to become anodes, while others are less active and are cathodic. Gold, for example, is very inactive and is cathodic and has about the lowest possible corrosion potential. The **galvanic series** of metals and alloys shows this difference in the activity of metals in sea water (Table 1). The galvanic series (also called the electromotive series) is so named because of the direct electric current produced (galvanism) when two dissimilar metals are immersed in an electrolyte. At the more active end, we find the metals that most readily become anodic are magnesium and zinc. Except for those metals at the extreme opposite ends (gold and magnesium) any given metal may be an anode or cathode when it is coupled with another metal in the series.

In the galvanic series the metals become increasingly active toward the anode end. For example, if potassium is placed in water, a violent reaction takes place, but gold or platinum in strong nitric acid will not even show evidence of attack. In the electromotive series, a metal may displace any metal below it. Iron can displace copper from solution with the copper ions acting as receivers for electrons in the corrosion of the iron. The displaced copper forms a coating on the corroded iron surface. This takes place when a solution of copper sulfide containing copper ions is placed on a clean surface of iron; iron atoms are detached (corroded) leaving the surface copper plated. In this way hydrogen ions act as electron accepters or receivers for any metal more active than hydrogen

Table 1 Galvanic Series of Metals and Alloys in Sea Water

Anode End (Highest Corrosion)	
Magnesium	Copper
Beryllium	Bronze
Aluminum	Copper-nickel alloys
Cadmium	Nickel
Aluminum alloys	Tin
Uranium	Lead
Manganese	Hydrogen
Zinc	Inconel
Plain carbon steel	Silver
Alloy steels	Stainless steel (passive)
Cast iron	Monel
Cobalt	Titanium
Stainless steel (active)	Platinum
Brass	Gold
	Cathode End (Least Corrosion)

in the galvanic series. Another example is the displacement of silver by copper with the silver ions becoming the electron acceptors and the copper being corroded; a silver coating is formed on the surface of the copper. This can be demonstrated by placing a strip of copper in a silver nitrate solution; a film of silver will then be deposited on the copper by ion transfer. The rate at which galvanic action and corrosion takes place depends on the degree of difference in electrical potential and the resultant current flow. If zinc is coupled with copper, a large current will flow, but if brass is coupled with copper, a very small current will flow and less reaction will be seen. A galvanic corrosion of magnesium that is in contact with a steel core can be seen in Figure 7.

Dezinctification

Dezinctification is a form of corrosion that often takes place in brass, cast iron, and other alloys. In dezinctification, the removal of zinc from brass continues throughout the alloy until the entire part has failed. Corrosion cells are set up between the zinc and the copper in the brass. The zinc remains in the electrolyte solution and the copper remains by plating on the brass. The final structure is very porous and brittle since it is made up of copper oxide and copper.

Cast iron can be completely corroded in that the iron goes into solution as ferrous ions, leaving mostly the graphite. This final condition of the iron is a very weak, soft, brittle and porous material.

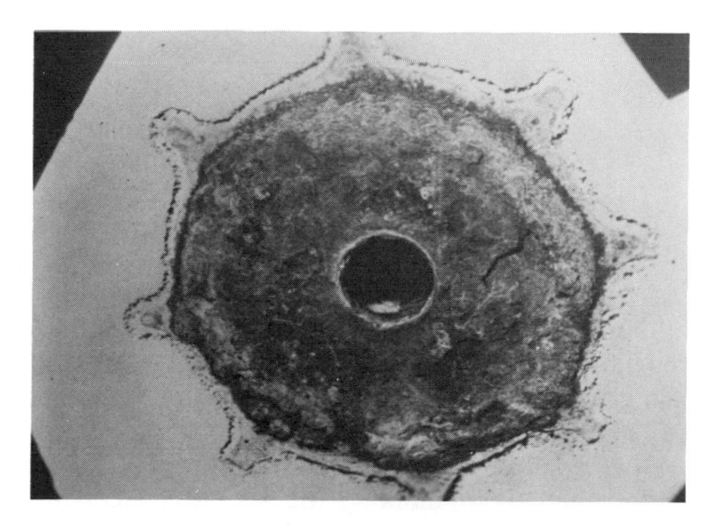

Figure 7. Galvanic corrosion of magnesium that surrounds a steel core (Courtesy The International Nickel Company, Inc.).

Rate of Corrosion

The speed at which corrosion proceeds depends on several factors: the type of corrosion product, the kind of electrolyte, galvanic current density, environmental condition (crevices, rate of electrolyte flow past the cathode, concentration cells), and the availability of oxygen as a depolarizer. The current density can be greater on small electrodes than on large ones, if a large cathode is coupled with a small anode. The current density for its size can be very high since the cathode provides a relatively larger area for depolarization of hydrogen by any available oxygen, thus allowing more current flow from the small anode and causing a large number of metal ions to leave the anode. If a small cathode is coupled to a large anode, however, the reverse is true; the cathode provides only a small surface for depolarization. This allows a small current density and very little corrosion occurs at the anode (Figure 8).

There is also a difference in rate of corrosion in steels having different alloying elements such as copper, nickel, or chromium steels, which have less tendency to corrode than most other alloy steels. However, as a general rule, all metals having almost no alloying elements or any other kinds of contaminating materials, such as pure iron or pure aluminum, tend to have very low corrosion rates. Wrought iron containing almost no carbon will tend to resist corrosion or rust more than carbon steel. Ancient wrought iron artifacts more than a thousand years old have resisted corrosion over the years. The iron tower at Delhi, India, and the wrought iron nails used by the Vikings in their ships are examples of this resistance to rusting. Other forms of protection used to inhibit the process of corrosion are organic and inorganic coatings, the use of cathode protection, and inhibitors. Inhibitors are usually liquids that are placed in the electrolyte or other corrosive medium to render it inert. They are sometimes used as a coating on metals.

Rates of corrosion are determined by practical tests in certain atmospheres (Figures 9*a* to 9*c*). The progress of corrosion is plotted on a graph showing loss of weight in relation to the amount of time the specimen was subjected to a given environment (Figure 10).

(a)

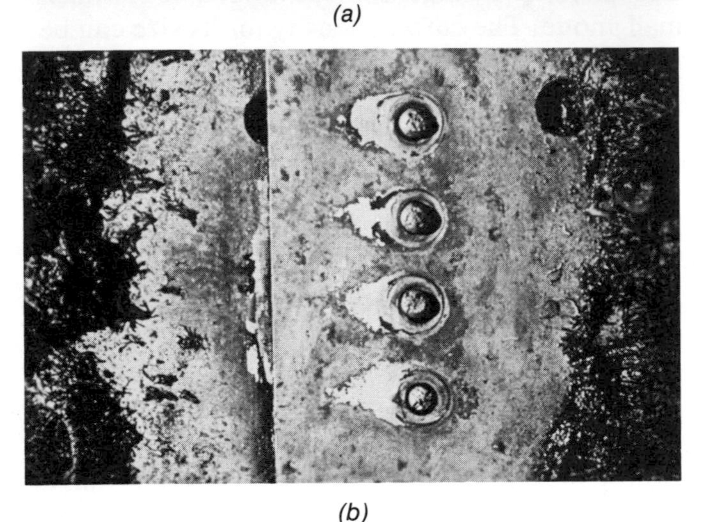

(b)

Figure 8. Influence of area relationship between cathode and anode illustrated by copper-steel couples after immersion in sea water. *(a)* Copper rivets with a small area in steel plates of large area have caused only slight increase in corrosion of steel. *(b)* Steel rivets with small area in copper plates of large area have caused severe corrosion of steel rivets. (Courtesy The International Nickel Company, Inc.).

Figure 9*a*. Corrosion of steels in marine atmosphere. *(Left)* Low copper steel; *(center)* ordinary steel; and *(right)* nickel-copper-chromium steel (Courtesy The International Nickel Company, Inc.).

Figure 9*b*. Mirror finish on Hastelloy alloy C after ten years of exposure in marine atmosphere (Courtesy The International Nickel Company, Inc.).

Concentration Cells

Concentration cells are created on a single piece of metal on which a deposit of sand or any material can produce a low oxygen area and a high oxygen area (Figure 11). Lapped metal on riveted or bolted joints is especially susceptible to this form of corrosion. Droplets of water can produce concentration cells (Figure 12). Cathodes are formed at the areas of high oxygen concentration, and anodes at areas of low concentration when corrosion occurs. This can be seen by observing a drop of salt water placed on a freshly polished steel surface. In only an hour or two, a ring of rust will form inside the drop (anode) while the outer edges (cathode) remain clear.

Stainless steel is protected from the environment with an invisible but effective film that can be strengthened by immersing the stainless steel in concentrated nitric acid. If stainless steel, however, becomes oxygen starved in any particular area by a washer or a pile of sand, the passivity

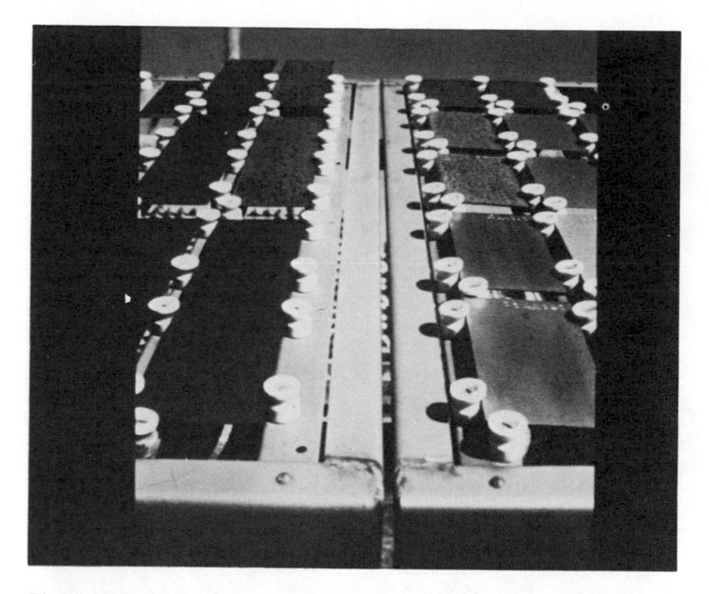

Figure 9c. The racks used in this corrosion experiment, made of monel-nickel-copper alloy, shows only a superficial tarnish after years of exposure in marine atmosphere (Courtesy The International Nickel Company, Inc.).

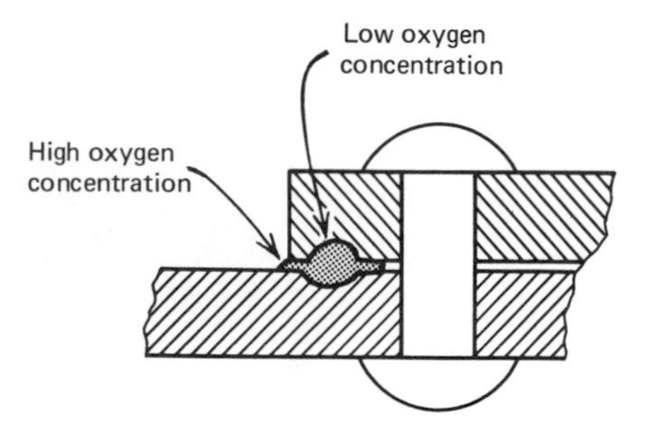

Figure 11. Concentration cell. The crevice in this concentration cell hinders the diffusion of oxygen and causes high and low oxygen areas. The low oxygen area is anodic. When the solution surrounding a metal contains more metal ions at one point than another, metal goes into solution where the ion concentration is low.

of the stainless steel can break down in these places. The area of the stainless steel that is freely exposed to the dissolved oxygen acts as a cathode; corrosion will take place, however, under these deposits or crevices where oxygen cannot penetrate; therefore when stainless steel is used around sea water or other corrosive media it is important to avoid such crevices or nitches where oxygen cannot penetrate. In these cases, stainless steel can corrode just like ordinary steel.

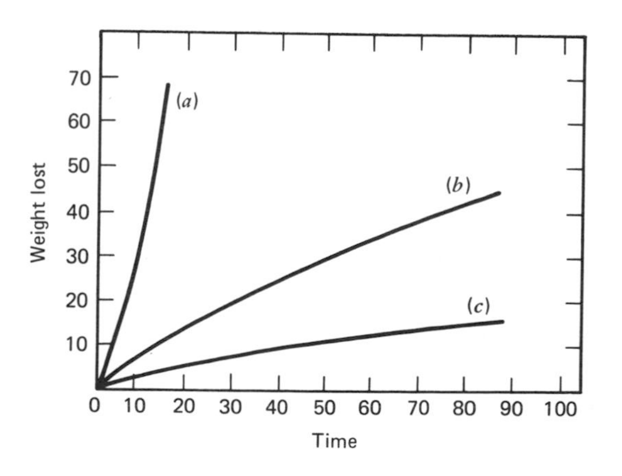

Figure 10. Progress of corrosion on steels with different resistance to atmospheric corrosion. *(a)* Steel containing very small amounts of copper. *(b)* Plain carbon steel. *(c)* Steel containing nickel, chromium, and copper.

Erosion Corrosion

Corrosion can also take the form of erosion in which the protective film, usually an oxide film, is removed by a rapidly moving atmosphere or media (Figure 13). Depolarization can also take place, for example, on the propellers of ships because of the movement through the water, which is the electrolyte. This causes an increased corrosion rate of the anodic steel ship's hull. Impellers of pumps are often corroded by this form of erosion corrosion in which metal ions are rapidly removed at the periphery of the impeller but are concentrated near the center where the velocity is lower.

Intergranular Corrosion

Another form of corrosion is intergranular corrosion. This takes place internally. Often the grain boundaries form anodes and the grains themselves form cathodes, causing a complete deterioration of the metal in which it simply crumbles when it fails

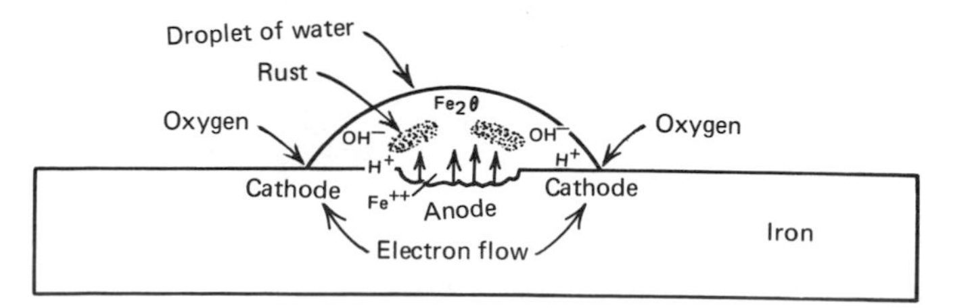

Figure 12. The action of corrosion in a droplet of water. Pitting corrosion is the result of this type of action.

Figure 13. Butane gas impinges on the internal passage of this conversion part in a carburetor on a gasoline engine and has badly eroded the metal. This is an example of erosion corrosion.

Figure 14. The very large grains are outlined in this zinc based die casting because of intergranular corrosion. The metal has deteriorated at the grain boundaries, which have become anodes.

(Figure 14). This often occurs in stainless steels in which chromium carbides precipitate into the grain boundaries. This lowers the chromium content adjacent to the grain boundaries, thus creating a galvanic cell. Most types of austenitic stainless steel when held at temperatures of 1500°F (815°C) and cooled slowly are subject to this kind of intergranular corrosion (Figure 15).

Pitting Corrosion

Differences in environment can cause a high concentration of oxygen ions. This is called cell concentration corrosion. Pitting corrosion is localized and results in small holes on the surface of a metal caused by a concentration cell at that point.

Stress Corrosion

Another form of corrosion is called stress corrosion. When high stresses are formed on metals in a corrosive environment, cracking can also be accelerated in the form of corrosion fatigue failure (Fig-

Figure 15. Section through type 310 stainless steel hot rolled plate in the as-welded condition, showing intergranular microcracking through the multiple-pass weld metal and the base metal (¾ size) (By permission, from *Metals Handbook*, Volume 7, Copyright American Society for Metals, 1972).

ure 16). It is a very localized phenomenon and results in cracking type of failures (Figure 17).

Protection against Corrosion

Cathodic protection is often used to protect buried steel pipelines and bronze ships' propellers. This is done by using zinc and magnesium sacrificial anodes that are bolted to the ship's hull or buried in the ground at intervals and electrically connected to the metal to be protected. In the case of the ship, the bronze propeller acts as a cathode, the steel hull as an anode, and the sea water as an electrolyte. Severe corrosion can occur on the hull as a result of galvanic action. The sacrificial anodes are very near the anodic end of the galvanic series and have a large potential difference between both the steel hull of the ship and the bronze propeller (Figures 18*a* and 18*b*). Both the hull and propeller become cathodic and consequently do not deteriorate. The magnesium anodes are replaced from time to time.

Selection of materials is of foremost importance. Even though a material may be normally resistant to corrosion, it may fail in a particular environment or if coupled with a more cathodic metal.

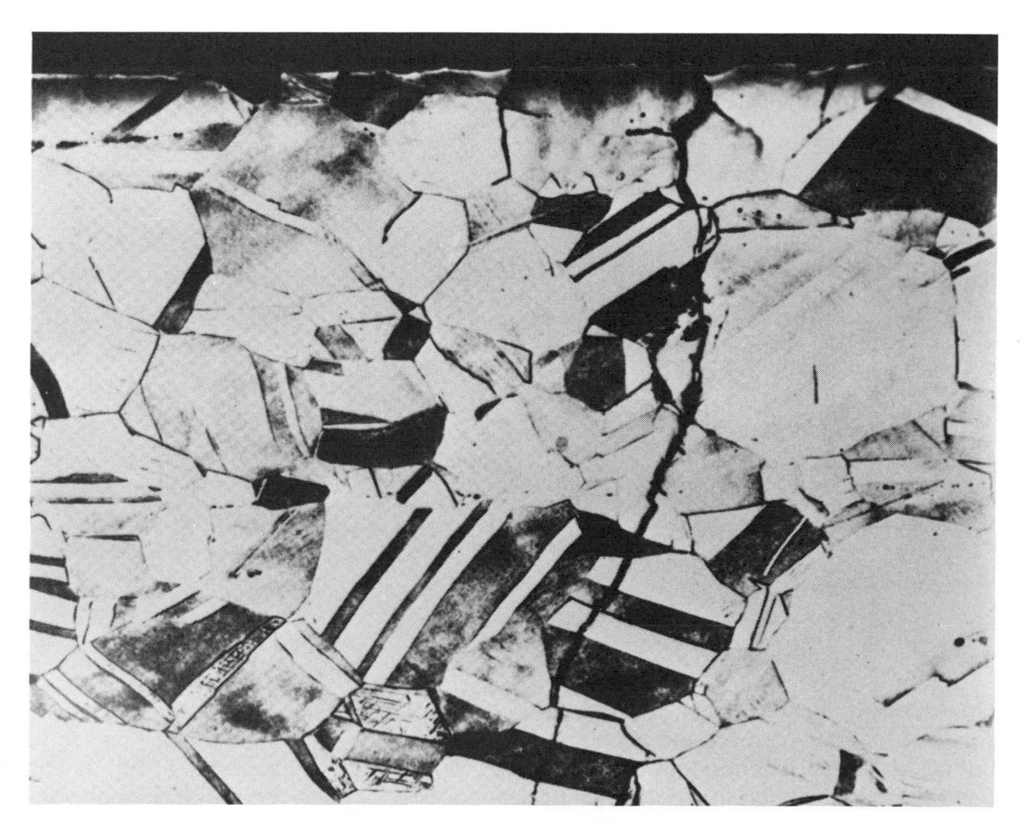

Figure 16. Typical transgranular corrosion-fatigue crack in alloy 260 (cartridge brass 70 percent). Note the lack of branching in the inner, or fatigue section of the crack (By permission, from *Metals Handbook,* Volume 7, Copyright American Society for Metals, 1972).

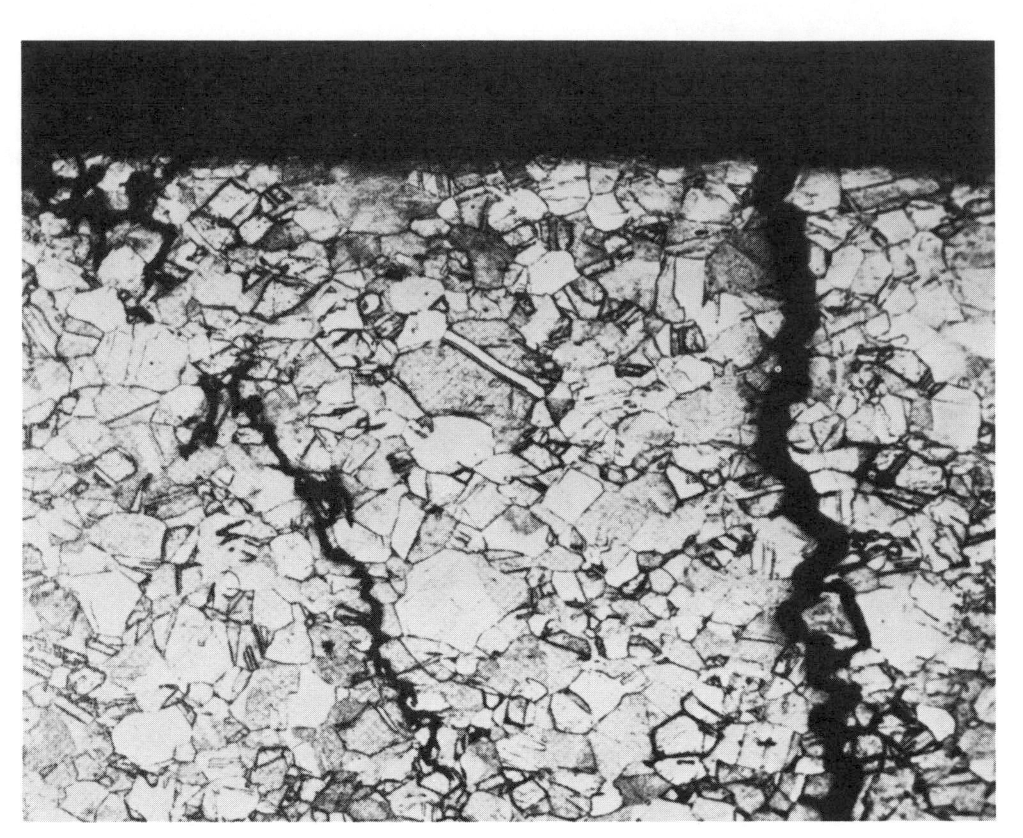

Figure 17. Typical intergranular stress-corrosion cracks in alloy 260 (cartridge brass 70 percent) tube that was drawn, annealed, and cold reduced 5 percent. The cracks show some branching (By permission, from *Metals Handbook*, Volume 7, Copyright American Society for Metals, 1972).

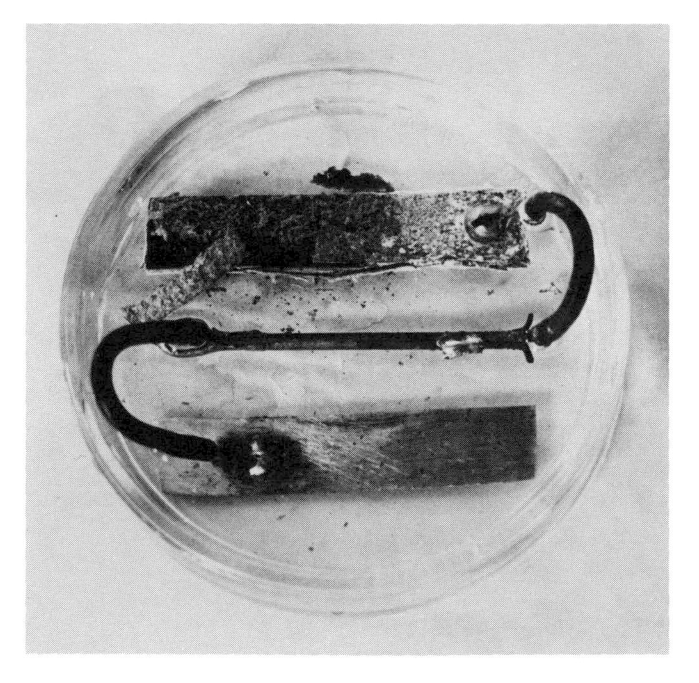

Figure 18*a*. In this experiment, a brass strip *(bottom)* is soldered to a nail *(center)* and a magnesium strip *(top)* bolted to the nail. The magnesium has started to deteriorate in the sodium chloride electrolyte.

Figure 18*b*. The sacrificial anode has deteriorated and separated from its connection. The iron nail has not corroded since it was cathodic.

Coatings are extensively used to prevent corrosion. They are classified as follows:

1 Anodic coatings (anodizing)
2 Cathodic coatings (electroplating, chrome, copper, nickel)
3 Organic and inorganic coatings
4 Inhibitive coatings (red lead, zinc chromate)
5 Inhibitors (placed in the electrolyte)

Metal coatings may be listed as follows:

1 Hot dip process (galvanizing)
2 Metal spraying
3 Metal cementation (diffusion of atoms into the surface of a metal part)
   a. Sherardizing (metal parts heated with zinc powder to 600 to 850°F (315 to 454°C)
   b. Chromizing
   c. Irighizing
4 Metal cladding (rolling a "sandwich" of metal with the outer layers of a corrosion resistant metal). For example, clad aluminum sheet on which a thin outer layer of pure aluminum covers the strong aluminum alloy that is not so resistant to corrosion.

Anodizing is the process of thickening and toughening the oxide film. On metals such as aluminum the thicker layer of oxides increases resistance to further corrosion. Anodized films not only provide a hard wear surface, but also provide a somewhat porous base for paint or other coatings.

The familiar electroplating process is similar to galvanic corrosion in that metal ions are detached from the anode and are deposited on the cathode to provide protection from corrosion. In most cases, the noble metals that are cathodic in the galvanic series are used for plating materials. Zinc plated steel, however, is used for roofing material and many other products, and tin plate is used extensively for food containers. If there is a break in the plating material, corrosion will begin at that point if the plating is cathodic to the base metal. If a break occurs in tin plate on iron, the tin is more cathodic than iron and therefore accelerates the corrosion of the iron at the break (Figure 19).

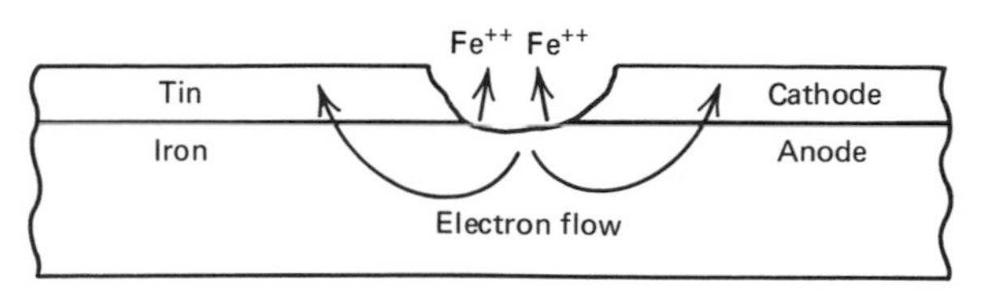

Figure 19. The action of corrosion on tin plate.

In air-free conditions, however, as when food is sealed in "tin" cans the tin plate is generally anodic to steel. Zinc is more anodic than iron and, if there is a break in the zinc plate on iron, the zinc will be a sacrificial metal in this case and will corrode instead of the iron (Figure 20). Large patches of zinc may corrode away on galvanized sheet iron before denuded areas become sufficiently wide to allow anodic areas to form on the steel. Only then will it begin to rust.

Chemical inhibitors used in the corrosive medium make it inert, that is, unable to transfer metal ions from the anode to the cathode. Some types are used as a protective coating on the surface of the metal. Sodium phosphate in water can be used to produce a passive (inactive) ferrous oxide (Fe_2O_3) film on steel.

Organic coatings such as paint, tar, grease, and varnishes prevent corrosion by keeping the corrosive atmosphere from contacting the surface. Inorganic coatings such as vitreous enamels or even mill scale also create a barrier, but one drawback is their brittleness. Corrosive attack can take place where the coating is chipped or broken.

Metal cementation or diffusion of a material into the surface of metal is done by applying heat. Carburizing of low carbon steel is one example of metal cementation. Zinc powder can be diffused into the surface of steel by heating both powder and steel together to a few hundred degrees Fahrenheit. This is contrasted to the hot dip zinc

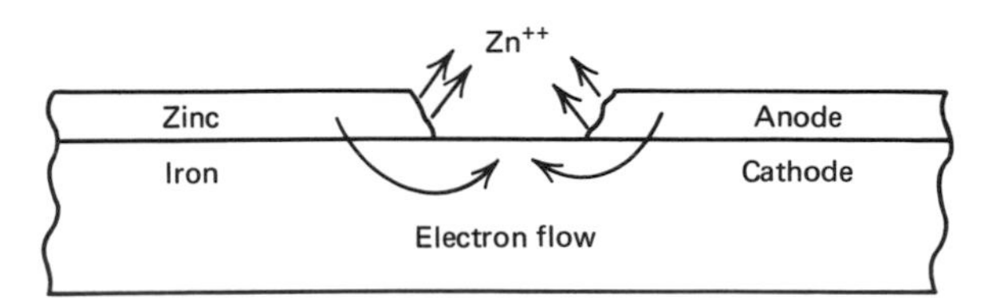

Figure 20. The action of corrosion on zinc plate (galvanized iron).

process or electroplating in that here the powder penetrates to some extent into the surface of the steel. Silicon, aluminum, chromium, and many other elements are diffused or cemented into steel and other metals.

★ Note

The study of galvanic corrosion as presented in this chapter is based on the current flow theory in which the flow of current is considered to be opposite to the electron flow and in which the anode deteriorates because of corrosion as the electrons leave it. In contrast, the electron theory as used in the study of electronics shows the cathode with a negative charge as the emitter of electrons and it deteriorates (as in a vacuum tube) while the anode with its positive charge (as the plate in a vacuum tube) does not.

Positive and negative signs are not used in order to avoid confusion about polarity. The terms anode and cathode as related to the current theory are in common use in industrial circles and probably will not be changed to the electron theory terminology for many years, if at all. The terms anodizing and cathodic protection are examples of common usage of current flow theory terminology.

Self-Evaluation

1. Corrosion may be classified in what two ways?
2. What is the process by which galvanic corrosion takes place?
3. Atoms may exist as positive or negative ions. For the most part, do metals become negative or positive ions?
4. Name three or more necessary conditions required for any galvanic corrosion to occur.
5. What role does hydrogen play in cathodic polarization? How does this affect the rate of corrosion?
6. What happens to cathode polarization when oxygen is dissolved in the electrolyte?
7. In the galvanic series, which would be more active (anodic), iron or copper?
8. Why will a galvanic couple having a large cathode and a small anode corrode much more rapidly than one having a large anode and a small cathode?
9. Some alloying elements such as copper in steel tend to reduce corrosion but, in general, in what conditions are metals least likely to corrode?
10. How can some metals like stainless steel, that are normally resistant to corrosion, rust under a washer, a crevice, or a pile of sand?

Worksheet 1

Objective Demonstrate the importance of oxygen in the corrosion of iron in water.

Material Two 500 ml (milliliter) Erlenmeyer flasks or equivalent, rubber tubing, glass tubing, oxygen tank, distilled water (deaerated by boiling), iron (low carbon steel) turnings, benzene, and acetone.

Procedure Pour 250 ml of distilled water in each of the Erlenmeyer flasks. Degrease the iron turnings by rinsing first with the benzene and then the acetone. Allow to dry and weigh two 50 gram samples and put one sample in each flask. Using a short length of glass tubing as a bubbler in one flask, connect it with the rubber tubing to the oxygen tank. Allow oxygen to bubble through one flask for a period of several hours.

Conclusion Observe the development of rust on the iron turnings. Which flask, the one with the still water or the one with the oxygenated water, shows rust forming on the turnings? Explain why you think this happened.

Worksheet 2

Objective Demonstrate the development of corrosion in crevices because of oxygen concentration differential.

Material One strip of AISI 410 chromium steel 2 × 6 in., two small pieces of wood 1 × 1 in., a strong rubber band, one 400 ml beaker or equivalent, and a 3 percent solution of sodium chloride, containing 3 or 4 ml of a 5 percent ferric chloride solution.

Procedure Clamp the two pieces of wood on opposite sides of the stainless steel strip with the rubber band and insert it into the 400 ml beaker. Fill the beaker with the sodium chloride solution and set it aside for a period of six weeks or more. It may be necessary to replace the solution about once or twice a week after corrosion has begun.

Conclusion After the test period, remove the sample of stainless steel and inspect under the crevice provided by the blocks of wood. Has any corrosion begun? If so, how can a normally corrosion resistant metal like stainless steel begin to rust in a crevice area?

Worksheet 3

Objective Illustrate ion displacement in a solution by a more active metal in the galvanic series.

Material Four small beakers or other glass containers, 5 percent solution of copper sulfate, 5 percent solution of silver nitrate, 10 percent solution of hydrochloric acid, and small strips of zinc, copper, and low carbon steel.

Procedure

1. Immerse a strip of copper in the 5 percent silver nitrate solution and note the deposition of silver on the copper surface. Copper displaces silver from solution because copper is more active (anodic) than silver in the galvanic series and suffers from corrosion or loss of copper ions in the process.
2. Immerse a strip of copper in the 10 percent hydrochloric acid and note the absence of any reaction. Hydrogen (in the acid) is more anodic than copper in the galvanic series and is not displaced by copper.
3. Immerse a strip of zinc in the 10 percent hydrochloric acid. Note the release of hydrogen as the zinc corrodes. In this case, the zinc is more active (anodic) in the galvanic series than hydrogen, which is displaced. Remove the zinc and note its etched or corroded condition.
4. Immerse a strip of carbon steel in the 5 percent copper sulfate solution and note the deposition of copper on the part of the strip that was in the solution. Steel is more active (anodic) than copper in the galvanic series and therefore displaces the copper from solution and at the same time is corroded an equivalent amount.

Conclusion Is ion displacement by a more active metal in an electrolyte solution similar to other forms of corrosion? In what ways?

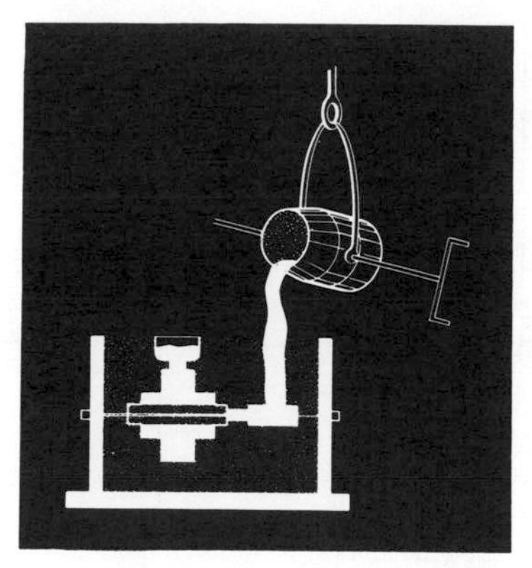

24 Casting Processes

The casting of molten metals into molds to produce useful items is one of the oldest methods of metal forming. Cast ornaments and tools over 4000 years old have been found from Ancient Egyptian, Assyrian, and Chinese cultures.

Molten metals such as iron, steel, and aluminum are cast into ingot molds and allowed to solidify before they are further processed. For the purpose of producing a needed or useful shape, casting is done by using many different techniques. These processes involve a large segment of the metals industry. The castings produced range from small intricate precision parts to massive machinery sections weighing many tons. In this chapter, you will be introduced to the various casting methods used by the industry today.

OBJECTIVES

After completing this chapter, you will be able to:

1. Identify and list the various types of casting processes.
2. Describe each casting process.

INFORMATION

Almost any metal can be cast into molds from its molten state. Iron, steel, aluminum, brass, bronze, and die cast metals are examples. (See Chapter 4 for identification of these metals.) The casting process requires a **pattern** (having the shape of the desired casting) and a **mold** made from the pattern. The mold must be made to withstand the heat of the molten metal, either of sand, plaster of paris, ceramic, or metal. Wood is most often used for the mold patterns, but metal and wax are also used. The different methods of casting are:

A. Sand casting
   1. Green sand molding
   2. Dry sand molding
      a. Core sand molding
      b. Shell molding

B. Centrifugal casting

C. Investment (lost wax) molding
   1. Solid investment casting (plaster molding)
   2. Shell investment casting

D. Permanent molds
   1. Book-type
   2. Deep-cavity

E. Die casting

Sand Casting

One type of sand mold, **green sand molding,** consists of moist sand with small amounts of clay and other substances. Green refers to the moisture content rather than to the color of the sand. The second type of sand casting is **dry sand molding,** in which a resin bond is mixed with the sand. By far, the greatest number of castings is made using a sand aggregate (additives and various particle sizes). Silica sand is combined with cereal, moisture, and sometimes a carbonaceous substance such as pitch by a process called mulling. After sand has been used for casting, it is usually reclaimed by crushing, screening, magnetic separation, and impinging of sand grains against a wear plate at high velocity. The fines are vacuumed away and the clean sand is transported back to hoppers to be used again. A simple casting, such as the one shown in Figure 1, can be made by sand casting processes. In these processes, the mold is made by packing or ramming sand around a pattern, and removing the pattern when the sand is sufficiently hard.

Patterns for green sand molding are usually made in two halves and are called **match plate patterns,** or **cope** (top half) (Figure 2) and **drag** (bottom half) (Figure 3) patterns. Loose patterns are simply placed on the mold board with the drag half of the flask, and the sand is rammed into place and struck off (Figure 4). A mold board is placed on the drag half, which is then rolled over so that the cope half of the pattern can be put into position over the drag half (Figure 5). The cope pattern is placed and sand is rammed into the mold; one or more holes are provided for pouring the metal. This hole, called the sprue, is connected to the cavity by a gate after the cope is removed. Other holes, called risers, are used as reservoirs to feed liquid metal to the casting as it shrinks while solidifying and as vents for escaping gases. Both the cope and drag patterns must be removed before the flask is put back together. All patterns must have a draft or tapered shape so they can be removed easily from the sand without disrupting the mold. Patterns must also be larger than the required size to allow for shrinkage when the molten metal solidifies as it cools. Pattern makers allow for shrinkage, depending on what metal is used for the casting; for example, $\frac{1}{8}$ inch per foot for cast iron. After the patterns are removed, and the new molds are put back together, the molten metal can be poured.

In dry sand molding, the specially prepared sand packs well when it is rammed into the mold, but it must be baked to drive off the excess mois-

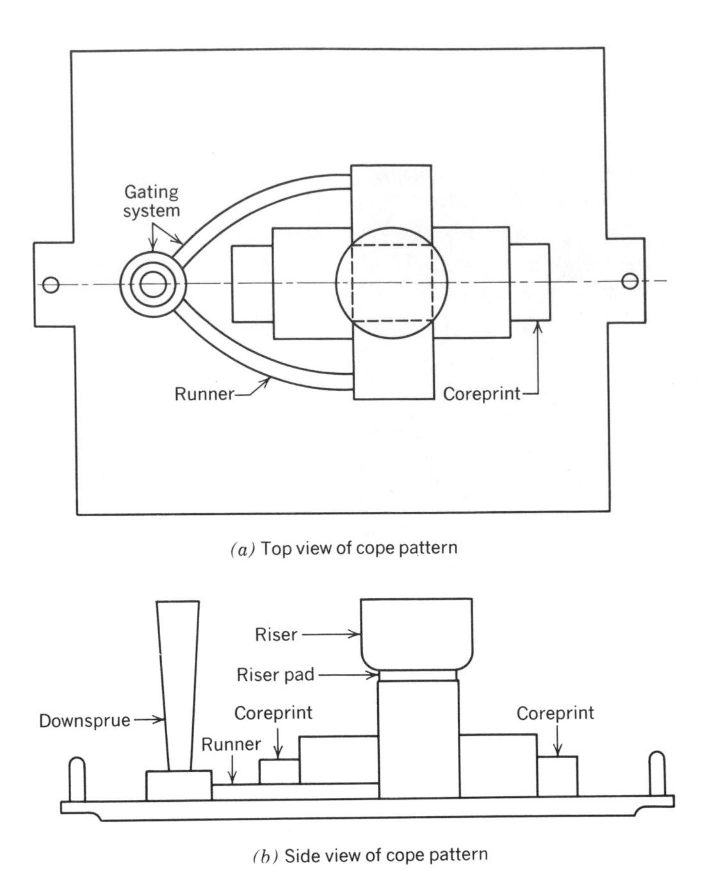

Figure 2. Cope match plate pattern. *(a)* Top view of cope pattern; *(b)* side view of cope pattern *(Machine Tools and Machining Practices)*.

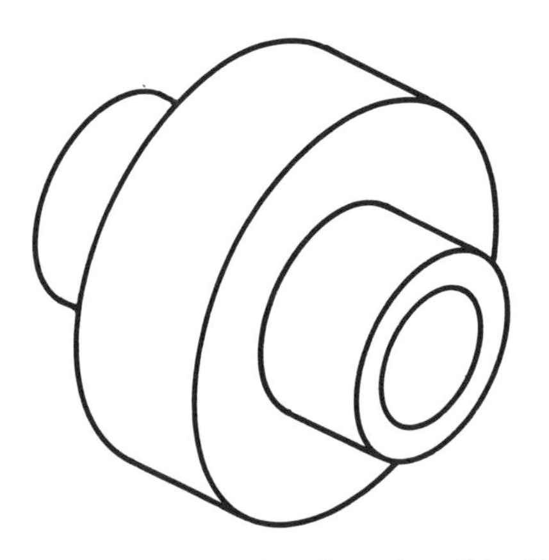

Figure 1. Completed casting (*Machine Tools and Machining Practices*).

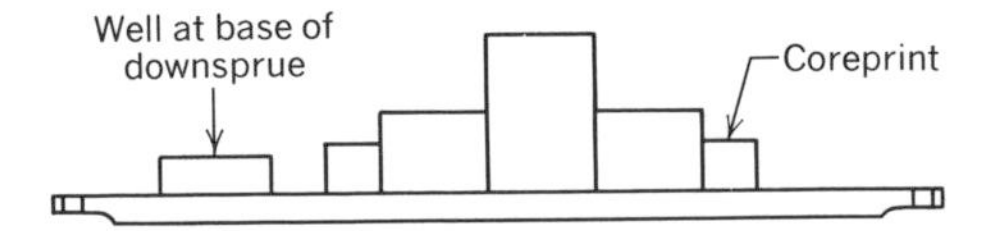

Figure 3. Drag match plate pattern *(Machine Tools and Machining Practices)*.

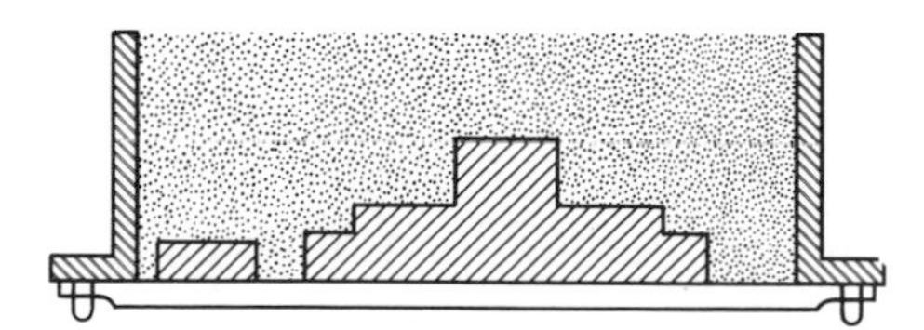

Figure 4. Sectional view of drag pattern in a flask with the sand rammed in place and struck off *(Machine Tools and Machining Practices)*.

ture and harden before the molten metal can be poured. Core sand is similar to that used in dry sand molding, and is used often when a core is needed to produce hollow parts. Core sand is clean, free of clay, and mixed with an organic binder such as linseed oil, cereal, or resins. After the cores are formed, they are hardened by baking in an oven. **Cores** are made by ramming sand into a simple core box (Figure 6), or by forcing sand into a mold using a core blowing machine. They are, like most sand molds, formed in halves, and then put together to be baked (Figure 7). The tedious process of oven drying cores has been, for the most part, replaced by the use of **no-bake sands,** a somewhat recent discovery. These sands are combined with a resin and a catalyst. Several types of resins are used such as furan and phenolic.

Some of the requirements for sand cores are:

1 Cores must be able to withstand high temperatures.
2 They must have a green bond strength so that they will hold together while damp.
3 They must have dry strength after baking.
4 They must have permeability to enable gases to escape through them.
5 They must be easily broken apart and removed from the finished casting.

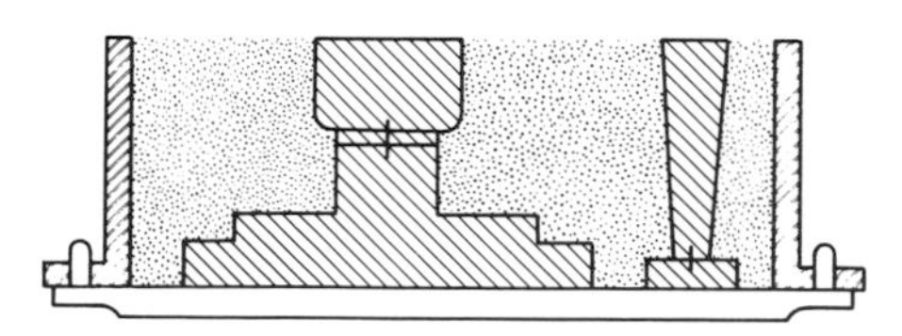

Figure 5. Sectional view of cope pattern in flask with the sand rammed in place and struck off *(Machine Tools and Machining Practices)*.

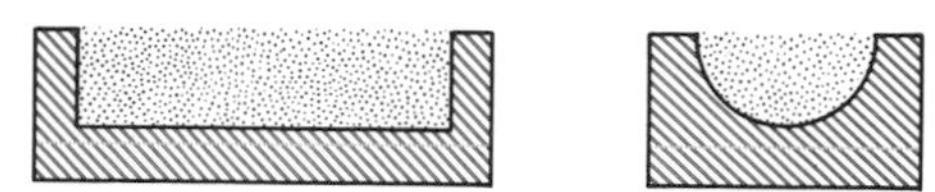

Figure 6. Sectional view of core box with core *(Machine Tools and Machining Practices)*.

When cores are positioned in a flask, they are held in place by a section of the pattern called the **core print.** The core print does not contribute to the final shape of the casting in any way. The cope and drag with core in place are shown in Figure 8, ready to make the casting.

Foundry Practice

The metallurgical properties of various cast irons was discussed in Chapter 8. Of all the types of cast iron it is gray cast iron that is most commonly used for casting purposes. However, large amounts of malleable, ductile, and some white cast irons are also produced. In most cases, molten cast iron is ladled (poured) into sand molds.

The **cupola furnace** (Figure 9) has been of primary importance in the melting of cast irons. Prior to the development of various kinds of electric furnaces, cast iron was melted in cupola furnaces that usually used coke for producing heat. When fossil fuels such as coke are burned, considerable fumes are released into the atmosphere; therefore in areas where air pollution is a critical problem, electric furnaces are most often used. Electric furnaces for melting cast iron can be either direct arc, electric induction, or electric resistance types.

The cupola furnace is a circular steel shell lined with a refractory material such as firebrick. It is equipped with a blower, air duct, and sand box with tuyeres for admitting air into the cupola. A coke bed is prepared and burned in with propane torches and a charge of pig iron (or scrap cast iron) and steel scrap is placed in the furnace. Limestone

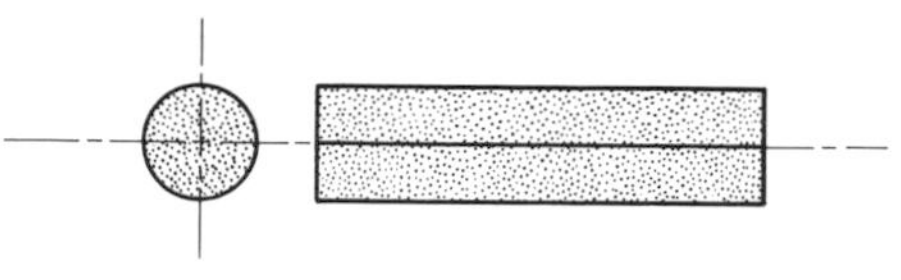

Figure 7. Two halves of core fastened together after being removed from core box and baked in oven *(Machine Tools and Machining Practices)*.

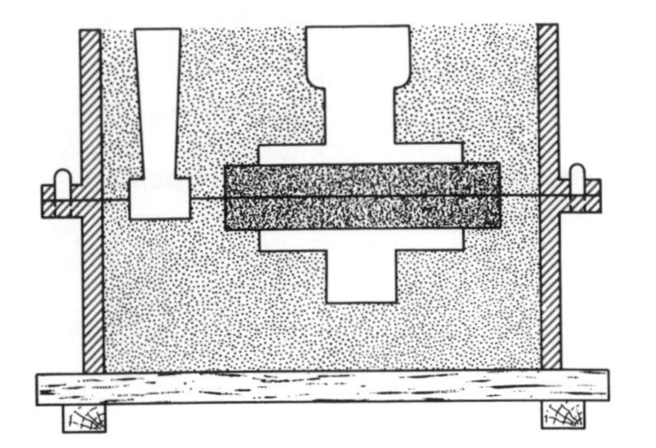

Figure 8. Sectional view of cope and drag with core in place on the mold board. At this point it is ready to make the casting as shown in Figure 1 *(Machine Tools and Machining Practices)*.

is used for a flux. The blower is turned on and the iron begins to melt at the top of the coke bed. The melting rate is determined by the diameter of the cupola. The molten iron is drawn off into a ladle after suitable metallurgical tests are made.

Metallurgical control of the molten cast iron is made at the furnace. Small wedges are poured in prepared sand molds and quickly cooled. The wedges are broken in cross section revealing the chill area (white iron). The depth of the chill area reveals the amount of graphitization. The tests reveal whether sufficient inoculation of silicon or magnesium (for nodular iron) has taken place.

Thin sections in castings likewise tend to cool

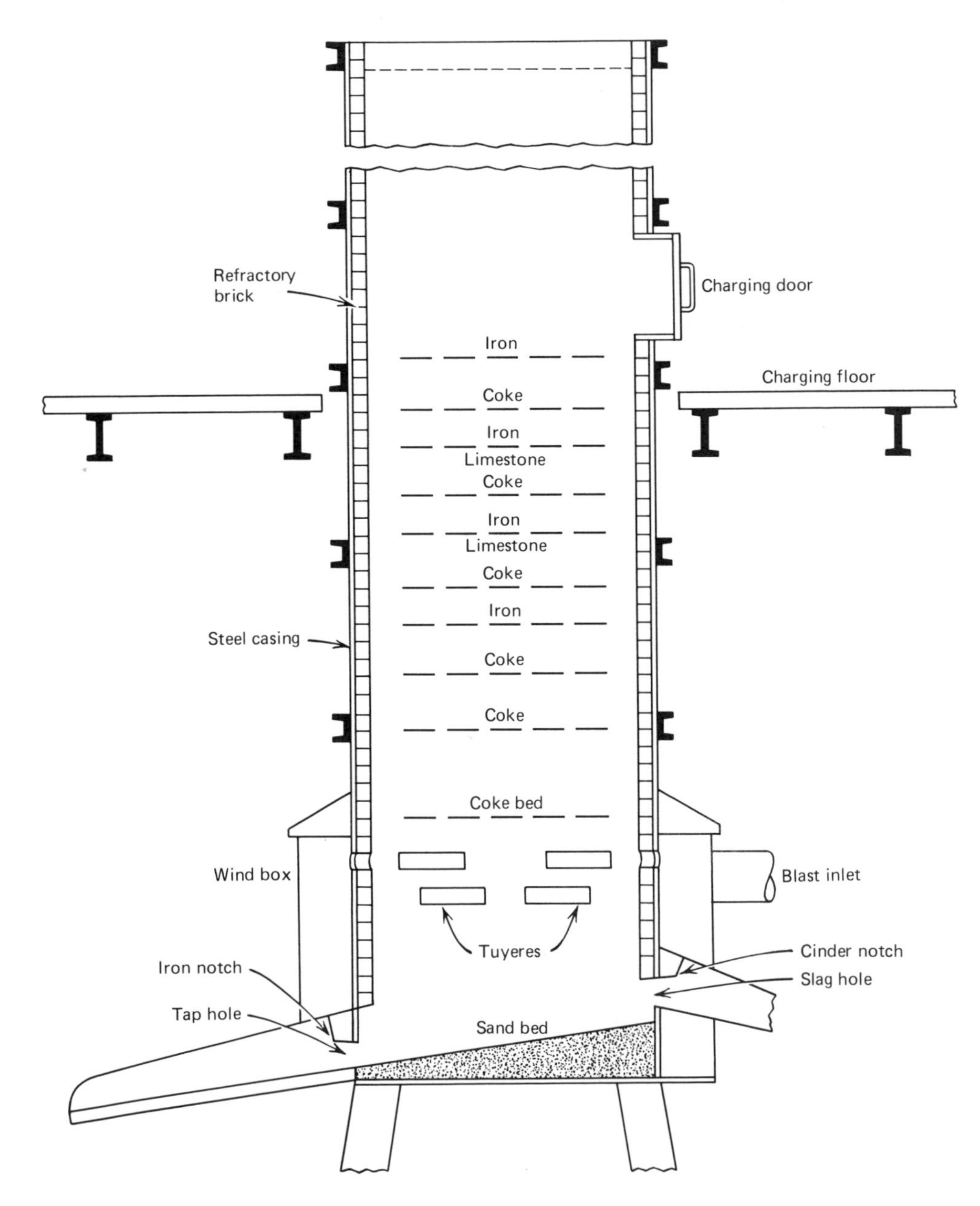

Figure 9. The cupola furnace.

more quickly than thick sections. Thinner areas in a casting will tend to be harder and contain less graphite (with smaller flakes) than do the thicker sections.

The green sand molds are poured from a ladle that was filled at the spout of the cupola or electric furnace. A crane or monorail transports the ladle to the molds on the pouring floors.

Gray iron is an easy alloy to cast as it possesses good fluidity; this allows intricate designs to be cast. Logical designs are necessary, however, to compensate for contraction, which can leave voids or shrinkage cavities. Large cross sections adjacent to thin sections should be avoided. A means of feeding extra molten metal into the mold as the iron shrinks is necessary to prevent shrinkage cavities. This is done with risers, which are reservoirs of molten metal. The pattern must also allow for progressive solidification toward the risers to produce sound castings. Impurities and gas pockets move to the last area of solidification that should be in the riser. Gray irons are more sensitive to differences in cooling rates than most alloys, and heavy sections of castings cool more slowly than lighter ones.

Chills are metal pieces placed in a mold to control the cooling rate of a casting. A wheel with a heavy rim section may crack in a flange or spoke during solidification if the rim section cools too slowly. A chill ring is therefore put into the mold to overcome this tendency. When cast iron is poured against a chill plate, the resultant product is a hard, abrasive-resistant iron that can be either partial or complete white iron.

Malleable iron is produced by prolonged heat treatment of white cast iron, but ductile (nodular) iron is made by inoculating a magnesium alloy into the ladle. Usually the magnesium alloy is placed in the ladle and the molten cast iron is poured over it. A nodular graphite is formed in the iron by this process in place of the flake graphite found in ordinary gray irons. The nodular graphite gives the iron more ductility.

The **shell molding** process, a type of sand mold casting, requires the use of metal patterns. The patterns are heated to 450°F (232°C), coated with a silicon release agent, and put into a dump box where a prepared sand is poured over them (Figure 10). The sand is mixed with a phenolic resin binder. Some of the sand adheres to the pattern and solidifies, forming a shell around it. The thickness of the shell is determined by the length of time the pattern is in contact with the sand, and ranges from $\frac{1}{8}$ to $\frac{1}{4}$ inch. The dump box is turned over and the loose sand is removed (Figure 11). The pattern and the adhering sand are placed in an oven and heated to a temperature of 600°F (316°C) for one or two minutes (Figure 12). The half shell is then removed from the pattern (Figure 13) and clamped together with its mating half to form the mold (Figure 14). When the mold is used for small or thin castings, the shells have sufficient strength to withstand the pressure of the molten metal. Heavier castings require the use of backup materials around the shells, usually sand, gravel, or metal shot. The advantages of shell molding over other forms of sand casting are that high precision, good finishes, and more complex shapes are possible, and less machining is needed.

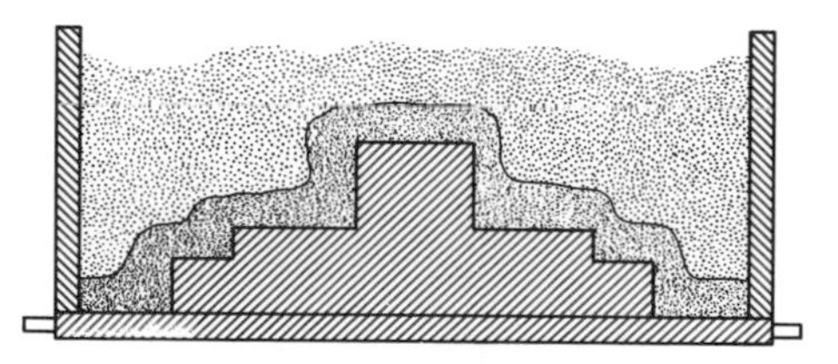

Figure 10. Heated metal pattern in dump box with sand *(Machine Tools and Machining Practices)*.

Centrifugal Casting

Centrifugal casting is a process in which molten metal is poured into a rapidly rotating mold. The liquid metal is forced outward by centrifugal forces to the mold cavity. Wheels, tubing, and pipe (Figure 15) are made by the centrifugal casting process. The centrifugal process also makes possible the casting of two dissimilar metals. For example, an

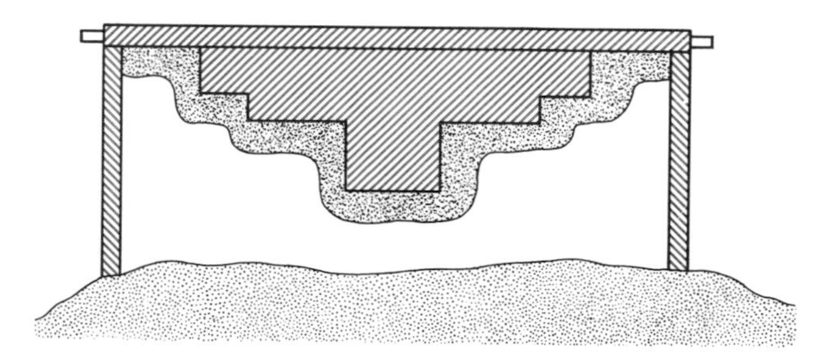

Figure 11. Dump box is turned, leaving some sand adhering to the pattern *(Machine Tools and Machining Practices)*.

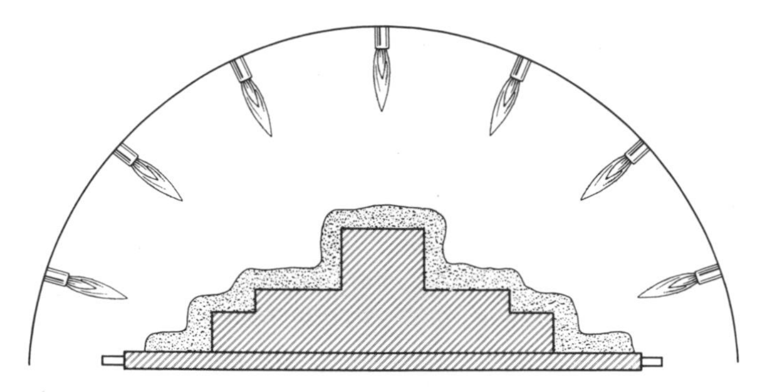

Figure 12. Pattern is placed in oven for one or two minutes to harden it *(Machine Tools and Machining Practices)*.

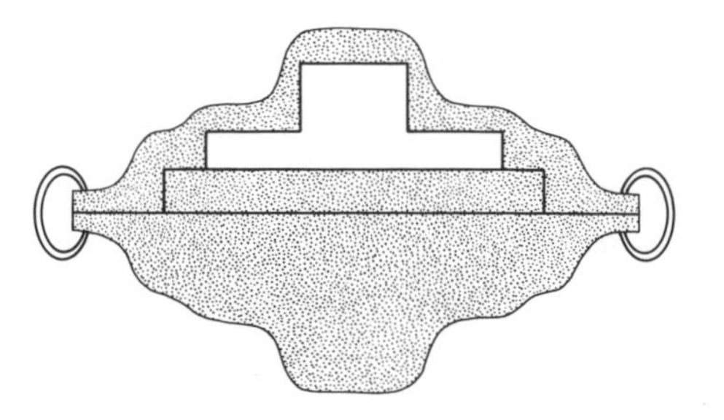

Figure 14. Mold halves are clamped together with core. The top half shell is shown as a sectional view *(Machine Tools and Machining Practices)*.

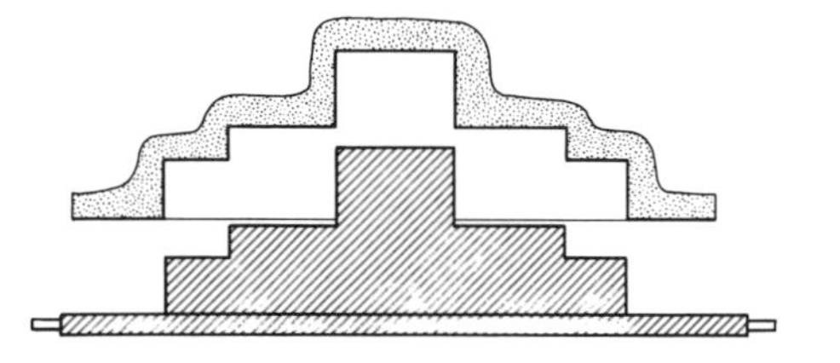

Figure 13. Half shell mold is stripped from pattern *(Machine Tools and Machining Practices)*.

outer surface of hard alloy can be poured, followed by an inner layer of softer metal. This would give the casting an outer wear surface while maintaining the machinability and weldability of the inner core.

Centrifuging of castings is a similar process, but with this method the entire mold or group of molds is rotated away from the center of rotation. Castings made by this process have superior mechanical properties and a uniform grain structure.

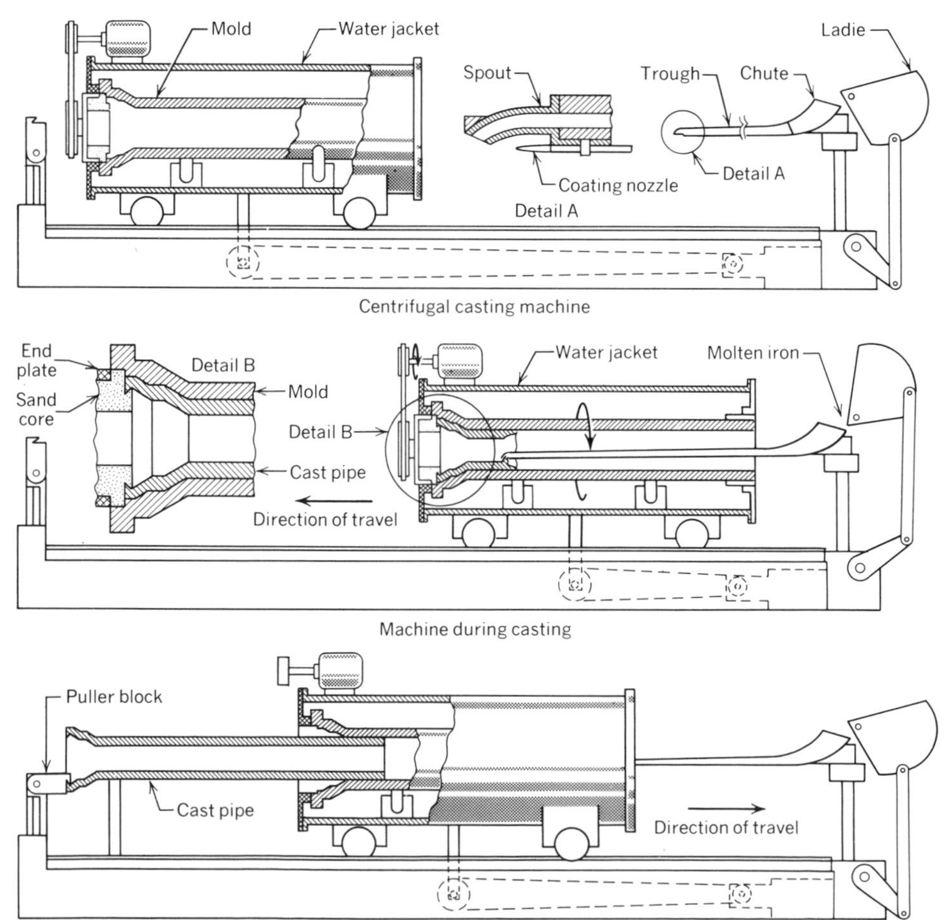

Figure 15. Machine for centrifugal casting of iron pipe in a rotating water-jacketed mold (By permission, from *Metals Handbook*, Volume 5, Copyright American Society for Metals, 1970).

The center area is the last to solidify so that the impurities work to the center where they are later removed by machining processes.

Investment Casting

One of the oldest methods of casting metals is that of **investment casting** or **lost wax process.** The pattern is made of wax and is used only once, but the wax is not really lost as it may be reused in another mold. Wax patterns, including their sprues and risers, are usually cast in metal molds (dies) or formed by injection molding. This method is used for industrial purposes, but unique (one-of-a-kind) patterns are often used for art metal castings.

Plaster of paris has long been used to make molds for solid investment castings (Figure 16). Plaster casting can only be used for the lower temperature metals such as aluminum, zinc, tin, and some bronzes. Metals having higher pouring temperatures such as steel and cast iron are precision cast in molds that are made by the **investment-shell** process. In this process the wax pattern is dipped into a slurry (thin liquid mixture of several substances) of refractory (resistant to high temperatures) material and dried. This process is repeated until a suitable shell up to $\frac{1}{4}$ inch thick has been built up. This mold is then heated in an oven to melt out the wax, which is collected and reused. The shell is heated in a furnace to 1600°F (871°C) and the metal is then poured. The mold cannot be reused since it must be destroyed in order to remove the casting. Figures 17*a* to 17*j* show the process from the wax pattern to the finished casting.

Investment casting is used for intricate shapes that would make the withdrawal of a normal pat-

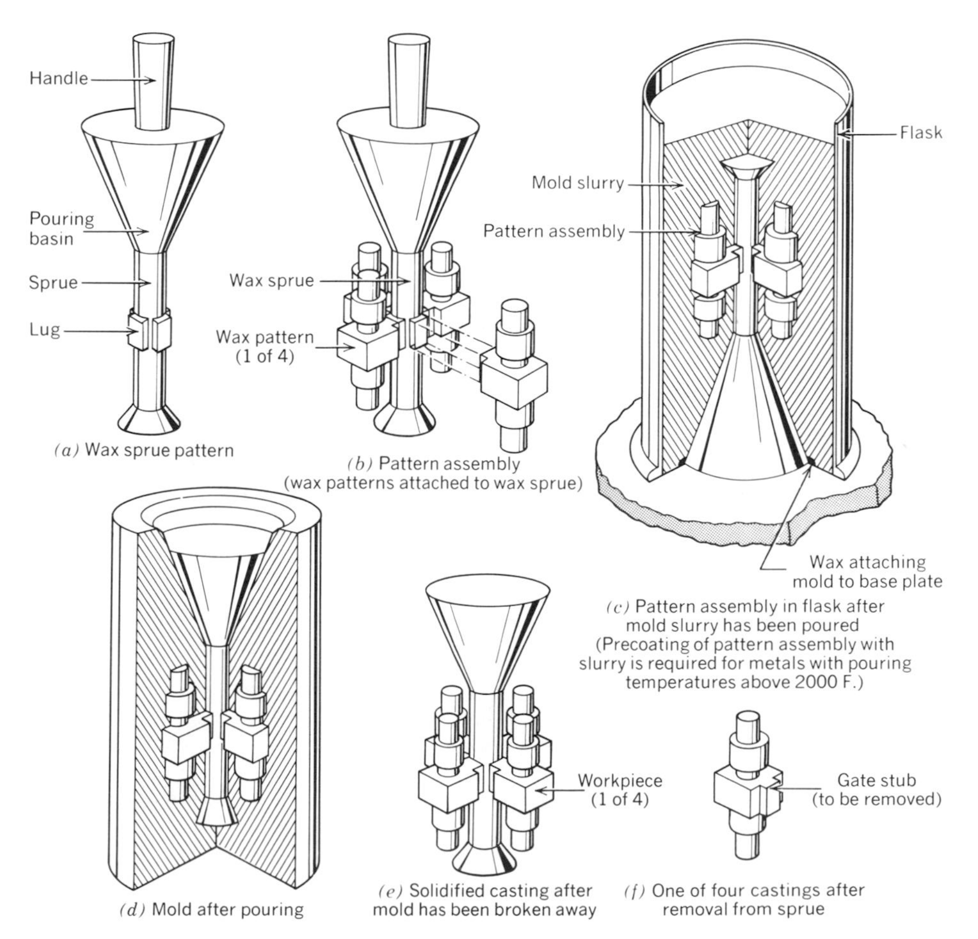

Figure 16. Steps in the production of a casting by the solid investment molding process, using a wax pattern (By permission, from *Metals Handbook*, Volume 5, Copyright American Society for Metals, 1970).

Figure 17*a*. First step in the production of a shell investment casting shows a wax replica of the lower half of valve assembly, together with die. Special wax is forced into die under high pressure (Investment Casting Corporation).

Figure 17*b*. Wax replica of upper half of valve assembly, made in a separate mold to provide for interchangeability of various vane configurations (Investment Casting Corporation).

Figure 17*c*. Two piece wax assembly shown prior to sealing (Investment Casting Corporation).

Figure 17*d*. Runners are attached to wax assembly, which provide the necessary gating and risers for sound castings (Investment Casting Corporation).

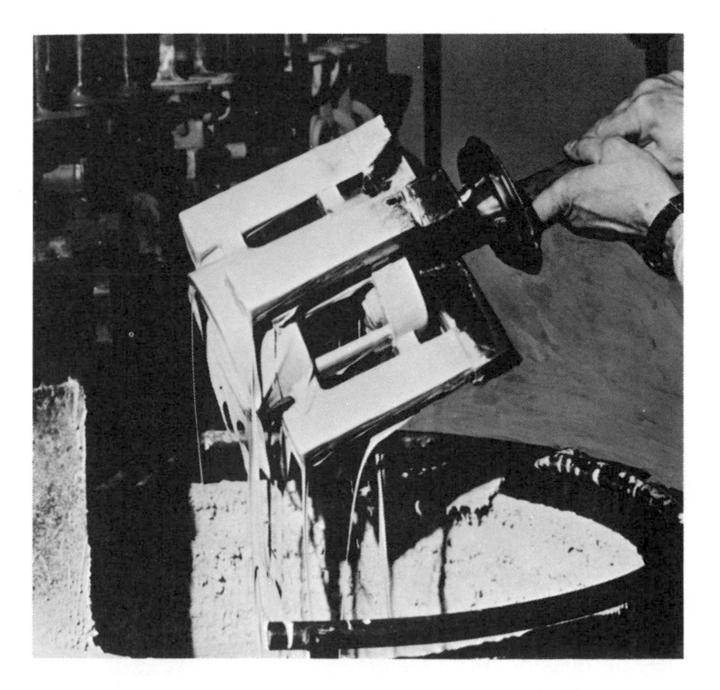

Figure 17*e*. Wax assembly is coated with first coat of ceramic slurry. As many as eight coats are applied (Investment Casting Corporation).

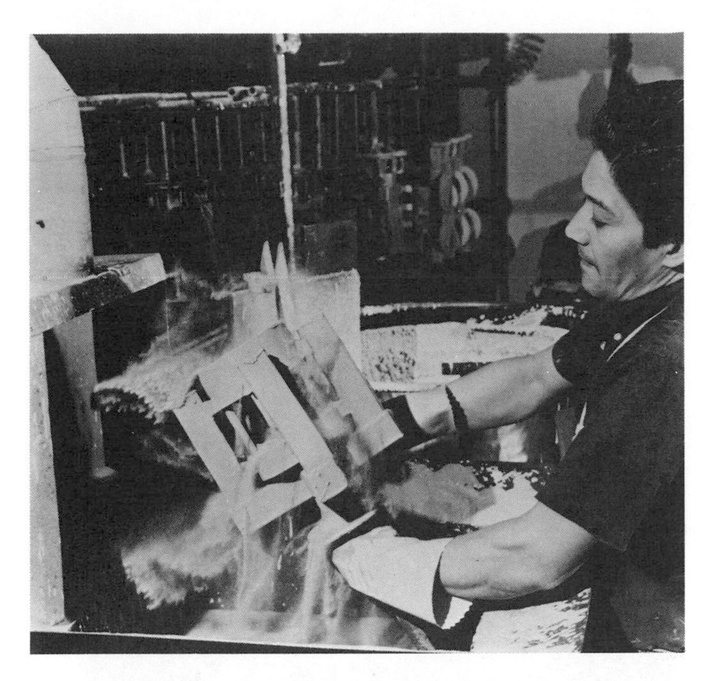

Figure 17*f*. Slurry ceramic is stuccoed with zircon sand and allowed to dry (Investment Casting Corporation).

Figure 17*h*. Ceramic slurries about to enter high pressure steam autoclave to remove majority of wax present in mold (Investment Casting Corporation).

tern impossible. Great detail and high precision with no parting line or seam is typical of this method. Since nonmachinable alloys can be cast with this precision casting, the need for machining is almost eliminated. Automotive and aerospace industries, for example, use the investment casting method for small parts (Figure 18).

Permanent Molding

Permanent molds (Figures 19 and 20) are made of metal, usually gray cast iron or steel. Graphite and other refractory materials are sometimes used for steel casting. The molds are usually machined to a rough shape and hand finished. A refractory wash

Figure 17*g*. Ceramic shells drying (Investment Casting Corporation).

Figure 17*i*. Induction heat metal being ladled into ceramic shell (Investment Casting Corporation).

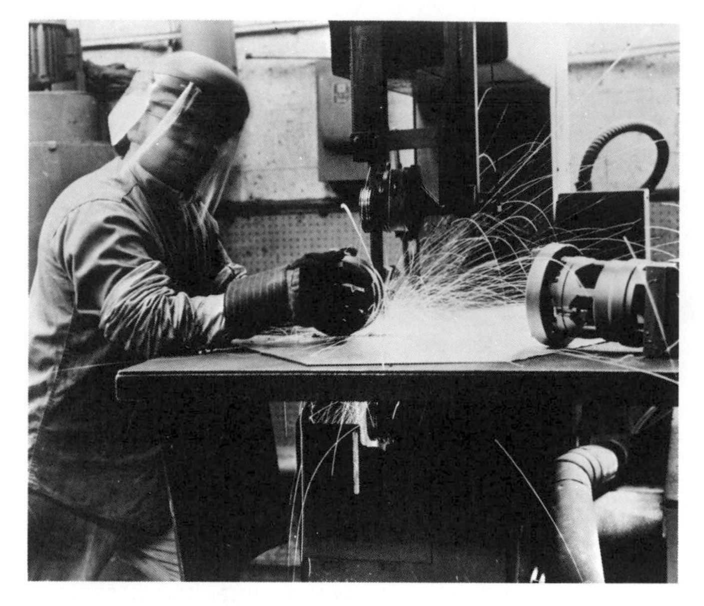

Figure 17*j*. After shakeout, castings are cut off on friction saw, finish ground, and subjected to quality assurance checks prior to shipment. X-ray, Zyglo, and Magnaflux facilities are available if required, in addition to dimensional checks (Investment Casting Corporation).

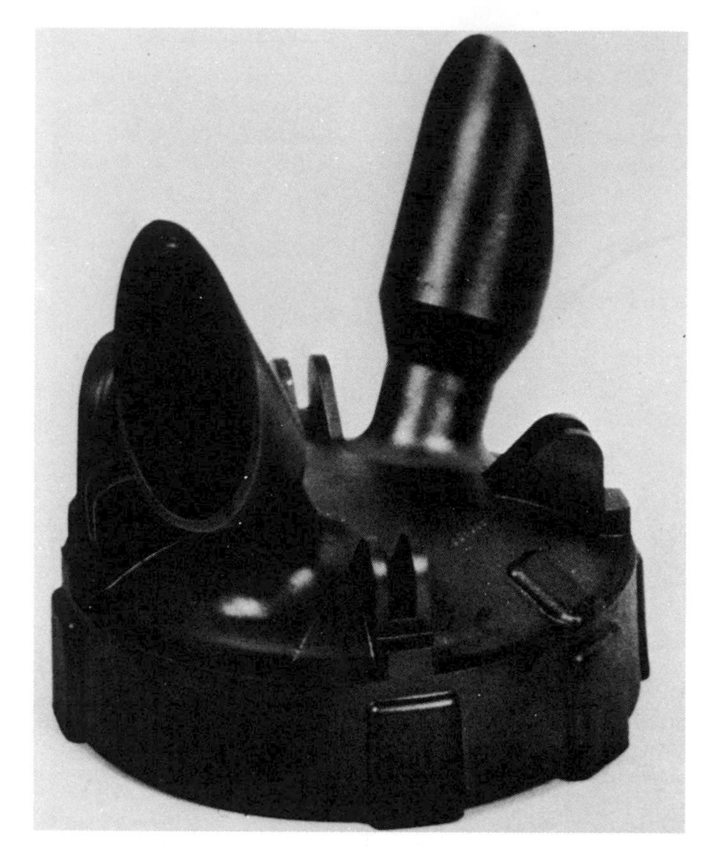

Figure 18. An experimental rocket part cast in Maraging steel by the investment casting process (Hitchiner Manufacturing Company, Inc.)

is applied to the mold prior to casting. As in sand casting, the cores are made of sand and they are not reused. Molten metal is poured into the mold from a ladle as in sand casting. The mold must be heated to and maintained at about 700°F (371°C) in order to produce good castings.

Permanent molding is a step between sand casting and die casting processes. Initial costs are low compared to die casting. Fairly high precision can be achieved, thus eliminating considerable machining time when compared to a sand casting. Permanent molds are used mostly for limited production runs since they can be reused for only a few thousand castings before they have lost their shape and must be scrapped.

Die Casting

Die casting differs from permanent molding and sand casting in that the metal is not poured into the mold, but is injected under high pressures from

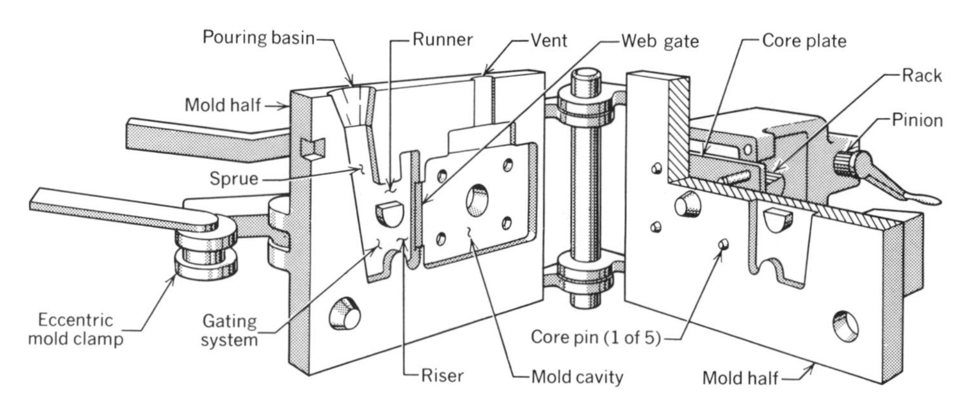

Figure 19. Book type manually operated mold casting machine, used principally with molds having shallow cavities (By permission, from *Metals Handbook*, Volume 5, Copyright American Society for Metals, 1970).

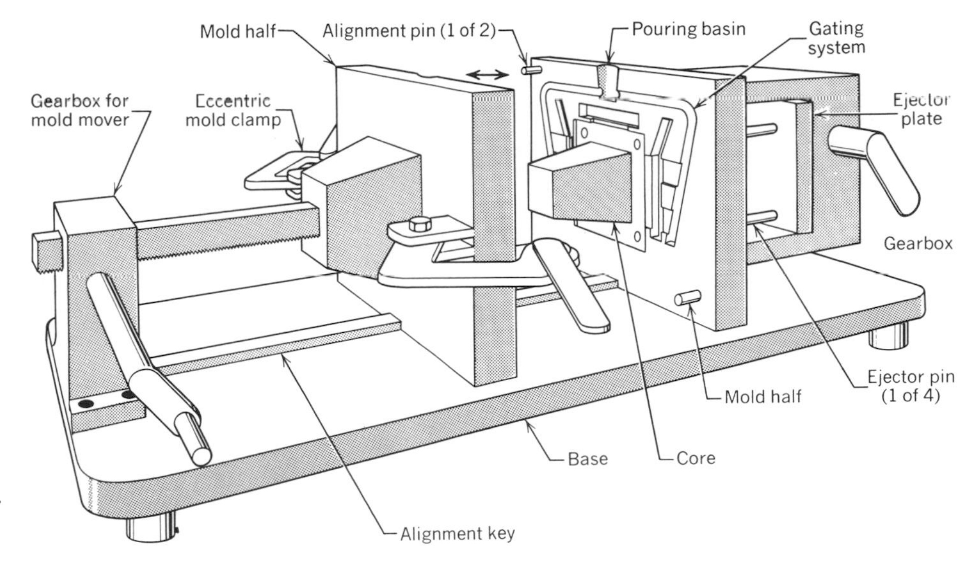

Figure 20. Manually operated permanent mold casting machine with straight-line retraction; required for deep-cavity molds (By permission, from *Metals Handbook*, Volume 5, Copyright American Society for Metals, 1970).

1000 to 100,000 psi. The two basic systems of die casting are the **hot-chamber** and the **cold-chamber** methods. Cold-chamber machines are used for casting aluminum, magnesium, copper-base alloys, and other high melting point alloys. Hot-chamber machines are used for casting zinc, tin, lead, and the low melting point alloys. Die casting machines are heavy and massive (Figure 21), generally hydraulically operated, and capable of exerting the hundreds of tons of force needed to hold the die halves together. This high pressure is necessary to keep the injected molten metal from leaking at the parting line.

A typical hot-chamber machine with the submerged plunger or gooseneck injector is illustrated in Figure 22. This type has an oil or gas fired furnace with a cast iron pot for melting and holding the metal. The plunger and cylinder are submerged in the molten metal that is forced through the gooseneck and nozzle into the die.

Cold-chamber machines (Figure 23) have to a large extent replaced the gooseneck type machines. One major difference is that cold-chamber machines have a separate melting and holding furnace. A sequence of steps of operation can be seen in Figure 24.

High production rates are made possible by using the die casting process, from 100 to 500 cycles an hour. Small thin sections can be produced and very high tolerances can be maintained. Surface finishes are usually so smooth that subsequent finishing or machining processes are not needed. The quality of the cast metal is better than that of sand castings. The metal mold or die cools the molten metal at a higher rate, thus producing a superior grain structure in the metal. Die castings are of typically small, high production parts (Figure 25).

Figure 21. Die casting machine (HPM Corporation).

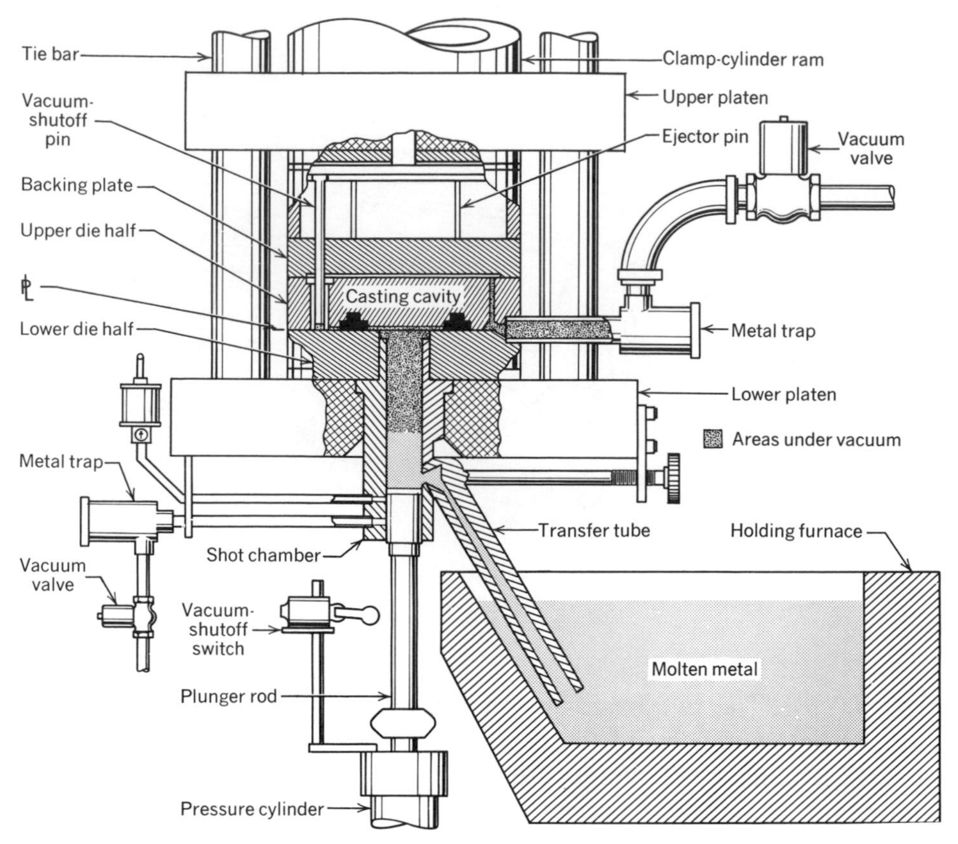

Figure 22. Hot-chamber die casting machine equipped with hood and hood seal for vacuum feeding and casting (By permission, from *Metals Handbook*, Volume 5, Copyright American Society for Metals, 1970).

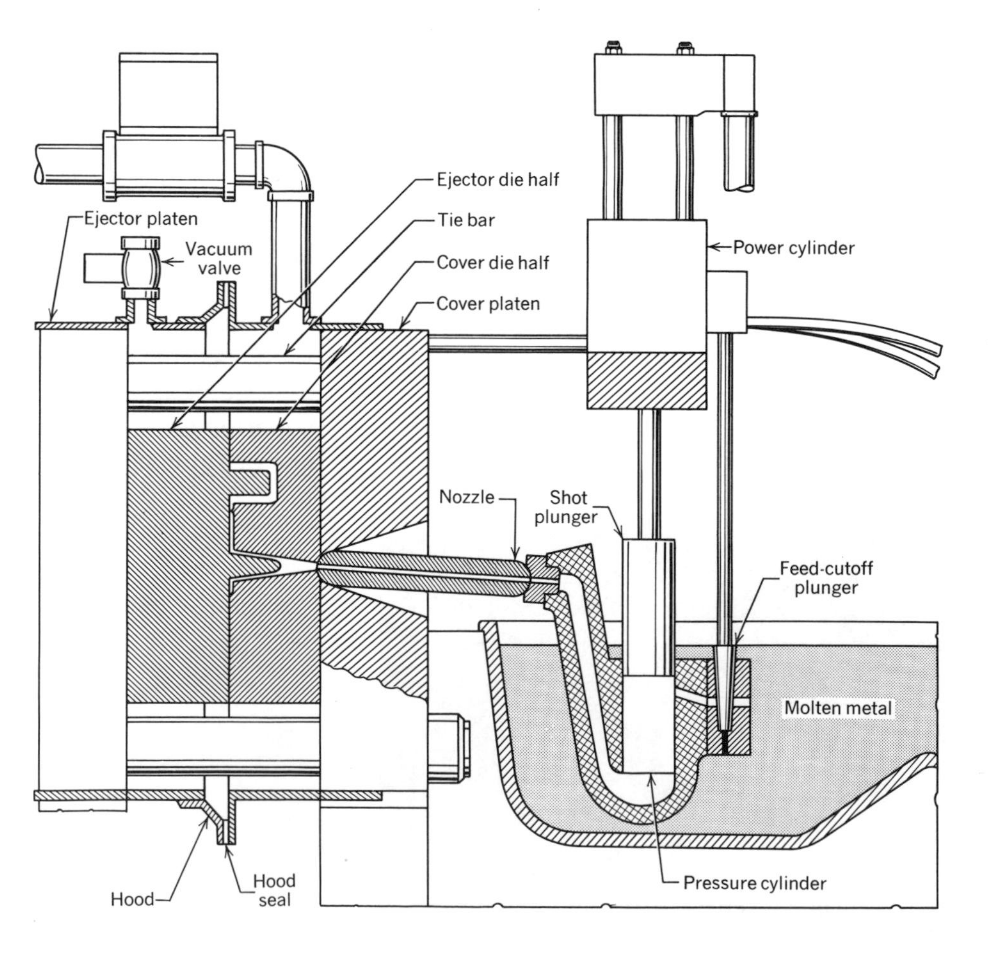

Figure 23. Principal components of a vertical cold-chamber die casting machine with the die parting line in the horizontal plane (By permission, from *Metals Handbook*, Volume 5, Copyright American Society for Metals, 1970).

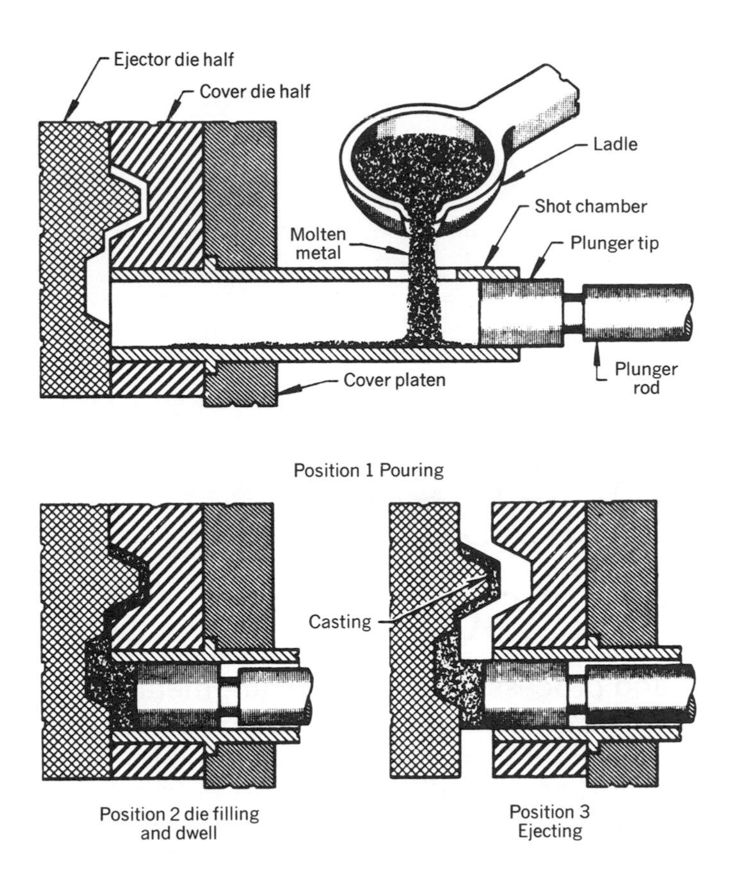

Figure 24. Operating cycle of a horizontal cold-chamber die casting machine (By permission, from *Metals Handbook*, Volume 5, Copyright American Society for Metals, 1970).

Figure 25. Die cast parts *(Machine Tools and Machining Practices)*.

Self-Evaluation

1. Sand casting may be divided into two types. Name them and explain the differences.
2. What kinds of materials are used to make patterns? Briefly describe the patterns used in sand molding.
3. Explan how cores are made and how they are used.
4. What is a ''chill''? Explain.
5. Briefly list the steps used in the shell molding process. Explain the types of materials needed.
6. How does centrifugal casting work? Name two advantages. What is centrifuging?
7. Can you use a wood or metal pattern for investment casting? Explain. List the steps necessary to produce an investment casting.
8. Name two advantages and one disadvantage of permanent molding.
9. How does die casting differ from permanent molding?
10. What major difference is found between the hot-chamber and cold-chamber machines? Name two or more advantages of die casting.

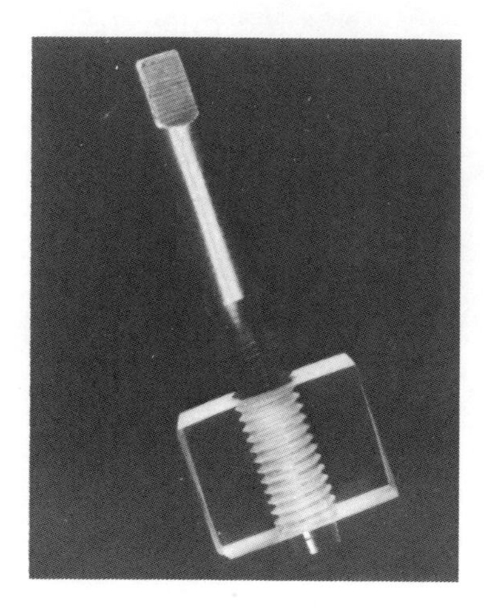

25 Plastics, Rubbers, Industrial Gases, and Oils

Not all materials used and processed in industry are metals. Ceramics, plastics, adhesives, and cements are some of the great variety of nonmetals that will be discussed in this and later chapters. Because industrial workers are often called upon to fabricate and make products of these materials, they should be aware of their behavior and characteristics. This chapter will familiarize you with some of these products.

OBJECTIVES

After you have completed this chapter, you will be able to:

1. Identify kinds of plastics, rubber, and other organic materials and some of their uses.
2. Identify some of the more common industrial gases and oils used in welding and machine shops.

INFORMATION

Organic Materials

Organic materials such as cotton, wood, plastics, rubber, and resins are typically **polymers.** Polymers are made from small molecules called **monomers**, and the monomer molecules are the building blocks from which the polymer chain is built (Figure 1). These building blocks of materials derive their names (monomer and polymer) from the Greek mono (one), poly (many) and meros (parts). Polymer chains are monomer cells that are combined to produce these strong, versatile materials (Figure 2). Polymerization is a chemical process in which many molecules are linked together to form one large molecule. Organic solids are not resistant to even moderately high temperatures, and they tend to break down when exposed to the ultraviolet rays of sunlight (Figure 3). However, chemical additives can improve this degradation; for example, acrylics resist the deleterious effects of sunlight very well.

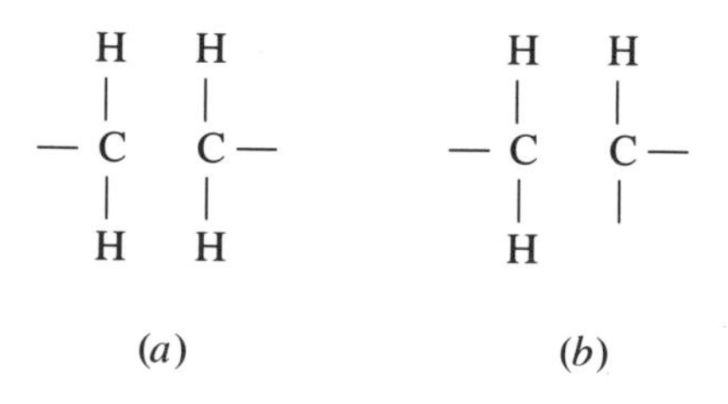

Figure 1. A monomer unit *(a)* can become a branch source unit *(b)* if it loses one or more atoms.

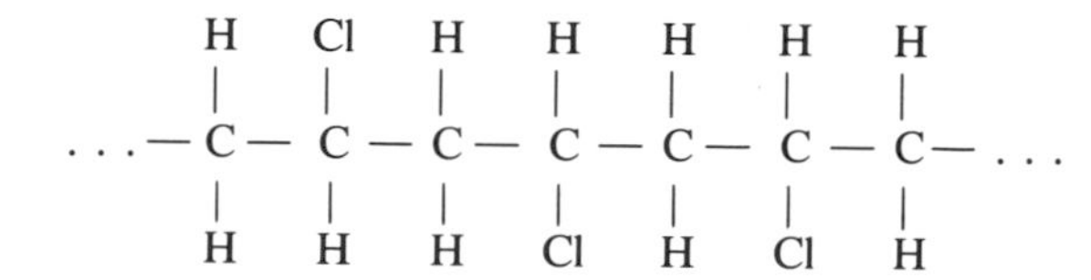

Figure 2. Giant chain molecules may be formed from monomers as in Figure 1 *(b)* by combining other atoms such as chlorine (Cl) to the oxygen and hydrogen atoms to produce rigid polyvinyl chloride (PVC) as shown here.

Fibers may be natural such as those derived from wood cellulose, cotton, or wool, or they may be synthetic such as nylon or Dacron®. Some are nonorganic in origin, for example, asbestos and fiberglass. All of them may be used as fabrics for making clothing or as reinforcing in other substances such as adhesives or automobile tires.

Plastics

Plastics may be divided into two groups: **thermosetting** and **thermoplastic.** Some of the most important thermosetting types are phenolics, polyesters, epoxies, silicones, and some types of urethanes. Plastics are often reinforced with other materials in the form of fillers or laminates. The strongest types are the laminates of phenolics or epoxies using fibers of glass, fabric (cotton or nylon), asbestos, or paper. These thermosetting plastics undergo a chemical change when compressed, heated, and cured to a final shape. They cannot be softened again with heat and are often used for casting electrical and mechanical parts.

Thermoplastic materials, on the other hand, soften with heat and harden when cooled and can be resoftened after the part is made. This kind of plastic is often used for "hot melt" injection molding. The most common thermoplastics are nylon, cellulose acetate, some types of urethanes, vinyls, acrylics, acetals, and teflon. Nylon is a useful material for gears, bearings, and cams, and is easily machined. The acrylic plastics have good machining qualities and are useful for optical purposes (Figure 4).

Machinability of Plastics

Nonmetallic materials are very poor conductors of heat; therefore, heat dissipation is concentrated in the tool and cutting edge. Surprisingly, conventional tool materials tend to break down on many soft materials since the temperature rise is sufficient to cause tool breakdown. When machining thermoplastics and other materials having low heat conductivity at high speeds, inadequate heat distribution may result in the melting of the workpiece and a dulled or damaged tool. Thermosetting materials do not soften when exposed to heat, but heat dissipation is still low. Filler materials, such as glass fiber, asbestos, paper, or cotton used for

Figure 3. Severe cracking on this rubber tire is due to long exposure to ultraviolet in sunlight.

Figure 4. Both the nylon rod on the left and the acrylic plastic sheet on the right have good machining qualities.

reinforcement, can be very abrasive and shorten tool life considerably.

Drills for plastics should have an included angle of 60 to 90 degrees and zero rake (Figures 5*a* and 5*b*). Special drills for plastics are made with large, polished flutes. Some types are carbide tipped.

Tools for plastics vary considerably according to their heat and abrasive resistance and the required rake and clearance angles. On some materials such as the acrylics, cutting tools tend to gouge and dig into the surface if a positive rake is used; therefore a zero rake is needed in these cases. High speed, carbide, ceramic, and diamond tipped tools are all used for machining plastics. These tools must always be kept very sharp. Diamond tipped tools can, in many cases, increase production 10 to 50 times over that of carbides when applied to the machining of nonmetallic materials. Table 1 gives the machining characteristics of some plastics and other nonmetallic materials.

Standard taps and dies may be used for threading in nonmetals, but they must be sharp. Threading, reaming, milling, and other machining operations on plastics are performed dry or with coolants such as a blast of air, soluble oil, a water-soap solution, or plain water.

Ultrasonic machining is also performed on plastic materials. With this method, taps or other shapes are inserted into the work to form the desired shape with little machining stress. Ultrasonic welding is also used to bond thermoplastic parts. Butt, scarf, and shear joints are bonded with high frequency vibrations that cause the material to melt between the surfaces of the joints.

Table 1 Machining Characteristics of Nonmetallic Materials

Material	Drill Point Angle	Cutting Tool	Single Point Rake Angles in Degrees	Cutting Speeds	Feed Rate in per rev.	Coolant/Lubricant Required
Nylon	118°	High speed	0–5 positive	250–500	0.002–0.005	Dry or soluble oil
Rubber	118°	High speed	0–hard positive	200–1000	0.010–0.025	Dry or water
Phenolics	118°	Carbide	10–15	700–900	0.003–0.006	Dry or water
Vinyls	60°	High speed	0	250–500	0.003–0.005	Dry or water
Acrylics	60°	High speed	0–3 negative	200-500	0.002–0.006	Dry or water
Teflon	90°	High speed	0	200–500	0.004–0.006	Dry, air, or soluble oil
Epoxies	60°	High speed or carbides	0–3 negative	300–700	0.004–0.006	Dry, air, or soluble oil
Acetals	118°	High speed	0	700–850	0.002–0.005	Dry or soluble oil

Source. *Machine Tools and Machining Practices,* 1977.

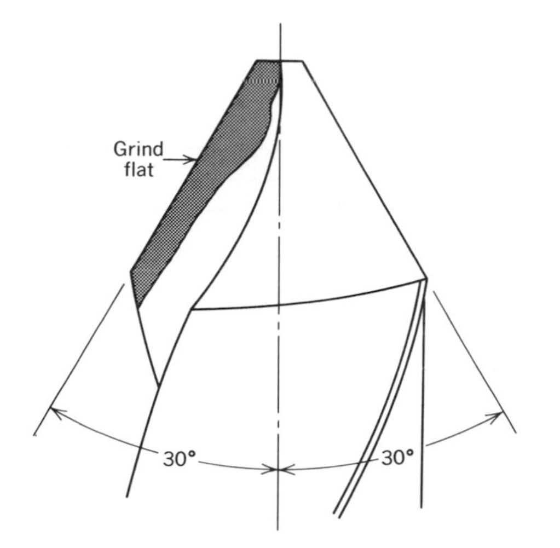

Figure 5a. A drill used for most plastics such as acrylics, fiber, and hard rubber should be sharpened as illustrated. The rake is ground with a zero angle *(Machine Tools and Machining Practices)*.

Other Fabrication Methods

Extrusion is a method of squeezing plastics into lengths of rods, pipes, films, sheets, and profile shapes. The extruder consists of a screw that rotates inside a heated barrel that forces molten plastic through a die.

Injection molding is a process in which molten plastic is forced into a cold mold by injection under pressure. The molded part quickly solidifies and the mold is opened to eject the part. The mold is then closed and the cycle continues. A great many of the plastic items that we use every day are made in this manner.

Blow molding is a method borrowed from the glassmaking industry and is used to make bottles and other containers. An external split mold is closed around a thick-walled heated tube of plastic. Compressed air is blown inside the plastic tube and it expands to fit the configuration of the mold. The mold is then opened, the bottle ejected,and the cycle is continued.

In **compression molding,** heated platens shaped to the desired form are used in a hydraulic press. The plastic sheet or other shape is inserted between the platens, and pressure from 1000 to 3000 psi is applied. The sheet takes on the form of the heated mold and is ejected when the press opens.

Casting is another way of shaping polymers. It is not possible to heat plastics to be sufficiently fluid for casting purposes, as is done with metals. A liquid polymer or monomer, or one dissolved in the other, is poured into the mold. Solidification is achieved either by the completion of polymerization or the crosslinking in polyester resins. One of the more utilitarian uses of the casting process is in the making of dentures. Casting is often used for art forms, decorative molding, and for potting (encapsulating) electrical components. Other forms of casting, however, are very important in industrial production. Photographic film and cellophane are made by feeding a solution of cellulose acetate onto a moving polished metal band.

Spinning of filaments of polymer to produce fibers is the source of such artificial clothing materials as Dacron®, nylon, and rayon. This process is actually a form of extrusion in which the liquid polymer is forced through a spinneret, a metal plate containing many small holes. Rayon is spun from cellulose acetate in an acetone solution, the filaments are dried in a heated chamber, and the solvent is driven off. Nylon and Dacron®, however, can be easily liquified by heat sufficiently to be forced through the spinneret.

Figure 5b. This nylon round rod was being drilled with an improperly shaped drill when it suddenly grabbed and split the part in several pieces.

Testing

Many of the same tests that are used for metal are used for plastics, such as tensile, hardness, and shock resistance. The Izod-Charpy test is used for the latter (Figures 6*a* and 6*b*). Other tests are electrical resistivity (dielectric strength), cold flow, softening temperature (heat resistance), flammability, dimensional and color change on aging, water resistance, acid, alkali and solvent resistance, and specific gravity.

Figure 6a. Plastic specimens are tested on this Izod-Charpy impact testing machine. (Tinius Olsen Testing Machine Company, Inc.)

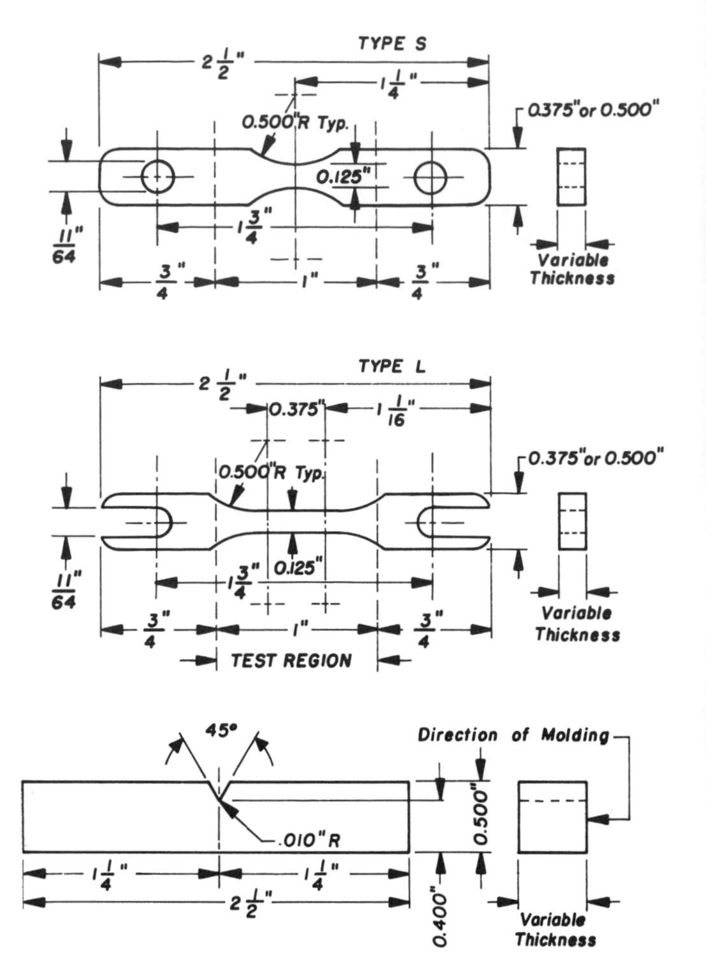

Figure 6b. Plastic specimens for tension impact tests (top two diagrams) should be made according to these specifications. The Izod test specimen for impact testing is shown below (Tinius Olsen Testing Machine Company, Inc.).

Rubber

Rubbers and elastomers may be either natural or synthetic. Natural rubber is obtained from a thick milky fluid (latex) that oozes from certain plants when they are cut. Most of the latex is derived from the para' rubber tree (Havea brasilinesis) in South America, southeast Asia, and Ceylon. Natural rubber in the form of cured latex is a sticky, gummy substance having limited usefulness.

The hot vulcanization process was discovered by Charles Goodyear in 1839. Goodyear combined latex rubber and sulfur in a mixture that he heated to the melting point of sulfur. This process made possible the many useful rubber products that we know so well. Rubber could be made hard or soft, very elastic, or semirigid.

Rubber is an extremely useful material. Modern society could scarcely do without such products as automobile tires, shock absorbers for machines, belting, wire covering, O-rings, and seals (Figure 7). Hard rubber (ebonite) has been used for many years as an industrial plastic material. Such products as battery boxes and combs are made of ebonite.

Synthetic rubbers were first developed in the early 1930s, commercialized in the 1940s, and were badly needed during World War II since natural rubber sources in southeast Asia were cut off. A great variety of synthetics has since been developed. Natural rubber is somewhat more resilient than synthetic rubbers after it is vulcanized (the cross linking of the polymer chains by sulfur atoms when heated). A disadvantage of natural rubber is that it swells when in contact with oil and sub-

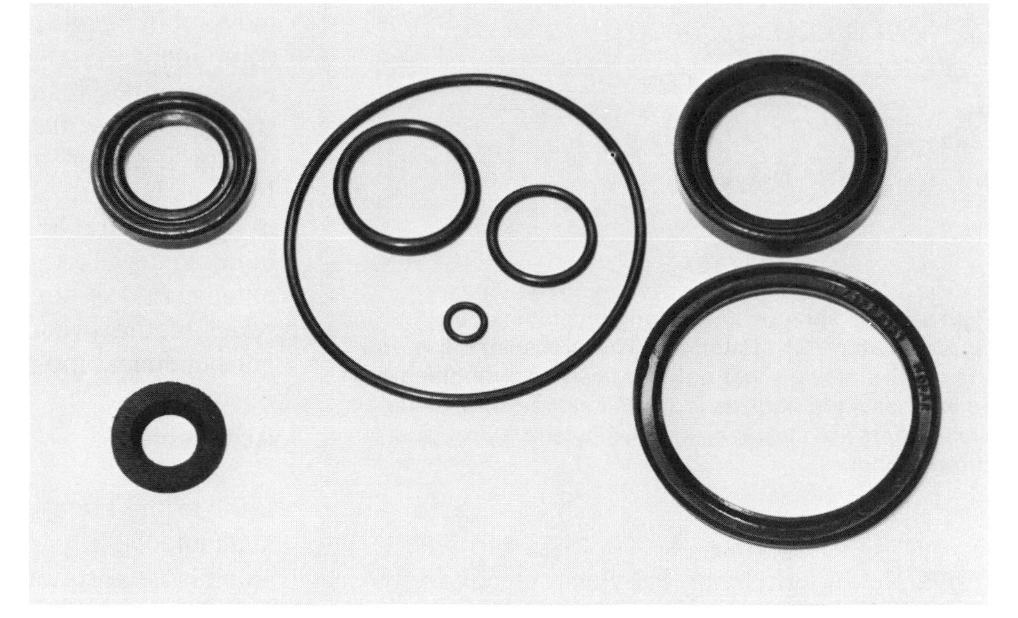

Figure 7. Many rubber products such as these O-rings and seals are used in hydraulic and other machinery.

sequently deteriorates; and when exposed to strong sunlight or heat, it becomes harder and more brittle, leading to cracking. Chemical additives such as antioxidants and stabilizers have greatly increased the life of soft vulcanized rubber. Plasticizers are sometimes added to polymers to change the properties and increase their usefulness.

Synthetic rubbers have a great variety of uses. Neoprene, for example, does resist oils and does not swell when in contact with them.

Petrochemicals

Many of today's plastic and rubber materials are derived from petroleum distillates, often called petrochemicals. Before 1920, gasoline and kerosene were obtained by the distillation of crude oil in a retort. The lighter elements, which were called napthas, were considered useless and were burned off to get rid of them. Later, it was discovered that many useful chemicals could be derived from these lighter elements. This was closely followed in the 1930s by the discovery of the process called catalytic cracking. This process uses heat to induce a chemical reaction with the help of catalysts, usually composed of refractory oxides of aluminum, silicon, and magnesium. High temperature boiling components of petroleum are broken down into lower boiling point components that are suitable for gasoline. The remaining gaseous material is made up of compounds having from one to four carbon atoms. Among these are ethylene, propylene, and the butylenes. Natural gas, from which the liquified gases butane and propane are produced is also a source of propylene, ethylene, hydrogen, and methane through the process of thermal cracking.

These petrochemicals can be formed into strong, tough plastics or rubbery materials through the amazing process of chemical synthesis. These are polymetric materials whose long, fibrous chain molecules are cross linked together to provide a strong, flexible bond (Figure 8).

Acetylene gas, made from the hydration of calcium carbide, is also a source of plastic and rubber materials. Calcium carbide is a synthetic substance, produced in an electric furnace from carbon (in the form of coke from coal or petroleum) and calcium oxide. Acetylene is also synthesized from natural gas by the high temperature cracking of natural gas. Probably the greater amount of acetylene gas used today is produced from methane rather than calcium carbide. From acetylene many products have been derived, but the most important of them are chloroprene and polychloroprene (or neoprene). Neoprene was the first synthetic rubbery material to become commercially successful.

From ethylene is derived ethyl alcohol, tetraethyl lead (a gasoline additive), ethyl benzene, styrene and polystyrene, ethylene glycol (antifreeze), and polyethylene (from which plastic squeeze bottles are made). Polyesters are also a

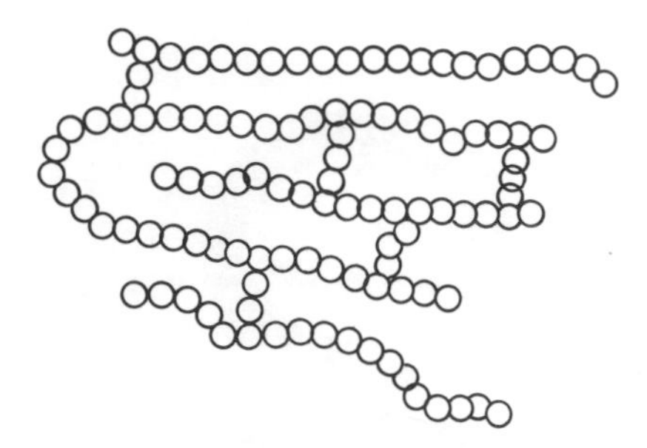

Figure 8. The spheres in this long chain molecule represent atoms or monomers. This cross-linking is only one of several ways that polymers can be strengthened. Some chains are rigid, as found in heat resisting materials, while others are elastic as in polyethylene, nylon, and cured rubber.

derivative of ethylene. Ethylene glycol forms a methyl ester from which polyester is subsequently made. It is a polyester from which fibers are made for clothing (Dacron®), magnetic tape for recording (Mylar®), and many other products.

Polystyrene is today one of the most important industrial plastics. It is a clear, glasslike product that dissolves in many solvents and is rather brittle. This thermoplastic is used where its clear, transparent qualities are useful, as in window panes and moldings. Polystyrene, though useful for many products, is also made into a foam with carbon dioxide gas, making a strong lightweight material known as styrofoam, which is used for insulation and packaging.

Cellulose is found in all plants, but is in its purest form in cotton. It is a polymer and is very useful in the form of cellulosic textiles such as rayon. When cellulose is treated with nitric acid, cellulose nitrate is formed. This product is well known as guncotton or nitrocellulose. It can be used as an explosive or a lacquer. The plastic called celluloid is made by compounding nitrocellulose with camphor and is used for lacquers in auto painting. Plasticizers are added to give it more resiliency.

Cellulose nitrate burns very rapidly and in the past, when movie film was made of it, created a fire hazard. Cellulose acetate was developed for the purpose of making nonflammable movie film. Since this material is soluble in acetone and other solvents, it has found many other uses, especially in the textile industry.

Epoxies are very strong but brittle thermosetting plastics. They may be cured with heat or at room temperature. Epoxy resins may be hardened by combining with a catalyst or they can be single component epoxies. These versatile plastics are used as adhesives or are fiber reinforced for added strength. They may be cast in a mold for tooling, jigs or dies, or for potting electrical hardware. Epoxies bond to almost any material and are used to make a durable surface on concrete floors or to bond materials to concrete. They are also used extensively as household adhesives. Table 2 lists some of the products that are derived from the petrochemical monomers.

Adhesives

Prior to the twentieth century, the only adhesives available were made from the hides and hooves of animals or parts of fish. Vegetable materials such as tree gums, starch, and casein were also used. None of these products provide a very strong bond, however, and they are easily loosened by

Table 2 Products from Petrochemicals

Products	Petrochemical Monomer (Raw Materials)
Antifreeze fluid, Mylar	Ethylene glycol
Butyl rubber	Isobutene
Epoxy resins, polycarbonates (Lexan®) acetone, phenol, bisphenol-A	Isopropylbenzine
Ethylene glycol antifreeze, Mylar (Dacron®) ethyl alcohol	Ethylene
Mylar, poly-para-xylene (Parylene®)	Xylenes
Neoprene rubber	Chloroprene
Nylon synthesis, neoprene, and polybutadiene rubber (source)	Butadiene
Orlon: chloroprene and neoprene rubber	Acetylene
Phenol, phenolic resins, styrene, and polystyrene	Benzene
Phenolic plastics	Toluene
Phenolic resins, epoxy resins	Phenol
Polypropylene, isopropylbenzine, isopropyl alcohol	Propylene
Polystyrene	Styrene
Polystyrene (nonsoluble)	Divinylbenzene
Production of acetylene	Methane
Synthetic rubber	Isoprene

the action of moisture. The synthetic resins in use today are each designed for specific uses. Some adhere well to soft, porous surfaces like wood, while others work best with metals or ceramic surfaces. Many are very resistant to deterioration from moisture, and bonding strengths are very high as compared to the old glues.

Sealants differ from adhesives in that they are designed to seal surfaces and joints from leakage. Although they are in a class by themselves, they are very similar to adhesives. They must adhere to the sealing surface though they usually do not have the high bonding strength of adhesives.

Because of their flexibility, structures that are bonded with adhesives resist damage from vibration better than riveted or bolted assemblies. The bond is made over the entire surface instead of a few places as in a riveted or nailed joint where each connector must resist high stresses that can result in cracking. Connectors such as bolts or rivets can be a cause of corrosion, but adhesives prevent electric current flow, because they normally have high electrical resistance, thus reducing corrosion.

Adhesives can be used to assemble unlike materials, such as glass and metal that have wide differences in their coefficients of expansion. One of the main disadvantages of adhesive joints is their relatively low bonding strength (1000 to 4000 psi) compared to welded or riveted structures. Usually, nondestructive testing or visual inspection is not possible except for ultrasonic testing, which makes routine inspection of bonded assemblies difficult. Also adhesives can be used only at relatively low temperatures, 200 or 300°F (93 or 149°C). A few adhesives have been developed that can be used at 600°F (315.5°C). These are mostly used on supersonic aircraft structures.

Bonding

In order for maximum bonding strength to be realized, absolute cleanliness of the part to be bonded is needed. Metals must be cleaned with degreasers, etching baths such as sulfuric acid, sand blasted, or anodized (as in the case of aluminum). Water is almost always a contaminant on the bonded surface; moisture should be removed as much as possible before bonding. Some of the several theories of adhesion are as follows:

1. The chemical bond theory in which electrostatic forces between the adhesive and the bonding surface (adherend) produce theoretical strengths that are much higher than any actual bond yet produced. You may remember from Chapter 7 that sodium chloride salt (NaCl) has an electrostatic bond that has a high theoretical strength, but because of natural dislocations, salt is a weak substance.

2. The inherent roughness theory is based on the principle of the roughness of the bonding surface producing a mechanical couple with the adhesive. According to this theory it would seem that a rougher surface would adhere with greater strength than a smooth one. However, with the development of epoxies and the super adhesives, the anaerobics or cyanoacrylates, very smooth metal surfaces can be joined with a tensile adhesion of 2000 psi or more within a few minutes. This adhesive is **anaerobic,** that is, it does not cure in the presence of air but by its absence between the contacting surfaces. The bond does not cure if it is more than 0.002 in. thick.

3. The surface energy theory in which surface tension of the bonding surface is greater than that of the liquid adhesive. This causes the adhesive to "wet" the surface and create a bond.

4. The polar theory states, in effect, that both materials of the adhesive and adherend must be similar in nature in order to bond to each other.

5. According to the weak boundary layer theory some weaknesses in bonding strength are caused by a volatile (unpolymerized) material that separates the surface of the adherend and the adhesive. This theory partly explains why the use of primers prior to applying the adhesive sometimes increases the bond strength.

None of these and other theories adequately explain all of the characteristics and problems encountered in bonding. Perhaps bonding is achieved through a complex combination of principles in which most or all of these theories are involved. Table 3 lists some of the kinds of adhesives and their uses.

Table 3 Adhesives and Their Uses

Name	Advantages	Disadvantages	Uses
Anaerobic cyanoacrylates (popularly called super glue)	High strengths (2000 psi) in a few minutes	High cost, thicknesses greater than 0.002 in. cure slowly or not at all, does not cure in presence of air	Production speed in bonding small parts, especially metal to metal and metal to plastics
Acrylic	Clear with good optical properties, ultraviolet stability	Moisture sensitive, high cost	Bonding plastics
Cellulose esters and ethers	Very soluble	Moisture sensitive	Bonding of organic substances such as paper, wood, and some fabrics
Epoxies: modified, nitrile, nylon, phenolic, polyamide, vinyl	Low temperature curing, adheres by contact pressure, good optical qualities for transmitting light, high strength, long shelf life	Cost, low peel strength in some types	Bonding microelectronic parts, aircraft structural bonding, bonding cutting tools and abrasives
Phenolics (bakelite)	Heat resistance, low cost	Hard, brittle	Abrasive wheels, electrical parts
Polyimides	Resistance to high temperature (600°F or 315.5°C)	Cost, high cure temperature (500 to 700°F or 260 to 371°C)	Aircraft honeycomb sandwich assemblies
Polyester (anaerobic)	Hardens without the presence of air	Limited adhesion, cost	Locking threaded joints
Polysulfide	Bonds well to various surfaces found in construction	Low strength	Caulking sealant
Polyurethane (rubbery adhesive)	Strong at very low temperatures, good abrasion resistance	Low resistance to high temperatures, sensitive to moisture	Flexible and rigid foams, protective coatings, vibration damping mounts, coatings for metal rollers and wheels
Rubber adhesives: Butadienes, butyl, natural (reclaim), neoprene, nitrile	High strength bonds with neoprene (4000 psi), Oil resistance (nitrile)	Low tensile and shear strength, poor solvent resistance	Adhesives for brake linings, structural metal applications, pressure sensitive tapes, household adhesives
Silicones	High resistance to moisture, can resist some temperature extremes	Low strength	Castable rubbers, vulcanized rubbers, release agents, water repellents, adhesion promoters, varnishes and solvent type adhesives, contact cements
Urea-formaldehyde resins	Low cost, cures at room temperatures	Sensitive to moisture	Wood glues for furniture and plywood
Vinyl: Polyvinyl-butyral-phenolic vinyl plastisols, polyvinyl acetate and acetals, polyvinyl chloride	High room temperature peel, ability to bond to oily steel surfaces (vinyl plastisols)	Sensitive to moisture	Bonds to steel (brake linings), safety glass laminate, household glue, water and drain pipe (PVC), household wrap

Some adhesives such as the epoxies require a curing agent or catalyst and others require heat for curing. Additives such as flexibilizers impart a resilience to the cured adhesive while fillers reduce the coefficient of expansion and thus the stress in the bond area. Fibers, when added, lend toughness to the adhesive. For example, fiberglass cloth and epoxy resins are used on the hulls of boats to give them a tough, smooth, leakproof surface. Auto body repairs are made with epoxy resins since these adhesives easily bond to metals. Commercial jet planes are bonded with adhesives in horizontal and vertical tail surfaces, wing panels, and ailerons. Nitrile rubber epoxy resins are used for bonding honeycomb structures for aircraft (Figure 9).

Two part epoxy kits are available for general use in which a small amount of each part of the epoxy is measured out on a 1:1 weight or volume measure. Abrasives are often bonded in grinding wheels; phenolic resins have been used for many years for this purpose.

The polysulfides, polyurethanes, and silicones are used as sealants for the caulking of buildings and boat hulls. Polyurethanes are also used as either flexible or rigid foams. It is also used as a rubbery adhesive. The silicones are used as castable rubbers for molds and for flexible adhesive joints.

Synthetic resins normally have a high resistance to the flow of electricity. When the passage of electric current through a joint or a synthetic resin part is required, conducting additives are used. Carbon black, graphite, or metals such as silver flake or copper powder are some of the additives that can be mixed with epoxies to make them electrically conductive. Gold and silver plating or other coatings on epoxy are sometimes used where higher current flows are required.

Adhesive bonds are tested for joint efficiency with destructive tests such as shear, tensile, compression, impact, and peel tests. Also cleavage (a variation of the peel test), creep, and fatigue tests are made on assemblies. Ultrasonic testing is used to some extent on bonded assemblies.

Industrial Gases

Compressed air may be considered as an industrial gas since it is used everywhere to operate air cylinders, pneumatic rock drills, to clean and ventilate, and to supply oxygen for combustion. Air is composed of oxygen, nitrogen, and some rare gases and has a pressure of 14.7 psi at sea level. Air contains moisture that condenses when it is compressed and then collects in air lines. This must be removed with water traps for most industrial air uses such as spray painting or operating air equipment.

Figure 9. Cutaway views of sandwich honeycomb structures used for aircraft.

Ammonia (NH_3) gas is used for nitriding, a type of case hardening of steel and for refrigeration. Nitrogen is an inert gas used for charging hydraulic accumulators.

Oxygen is used where heating or burning is required as in furnaces, cutting torches, and rockets. Oxygen can be dangerous since oils, greases, and other hydrocarbons will explode in the presence of pressurized oxygen.

Helium is also an inert gas used as a refrigerant, in balloons and blimps, and for welding. **Argon** is an inert gas that is used in electric lamps and for welding. **Carbon dioxide** (CO_2), when frozen, is the familiar ''dry ice.'' It is used in fire extinguishers, soft drinks, and as a shield gas for welding. Carbon dioxide is not an inert gas, and it disassociates into oxygen and carbon monoxide when heated by welding. It is therefore used for welding metals such as steel and iron that are not so affected by the presence of oxygen. It is not suitable for welding aluminum, titanium, and copper.

Fuel gases such as hydrogen, acetylene, natural gas, propane, and butane are burned with oxygen (O_2) or in air for heating purposes. The highest temperature is obtained by burning acetylene in oxygen (6200°F or 3427°C). For information on the many (more than 100) industrial gases, their uses, and toxicity obtain a good reference book such as the *Matheson Gas Data Book,* The Matheson Company, Inc., 1966.

Industrial Oils

Without lubricants such as oils and greases, machines would all cease to operate within a very short time. Friction takes place as the result of contacting moving surfaces even when they appear to be smooth. Abrasive wear is caused by contact between two metals that move against each other. Sometimes a foreign nonmetallic material such as sand or grit gets between the surfaces and causes rapid wear. Erosion (by contact with moving liquids or gases) is a form of wear that is often seen in pipe fittings and valve faces.

Wear is usually the result of the constant rubbing of the contact areas that may either break off small particles or displace the metal to flatten and make it smoother. This is often what happens when bearings are ''run in'' when they are new. A thick film (hydrodynamic lubrication) completely eliminates metallic contact (Figure 10) and wear is kept at a minimum. Boundary layer or adhesive film lubrication (Figure 11) is often used when it is not possible to maintain hydrodynamic lubrication.

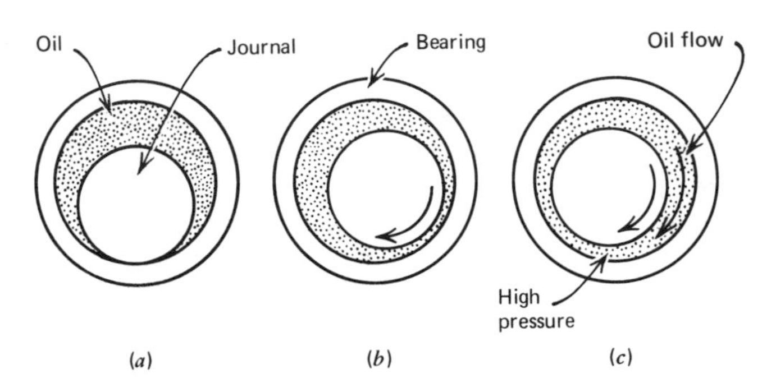

Figure 10. The journal is at rest at *(a)* and is in metal-to-metal contact. At *(b)*, as the journal begins to turn, it rolls upward to the right, but after adherent oil clinging to its surface begins to build up pressure and a separating oil film between the journal and the bearing, the journal moves to the left as shown at *(c)*. Bearing friction is largely dependent on the force necessary to shear the oil film.

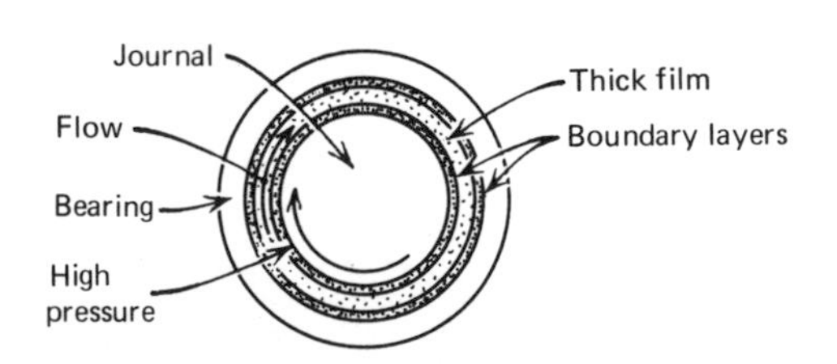

Figure 11. Thick film or boundary lubrication almost always provides a lubricating surface even when the parts are not moving. The boundary layers do not move to any great extent because of adhesion, and the shear is confined to the center area of the thick film. The oil flow is always slower than the movement of the journal depending on the amount of shear.

The addition of graphite, molybdenum disulfide, phosphides, or similar high pressure lubricants to the oil or grease will provide the boundary layer that reduces friction. Hypoid oils and greases were developed for the automotive industry to provide boundary lubrication for the intense pressures generated in automobile differentials.

Petroleum oils are universally used for lubrication. They all contain some sulfur, but the best types of oils contain the least amount of sulfur; however, they do contain paraffin, a saturated hydrocarbon that forms long polymer chains. Synthetic oils have been developed that resist higher temperatures and do not change in viscosity with temperature changes as much as petroleum oils do.

Oils for hydraulic machinery are specially prepared with antifoaming agents, controlled viscosity, corrosion inhibitors, and a high flash point so that they will not burn at lower temperatures. Both petroleum and synthetic oils are used in industrial hydraulic machines.

Self-Evaluation

1. Name two conditions that cause the breakdown of organic materials.
2. Plastics may be divided into two major groups. Name them.
3. Which of the two groups of plastics are the kind that can be resoftened by heat after the part is made?
4. What is the source of natural rubber?
5. What is the source of synthetic rubbers such as neoprene, isoprene, butadiene, and chloroprene?
6. The first synthetic rubbery material used commercially was derived from acetylene gas. What was it called?
7. Name three derivatives of ethylene.
8. Why do conventional tool materials often break down when machining plastics?
9. Before the age of synthetics, what materials were used for adhesives, pastes, and glues?
10. Name two advantages of adhesives over riveted or bolted joints?
11. How should a bonded surface be prepared?
12. The so-called super glues, the cyanoacrylates, can join two smooth metal surfaces with a single drop for each square inch and almost immediately gain a bond strength of 2000 psi. However, it will not even harden in the presence of air. Why is this?

13 Sealants are usually rubbery, flexible materials. Name one or more synthetic materials that are used as sealants.

14 Name three fuel gases and two inert gases used for industrial purposes.

15 Hydrodynamic lubrication reduces wear because the thick film lubricant keeps the moving parts separated. When the operating pressure is so intense that the thick film cannot be maintained, wear is the result. What kind of lubrication would eliminate rapid wear in this case?

This chapter has no Posttest.

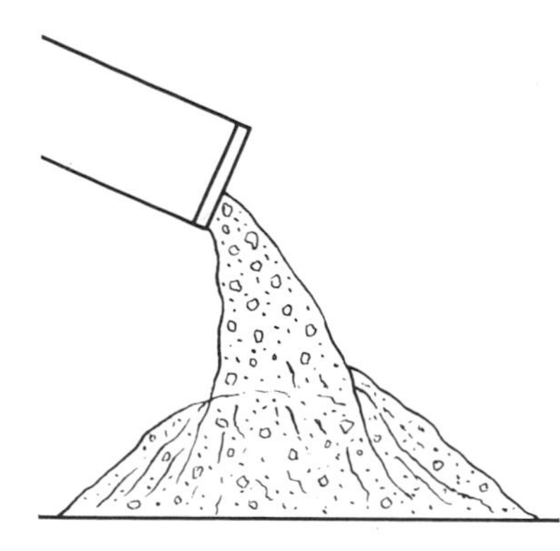

26 Concrete, Ceramics, and Related Materials

From the dawn of history, ceramic materials have been most useful to mankind. Indeed, they were the first materials used and developed. Primitive man fashioned tools from hard stones and pointed spears and arrows with sharp bits of stone. Ancient buildings were constructed of massive stones or clay bricks dried in the sun. Later, when it was discovered that clay could be hardened and glazed by firing, hard brick and pottery were made and are still in use today. This chapter will introduce you to these useful materials.

OBJECTIVE

After completing this chapter, you will be able to:
Describe the composition, characteristics, and uses of several inorganic materials.

INFORMATION

Ceramic Materials

Ceramic materials include clay, glass, silica, graphite, asbestos, limestone, and Portland cement (from which concrete is made). Fired earthenware is formed on a potter's wheel (Figure 1) and fine china is made from porcelain (a pure form of kaolin) clays. All of these inorganic materials are held together by ionic and covalent bonds that are more rigid than the metallic bond. Since these bonds, however, are very strong, the ceramic materials have greater heat and chemical resistance than organic materials. Ceramics are also usually good electrical and thermal insulators.

Local stress concentrations may exceed the bond strength, however, and cause a brittle failure. These materials, unlike metals, have few slip planes to absorb local stresses. This is why ceramics are very brittle materials. Ceramic materials have a very high compression strength, but low tensile strength. This feature makes them useful for load bearing and supporting structures in building construction.

Ceramic products have been machined or ground for many years. Marble and slate are examples. Many ceramics cannot be machined by traditional methods. Ultrasonic, abrasive jet, electron beam, and laser beam are applicable machining processes for ceramic materials.

Refractories

A refractory material can withstand high temperatures without breaking down (spalling or melting). Refractory brick used in furnaces is a very common example, and without refractories modern

Figure 1. The process of forming (throwing) a clay pot on a potter's wheel. *(a)* Clay is being prepared by wedging (kneading).

steel making would not be possible (Figure 2). Refractory materials can be cast or built of brick, which is made when fire clay or other material is used as mortar to bond the fire brick together.

The silicate structures commonly found in most rocks and clays (when fired) give them their characteristic properties of hardness and stability. Silica (silicon dioxide, SiO_4) combines in the unit structure of a tetrahedron (Figure 3). When linked together as a chain, the units become Si_4O_7 (Figure 4); this is the form in which it is usually found in minerals and rocks. Combined with water and alumina (aluminum oxide, Al_2O_3), the resultant aluminum silicate is the basic component of the clays, which are complex silicates containing attached water molecules.

Graphite is also an excellent refractory material since it cannot be spalled (that is, pieces cannot be broken off by thermal shock) because of its high thermal conductivity. Most refractories such as fire brick can withstand temperatures slightly higher than 3000°F (1649°C) before they break down. Graphite tends to oxidize in the presence of air and can be used up to 6000°F (3316°C). Amorphous carbon and graphite are both forms of the element carbon. Usually, carbon is mixed with a binder such as clay and is formed by molding and pressure. Both carbon and graphite exhibit low tensile to high compressive strength ratios. They also have the unusual property of gaining in strength as the temperature rises. After being baked in an oven to harden and strengthen it, the carbon is either ground or machined. Graphite is relatively easy to machine, but carbon requires carbide, ceramic, or diamond tools because it is very abrasive.

The abrasive silicon carbide can also be used as a high temperature refractory, but it is quite

(b) Lump of clay is being molded into a round shape.

(c) Hole is formed into the center and the sides are being raised.

expensive to use in this way. Refractory bricks containing large quantities of chromium oxide, referred to as chromite refractories, are especially well suited for high temperature use in steel melting furnaces. Magnesite brick, composed largely of magnesium oxide, is also used for this purpose. Insulating fire bricks are made from ordinary fire clay, but to provide porosity, the clay is combined with sawdust or coke, which is burned out when the brick is fired.

Abrasives, such as aluminum oxide and silicon carbide, are also ceramic materials. Most abrasives in use today are manufactured synthetics rather than the natural abrasives such as emery that were used in the past. Aluminum oxide abrasives are produced from bauxite ore that is calcined and placed in an electric furnace with steel turnings and coke. A fused ingot of aluminum oxide is formed. The ingot is allowed to cool; it is then crushed and screened for grade size.

Silicon carbide is also produced in an electric furnace. Silica, coke, sawdust (to produce porosity), and salt are charged in the furnace. A silicon carbide ingot is formed, which is allowed to cool

(d) Pot is now in final stages of finishing and will be dried and later fired.

Figure 2. Fire bricks are to be used for lining a furnace.

and is then crushed, cleaned, screened, and graded for grit size. Grit (grain) size refers to the particle size that will pass through a screen or grid with a given number of holes; that is, a number 100 grit size will pass through a 100 mesh screen that has 100 meshes per linear inch; the size of this particle is 0.010 inch in diameter. Particle sizes range from very coarse (numbers 6 to 12), coarse (numbers 14 to 24), medium (numbers 30 to 60), fine (numbers 70 to 120), very fine (numbers 150 to 240), and flour size (numbers 280 to 600).

Some manufactured refractory materials such as aluminum oxide are machined with diamond tools before firing. Silicon, a metalloid (which is a nonmetal that has some characteristics of a metal or forms an alloy with one), is very hard and abrasive and can be successfully machined with a polycrystalline diamond tool (Figure 5).

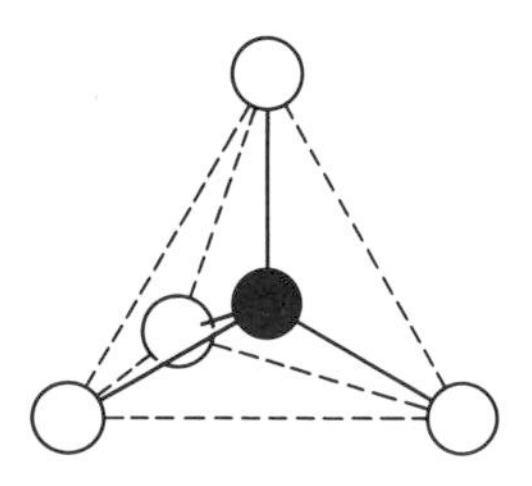

Figure 3. Unit structure of silicon dioxide (SiO_4). The darkened circle is silicon and the plain circles represent oxygen atoms.

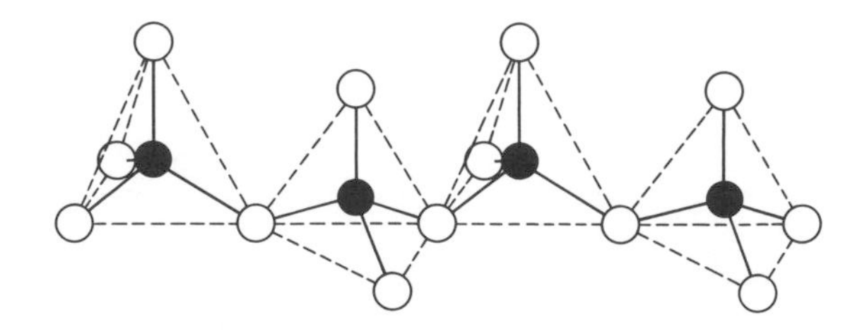

Figure 4. Silicate units can link through any of the oxygen atoms as shown. This is one way in which a chain of silicate tetrahedra can be joined. The dark circles represent silicon atoms and the plain circles are oxygen atoms.

Clays and Enamels

Several types of clays are mined. Kaolin (a white clay that is mostly composed of alumina and silica) is used for earthenware and fine china. Ordinary clays are used in making brick, in building construction, and for fire brick. These ordinary clays are composed of alumina and silica in various proportions with other impurities present, such as iron oxide (which gives it a red color) and man-

Figure 5. Megadiamond insert turns rebonded fused silica cylinder (Courtesy, Megadiamond Industries).

ganese oxide. The impurities lower their melting point.

The bonding of silica and alumina atoms forms a chain that develops tiny, flat plates of clay that tend to flow over one another like a deck of cards when they are wet; this gives clay its plastic character. When dried, the plates become linked together solidly as in unfired clay objects, and when they are fired, they fuse together into a hard, glasslike substance. The glassy appearance is due to both a liquid and solid phase that forms between the chains when they are fired. This is also true of the porcelains that are made from almost pure kaolin.

Porcelain enamels that are composed of quartz, felspar, borax, soot ash, and other elements are used to coat such articles as stoves, refrigerators, pots, and pans. This material can be applied to iron, steel, or aluminum for wear and decorative appearance.

Figure 6*b*. Large ball mills, such as this one, are used for raw grinding basic materials and clinker in the production of cement at the Oregon Portland Cement Company.

Concrete and Mortars

Silica, when combined with sodium and water, becomes sodium silicate or water glass which, when dried, forms a silica gel. This material was once used for bonding grinding wheels. Similarly, Portland cement, used for pavements, foundations, and other construction, contains a large proportion of calcium silicate; it also forms a silica gel when combined with water and allowed to harden.

Figure 6*a*. Powdered cement materials are fired in this large rotating kiln producing a clinker at the Oregon Portland Cement Company.

Portland cement is basically composed of clay and limestone with small amounts of iron oxide (Fe_2O_3) and dolomite ($MgCO_3$). These materials are ground to a powder and fired in a large rotary kiln. The resulting "clinker" is again ground to a fine powder as a final operation (Figures 6*a* and 6*b*). Cements are ground to about 44 microns particle size, or 325 mesh. The cement powder is shipped in bulk or in sacks that weigh about 100 lbs each.

Mortars for brick construction are composed of sand, hydrated lime, and Portland cement. Concrete is a mixture of Portland cement, water, and a suitable aggregate such as gravel and sand. Expanded, lightweight aggregates such as pearlite or vermiculite are sometimes used in concrete where weight is a factor. Concrete aggregate is composed of graded sizes of clean (washed) gravel and coarse sand to fill the voids (spaces between the gravel). Sand may have a particle or grain size that varies from 100 mesh to $\frac{3}{16}$ inch. Gravel is measured by the average aggregate diameter, which ranges from $\frac{3}{16}$ to $1\frac{1}{2}$ inches and occasionally up to $2\frac{1}{2}$ to 3 inches.

Several proportioning methods are used to determine the concrete mixture either by volume or by weight. An approximate method used for small batches of concrete is to refer to the cement content as a unit followed by fine aggregate (sand) and coarse aggregate (gravel). A 1:2:3 mixture would contain one part cement, two parts sand, and three parts gravel. A proportioning method used, when a large quantity of concrete is involved, uses the 94 lb net weight (about 1 cubic foot) bag of cement as a measure for the cement. The number of bags or sacks of cement used per cubic yard of aggregate serves as a measure. A five sack mixture is considered to be a standard mix for most purposes.

Proportioning is also done according to the weight of the aggregate. With this method, the bulk density of the gravel and sand must be known in order to determine the correct proportion of each. The gravel forms, more or less, a group of spheres with voids or holes between them (Figure 7). Sand is also made up of smaller spheres with voids. Ideally the cement fills the voids in the sand and the sand-cement mixture fills the voids in the gravel.

A method of expressing various mixtures used for premixed concrete is based on its expected ultimate compression strength when it is completely cured. Since a standard mix, for example, should reach at least 2000 psi, a 2000, 2400, or 3000 psi mix might be requested for a particular job. Compression tests are used to determine concrete strength after the curing period. Standard mixes for concrete sidewalks, foundations, and driveways should have a breaking strength of 2400 to 3000 psi and a high strength concrete up to 7000 psi. Additives, such as chemicals to speed the hardening or setting rate or to entrain or homogenize the entrapped air, are sometimes added when they are needed for a specific job.

The strength of concrete is determined largely by the proportion of cement used per a cubic yard of aggregate; a standard five sack mix is more economical than a six sack mix and is used for most purposes; however, a six or seven sack mix results in a higher strength concrete because of the extra cement that is added. Strength is also dependent to some extent upon the amount of mixing water used; too much water produces a weak concrete. A slump test is sometimes used to determine water content (Figures 8*a* to 8*c*). Most concrete, for testing purposes, reaches its highest strength in 28 days, but for all practical purposes, it continues to grow in strength for 25 years after pouring if it is in a damp environment.

The sharpness of the sand aggregate also affects the final strength of the concrete to a considerable degree. Sharp glacial sand, found in the Midwest where ancient glaciers formed sand, produces a very high strength concrete, but rounded coastal or river sands produce a weaker concrete if all other factors are the same. Beach sand is a very poor aggregate for concrete because it is not only rounded but also contains salt, which slows down the curing of concrete and prevents the concrete from attaining its full hardness. Any humus or dirt of any kind in the aggregate will also lower the ultimate compression strength of concrete.

Concrete is used extensively in the construction of building foundations, as pavements for streets and highways, bridges, and building construction (Figure 9). Since concrete has a low tensile strength compared to its high compression

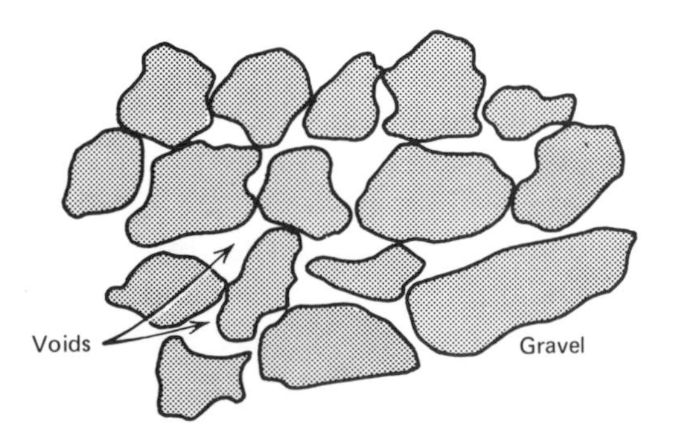

Figure 7. Gravel showing voids. Sand and cement fill the voids to make concrete.

Figure 8*a*. Mold used to make slump test.

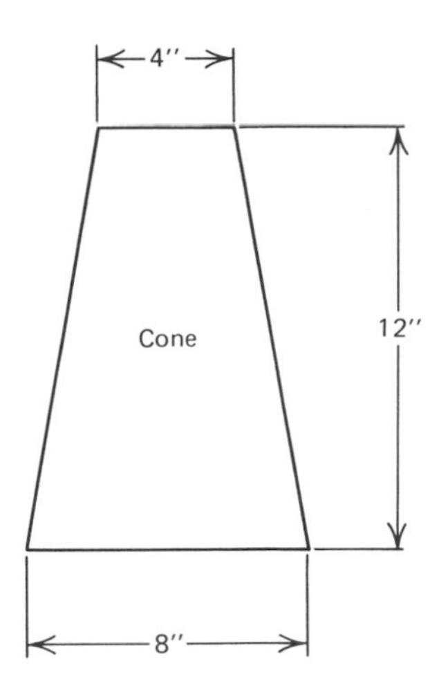

Figure 8b. The dimensions of the mold in Figure 8a are given here. The concrete mixture to be tested is placed in the mold and rammed with a steel rod. The excess at the top is struck off and the mold removed vertically.

Figure 9. Concrete foundations are being poured for a large building.

strength, concrete load support beams must be of massive size. Prestressing concrete beams provides the tensile strength of steel wires. Long, slender, more graceful beams and concrete structures are made possible with this technique (Figure 10). The steel wires do not corrode because they are buried in the concrete where oxygen cannot penetrate.

Another form of concrete construction is sometimes called "Gunite," a method of using compressed air to blow a specially prepared concrete mixture onto a surface. It is used to form a protective covering on steel construction and to line drainage ditches and the inside of tunnels. Tunnels are lined by blowing the concrete mix onto the inside surface that is covered with a steel mesh and reinforcing bars. The concrete is mixed with a high slump and a consistency that allows it to "stick" when blown to a surface. Probably the most common use of this method is for building backyard swimming pools (Figure 11). Some advantages of this way of placing concrete are that it produces a very high strength, dense concrete and it facilitates the construction of artistic and complex shapes without the use of expensive forms.

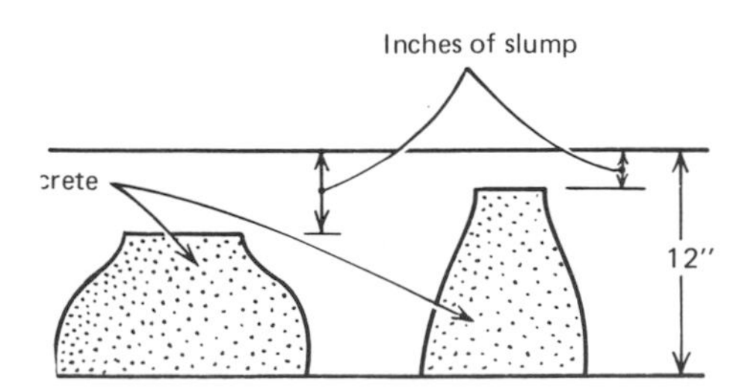

Figure 8c. After the mold has been removed, the concrete will slump (flatten out) an amount that depends on its consistency, which is controlled mostly by mixing water.

Figure 10. Graceful structures such as this curved concrete bridge are made possible by the technique of prestressing concrete beams.

Figure 11. Placing concrete in a backyard swimming pool with the gunite process.

Sewer and water pipes are made of concrete and a useful cement-asbestos concrete is used to make "transite" pipe. Transite pipe is sometimes machined on lathes to form fitted joints. The dust from machining this material can cause lung irritation and should be removed with an exhaust fan. The dust is also very abrasive on tools and machine ways must be protected to prevent damage.

Glass

Quartz is composed of silicon dioxide (silica) from which the amorphous structure known as glass is made after the silica is combined with sodium or lime. This combination causes the glass to solidify at a lower temperature than quartz and, instead of forming the crystalline structure of quartz, it solidifies directly into an amorphous or random lattice structure. Silica (in alumino-silicate structures, sand, and many other compounds) is one of the most plentiful elements in the earth. This is why glass is a relatively inexpensive material.

Glass is a very strong substance. A glass plate, for example, will support a heavy load without taking a permanent set (deforming). However, since glass is also a very brittle and notch sensitive material, a single scratch could cause a sudden failure when loaded in tension. Glass that is spun into fiber, known as fiberglass, is used with other components such as adhesives and resins to produce very tough materials that are not brittle or notch sensitive.

Composites

In the production of many materials, a great effort is often made to maintain their purity and eliminate any foreign matter. A composite is a material in which a stronger, sometimes fibrous material is usually combined with another to reinforce or strengthen the resultant mass. This is not an alloy but a mixture of two entirely differing materials. An example of a composite is the use of fiberglass to strengthen epoxies. Other organic and inorganic materials may be strengthened in this way.

Some materials are extremely strong when formed in small filaments or "whiskers" that are "grown" or formed by special techniques. Whiskers of sapphire (alumina), boron, or graphite are extremely strong (millions of pounds per square inch) and, when they are combined with metals such as aluminum, nickel, and titanium, greatly increase their tensile strength. Composites are often abrasive and tend to cause wear on tools when they are machined.

Self-Evaluation

1. Why are ceramic materials such as glass, fired clays, and stones rigid and brittle?
2. In what way do ceramic materials have high strength?
3. What is a refractory material and what is it used for?
4. Name the two principal ingredients in clay.
5. Abrasives such as aluminum oxide and silicon carbide are both produced in an open hearth furnace. True________ False________
6. Portland cement contains sharp sand and gravel or vermiculite. True________ False________
7. The strength of concrete is determined mostly by the properties of cement and mixing water plus the sharpness and cleanliness of the aggregate. True________False________
8. How can slender beams be used in constructions that are made of a substance such as concrete that has a low tensile strength?
9. What is glass made of?
10. Define a composite material.

This chapter has no Posttest.

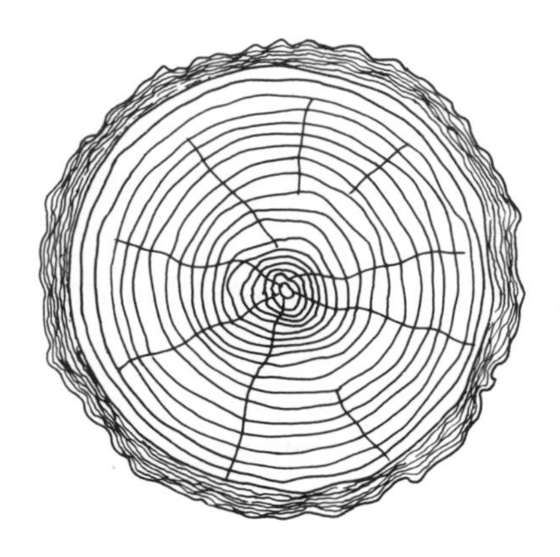

27 Wood and Paper Products

Trees, or timber as it is called by the lumber industry, are one of our most valuable renewable resources. The homes we live in, the paper we write on, furniture, and many other products depend on our northern, southern, and western forests. You will learn in this chapter about the many operations required to process the raw timber into useful products.

OBJECTIVES

After completing this chapter, you will be able to:

1. Describe the structure of wood and how it is processed to make lumber and plywood.
2. Explain the process of making paper.

INFORMATION

Timber, especially standing timber, refers to a stand of commercially valuable growing trees. Forest products consist of anything made of the woody parts of trees such as boards, planks, timbers (wood beams), plywood, and paper.

In North America, there are basically two kinds of forests that are harvested for commercial purposes: softwoods and hardwoods. Both hardwoods and softwoods are used in our civilization in a variety of applications. Hardwoods are typically used for furniture making. Conifers (softwoods) are used extensively for lumber products. Large stands of forests in the southern and western United States are harvested for lumber, plywood, and particle board (Figure 1).

Hardwoods

Deciduous trees shed their leaves each year and are classed as hardwoods. Oaks, maples, and birch are in this group and grow in the northern forests from Minnesota east and south to Virginia. Sweet gum, ash, and tupelo are found in Texas and the south. Hickory, oak, black walnut, and basswood grow in the prairie states. Balsa, although it is very light in weight, soft, and porous, is classed as a hardwood since it comes from a deciduous tree.

Softwoods

Coniferous trees have cones and needles instead of flat leaves and are sometimes called "evergreen" trees; that is, they do not lose their needles in the fall but keep them throughout the year. The coniferous trees are softwoods such as the southern and northern pines; western softwoods include the Douglas fir, white fir, pines, hemlocks, and spruces.

Softwoods such as fir, pine, and spruce often have greater hardness and strength than some hardwoods. The sitka spruce, for example, is classed as a softwood since it has needles, but it is

Figure 1. A stand of Douglas fir trees that will be harvested for lumber and other products (Weyerhaeuser Company).

considered to have the highest strength to weight ratio of any wood. It is noteworthy that sitka spruce was once used for structural frames in aircraft. It is also used to build musical instruments because of its resonant qualities.

Structure of Wood

Growth takes place in trees in the cambium layer, which is the only living part of the tree. Nutrients and moisture are carried to the leaves through this outer layer. As the tree grows, rings of inner cambium cells become inactive in winter, lose their moisture, and become sapwood. The sapwood eventually becomes part of the lifeless core or heartwood, which is the useful part of the tree (Figure 2). These growth rings, called annular rings, indicate the age of a tree since a new growth ring is formed every spring. The heartwood is composed of a cellular structure made of fibrous cellulose. These cells are held together by a natural adhesive called lignin (Figure 3).

Lumber Production

When these trees are ready for harvesting, they are felled (Figure 4) and trucked to the mill (Figure 5). The logs, while awaiting processing, are not allowed to dry out as this would cause cracking and splitting. The logs are either stored in a mill pond

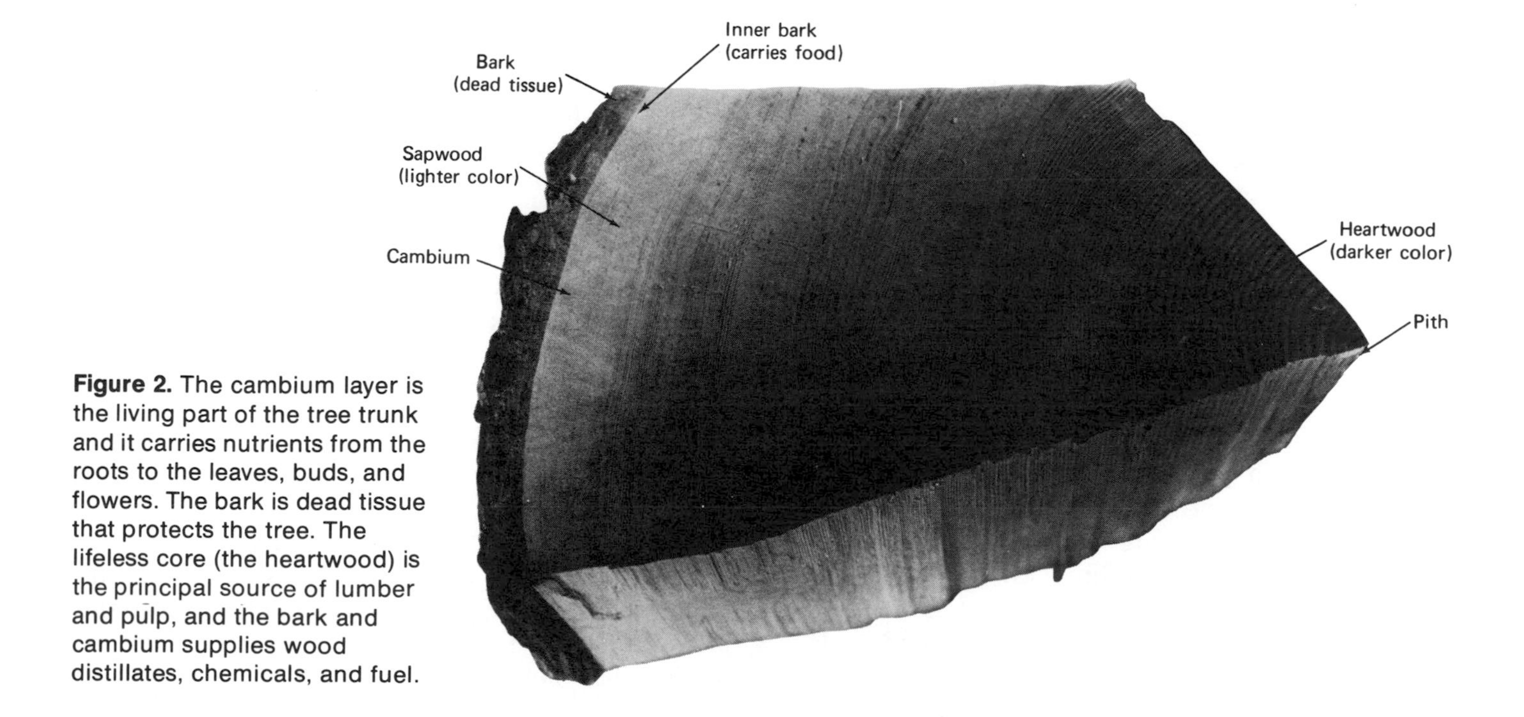

Figure 2. The cambium layer is the living part of the tree trunk and it carries nutrients from the roots to the leaves, buds, and flowers. The bark is dead tissue that protects the tree. The lifeless core (the heartwood) is the principal source of lumber and pulp, and the bark and cambium supplies wood distillates, chemicals, and fuel.

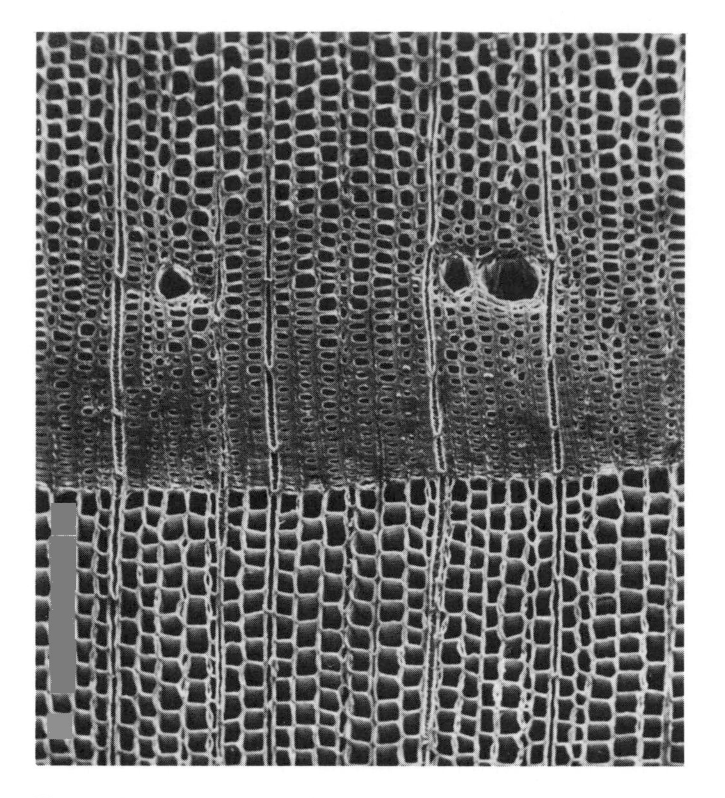

Figure 3. Transverse surface of a conifer or softwood (Douglas fir) illustrating the differences in cell shape and wall thickness of earlywood and latewood fibers. The latter are thicker-walled and flattened radially. Three vertical resin ducts are also seen here in cross section (100×). (Scanning electron micrograph courtesy R. A. Parham of The Institute of Paper Chemistry, Appleton, Wisconsin.)

(Figure 6) or in a cold deck where they are kept wet with sprinklers (Figure 7). The logs are first sawed into large slabs on the head rig (Figure 8) from which various lumber products (boards, planks, or timbers) are made. Figures 9*a* to 9*c* show the various cuts that can be made to utilize the log in the best ways. The lumber is further cut to usable sizes on other sawing machines. It is then carried along on a green chain (a set of moving chains that carries newly sawed lumber slowly along a long table). Here it is graded or selected for different uses.

Wood can have defects such as knots, dry rot, cracks, or pitch pockets. This is the reason that lumber must be graded (Figure 10). Softwood lumber is graded into three major categories: yard lumber, shop lumber, and timbers. Yard lumber is divided into select, common, and dimension lumber. The select (finish) lumber is graded A, B, C, and D according to the appearance of the board. Select grades are used for window frames, doors, and trim since they can have no defects and must be clear of knots and other blemishes.

The common grades of yard lumber are divided into five numbered grades. Number 1 grade common is free from structural defects having no loose knots. Number 2 has some structural defects. Number 3 can have loose or open knots and can be twisted or warped. These grades are used for construction. Numbers 4 and 5 are not acceptable for construction specifications since they may have pitch or dry rot.

Shop lumber is used for such purposes as furniture parts, cabinets, and window sashes. Timbers are used for structural purposes. A four inch or larger dimension is considered to be a timber.

The lumber, after it has been graded, is either dried in the open air in stacks or is kiln dried by being subjected to heat in an enclosed area. This operation tends to shrink the wood to some extent. This shrinkage can cause warping depending on how the board was sawed from the log (Figures 11*a* and 11*b*). Rough-sawed lumber is then finished by planing in a machine with whirling knives. By-products of sawing and planing in lumber mills are sawdust and chips. These can be processed into particle board or pressed logs for burning (Figure 12). Wood products are sometimes immersed in chemicals such as wood preservatives to prevent dry rot and fire retardants.

A rather unique recent development of the wood products industry is the laminated construction of beams that makes possible graceful shapes (Figure 13) of strong, light weight trusses that are used to support roofs in building construction. Flat boards are laid together and clamped in fixtures that form the shape of the beam. Strong adhesives are used to bond the boards together. Many of the weaknesses of sawed beams (cut from a single log) are eliminated with the lamination method. Knots, cracks, and rot can severely weaken a sawed beam, but in laminated beams the best boards, free from defects, are used on the top and bottom sides of the beam where the stress is the greatest.

Plywood is made by gluing veneer (thin slices of wood) together in alternate layers in which the direction of the grain is placed at right angles to the previous layer. Since wood is stronger in one direction than the other, crossing the grains greatly

Figure 4. A tree being cut down (felled) in a Douglas fir forest (Weyerhaeuser Company).

increases its strength. This criss-cross stack of veneer and adhesive is then placed in a hot press that has a series of heated plates or platens. Pressure is applied until the glue has hardened (Figure 14). The plywood sheets are then removed from the hot press, trimmed to size, and then sanded to give them a smooth, finished surface. The thickness of the finished plywood sheets depends on the number of plies or layers of veneer. Veneers are sliced from logs by either rotating them against a knife (Figure 15) or by slicing them with a reciprocating motion (flat cut) across the log (Figure 16).

Various hardwood veneers are used in the manufacture of furniture, and special cuts are needed for this purpose to provide a satisfactory wood grain, which can be either the flat cut or

Figure 5. Logs being trucked to the mill (Weyerhaeuser Company).

Figure 6. Logs stored in a mill pond (Weyerhaeuser Company).

Figure 7. Pile of logs called a cold deck is waiting to be processed in the mill (Weyerhaeuser Company).

rotary cut. Softwoods, such as fir or pine, are usually cut into veneer by the rotary method for making plywood. Plywood is mostly used in building construction (Figure 17). Like lumber, plywood is also graded by a classification system.

Hardboard and Particle Board

Hardboard and particle board are made from wood fibers such as sawdust or wood chips. Hardboard is bonded together by its own natural adhesive (lignin) while under heat and pressure. Adhesives are added to particle board to bond the fibers together when they are compressed. Both processes are somewhat similar, but hardboard is stronger and more dense than particle board.

In 1924, William H. Mason discovered the process of making hardboard, which was then produced under the name "masonite." Standard hardboard is a dense and stable material that is useful for making furniture parts, drawer bottoms, pegboard, and many other products. Tempered hardboard is treated with chemicals to make it more dense and water resistant. It is often made with a colored shiny surface for cabinets and bathroom walls.

Particle board is made of wood chips or sawdust mixed with an adhesive and formed into a mat (Figure 18). Ten or twenty mats are then placed in a hot press between platens and put under pressure

Figure 8. Logs are first sawed on the head rig. Other machines cut these slabs into smaller boards (Weyerhaeuser Company).

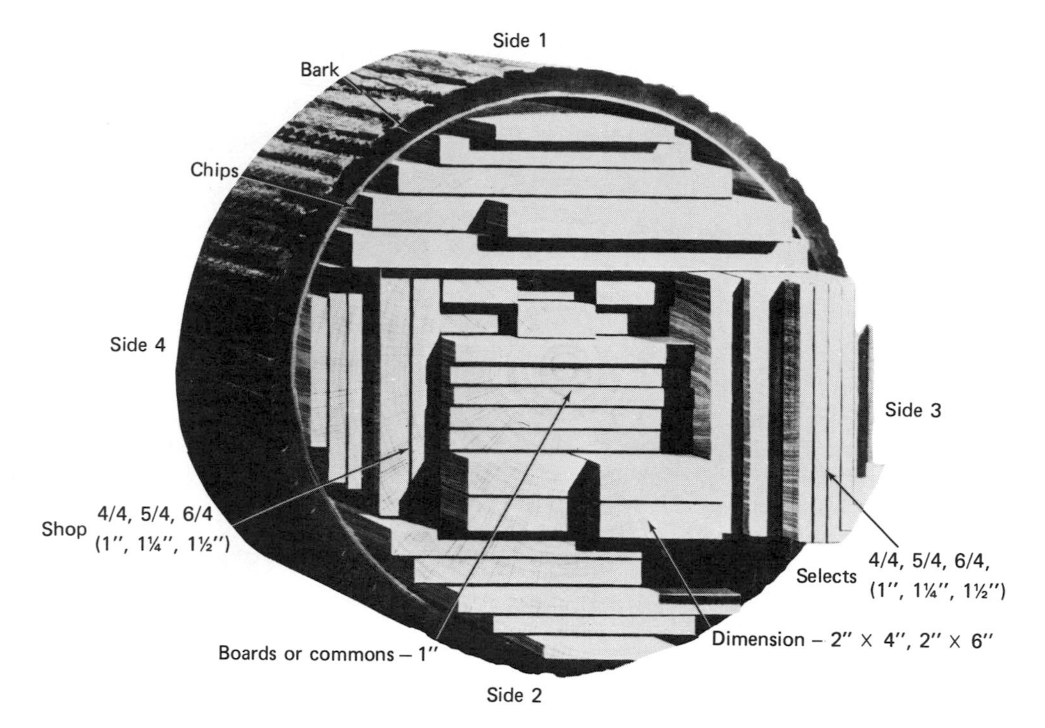

Figure 9*a*. The sawyer opened the log marked Side 1 and sawed off five thicknesses before turning the log and sawing Side 2. He then removed five pieces of lumber before turning the log and sawing Side 3. After removing eight cuts, he rotated the log and took five cuts from Side 4. By this time, he had reduced the log to a timber 12 inches square and had the choice of reducing it further or transferring the timber to other saws in the mill (Photo courtesy Western Wood Products Association).

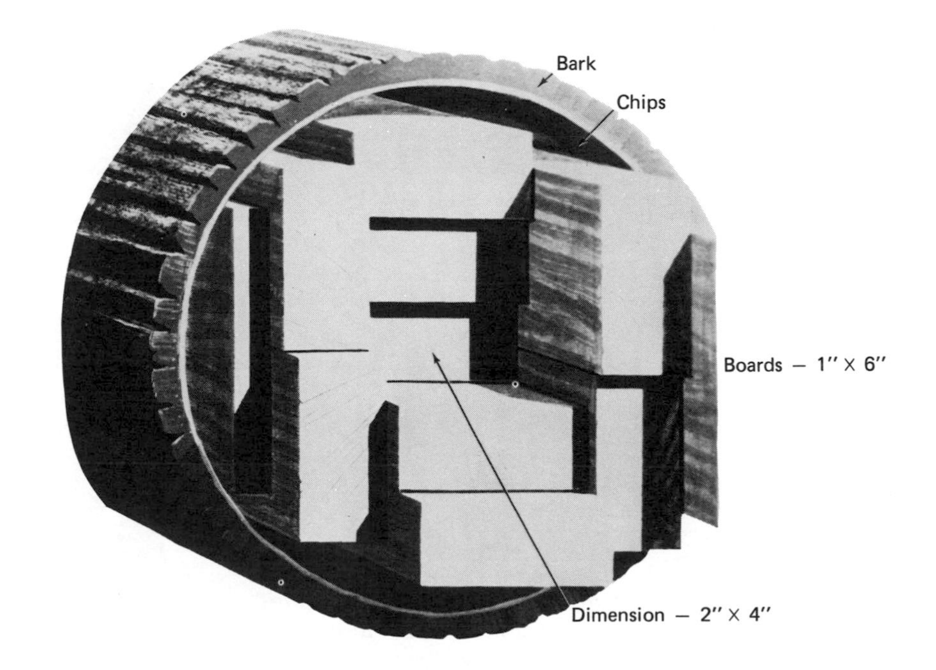

Figure 9*b*. This method is used by mills specializing in sawing small logs. The whole log passes in one motion through a series of circular or band saws and wood chippers, which reduce it to 2-inch or 4-inch thick pieces. These pieces are then turned flat and transported through a series of saws which cut them in one operation into 2 × 2's, 2 × 3's, or 2 × 4's. Such pieces, when cut 10 feet or shorter, may be specially graded and called studs (Photo courtesy Western Wood Products Association).

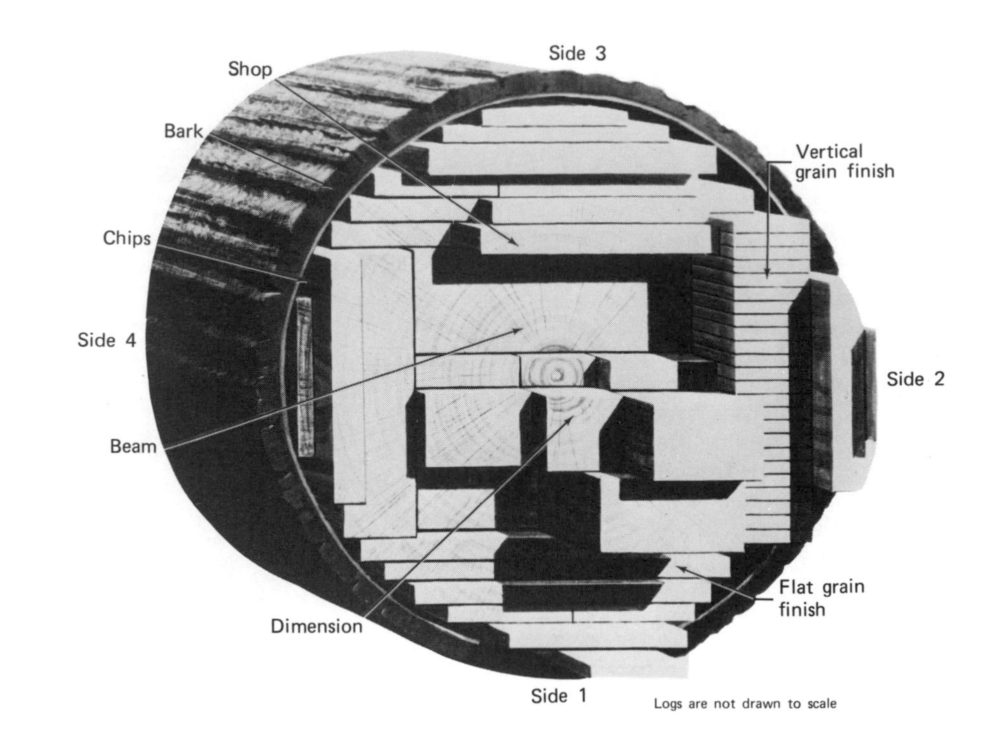

Figure 9c. This method is called "sawing around the log." The sawyer took six cuts from Side 1, then turned the log to begin sawing Side 2. He followed the same procedure with Sides 3 and 4 until a 16-inch square timber remained. He then had the same choices available as he had with Figure 9*a* to reduce it as shown or market it as a timber 16 inches square (Photo courtesy Western Wood Products Association).

Figure 10. Lumber being graded (Weyerhaeuser Company).

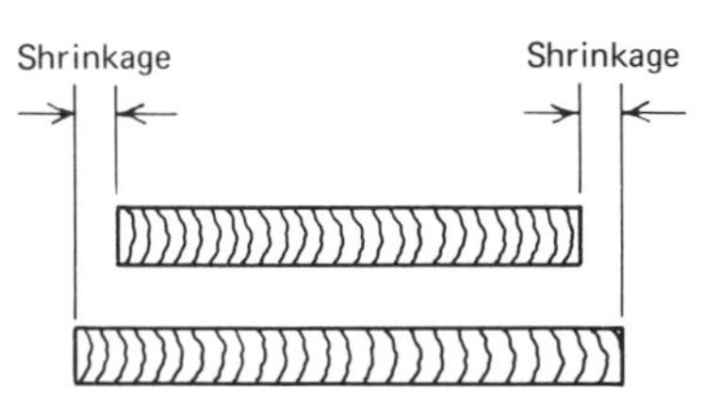

Figure 11*b*. Shrinking without warping is the major advantage of quartersawing (sawing perpendicular to the annular rings).

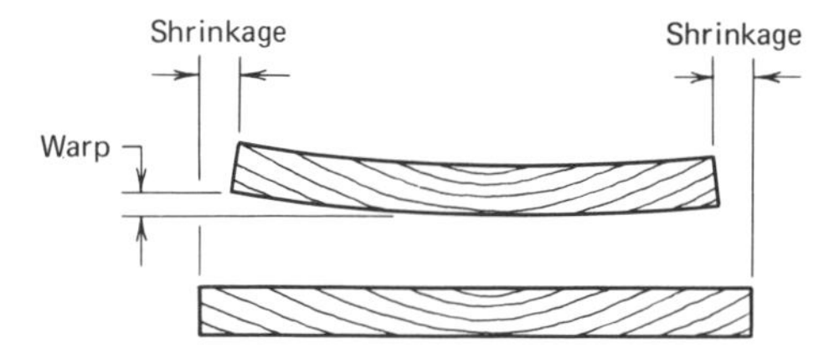

Figure 11*a*. Warping is a disadvantage with plain sawed lumber. Wood products always shrink when dried.

Figure 12. Sawdust is pressed to make logs for fireplace use (Weyerhaeuser Company).

Figure 13. These graceful arched beams in this construction project are laminated and glued. The trusses are still covered with paper to protect them from the elements while construction is in process.

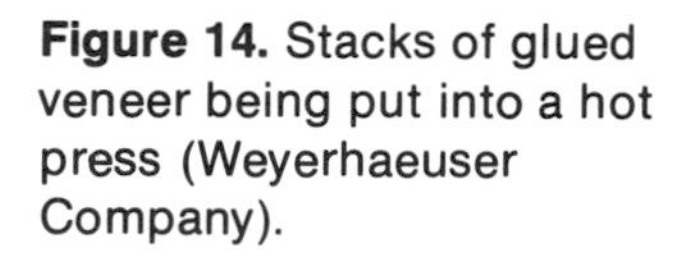

Figure 14. Stacks of glued veneer being put into a hot press (Weyerhaeuser Company).

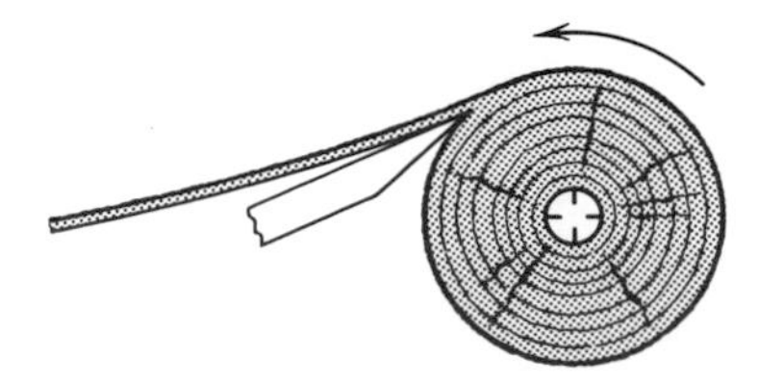

Figure 15. Rotary cut method of making veneer.

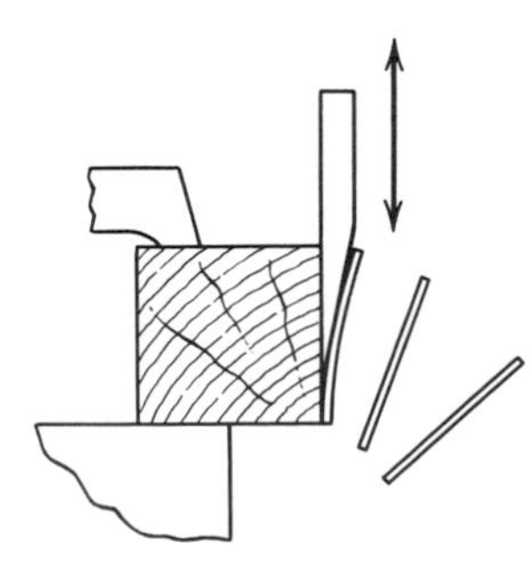

Figure 16. Flat cut method of making veneer.

Figure 17. Lumber and plywood are used extensively for building construction.

for 20 to 30 minutes (Figure 19). When the boards are removed, a somewhat dense, hard material is formed (though it is not as dense as hardboard). Particle board is used extensively for furniture making and building construction. Thin plastic simulated wood grain surface and other textures are sometimes cemented to the surface of particle board for furniture and wall panels.

Paper

Our modern civilization could scarcely get along without paper. Papermaking is one of the world's oldest industries. Ancient Egyptians made a kind of writing material from a native grass called papyrus. The art of papermaking seems to have been discovered by the Chinese and from there spread to Europe. Many materials such as linen or cotton rags, wood, and other vegetable fibers have been used to make paper. Perhaps the idea of papermaking derived from observing the wasps that build their nests of wood fiber and secretions from their bodies. Paper is used for the printing and publishing of books, magazines, newspapers, building materials, many kinds of containers, and other products.

Softwoods, such as spruce, balsam fir, hemlock, and southern pine, are usually used as pulpwood for papermaking, although hardwoods are sometimes used. The logs are cleaned and the bark, pitch, and foreign materials are removed. They are then cut into suitable lengths. The clean wood is conveyed to a chipper or a grinder, and the particles are placed in a digester to remove the lignin.

Figure 18. A mat of sawdust and adhesive mixture is formed on a large metal plate called a caul. The cauls move down a conveyor and are placed in a charging machine ready to be inserted in the hot press (Bohemia Inc.).

Figure 19. The hot press compresses the mat (12 mats in this hot press at one time) with hydraulic pressure. The mats are between heated platens. This forms a somewhat dense board that will then be trimmed to size and stored (Bohemia, Inc.).

There are four major processes for pulping wood for paper manufacture. One of these is known as groundwood or the mechanical process. The wood is ground on a rough grindstone to produce a fine fiber that is usually steamed to soften it. Although this does not produce a chemical change, the paper made from it is not stable, causing it to discolor, and it often disintegrates in use. The other three processes are chemical in nature. They are the sulfite process, the soda process, and the sulfate process (also called the Kraft process).

Sulfite Process In the sulfite process, the chips are cooked in a digester in an acid liquor under pressure. The cooking process in the bisulfite of lime liquor removes most of the noncellulose material in the wood. The chief advantage of the sulfite process is that the pulp can be easily bleached to produce white writing and printing paper.

Soda Process Caustic soda is used as a solution in the digesters to remove the unwanted lignin from the wood pulp. This process is used for short fiber woods such as poplar and other deciduous trees.

Sulfate Process This process is similar in some ways to the soda process in that an alkali (sodium sulfide produced from sodium sulfate) is used in place of part of the caustic soda in the cooking liquor. The sulfate process, or Kraft process, makes possible the use of fibrous substances that are difficult to break down by other processes. The Kraft process takes less time than the other methods and produces a pulp of exceptional strength (Figure 20). Sulfate pulp is manufactured from pine in the southern United States, spruce and eastern hemlock in the northeast, and western hemlock in the Pacific coast states. Other materials used are straw, jute, rags, and waste paper. These materials are often used to make cheap, brown paper for corrugated cartons and other containers. The Kraft process is mostly used in the south and far west.

The Paper Machine

In the Fourdrinier machine (named for its inventor), a fiber mat is continuously formed, which is

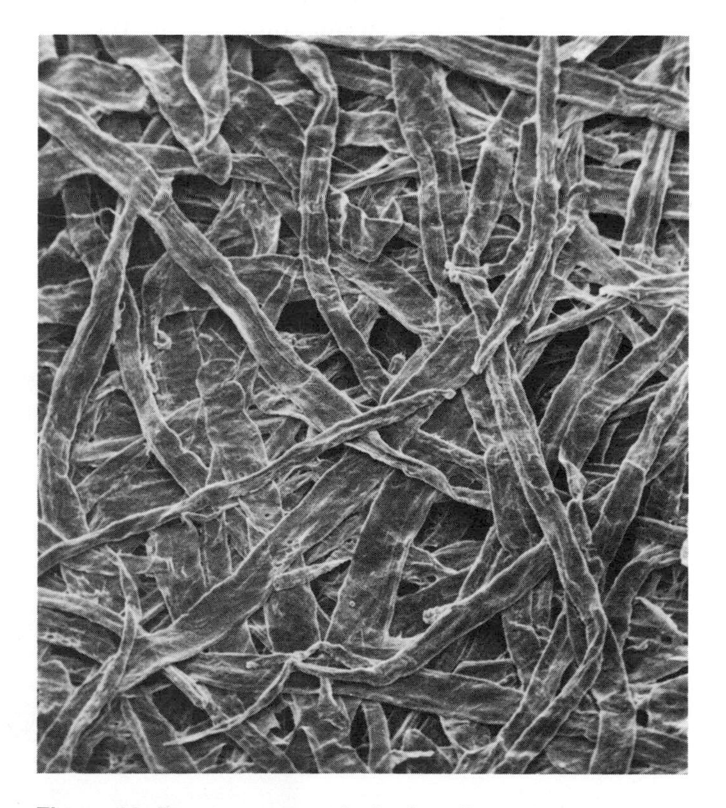

Figure 20. Paper composed of primarily earlywood fibers of southern yellow pine. They are thin walled and collapse readily into ribbons to form a dense sheet. (120 ×) (Scanning electron micrograph courtesy R. A. Parham of The Institute of Paper Chemistry, Appleton, Wisconsin.)

Figure 21. Fiber mat in the Fourdrinier machine (Weyerhaeuser Company).

dried, then rolled between a series of heavy steel rolls, and finally made into a thin, hard sheet (Figure 21). The Fourdrinier machine is the type of paper machine most used to form the pulp into a continuous ribbon on an endless belt of bronze wire cloth. As the pulp moves forward, water drains from it. The pulp is then squeezed in a series of rolls and is further dried on steam heated rolls. The finishing rolls further press the paper into a dense product of controlled thickness. The finished paper is then wound on large rolls for shipment (Figures 22*a* and *b*).

Figure 22*a*. At the end of the machine the paper is wound up on huge rolls (Weyerhaeuser Company).

Figure 22*b*. Rolls of paper are stored ready to be shipped (Weyerhaeuser Company).

Self-Evaluation

1 What are the two classes of wood used for lumber products?

2 Name several coniferous trees.

3 In which part of the tree does the growth take place?

4 How can the age of a tree be determined?

5 Describe the major processes in a sawmill required to make finished lumber.

6 Name two by-products of the sawmill and two uses.

7 Why is plywood stronger than ordinary sawed lumber?

8 What is the reason that lumber is graded and a classification system used?

9 What are hardboards and particle boards made of and used for?

10 Briefly describe how paper is made.

This chapter has no Posttest.

APPENDIX 1
Tables

Table 1 Heat Treatments for Various Tool Steels

Carbon Tool Steel
AISI-SAE Type W1

Atlas X-12
Best Carbon
B-F Extra
Blue Label
Carbon
Cutlery
Diamond S
Extra
G. W. Extra
H & R Carbon
Pompton
Regular
Special A.S.V.
Standard

Temp.	R.C.
As-Q	67-68
150°C	64-65
175	63-64
200	61-62
230	60-61
260	59-60
290	57-58
315	53-55
370	50-52
425	46-48
480	40-42
540	33-36

Analysis: C .60-1.40
Harden: 770°-815°C
Quench: Brine/Water

High Carbon High Chromium Oil
AISI-SAE Type D3

Alloy C
Atlas NN
CNS-2
Crocar
H & R K
Hampden
Hicro 200
Huron
Lehigh S
Neor
Republic HC
Superdie
Superior No. 1
Vi-Chrome

Temp.	R.C.
As-Q	64-66
150°C	63-64
175	62-63
200	61-62
230	60-62
315	59-60
370	58-60
510	57-59
540	56-57
550	52-54
565	50-53

Analysis: C 2.25 Cr 12.00
Harden: 925°-980°C
Quench: Oil/Salt/Air

Low Manganese, Cr-W Oil Hardening
AISI-SAE Type O1

Atlan
BTR
B 76
Carpenter O-1
Choyce 77
Fisco Oilhard
Invaro No. 1
Ketos
Kiski
Oil Hardening
Oneida
Ready-Mark
Utex
Veri Best Drill Rod

Temp.	R.C.
As-Q	64-65
150°C	63-64
175	62-63
190	61-62
200	59-61
230	58-59
260	57-58
315	55-56
370	51-53
425	48-50
480	46-47
540	42-43

Analysis: C .90 Mn 1.20 Cr .50 W .50
Harden: 790°-815°C
Quench: Oil/Salt

5% Chromium Air Hardening
AISI-SAE Type A2

Airkool
Airvan
Aircrat
Airtem
Concord
Crest A.H.
Dumore
Econo No. 5
E-Z-Die Smooth Cut
Hardnair
Krovan
Pittsburgh
Precision Ground
Simonds Airtrue

Temp.	R.C.
150°C	63-65
150	63-64
175	62-63
200	60-62
260	60-61
315	59-60
370	58-59
425	57-58
480	56-57
540	54-56
565	52-54
590	50-52
650	43-45

Analysis: C 1.00 Cr 5.00 Mo 1.10
Harden: 925°-1025°C
Quench: Air

Table 1 Heat Treatments for Various Tool Steels *(Continued)*

Manganese Oil Hardening
AISI-SAE Type O2

Deward
H Brand
H & R Oil-Hardening
Prescott
Ry-Alloy
Ry-Alloy Drill
Republic Arrestite
Simonds 864
S.O.D.
Special Oil-Hardening
Stentor

Temp.	R.C.
As-Q	64-65
150°C	63-64
175	62-63
190	61-62
200	60-61
230	59-60
260	58-59
290	56-58
315	55-56
370	51-53
425	47-49
540	42-43

Analysis: C .90 Mn 1.60
Harden: 770°-800°C
Quench: Oil/Salt

Silicon-Manganese Shock Resisting
AISI-SAE Type S5

Achorn USI Steel
Atsil
Bedco Alloy
BTF Alloy
Champion 255
Chimo
Duro Chip
FB-S5 Tufkut
Fisco Omega
H & R No. 8
Lanark
Ludlum 602
Omega
Rocket
Silicon Alloy

Temp.	R.C.
As-Q	61-63
150°C	59-61
175	58-60
200	57-59
230	56-59
260	55-58
315	54-57
370	53-55
425	50-52
480	46-50

Analysis: C .55 Si 2.00 Mn .80 Mo .40
Harden: 870°-910°C
Quench: Oil/Salt

High Speed Steel
AISI-SAE Type T1

Blue Chip
Blue Streak
Cannon
Clarite
F.C.C. No. 41 M
Gold Anchor
H & R No. 1
H.S.C. 18-4-1
Kutkwik
Record Superior
Red Label
Spartan
Superior Ark H.S.
Supremus

Double Tempered Temp.	R.C.
As-Q	63-65
540°C	63-64
550	63-64
565	63-65
580	62-64
590	61-63
620	57-60
650	58-60

Analysis: C .75 W 18.00
Harden: 1250°-1290°C
Quench: Air/Oil/Salt

High Speed Steel
AISI-SAE Type M2

Achorn
Bedco M-2
Blue Streak Moly
Braemow
DBL-2
FB-M2
H & R No. 57
Mocarb
Molite
Moly
Mustang
Rex M2-S
Sixix
Star Mo-M2

Double Tempered Temp.	R.C.
As-Q	64-66
540°C	64-66
550	64-66
565	64-66
580	62-64
590	61-63
620	59-61
650	54-57

Analysis: C .83 W6.50 Mo 5.00
Harden: 1175°-1230°C
Quench: Air/Oil/Salt

Table 1 Heat Treatments for Various Tool Steels *(Continued)*

High Carbon High Chromium Air
AISI-SAE Type D2

Atlan HCC
Atmodie
Atmodie Smoothcut
Airdi 150
Beaver
Darwin No. 1 FM
Hyco-1
High Production
Lehigh H
Olympic F M
Ontario
Republic 404
Superior No. 3
White Label

Temp.	R.C.
As-Q	63-64
150°C	62-63
175	61-63
200	60-62
230	60-61
260	59-60
340	58-59
510	58-60
540	57-58
550	56-57
565	54-56
590	48-52

Analysis: C 1.50 Cr 12.00 Mo .80
Harden: 980°-1025°C
Quench: Air

Tungsten Chisel
AISI-SAE Type S1

Achorn UBC Steel
Bengal
Brown Label
Bull Dog
Cyclops S 1
Falcon—6
Falcon—4
Ideor
Macco Foolproof
Maxtuff
Par-Exc
Pure-Ore Chiz-Alloy
Seminole Hard
Super Shock

Temp.	R.C.
As-Q	58-61
150°C	58-61
200	57-60
260	56-59
315	56-58
370	54-56
425	52-54
480	50-53
540	47-50
590	44-46
650	40-43

Analysis: C .50 Cr 1.35 W 2.50
Harden: 900°-980°C
Quench: Oil/Salt

Manganese Air Hardening
AISI Types A4, A5, A6

Air 4
Airloy
A.S. No. 121
A.S. Ack Low
Apache
CM Air-Hardening
H & R A-4
Macco AL-4
Milnair 4
Nutherm
Pure-Ore Air-Eez
Tempair
Uni-Die Smooth Cut
Vega Mold Steel

Temp.	R.C.
As-Q	63-64
150°C	61-63
175	61-62
200	60-61
260	58-59
315	55-57
370	54-55
425	52-53
480	50-52

Analysis:
A4 C 1.00 Mn 2.00 Cr 2.00 Mo 1.10
A5 C 1.00 Mn 3.00 Cr 1.00 Mo 1.00
A6 C .70 Mn 2.00 Cr 1.00 Mo 1.30
Harden: A4 815°-870°C
A5 790°-845°C
A6 830°-870°C
Quench: Air

Tungsten Fast Finishing
AISI Type F2

Atlas XXX
B. F. Fast Finishing
BFS
Celero
Coppco Fast Finishing
Double Special
E-E
F.F.F.
Fast Finishing
Finishing
GW-350
K.W.
Penco No. 20
Special Finishing

Temp.	R.C.
As-Q	66-68
150°C	64-66
175	63-65
200	63-65
230	61-63
260	61-62
315	59-61
370	55-56

Analysis: C 1.30 W 3.75
Harden: 790°-870°C
Quench: Brine/Water

THE ABOVE FIGURES ARE AVERAGE AND FOR 1″ DIAMETER ROUNDS, ADEQUATELY QUENCHED TO OBTAIN FULL HARDNESS.

Table 1 Heat Treatments for Various Tool Steels *(Continued)*

High Speed Steel
AISI-SAE Type M30

Como
Electrite Lacomo
H & R Super Molyhi
Super Hi-Mo
Super LMW
Super Motung
8-N2 Cobalt

Temp.	R.C.
As-Q	65-66
540°C	66-67
550	66-67
565	65-66
580	64-65
590	63-64
620	61-62
650	57-58

Analysis: C .80 Cr 4.00 Mo 8.25 Co 5.00
Harden: 1200°-1245°C
Quench: Air/Oil/Salt

Stainless Steels
403-410-416 420 440C

Temp.	R.C. (403-410-416)	R.C. (420)	R.C. (440C)
As-Q	38-45	52-55	58-61
175°C	38-43	52-55	58-60
230	38-43	50-52	56-58
315	38-43	48-50	56-58
425	38-43	48-50	56-58
420	38-43	45-47	56-58
510	38-42	44-46	53-55
525	30-40	42-45	53-55
540	26-35	40-42	52-55
550	25-34	35-38	50-52
620	20-25	33-35	46-48
590	B95-100	28-32	40-43

Harden:
403/410/416—925°-1010°C
420—980°-1040°C
440C—1010°-1065°C
Quench: Air/Oil/Salt

5% Chromium Hot Work
AISI-SAE Types H11, H12, H13

A. S. Cromat V
Crodi
Cromo-W V
Coppco H. W. No. 1
G. W. 99
H. & R. Hot Work No. 6
H W S
Howard A
Hotform No. 2
Macco M. L.
M. G. R.
Redstone 12
Republic 10-H-W
Thermotem 11

Double Tempered Temp.	R.C.
As-Q	54-55
540°C	54-56
550	53-54
565	51-52
570	48-49
580	47-48
590	44-45
605	42-43
620	40-41
635	38-39
650	36-37

Analysis:
H11 C .35 Cr 5.00 Mo 1.50 V .40
H12 C .35 Cr 5.00 Mo 1.50 W 1.30 V .40
H13 C .35 Cr 5.00 Mo 1.25 V 1.00
Harden:
H11 995°-1040°C
H12 995°-1025°C
H13 995°-1025°C
Quench: Air

9% Tungsten Hot Work
AISI-SAE Type H21

Air Hardening No. 30
B.B.
B-44J
C.L.W.
D. C. 66
D.N.V.
E. I. S. 73
F.C.C. No. 14
H & R Hot Work No. 2
Halcomb H. W.
Macco P-175
Nitung
Nut Piercer
Redstone

Temp.	R.C.
As-Q	52-54
540°C	52-55
565	52-55
590	49-52
620	48-51
635	47-50
660	45-48
675	39-43
705	34-38

Analysis: C .35 Cr 3.50 W 9.50
Harden: 1090°-1200°C
Quench: Air/Oil/Salt

Note: Steels listed are representative groups of brands as used in various sections of the United States and are not to be construed as recommendations for specific brands. All analysis percentages are averaged from representative specifications from steel producers.

Source. "Tempering Chart for Tool Steels," *Lindberg, Sola Basic Industries,* Chicago, Illinois, 1973.

Table 2 Heat Treatments for Various Production Steels

Alloy Steels (Water Hardening 0.30%C)

Temp.	1330	2330	3130
205°C	47RC	47RC	47RC
260	44	44	44
315	42	42	42
370	38	38	38
425	35	35	35
480	32	32	32
540	26	26	26
590	22	22	22
650	16	16	16

Heat Treatment:
Normalized at 900°C
Water quenched from 800°-815°C
Tempered—2 hours

Alloy Steels (Water Hardening 0.30%C)

Temp.	4130	5130	8640
205°C	47RC	47RC	47RC
260	45	45	45
315	43	43	43
370	42	42	42
425	38	38	38
480	34	34	34
540	32	32	32
590	26	26	26
650	22	22	22

Heat Treatment:
Normalized at 885°C
Water quenched from 800°-855°C
Tempered—2 hours

Alloy Steels (Oil Hardening 0.40%C)

Temp.	1340	3140
205°C	57RC	55RC
260	53	52
315	50	49
370	46	47
425	44	41
480	41	37
540	38	33
590	35	30
650	31	26

Heat Treatment:
Normalized at 870°C
Oil quenched from 830°-845°C
Tempered—2 hours

Alloy Steels (Oil Hardening 0.40%C)

Temp.	4340	4640	8740
205°C	55RC	52RC	57RC
260	52	51	53
315	50	50	50
370	48	47	47
425	45	42	44
480	42	40	41
540	39	37	38
590	34	31	35
650	31	27	22

Heat Treatment:
Normalized at 870°C
Oil quenched from 830°-855°C
Tempered—2 hours

Alloy Steels (Oil Hardening 0.50%C)

Temp.	4150	5150	6150
205°C	56RC	57RC	58RC
260	55	55	57
315	53	52	53
370	51	49	50
425	47	45	46
480	46	39	42
540	43	34	40
590	39	31	36
650	35	28	31

Heat Treatment:
Normalized at 870°C
Oil quenched from 865°-870°C
Tempered—2 hours

Alloy Steels (Oil Hardening 0.50%C)

Temp.	8650	8750
205°C	55RC	56RC
260	54	55
315	52	52
370	49	51
425	45	46
480	41	44
540	37	39
590	32	34
650	28	32

Heat Treatment:
Normalized at 870°C
Oil quenched from 815°-845°C
Tempered—2 hours

Source. "Production Steels, Analysis and Heat Treatment," *Lindberg, Sola Basic Industries,* Chicago, Illinois, 1973.

Carbon Steels (Water Hardening 0.30-0.50%C)

Temp.	1030	1040	1050
205°C	50RC	51RC	52RC
260	45	48	50
315	43	46	46
370	39	42	44
425	31	37	40
480	28	30	37
540	25	27	31
590	22	22	29
650	95RB	94RB	22

Carbon Steels (Water Hardening 0.60-0.95%C)

Temp.	1060	1080	1095
205°C	56RC	57RC	58RC
260	55	55	57
315	50	50	52
370	42	43	47
425	38	41	43
480	37	40	42
540	35	39	41
590	33	38	40
650	26	32	33

Carbon Steels (Water Hardening 0.40%C)

Temp.	1137	1141
205°C	44RC	49RC
260	42	46
315	40	43
370	37	41
425	33	38
480	30	34
540	27	28
590	21	23
650	91RB	94RB

Table 2 Heat Treatments for Various Production Steels *(Continued)*

Heat Treatment:	Heat Treatment:	Heat Treatment:
Normalized at 900°C.	Normalized at 885°C	Normalized at 900°C
Water quenched from 830°-845°C	Water quenched from 800°-815°C	Water quenched from 830°-855°C
Tempered—2 hours	Tempered—2 hours	Tempered—2 hours

THE ABOVE FIGURES ARE AVERAGE AND FOR 1″ DIAMETER ROUNDS,
ADEQUATELY QUENCHED TO OBTAIN FULL HARDNESS.

Note: Oil hardening steels can often be quenched in hot salt or hot oil (martempering) to minimize distortion, cracking hazards and residual stresses.

Table 3 Heat Treatments for Case Hardening Steels

Case Hardening Steels

The low carbon steels used for case hardening in pack or gas carburizing must be properly treated to develop desired case and core properties. Various combinations of heating and quenching to refine structure are illustrated below:

	1015		1018		1022		C-1117		2317		2515		3120	
	Co (RB)	Ca (RC)	Co (RC)	Ca (RC)	Co (RB)	Ca (RC)	Co (RC)	Ca (RC)	Co (RC)	Ca (RC)	Co (RC)	Ca (RC)	Co (RC)	Ca (RC)
Carburized and oil quenched from 925°C	85	62	14	62	87	62	10	65	34	62	36	62	30	62
Reheated and oil quenched from 800°C	82	61	12	61	85	61	10	64	32	62	35	62	31	62
Double oil quenched from 800°C and 775°C	82	60	12	60	85	60	11	63	32	62	35	62	30	63

	4320		4620		4815		5120		6120		8620	
	Co (RC)	Ca (RC)	Co (RC)	Ca (RC)	Co (RC)	Ca (RC)	Co (RC)	Ca (RC)	Co (RC)	Ca (RC)	Co (RC)	Ca (RC)
Carburized and oil quenched from 925°C	38	64	28	62	33	62	33	62	35	62	30	62
Reheated and oil quenched from 815°C	38	63	29	64	32	63	29	63	27	63	27	63
Double oil quenched from 815°C and 775°C	38	63	28	63	32	63	30	63	25	62	24	62

NOTE:
All hardness readings are RC unless otherwise indicated.

Ca = Case
Co = Core
RB = Rockwell B-Scale
RC = Rockwell C-Scale

THE ABOVE FIGURES ARE AVERAGE AND FOR 1″ DIAMETER ROUNDS,
ADEQUATELY QUENCHED TO OBTAIN FULL HARDNESS.

Source. "Production Steels, Analysis and Heat Treatment," *Lindberg, Sola Basic Industries,* Chicago, Illinois, 1973.

Table 4 Hardness Conversion Table (Approximate Value)

C 150 kg	A 60 kg	15-N 15 kg	B 100 kg	15-T 15 kg	Diamond Pyramid Hardness 10 kg	Knoop Hardness 500 gr & Over	Brinell Hardness 3000 kg	G 150 kg	Tensile Strength Approx. 1000 lbs. per square inch Inexact and Only for Steel
70	86.5	94.0	—	—	1076	972	—	—	
69	86.0	93.5	—	—	1004	946	—	—	
68	85.5	—	—	—	942	920	—	—	
67	85.0	93.0	—	—	894	895	—	—	
66	84.5	92.5	—	—	854	870	—	—	
65	84.0	92.0	—	—	820	846	—	—	
64	83.5	—	—	—	789	822	—	—	
63	83.0	91.5	—	—	763	799	—	—	
62	82.5	91.0	—	—	739	776	—	—	
61	81.5	90.5	—	—	716	754	—	—	
60	81.0	90.0	—	—	695	732	614	—	
59	80.5	89.5	—	—	675	710	600	—	
58	80.0	—	—	—	655	690	587	—	
57	79.5	89.0	—	—	636	670	573	—	
56	79.0	88.5	—	—	617	650	560	—	—
55	78.5	88.0	—	—	598	630	547	—	301
54	78.0	87.5	—	—	580	612	534	—	291
53	77.5	87.0	—	—	562	594	522	—	282
52	77.0	86.5	—	—	545	576	509	—	273
51	76.5	86.0	—	—	528	558	496	—	264
50	76.0	85.5	—	—	513	542	484	—	255
49	75.5	85.0	—	—	498	526	472	—	246
48	74.5	84.5	—	—	485	510	460	—	237
47	74.0	84.0	—	—	471	495	448	—	229
46	73.5	83.5	—	—	458	480	437	—	221
45	73.0	83.0	—	—	446	466	426	—	214
44	72.5	82.5	—	—	435	452	415	—	207
43	72.0	82.0	—	—	424	438	404	—	200
42	71.5	81.5	—	—	413	426	393	—	194
41	71.0	81.0	—	—	403	414	382	—	188
40	70.5	80.5	—	—	393	402	372	—	182
39	70.0	80.0	—	—	383	391	362	—	177
38	69.5	79.5	—	—	373	380	352	—	171
37	69.0	79.0	—	—	363	370	342	—	166
36	68.5	78.5	—	—	353	360	332	—	162
35	68.0	78.0	—	—	343	351	322	—	157
34	67.5	77.0	—	—	334	342	313	—	153
33	67.0	76.5	—	—	325	334	305	—	148
32	66.5	76.0	—	—	317	326	297	—	144
31	66.0	75.5	—	—	309	318	290	—	140

Table 4 Hardness Conversion Table (Approximate Value) *(Continued)*

C 150 kg	A 60 kg	15-N 15 kg	B 100 kg	15-T 15 kg	Diamond Pyramid Hardness 10 kg	Knoop Hardnes 500 gr & Over	Brindell Hardness 3000 kg	G 150 kg	Tensile Strength Approx.
30	65.5	75.0	—	—	301	311	283	92.0	136
29	65.0	74.5	—	—	293	304	276	91.0	132
28	64.5	74.0	—	—	285	297	270	90.0	129
27	64.0	73.5	—	—	278	290	265	89.0	126
26	63.5	72.5	—	—	271	284	260	88.0	123
25	63.0	72.0	—	—	264	278	255	87.0	120
24	62.5	71.5	—	—	257	272	250	86.0	117
23	62.0	71.0	—	—	251	266	245	84.5	115
22	61.5	70.5	—	—	246	261	240	83.5	112
21	61.5	70.0	100	93.0	241	251	240	82.5	116
20	61.0	69.5	99	92.5	236	246	234	81.0	112
—	60.0	—	98	—	—	241	228	79.0	109
—	59.5	—	97	92.0	—	236	222	77.5	106
—	59.0	—	96	—	—	231	216	76.0	103
—	58.0	—	95	91.5	—	226	210	74.0	101
—	57.5	—	94	—	—	221	205	72.5	98
—	57.0	—	93	91.0	—	216	200	71.0	96
—	56.5	—	92	90.5	—	211	195	69.0	93
—	56.0	—	91	—	—	206	190	67.5	91
—	55.5	—	90	90.0	—	201	185	66.0	89
—	55.0	—	89	89.5	—	196	180	64.0	87
—	54.0	—	88	—	—	192	176	62.5	85
—	53.5	—	87	89.0	—	188	172	61.0	83
—	53.0	—	86	88.5	—	184	169	59.0	81
—	52.5	—	85	—	—	180	165	57.5	80
—	52.0	—	84	88.0	—	176	162	56.0	78
—	51.0	—	83	87.5	—	173	159	54.0	77
—	50.5	—	82	—	—	170	156	52.5	75
—	50.0	—	81	87.0	—	167	153	51.0	74
—	49.5	—	80	86.5	—	164	150	49.0	72
—	49.0	—	79	—	—	161	147	47.5	—
—	48.5	—	78	86.0	—	158	144	46.0	—
—	48.0	—	77	85.5	—	155	141	44.0	—
—	47.0	—	76	—	—	152	139	42.5	—
—	46.5	—	75	85.0	—	150	137	41.0	—
—	46.0	—	74	—	—	147	135	39.0	—

The values in this table correspond to the values shown in the corresponding joint SAE-ASM-ASTM Committee on Hardness Conversions as printed in ASTM E 48, Table 3.

Source. "Hardness Conversion Table," *Lindberg, Sola Basic Industries,* Chicago, Illinois, 1973.

Table 5 The Periodic Table of the Elements

H 1																	H 1
Light Metals		Heavy Metals										Nonmetals					Inert gases
IA	IIA											IIIB	IVB	VB	VIB	VIIB	He 2
Li 3	Be 4	←Brittle metals→					←Ductile metals→				Low melt-ing	B 5	C 6	N 7	O 8	F 9	Ne 10
Na 11	Mg 12	IIIA	IVA	VA	VIA	VIIA	VIII			IB	IIB	Al 13	Si 14	P 15	S 16	Cl 17	Ar 18
K 19	Ca 20	Sc 21	Ti 22	V 23	Cr 24	Mn 25	Fe 26	Co 27	Ni 28	Cu 29	Zn 30	Ga 31	Ge 32	As 33	Se 34	Br 35	Kr 36
Rb 37	Sr 38	Y 39	Zr 40	Nb 41	Mo 42	Tc 43	Ru 44	Rh 45	Pd 46	Ag 47	Cd 48	In 49	Sn 50	Sb 51	Te 52	I 53	Xe 54
Cs 55	Ba 56	* 57-71	Hf 72	Ta 73	W 74	Re 75	Os 76	Ir 77	Pt 78	Au 79	Hg 80	Tl 81	Pb 82	Bi 83	Po 84	At 85	Rn 86
Fr 87	Ra 88	† 89															

*	La 57	Ce 58	Pr 59	Nd 60	Pm 61	Sm 62	Eu 63	Gd 64	Tb 65	Dy 66	Ho 67	Er 68	Tm 69	Yb 70	Lu 71
†	Ac 89	Th 90	Pa 91	U 92	Np 93	Pu 94	Am 95	Cm 96	Bk 97	Cf 98	Es 99	Fm 100	Md 101	102	Lw 103

In general, it can be stated that the properties of the elements are periodic functions of their atomic numbers. It is in the vertical columns that the greatest similarities between elements exist. There are also some similarities between the A and B groups on either side of the VIII groups. For example, the Scandium group, IIIA, is in some respects similar to the Boron group, IIIB. The most important properties in the study of metallurgy that vary periodically with the atomic numbers are crystal structure, atomic size, electrical and thermal conductivity, and possible oxidation states.

Besides the relationship of the elements in the vertical groups of the transition metals, there are resemblances in several horizontal triads: iron, cobalt, and nickel (all three are ferromagnetic); ruthenium, rhodium, and palladium; and osmium, iridium, and platinum.

APPENDIX 2
Answers to Self-Evaluation

Chapter 1 Extracting Metals from Ores

SELF-EVALUATION ANSWERS

1 Iron ore is essentially iron oxide. When the oxygen is removed, thc iron remains. This is done by burning carbon (coke) in the presence of the ore. The carbon monoxide that is formed combines with oxygen in the ore to become carbon dioxide.
2 No. It is either used directly for steel making or for producing refined cast irons.
3 Iron ore, coke, and limestone.
4 No. The carbon content of pig iron (3 to 4.5 percent) renders it too weak and brittle for many uses. Steel with its lower carbon content (0.05 to 2 percent) is more ductile and has a higher tensile strength.
5 Ore dressing concentrates the ore so that a higher percentage of iron is shipped to the smelter.
6 Basically, the carbon must be removed from the pig iron and also such unwanted elements as sulfur and phosphorus. Some sulfur may be retained by the addition of manganese, which forms the compound, manganese sulfide, which is not as undesirable in the steel as iron sulfide. Free machining (resulfurized) steels contain manganese sulfide.
7 The electric furnace can produce a highly controlled, high grade steel product. In areas having little or no pig iron production, but with considerable scrap available, the electric furnace can operate quite well, especially if cheap hydroelectric power is available.
8 "Killed" steel is deoxidized in the furnace giving it more uniformity and less porosity than "rimmed" steels.
9 The Bessemer process blows cold air through molten pig iron to burn out the carbon in the iron. In the basic oxygen process, oxygen is blown down onto the surface of the molten pig iron. This method is not limited to pig iron alone, but can also process steel scrap.
10 Electric. Very high quality steel can be made by this process as a close control can be maintained on the proportions of the alloying elements.

11 A roasting process. Matte.
12 By the electrolytic process.
13 Bauxite.
14 Cryolite.
15 The oxygen is released as a gas and the metallic aluminum sinks to the bottom of the bath where it is siphoned off and cast into ingots.

Chapter 2 The Manufacture of Steel Products

SELF-EVALUATION ANSWERS

1 The ingot is heated to bring it up to rolling temperature, 2200°F (1204°C), and to give it a uniform temperature throughout its mass.
2 Rolling breaks down the coarse, weak columnar grains in the ingot and they reform into smaller but even grains. This recrystallizing process makes the steel stronger.
3 Mild steel bar contains up to 0.20 percent carbon.
4 Cold finished steel is stronger than equivalent hot rolled steel because the grains are permanently deformed and elongated in the direction of rolling. If it is in an annealed condition, the grains are restored to their former state and the metal is no longer much stronger than hot rolled steel.
5 Hot rolled steel is covered with black mill scale. Cold finished steels typically have a bright metallic finish.
6 Hot formed metals tend to take on a fibrous quality like wood grain. When a tool or a machine part is forged, this grain is shaped with the irregularities instead of being cut through as with machining.
7 Seamless pipe is made from a single billet that is pierced so that it is all one piece with no welds. Butt welded pipe is made from steel strip and the edges are joined by pressure welding all along its side.
8 Small pipes and tubes are rolled from strip and electric resistance welded. This material does not have to be heated first as with butt welded pipe. Automatic fusion welding joins pipes of large diameter.
9 The rough cuts should all be made first because of the residual stresses in these products. Forged and hot rolled bars tend to warp when these stresses are removed by machining processes and, if a finish cut is taken first, it will warp out of true position and it cannot be corrected by more cutting.
10 Steel rod is drawn through a series of increasingly smaller dies. Occasionally this process must be interrupted to anneal the wire before drawing can continue.

Chapter 3 Identification and Selection of Iron and Steel

SELF-EVALUATION ANSWERS

1 Carbon and alloy steels are designated by the numerical SAE or AISI system.
2 The three basic types of stainless steels are: martensitic (hardenable) and ferritic (nonhardenable), both magnetic and of the 400 series, and austenitic (nonmagnetic and nonhardenable, except by work hardening) of the 300 series.
3 The identification for each piece would be as follows:
   a AISI C1020 (CF) is a soft, low carbon steel with a dull metallic luster surface finish. Use the observation test, spark test, and file test for hardness.
   b AISI B1140 (G and P) is a medium carbon, resulfurized, free machining steel with a shiny finish. Use the observation test, spark test, and machinability test.
   c AISI C4140 (G and P) is a chromium-molybdenum alloy, medium carbon content with a polished, shiny finish. Since an alloy steel would be harder than a similar carbon or low carbon content steel,

a hardness test should be used such as the file or scratch test to compare with known samples. The machinability test would be useful as a comparison test.

d AISI 8620 (HR) is a tough low carbon steel used for carburizing purposes. A hardness test and a machinability test will immediately show the difference from low carbon hot rolled steel.

e AISI B1140 (Ebony) is the same as the resulfurized steel in *b*, only the finish is different. The test would be the same as for *b*.

f AISI C1040 is a medium carbon steel. The spark test would be useful here as well as the hardness and machinability tests.

4 A magnetic test can quickly determine whether it is a ferrous metal or perhaps nickel. A chemical test can tell you if it is a stainless steel. If the metal is white in color, a spark test will be needed to determine whether it is a nickel casting or one of white cast iron, since they are similar in appearance. If a small piece can be broken off, the fracture will show whether it is white or gray cast iron. Gray cast iron will leave a black smudge on the finger. If it is cast steel, it will be more ductile than cast iron and a spark test should reveal a smaller carbon content.

5 01 refers to an alloy type of oil hardening (oil quench) tool steel. W1 refers to a water hardening (water quench) tool steel.

6 The 40 inch long, $2\frac{7}{16}$ inch diameter shaft weighs 1.322 pounds per inch. The cost is 30 cents per pound.
$1.322 \times 40 \times 0.30 = \$15.86$ cost of the shaft

7 **a** No
b Hardened tool steel or case hardened steel

8 Austenitic (having a face centered cubic unit cell in its lattice structure). Examples are chromium, nickel, stainless steel, and high manganese alloy steel.

9 Nickel is a nonferrous metal that has magnetic properties. Some alloy combinations of nonferrous metals make strong permanent magnets; for example, the well-known Alnico magnet, an alloy of aluminum, nickel, and cobalt.

10 Some properties of steel to be kept in mind when ordering or planning for a job would be:
Strength
Machinability
Hardenability
Weldability (if welding is involved)
Fatigue resistance
Corrosion resistance (especially if the piece is to be exposed to a corrosive atmosphere)

11 Chromium. Nickel.

12 A general rule for machining stainless steel is to use slow speeds and heavy feeds.

13 Use a smaller percentage of thread, such as a 50 percent thread instead of a 75 percent thread.

14 White or chilled cast iron.

15 There could be hard spots in the heat affected zone of the weld. The hard carbides that are formed as a result of the heat of welding and sudden cooling in the base metal.

16 The proper carbide tool selection should be made for the job and the correct feeds and speeds determined. A very rigid setup is necessary when hard materials are machined.

Chapter 4 Identification of Nonferrous Metals

SELF-EVALUATION ANSWERS

1 Advantages: Since aluminum is about one-third lighter than steel, it is used extensively in aircraft. It also forms an oxide on the surface that resists further corrosion.
Disadvantages: The initial cost is much greater. Higher strength aluminum alloys cannot be welded.

2 The letter "H" following the four digit number always designates strain on work hardening. The letter "T" refers to heat treatment.

3 Magnesium weighs approximately one-third less than aluminum and is approximately one-quarter the weight of steel. Magnesium will burn in air when finely divided.
4 Copper is most extensively used in the electrical industries because of its low resistance to the passage of current when it is unalloyed with other metals. Copper can be strain hardened or work hardened and certain alloys may be hardened by a solution heat treatment and aging process.
5 Bronze is basically copper and tin. Brass is basically copper and zinc.
6 Nickel is used to electroplate surfaces of metals for corrosion resistance, and as an alloying element with steels and nonferrous metals.
7 All three resist deterioration from corrosion.
8 Alloy.
9 Tin, lead, and cadmium.
10 Die cast metals, sometimes called "pot metal."
11 Wrought aluminum is stronger.
12 Large rake angles (12 to 20 degrees back rake). Use of a lubricant. Proper cutting speeds.
13 No. If a fire should start in the chips, the water based coolant will intensify the burning.
14 The rake should be zero on all cutting tools for brasses or bronzes.
15 Tungsten is combined with carbon to form a tungsten carbide powder that is compressed into briquettes and sintered in a furnace. The resultant tungsten carbide cutting tool is extremely hard and will resist the high temperatures of machining.

Chapter 5 Using Rockwell and Brinell Hardness Testers

SELF-EVALUATION ANSWERS

1 Resistance to penetration is the one category that is utilized by the Rockwell and Brinell testers. The depth of penetration is measured when the major load is removed on the Rockwell tester and the diameter of the impression is measured to determine a Brinell hardness.
2 As the hardness of a metal increases, the strength increases.
3 The "A" scale and a brale marked "A" with a major load of 60 kgf should be used to test a tungsten carbide block.
4 It would become deformed or flattened and give an incorrect reading.
5 No. The brale used with the Rockwell superficial tester is always marked or prefixed with the letter "N."
6 False. The ball penetrator is the same for all the scales that use the same diameter ball.
7 The diamond spot anvil is used for superficial testing on the Rockwell tester. When used, it does not become indented, as is the case when using the spot anvil.
8 Roughness will give less accurate results than would a smooth surface.
9 The surface "skin" would be softer than the interior of the decarburized part.
10 A curved surface will give inaccurate readings
11 A 3000 kg weight is used to test steel specimens on the Brinell tester.
12 A 10 mm steel ball is usually used on the Brinell tester.

Chapter 6 The Mechanical and Physical Properties of Metals

SELF-EVALUATION ANSWERS

1 Creep occurs within the elastic range of metals.
2 Creep failures occur over a period of time; the rate of creep increases as the temperature increases.

3 A metal that is ductile above its transition temperature behaves as a brittle metal below the transition temperature and loses much of its toughness.
4 Hardness, strength, and modulus of elasticity increase with a decrease in temperature.
5 Nickel will lower the transition temperature when alloyed with steel.
6 Resistance to penetration may be measured by the Rockwell or Brinell testers; elastic hardness may be measured with a scleroscpe and abrasive resistance may be tested in the shop to some degree with a file.
7 Tensile, compressive, and shear.
8 Unit stress $= \frac{\text{Load}}{\text{Area}} = \frac{40{,}000}{4} = 10{,}000$ psi
9 Ductility is the ability of a metal to deform permanently under a tensile load.
10 Malleability is the ability of a metal to permanently deform under a compressive load.
11 Fatigue strength can be improved by eliminating sharp undercuts, deep tool marks, and other forms of stress concentration.
12 Electrical and thermal conductivity are related. A metal that is a good conductor of electricity is also a good conductor of heat.
13 A pure metal (unalloyed) will conduct best.
14 The rate of thermal expansion is expressed in inches (of dimensional change) per inch (of material length or diameter) per degree F. The value for each material is called the coefficient of thermal expansion.
15 A machinist works with dimensional tolerances of a few thousandths or ten-thousandths of an inch and the temperature rise caused by machining friction often exceeds these tolerances. Warping of parts and surface cracking when grinding can also cause damage and scrapping of parts.

Chapter 7 The Crystalline Structure of Metals

SELF-EVALUATION ANSWERS

1 Atoms are composed of a nucleus consisting of positively charged protons and neutrons. Electrons are negative in charge and revolve in paths called shells. The outer shell or valence electrons are important in determining chemical and physical properties.
2 The electron cloud of free valence electrons creates a mutal attraction for the metal atoms. This free movement accounts for their high electrical and thermal conductivity, plasticity, and elasticity.
3 **a** Body-centered cubic (BCC)
b Face-centered cubic (FCC)
c Close-packed hexagonal (CPH)
d Cubic
e Body-centered tetragonal
f Rhombohedral
4 Dendrite.
5 No. Each grain grows from its own nucleus in independent orientation; this means that the lattice structure of adjacent grains jam together in a misfit pattern.
6 Austenitized carbon steel is a solid solution of carbon in the interstices of the iron. When the steel is quenched, the FCC attempts to transform into BCC, but BCC can contain almost no carbon in its interstices. The result is a hard, elongated body-centered tetragonal cubic structure.
7 A material that changes its crystal structure is called allotropic. Iron changes from body-centered cubic to face-centered cubic as the temperature rises. It is therefore an allotropic element.
8 Fine grained steels are preferred over coarse grain for almost every application. Coarse grain has increased hardenability; that is, it will carburize faster than fine grain steels.
9 Deformation takes place when slip occurs along slip lines, which are rows of atoms shifting to fill vacancies and other distortions of the lattice.
10 A small piece is removed with a metallurgical abrasive saw using a coolant so that the piece won't be overheated. Since the metal piece is too small to hold in the hand to polish, it must be encapsulated in

plastic. The specimen is ground on successively finer grits of abrasive paper after which it is polished on a rotating table with a polishing compound.

Chapter 8 Phase Diagrams and the Iron-Carbon Diagram

SELF-EVALUATION ANSWERS

1 Check Figure 4 in this chapter for the correct answers.
2 It is a short horizontal section on the line.
3 Lowest transition temperature. Eutectic is the lowest melting temperature. Eutectoid is the lowest transition temperature from one solid phase to another.
4 A_3 shows the beginning of transition from austenite to ferrite. A_1 shows the completion of austenite transition to ferrite and pearlite. A_{cm} shows the limit of carbon solubility in austenite.
5 Cementite. It appears dark when in the form of lamellar pearlite, but white when massive such as in grain boundaries or in white cast iron.
6 It is FCC. Iron with a high solubility for carbon. Austenite normally exists at temperatures between 2700°F (1482°C) and 1330°F (721°C) depending on carbon content.
7 Ferrite. It appears light under the microscope. It dissolves very little carbon, about 0.008 percent at room temperature and 0.025 percent at 1330°F (721°C).
8 Pearlite. It looks like a fingerprint. It is composed of alternating layers of ferrite and cementite.
9 Ferrite. 1330°F (721°C).
10 The A_1 and $A_{3,1}$. Ferrite forms austenite at the A_3 line.
11 No.
12 It moves it to the left or decreases the carbon content of the eutectoid.
13 Cast iron can be identified by the forms in which graphite is found (flakes and ball); white cast iron is composed of pearlite and massive white cementite.
14 Nickel or manganese.
15 White cast iron may be produced by lowering the silicon content or by chilling the surface of the casting.

Chapter 9 Hardening and Tempering of Carbon Steel

SELF-EVALUATION ANSWERS

1 No hardening would result as 1200°F (649°C) is less than the lower critical point and no dissolving of carbon has taken place.
2 There would be almost no change. For all practical purposes in the shop these low carbon steels are not considered hardenable.
3 They are shallow hardening, and liable to distortion and quench cracking because of the severity of the water quench.
4 Air and oil hardening steels are not so subject to distortion and cracking as W1 steels and are deep hardening.
5 1450°F (788°C). 50°F (10°C) above the upper critical limit.
6 Tempering is done to remove the internal stresses in hard martensite, which is very brittle. The temperature used gives the best compromise between hardness and toughness or ductility.

7 Tempering temperature should be specified according to the hardness, strength, and ductility desired. Mechanical properties charts give this data.
8 525°F (274°C). Purple.
9 600°F (315°C). It would be too soft for any cutting tool.
10 Immediately. If you let it set for any length of time, it may crack from internal stresses.

Chapter 10 Annealing, Normalizing, and Stress Relieving

SELF-EVALUATION ANSWERS

1 Medium carbon steels that are not uniform, have hardened areas from welding, or prior heat treating need to be normalized before they can be machined. Forgings, castings, and tool steel in the as-rolled condition are normalized before any further heat treatments or machining is done.
2 1550°F (843°C). 50°F (10°C) above the upper critical limit.
3 The spheroidization temperature is quite close to the lower critical temperature line, about 1300°F (704°C).
4 The full anneal brings carbon steel to its softest condition as all the grains are reformed (recrystallized), and any hard carbide structures become soft pearlite as it slowly cools. Stress relieving will only recrystallize distorted ferrite grains and not the hard carbide structures or pearlite grains.
5 Stress relieving should be used on severely cold worked steels or for weldments.
6 High carbon steels (0.8 to 1.7 percent carbon).
7 Process annealing is used by the sheet and wire industry and is essentially the same as stress relieving.
8 In still air.
9 Very slowly. Packed in insulating material or cooled in a furnace.
10 Since low carbon steels tend to become gummy when spheroidized, the machinability is poorer than in the as-rolled condition. Spheroidization sometimes is desirable when stress relieving weldments on low carbon steels.

Chapter 11 I-T Diagrams and Cooling Curves

SELF-EVALUATION ANSWERS

1 About 50°F (10°C) above the A_3 or $A_{3,1}$ lines.
2 Martensite is produced by rapid quenching from the austenitizing temperature to the Mf or near room temperature. Time is a major consideration.
3 The Ms temperature is the point at which martensite begins to form, and the Mf is the point where it is at 100 percent transformation.
4 The critical cooling rate is the time necessary to undercool austenite below the M temperature to avoid any transformation occurring at the nose of the S-curve.
5 The microstructure should be partly fine pearlite and partly martensite.
6 An increase in carbon content moves the nose of the S-curve to the right.
7 Hardening and tempering.
8 This could be caused by the difference in internal and external cooling rates. Changing to an oil or air hardening steel may correct the problem.
9 It can be observed that a cooling curve will always cut through the nose of the S-curve no matter how fast the part is quenched. It is therefore evident that little or no martensite may be produced.
10 A steel that is hardenable and is deep hardening shows an S-curve that is moved to the right.

Chapter 12 Hardenability of Steels and Tempered Martensite

SELF-EVALUATION ANSWERS

1 The jominy end-quench hardenability test.
2 A 1 inch diameter specimen about 4 inches long is heated to the quenching temperature and placed in a jet of water so that only one end is cooled without getting the sides wet. When the whole specimen has cooled, flat surfaces are ground on the sides and Rockwell C scale readings are taken at $\frac{1}{16}$ inch intervals.
3 The rate of cooling and hence the microstructure as shown on the I-T diagram are related to the depth of hardening as shown by the jominy end-quench graph.
4 Circulation of the cooling medium in all cases increases depth of hardening.
5 Rc67 is about as hard as any carbon steel will get. Steels with less than 0.83 percent carbon will not get this hard.
6 Coarse pearlite.
7 Austempering is isothermal quenching in the lower bainite region of the S-curve. Since no martensite is formed, a tougher, more ductile microstructure is the result; it is superior to a quenched and tempered product of the same hardness.
8 The best time to temper is immediately after quenching, as soon as the piece is cool enough to be handheld.
9 The blue brittle tempering range is found between 400°F (204°C) and 800°F (427°C). Only some alloy steels show loss of notch toughness at higher tempering ranges. This is called temper brittleness. It can be avoided by quenching from the tempering temperature.
10 You can predict the hardness that a tempered part will be by consulting a mechanical properties chart for that particular grade of steel.

Chapter 13 Heat Treating Equipment and Procedures

SELF-EVALUATION ANSWERS

1 Electric, gas, oil fired, and pot furnaces.
2 The surface decarburizes or loses surface carbon to the atmosphere as it combines with oxygen to form carbon dioxide. An oxide scale forms on the surface.
3 Dispersion of carbon atoms in the solid solution of austenite may be incomplete and little or no hardening in the quench takes place as a result. Also, the center of a thick section takes more time to come to the austenitizing temperature than a thin section.
4 Circulation or agitation breaks down the vapor barrier. This action allows the quench to proceed at a more rapid rate and it avoids spotty hardening.
5 By furnace.
6 They run from the surface toward the center of the piece. The fractured surfaces usually appear blackened. The surfaces have a fine crystalline structure.
7 **a** Overheating.
b Wrong quench.
c Wrong selection of steel.
d Poor design.
e Time delays between quench and tempering.
f Wrong angle into the quench.
8 **a** Controlled atmosphere furnace.

b Wrapping the piece in stainless steel foil.
c Covering with cast iron chips.
9 a Tungsten high speed.
b Molybdenum high speed.
10 An air hardening tool steel should be used when distortion must be kept to a minimum.
11 Since the low carbon steel core does not harden when quenched from 1650°F (899°C), it remains soft and tough, but the case becomes very hard. No tempering is therefore required as the piece is not brittle all the way through as a fully hardened carbon steel piece would be.
12 A deep case can be made by pack carburizing or by a liquid bath carburizing. A relatively deep case is often applied by nitriding or by similar procedures.
13 No. The base material must contain sufficient carbon to harden by itself without adding more for surface hardening.
14 Three methods of introducing carbon into heated steel are roll, pack, and liquid carburizing.
15 Nitriding.

Chapter 14 Heat Treating of Nonferrous Metals

SELF-EVALUATION ANSWERS

1 A 5
B 2
C 4
D 3
E 1
2 The solid solution becomes supersaturated and the copper forms into highly dispersed globules of copper aluminide.
3 Solution heat treatment and quench and precipitation heat treatment or aging.
4 Aging causes the copper aluminide particles to act as keys to lock up the slip planes. This lowers the ductility and raises the strength.
5 Pure aluminum is more corrosion resistant than alloyed aluminum; for this reason, a thin layer of pure aluminum is clad to the sheet of alloy.
6 The aluminum is simply heated to about 650°F (343°C) and allowed to cool in air.
7 By the amount of cold working done prior to annealing. A highly strained metal will produce the finest grain size.
8 Beryllium copper, aluminum bronze, copper-nickel-silicon, copper-nickel-phosphorus, chromium copper, and zirconium copper.
9 A muffle (carbonizing atmosphere) furnace or bright annealing can be used to avoid oxide scale.
10 Alloying elements tend to cause titanium to become CPH (alpha) or BCC (beta). Some alloys are alpha-beta and they are highly heat treatable.

Chapter 15 The Effects of Machining on Metals

SELF-EVALUATION ANSWERS

1 The temperature of the chip and the kind of cutting, relative speeds for the tool material.
2 At lower speeds, negative rake tools usually produce a poorer surface finish than positive rakes.
3 It is called a built-up edge and it causes a rough, ragged cutting action that produces a poor surface finish.

4 Thin uniform chips indicate the least surface disruption.
5 No. It slides ahead of the tool on a shear plane, elongating and altering the grain structure of the metal.
6 Negative rake.
7 Higher cutting speeds produce better surface finishes. There is less disturbance and disruption of the grain structure at higher speeds, also.
8 Surface irregularities such as scratchy tool marks, microcracks, and poor radii can shorten the working life of the part considerably by causing stress concentration that can develop into a metal fatigue failure.
9 The property of hardness is related to machinabilty. Machinists sometimes use a file to determine hardness.
10 Sharp shoulders create stress concentration that leads to fatigue failures. A smooth fillet radius should be used whenever possible to avoid stress raisers.

Chapter 16 Metallurgy of Welds: Carbon Steel

SELF-EVALUATION ANSWERS

1 The weld zone bears a similarity to the ingot as it solidifies into columnar structures. It also compares to a small electric melting furnace. Peening of welds is like hot and cold rolling of metals.
2 Solid phase welding, fusion joining, and liquid-solid phase welding.
3 Diffusion of some of the carbon into the weld melt will take place causing hard carbides to form when cooling rates are rapid.
4 The weld zone, fusion zone, adjacent zones, and the heat affected zone.
5 In the weld zone.
6 Oxy-acetylene welding.
7 Martensite or hard carbides can form with high cooling rates in welds.
8 The heating action of subsequent passes tends to normalize and soften the hard zones in the previous pass.
9 Cracking, porosity, and embrittlement (from entrapped gases or formation of martensite).
10 By limiting or inhibiting the formation of martensite.
11 By removing them through tempering, spheroidizing, or annealing.
12 The tough, hard, distorted grains in cold rolled steel are recrystallized in the heat affected zone to a softer, larger grain that is not as strong as it was in the original stressed condition.
13 Large masses cause more rapid cooling rates.
14 To protect the weld zone from contaminators of the atmosphere (oxygen and nitrogen), both of which weaken the weld, causing porosity and hard zones.
15 The weld.

Chapter 17 Metallurgy of Welds: Alloy Steel

SELF-EVALUATION ANSWERS

1 The hardenability of steels, and consequently of welds, and of heat affected zones is increased with the addition of alloying elements.
2 High postheat (tempering) temperatures tend to promote decomposition of the retained austenite into fine carbides, thus increasing the hardness and brittleness of the weld or heat affected zone.
3 Complex carbides tend to form, causing brittleness and weakening of the steel in the grain boundaries of the large austenite grains.

4 Vanadium.
5 The brittleness in the HAZ of martensitic stainless steel is caused by the formation of carbides when rapid cooling takes place, but excessive grain growth in the HAZ of the ferritic stainless steel is the cause of brittleness.
6 a Preheat and postheat are usually needed since the HAZ of tool steel welds tends to become brittle.
  b The correct filler metal that is similar to the base metal in carbon and alloy content should be used.
  c Full annealing is necessary after welding if any machining is to be done. This is then followed by hardening and tempering procedures.
7 Standard ASTM specifications.
8 Hot cracks are usually located lengthwise down the center of a weld or in the crater at the end of a weld. Hot cracking can be caused by a too rapid cooling rate or by stresses in the weld zone at high temperatures.
9 Carbon can be picked up in the root pass from oil or grease causing embrittlement and cracking. All of these materials can cause hydrogen embrittlement in welds.
10 The overlay tends to spall (break off in chunks) when this procedure is followed.

Chapter 18 Metallurgy of Welds: Cast Iron

SELF-EVALUATION ANSWERS

1 No. All except white cast iron.
2 Repair of casting defects in foundries and of worn or broken castings.
3 They contain too much carbon.
4 Almost all the welding processes can be used.
5 Gray cast iron has a lower tensile strength than the weld metal (unless the weld is also cast iron) and, when the weld metal contracts while cooling, very high stresses are created in the base metal.
6 The major advantage is that preheat is not used where large castings are involved. The major disadvantage is the weakness of such welds due to microcracking near the junction zone of the weld.
7 Preheating and postheating can eliminate the hard fusion zone. It can be avoided by removing sufficient base metal prior to welding so that machining is done only on the weld itself.
8 The graphite in gray cast iron can be smeared over the surface to be welded by grinding procedures. Weld grooves should be prepared by chipping or sawing.
9 By studding procedures.
10 Gas welding with a cast iron rod raises the malleable iron above the transformation temperature and it reverts back to white cast iron.

Chapter 19 Metallurgy of Welds: Nonferrous

SELF-EVALUATION ANSWERS

1 It is their great affinity for oxygen that causes the rapid formation of oxides.
2 Because of their high thermal conductivity.
3 Aluminum should be cleaned prior to welding for two reasons: to remove the oxide film and to remove contaminants.
4 Recrystallization and grain growth is increased by strain hardening. These larger, more ductile grains cause the heat affected zone to have a lower tensile strength than the rest of the base metal.
5 The alloying elements (such as copper in 2024) tend to form compounds and precipitate to the grain boundaries due to the heat of welding. This, of course, produces an exceedingly weak, brittle weld that

can be restored to the original strength only by postweld solution heat treatment and by aging the entire weldment.

6 Oxygen in copper welds tends to form oxides that migrate to the grain boundaries, thus severely weakening the metal.

7 Quenching is not necessary in these alloys. All that is needed is a simple annealing procedure from 1100 to 1600°F (593 to 871°C). This step will harden the weldment uniformly and provide a higher corrosion resistance.

8 Nickel alloy electrodes such as monel are used; in addition, austenitic stainless steel, which contains high percentages of nickel, is used for joining dissimilar metals.

9 These metals react violently with oxygen and readily burn to form oxides. They can either be welded in a vacuum chamber or by the TIG process using an inert gas.

10 By keeping the weld area very clean, avoiding oils and other contaminants; even fingerprints can cause embrittlement.

Chapter 20 Powder Metallurgy

SELF-EVALUATION ANSWERS

1 The automotive industry.

2 Porous products such as filters and self-lubricating bearings, friction materials for brake linings, cutting tools, creep resistant alloys for jet engines, gears, cams, magnets, and metals that are difficult to process such as tungsten.

3 a The powder is pressed into a green compact or briquette shape.
 b The briquette is placed in an oven and sintered at a temperature just under its melting point.
 c When more precision is required, a resizing operation in a press is carried out.

4 The green compact is "cold welded" by pressure alone, but the bond is very weak. The sintering process forms metallurgical bonds similar to that of fusion welding.

5 Some advantages: small parts can be economically produced by P/M with no scrap loss. Porous materials and metal-nonmetal combinations are possible with this method. All of the normal machining, plating, and heat treating processes can be used on P/M metal.
Some disadvantages: P/M parts have low plasticity and lowered impact strength and ductility. They also have a lower resistance to corrosion than conventionally formed metals.

Chapter 21 Nondestructive Testing

SELF-EVALUATION ANSWERS

1 Nondestructive testing is used to inspect for cracks or other defects in various metal and nonmetal parts without damaging the part for its intended use.

2 Magnetic-particle inspection can only be used for ferromagnetic metals such as iron and steel.

3 The specimen must be magnetized lengthwise in order to find a crosswise crack. A circular magnetic field is needed to find a lengthwise crack.

4 Fluorescent penetrant systems make use of capillary action to draw the penetrating solution into the flaws. The surplus penetrant is removed and a developer is applied to enhance the outline of the defect. The inspection is carried out under black light, which causes the penetrant to fluoresce. This method may be used on almost any dense material.

5 Dye penetrants may be used in remote areas where no power is available for a black light source. Portability, low cost, and ease of application are also advantages.

6 The pulse-echo system and the through transmission system. The pulse-echo system uses one trans-

ducer that both sends and receives the pulse. The pulse is reflected and echoed back from the other side or from a flaw. The through transmission system uses a transducer on each side of the test piece. The signal passes through the material and is modified by any flaws present.

7 Internal defects, cracks, porosity, laminations, thicknesses, and weld bonds can be detected with ultrasonic testing systems.

8 X rays have the ability to pass through solids. A solid steel casting, for example, will have an X ray source on one side of the area to be inspected and a photographic plate on the other. A void or crack will appear darker on the negative.

9 Gamma radiation is omnidirectional — that is, it goes in all directions. Therefore, a hollow object may be inspected on its entire surface with one exposure. Radium or another radioactive substance is used to produce gamma rays, and extreme care must be taken with its use.

10 Eddy current techniques are used for sorting materials of various alloy compositions and with differing heat treatments. It can be used for detecting seams or variations in thicknesses, mass, and shape. It is limited to testing only materials that conduct electricity.

Chapter 22 Service Problems

SELF-EVALUATION WORKSHEET ANSWERS

1 (Figure 31) This fatigue failure is an example of a two-way bending load in which fatigue cracks began on opposite sides and worked toward a sudden fracture along a straight line in the center. The shaft could have been used in an application where it was fastened securely to prevent it from turning while a pulley or sprocket having an intermittent one-direction loading from the belt or chain turned on it. A correction of the problem could be case hardening or the use of a fatigue resistant material.

2 (Figure 32) Star pattern break, typical of torsional failures at the end of splines. This part was a drive line from a truck. No correction. Cause of failure was overstress beyond its margin of safety.

3 (Figure 33) Somewhat unusual torsional fatigue cracking parallel to the shaft axis. Here the fatigue crack took the form of a spiral and did not terminate in a brittle fracture as often happens in transverse fatigue cracking. This was probably a resulfurized, fibrous type of shaft; an alloy shaft may correct the problem.

4 (Figure 34) Longitudinal shear causing failure from splitting along fibrous type of steel. Again use of an alloy type shaft could correct the problem. Fibrous-type shafting should not be used where high torque is involved.

5 (Figure 35) Star-type pattern showing cracking from the inside corners of the square spline to the center of the shaft, which ultimately failed transversely. Perhaps an involute (more rounded roots) spline instead of a square spline would correct this problem.

6 (Figure 36) One-way bending load, single-sided concave with high overstress, but no stress concentration (no crack or groove to start failure). This bar was suspended in a cantilever fashion with intermittent loading hanging on one end. The load was higher than the fatigue strength of the material. The solution would be to make the part larger to increase the fatigue strength or change the design to distribute stresses more evenly.

7 (Figure 37) Failure of bearing insert by overheating and by the tearing away of bearing material. Cause was loss of lubrication. Solution is to always maintain surveillance on lubricating systems for machinery.

8 (Figure 38) Wear. The hypoid pinion gear from a truck on the right shows considerable wear, but not to failure as happened to the pinion at the left. The kind of wear shown on the right is largely unavoidable where extremely heavy loading must take place as in truck differentials. The complete failure as on the left could have been caused by loss of lubricants. The solution is the replacement of parts.

9 (Figure 39) A concentric fatigue failure having low overstress and high stress concentration. This failure occurred at the base of a sharp shoulder with no radius; this caused a high concentration of

stresses and began cracking around the circumference of the shaft. The small brittle section where final sudden failure occurred indicates a low overstress; that is, the loading of the shaft was comparatively light for its strength. The load, however, must have been cyclic (on and off) to cause this failure. A solution would be to use a resulfurized type of replacement shaft that resists transverse fatigue.

10 (Figure 40) Wear in a vane-type hydraulic pump cam ring. The cam ring should be very smooth, and wear to this extent would cause pump failure and excessive noise. Solution is the replacement of the part. This kind of wear can have a number of causes such as dirty oil or cavitation (the pump not getting sufficient oil because of dirty strainers).

This chapter has no Posttest.

Chapter 23 Corrosion in Metals

SELF-EVALUATION ANSWERS

1 Corrosion may be classified as direct oxidation and galvanic action.
2 An electrochemical process in which the metal is changed to the ionic state in an electrolyte.
3 Metals usually become positive ions while nonmetals become negative ions.
4 For galvanic corrosion, there must be an anode and a cathode electrically connected, and an electrolyte. The anode and cathode may be similar or dissimilar metals and an electric current must flow.
5 In more neutral electrolytes, an accumulation of hydrogen gas is formed on the cathode surface that prevents or slows down corrosion.
6 Oxygen reacts with the layer of hydrogen on the cathode to form water, accepting negative charges of electrons that allow the release of metal ions from the anode and the resultant corrosion.
7 Iron.
8 The current density is greater with the large cathode, small anode couple because of the relatively larger area for depolarization available to any oxygen atoms.
9 Metals are least likely to corrode when they are in their pure state or have few contaminants.
10 Any thing that can prevent the normally present oxygen from contacting the metal, will set up an anode area surrounded by the high oxygen concentration, which is the cathode. The anode becomes activated (stainless steel is normally passivated and cathodic) and is corroded in time.

Chapter 24 Casting Processes

SELF-EVALUATION ANSWERS

1 The two types of sand casting are green sand molding and dry sand molding. Moist sand with a small amount of clay and other additives make up the green sand mold. Dry molding sand, often used for making cores, has no clay, but contains linseed oil or organic resins.
2 Wood, metal, and wax are common materials used for making patterns. Patterns for sand molding are made of wood or metal and must be tapered so they can be easily removed from the mold. They also must have shrinkage allowance. Patterns may either be simple cope and drag or match plate types.
3 Core sand is rammed into a "core box" or mold and the green cores are baked in an oven to harden them. They are set into the pattern and are locked into the sand mold by the core print. The core remains in the mold when the pattern is removed.
4 A "chill" is usually a metal section embedded in the mold to increase or control the cooling rate of the casting.
5 The steps in the shell molding process are:
   a Heated metal patterns are brought into contact with a prepared sand mix in a dump box.

b The adhering sand and pattern are placed in an oven and heated to about 500°F (315.5°C) for two minutes.

c The shell is removed and clamped to its mating half to form a complete mold.

The sand is mixed with a phenolic resin binder and a silicon release agent is used on the pattern.

6 Molten metal is thrown out to the mold cavity by centrifugal forces when casting pipe, wheels, and other products by this process. The castings are more uniform and of a higher quality than that of gravity casting. Also, two different metals may be cast in layers. Centrifuging is the filling of a mold or molds located near the outside of a rotating body.

7 No. The pattern cannot be removed by separating mold halves as with sand casting since it is a one piece mold.

The steps necessary to produce an investment casting are:

a The wax pattern is made, including the sprue and riser.

b Plaster is cast around the wax or a refractory slurry is used. The pattern is then dipped into the slurry and dried repeatedly until a thick shell has formed on the pattern.

c The mold is heated to drive out the wax and the shell is fired to harden it.

d The molten metal is poured and, when it has solidified, the mold shell is broken off the casting.

8 Higher production rates and higher precision are possible with permanent molds as compared to sand casting. One disadvantage of permanent molding is that it is suitable for only limited production runs as the mold deteriorates after a few thousand castings.

9 The molten metal is not poured into the mold as in permanent molding, but is injected under very high pressures into the mold cavity.

10 The cold-chamber machine has a separate melting and holding furnace. The hot-chamber machine is characterized by the plunger and gooseneck in the melting pot located on the machine. Two advantages of die casting are:

a High production rates are made possible by this process.

b High finish and precision are obtained in the finished product.

Chapter 25 Plastics, Rubbers, Industrial Gases, and Oils

SELF-EVALUATION ANSWERS

1 High temperatures and ultraviolet light from sunlight.

2 Thermosetting and thermoplastic.

3 Thermoplastic.

4 Latex from plants such as the para-rubber tree is the source of natural rubber.

5 Petrochemicals and acetylene.

6 Neoprene.

7 Ethyl alcohol, tetraethyl lead, ethylene glycol, styrene, polystyrene, polyethylene, ethyl benzine, and nylon.

8 Plastics and other nonmetals are poor conductors of heat, and the temperature rise at the cutting surface overheats the tool causing failure.

9 The materials used were hides and hooves of animals, casein, starch, and tree gums.

10 Bonded structures resist vibration damage better, and they are less likely to have corrosion.

11 It should be thoroughly cleaned by sand blasting, etching, or with a degreaser and all moisture removed.

12 The bond in these adhesives is catalyzed by the contacting surfaces; the adhesive is anaerobic and will not cure in the presence of air or if it is over 0.002 inch thick.

13 The polysulfides, polyurethanes, and silicones are used as sealants.

14 Hydrogen, acetylene, natural gas, propane,and butane are fuel gases, and nitrogen, helium, and argon are inert gases.

15 Boundary layer or adhesive film lubrication such as the hypoid oils and greases.

This chapter has no Posttest.

Chapter 26 Concrete, Ceramics, and Related Materials

SELF-EVALUATION ANSWERS

1 They usually have ionic or covalent bonds, which are rigid, and they have few slip planes to absorb local stresses, which may exceed the bond strength and cause a brittle failure.
2 Ceramic materials have high compression strength but low tensile strength.
3 A refractory material such as clay brick can withstand high temperatures without breaking down or melting. Refractories are used to contain molten metals and for furnace linings.
4 Silica and alumina.
5 False.
6 False.
7 True.
8 The beams are prestressed with steel wires, which impart a higher tensile strength to the concrete.
9 Glass is made by combining molten silica (quartz) and lime or sodium. This combination solidifies into the noncrystalline (amorphous) structure of glass.
10 A composite is a material that is combined with another different material to reinforce or strengthen the resultant mixture.

This chapter has no Posttest.

Chapter 27 Wood and Paper Products

SELF-EVALUATION ANSWERS

1 The softwoods (coniferous) and the hardwoods (deciduous) are the two classes.
2 Pines, firs, spruces, and hemlock are coniferous trees.
3 Growth takes place in the cambium layer.
4 The age of a tree can be determined by counting its annular growth rings.
5 The log is first sawed into rough slabs in a large circular or band saw called a "head rig." The slabs are then sawed into lumber, which is graded and then dried. The dry boards may then be planed to provide a finished surface.
6 Sawdust and wood chips are by-products that are used for making hardboard, particle board, and fireplace logs.
7 The grains are all in one direction in wood, making it prone to splitting. Wood is simply weak in one axis and strong in the other. Plywood utilizes this strength by crossing each layer or ply alternately and gluing them together.
8 Wood is not a uniform material, having such defects as knots, cracks, pitch pockets, and rot. Boards must be selected according to their intended use.
9 They are made of sawdust and wood chips. They are used for panels in construction, furniture making, and pegboard.
10 A pulp is made by grinding the fibers and soaking them in a vat. The pulp is formed into a mat that is pressed, dried, and squeezed to form a hard, dense sheet.

This chapter has no Posttest.

Glossary

Acetylene (C_2H_2). A colorless gas having a characteristic odor that is very soluble in acetone. It is used with oxygen as a welding gas to produce a very high temperature flame.
Acicular. Needle-shaped particles or structures as found in martensite.
Acids, bases, and salts. Acids are electrolytes that furnish hydrogen ions, which allow any acid to react with water. Bases are electrolytes that furnish hydroxyl ions. Salts are electrolytes that furnish neither hydrogen nor hydroxyl ions. Salts are formed by a process called neutralization, the combination of equivalent weights of an acid and a base.
Adhesives. Materials or compositions that enable two surfaces to join together. An adhesive is not necessarily a glue, which is considered to be a sticky substance, since many adhesives are not sticky.
Admixture. In welding, the combining of a base metal with a filler metal to alter the characteristics of either.
Aggregate. Small particles as powders used for powder metallurgy which are loosely combined to form a whole; also sand and rock as used in concrete.
Aging. The process of holding metals at room temperature or at a predetermined temperature for the purpose of increasing their hardness and strength by precipitation; aging is also used to increase dimensional stability in metals such as castings.
AISI. Abbreviation for *American Iron & Steel Institute*.
Allotropy. The ability of a material to exist in several crystalline forms.
Alloy. A substance that has metallic properties and is composed of two or more chemical elements of which at least one is a metal.
Amorphous. Noncrystalline, a random orientation of the atomic structure.
Anistropy. A material having different physical properties in different directions. Rolled steel is strongest in the direction of rolling.
Anneal. To heat and then to cool (as steel) usually for softening and making the metal less brittle; it also refers to treatments intended to alter the mechanical or physical properties to produce a definite microstructure.
Anodizing. To subject a metal to electrolytic action as the anode of a cell in order to coat it with a protective or decorative film; used for nonferrous metals.
ASTM. Abbreviation for *American Society for Testing Materials*.
Austempering. A heat treating process consisting of quenching a ferrous alloy at a temperature above the transformation range in a medium such as molten lead; the temperature of the quenching medium is maintained below that of pearlite and above that of martensite formation to produce a tough, hard microstructure.
Austenite. A solid solution of iron and carbon and sometimes other elements in which gamma iron characterized by a face-centered cubic crystal structure is the solvent.
Austenitizing. The process of forming austenite by heating a ferrous alloy above the transformation range.
AWS. Abbreviation for *American Welding Society*.
Bainite. A structure in steel named after E. C. Bain that forms between 900°F (482°C) and the Ms temperature. At the higher temperatures, it is known as upper or feathery bainite. At the lower temperatures, it is known as lower or acicular bainite and resembles martensite.
Base metal. Also called parent metal, which is one of the parts to be welded together.
Bonding. Chemical bonds of metals or other compounds are related to the valence of the atoms. Bonding in welding is the fusion of two metals; in adhesives, bonds are either mechanical or chemical.
Brittleness. The property of materials that will not deform under load, but tend to break suddenly; for example, cast iron and glass are brittle. Brittleness is the property opposite to plasticity.
Carburizing. A process that introduces carbon into a heated solid ferrous alloy by having it in contact with a carbonaceous material. The metal is held at a temperature above the transformation range for a period of time. This is generally followed by quenching to produce a hardened case.
Case hardening. A process in which a ferrous alloy is hardened so that the surface layer or case is made considerably harder than the interior or core. Some case hardening processes are carburizing and quenching, cyaniding, carbonitriding, nitriding, flame hardening, and induction hardening.
Casting. A process of producing a metal object by pouring molten metal into a mold.
Cast iron. Iron containing 2 to 4½ percent carbon, silicon, and other trace elements. It is used for casting objects into molds. Cast iron is somewhat brittle.
Catalyst. An agent that induces catalysis, which is a phenomenon observed in chemical reactions in which the reaction between two or more substances is influenced by the presence of a third substance (the catalyst) that remains unchanged throughout the reaction.
Caustic. A chemical base material; caustic soda, for

example, is a solution of sodium hydroxide used to clean metal surfaces or as an enchant in microscopy.
Cellulose. A polysaccharide of glucose units that constitutes the chief part of the cell walls of plants. For instance, cotton fibers are over 90 percent cellulose, and is the raw material of many manufactured goods such as paper, rayon, and cellophane. In many plant cells, the cellulose wall is strengthened by the addition of lignin, forming lignocellulose.
Cementite. Also known as iron carbide, a compound of iron and carbon (Fe_3C).
Cold drawing. Reducing the cross section of a metal bar or rod by drawing it through a die at a temperature below the recrystallization range, usually room temperature.
Cold rolling. Reducing the cross section of a metal bar in a rolling mill below the recrystallization temperature, usually room temperature.
Cold working. Deforming a metal plastically at a temperature below its lowest recrystallization temperature. Strain hardening occurs as a result of this permanent deformation.
Compressive strength (ultimate). The maximum stress that can be applied to a brittle material in compression without fracture.
Compressive strength (yield). The maximum stress that can be applied to a metal in compression without permanent deformation.
Corrosion. A gradual electrochemical attack on a metal caused by galvanic action in an electrolyte such as moisture.
Corrosion embrittlement. The embrittlement caused in certain alloys that are susceptible to intergranular corrosion attack when they are exposed to a corrosive environment.
Corrosion resistance. The ability of some metals to form an oxide surface skin that resists further corrosion. Many nonferrous metals and stainless steel have this capability.
Creep. Slow plastic deformation in steel and most structural metals caused by prolonged stress under the yield point at elevated temperatures.
Critical temperature. The preferred term used by metallurgists is transformation point. The lower A_1 and the upper A_3 points are the boundaries of the transformation range in which ferrite transforms into austenite.
Crystal unit structure or unit cell. The simplest polyhedron that embodies all the structural characteristics of a crystal and makes up the lattice of a crystal by indefinite repetition.
Cyanogen $(CN)_2$. A colorless gas of characteristic odor. It forms hydrocyanic and cyanic acid when in contact with water. Cyanogen compounds are often used for a case hardening.
Decarburization. The loss of carbon from the surface of a ferrous alloy as a result of heating it in the presence of a medium such as oxygen that reacts with the carbon.
Deformation. Alteration of the form or shape as a result of the plastic behavior of a metal under stress.
Dendrite. A crystal characterized by a treelike pattern that is usually formed by the solidification of a metal. Dendrites generally grow inward from the surface of a mold.
Density. The density of a body is the ratio of its mass to its volume. For solids the method to determine density is to measure the buoyant force upon the specimen of liquid of known density in which it is immersed or by determining the volume of displacement from a specific gravity bottle.
Deoxidizer. A substance used to remove oxygen from molten metals, for example, ferrosilicon in steel making.
Diffusion. The process of intermingling atoms or other particles within a solution. In solids, it is a slow movement of atoms from areas of high concentration toward areas of low concentration. The process may be (a) migration of interstitial atoms such as carbon, (b) movement of vacancies, or (c) direct exchange of atoms to neighboring sites.
Dezinctification. Corrosion of a metal alloy that contains zinc; it involves the loss of zinc and a deposit on the surface of one or more of the less active components; in the corrosion of brass this component is usually copper.
Drawing. A term sometimes used for the process of tempering hardened steel. Also used for metal forming in presses and forming wire.
Ductility. The property of a material to deform permanently, or to exhibit plasticity without rupture, while under tension.
Elasticity. The ability of a material to return to its original form after a load has been removed.
Elastomer. Any of various elastic substances resembling rubber.
Electrolysis. The producing of chemical changes by passing an electric current through an electrolyte.
Electrolyte. A nonmetallic conductor in which electric current is carried by the movement of ions.
Electroplating. Coating an object with a thin layer of some metal through electrolytic deposition.
Equilibrium. A condition of balance in which all the forces or processes present are counterbalanced by equal and opposite forces or processes where the condition appears to be one of rest rather than change.
Eutectic. The alloy composition that freezes at the lowest constant temperature causing a discrete mixture to form in definite proportions.
Eutectoid. In binary (double) alloy systems, it is a mechanical mixture of two phases forming simultaneously from a solid solution as it cools through the eutectoid (A_1 in steels) temperature.
Fatigue in metals. The tendency for a metal to fail by breaking or cracking under conditions of repeated cyclic stressing that takes place well below the ultimate tensile strength.
Fatigue strength. The amount of stress that can be applied to a metal without failure while it is subjected to ten million or more cycles of load reversals. In mild steels, the fatigue strength is about 50 percent of the tensile strength.
Ferrite. A magnetic form of iron. A solid solution in which alpha iron is the solvent, characterized by a body-centered cubic crystal structure.

Fiber. (a) The directional property of wrought metals revealed by a woody appearance when fractured; (b) A preferred orientation of metal crystals after a deformation process such as rolling or drawing; (c) Cellulosic plant cells used for manufacturing paper and other products.
Fluxes. A solid or gaseous material that is applied to molten metal in order to clean and remove the oxides.
Forging. The shaping of metal by hammering or pressing. While forging may be used to shape malleable metals in the cold state, the application of heat increases plasticity and permits greater deformation without inducing undue strain in the metal.
Fracture. A ruptured surface of metal showing a typical crystalline pattern. Fatigue fractures, however, often display a smooth, clam-shell appearance.
Fusion. The merging of two materials while in a molten state.
Gangue. The commercially undesirable portion of an ore that must be removed before the ore is processed into a metal.
Grain. Individual crystals in metals.
Grain growth. An increase in the grain size of metal that often results from overheating.
Graphite. An allotropic form of carbon.
Hardenability. The property that determines the depth and distribution of hardness in a ferrous alloy induced by heating and quenching.
Hardening. The process of increasing the hardness of a ferrous alloy by austenitizing and quenching; also the process of increasing the hardness of some stainless steels and nonferrous alloys by solution heat treatment and precipitation.
Hardness. The property of a metal to resist being permanently deformed. This is divided into three categories: the resistance to penetration, abrasion, and elastic hardness.
Heat sink. A large mass of metal having a high thermal conductivity used to stabilize the temperature of a part held in contact with it. In welding, the mass of the base metal often acts as a heat sink to the weld metal.
Hot rolling. A process of forming metals between rolls in which the metals are heated to temperatures above the transformation range.
Hot-short. Brittleness in hot metal. The presence of excess amounts of sulfur in steel causes hot-shortness.
Hydrogen brittleness. A condition of low ductility resulting from absorption of hydrogen in a metal at high temperature. The hydrogen often causes intergranular holes and porosity along with some cracking after the metal has cooled.
Impact test. A test in which small notched specimens are broken in an Izod-Charpy machine. This test determines the notch toughness of a metal.
Inclusions. Particles of impurities that are usually formed during solidification and are usually in the form of silicates, sulfides, and oxides.
Inert gas. Noble gases such as helium or argon that are not reactive with any other elements to any great extent.
Ingot. A large block of metal that is usually cast in a metal mold and forms the basic material for further rolling and processing.
Intergranular corrosion. A type of galvanic corrosion that progresses along the grain boundaries of an alloy. The grain boundaries become anodic to the grains and deteriorate, usually causing failure of the part.
Intergranular cracking. This is a crack that forms along the grain boundaries and not through the grains.
Interstitial lattice structure. A crystalline lattice containing smaller atoms of a different element within its interstices (voids or holes between the atoms and the lattice).
Ion. An electrostatically charged atom due to losing or gaining one or more valence electrons.
Iron. The term iron always refers to the element Fe and not cast iron, steel, or any other alloy of iron.
Isothermal transformation (I-T). Transformation that takes place at a constant temperature.
Jominy end-quench test. A test to determine the hardenability of alloy and tool steels.
Kaolin. A fine white clay that is used in ceramics and refractories composed mostly of kaolinite, a hydrous silicate of aluminum. Impurities may cause various colors and tints.
Killed steel. Steel that has been deoxidized with agents such as silicon or aluminum to reduce the oxygen content. This prevents gases from evolving during the solidification period.
Lamellar. An alternating platelike structure in metals (as in pearlite).
Lattice, space. A term that is used to denote a regular array of points in space. For example, the sites of atoms in a crystal. The points of the three-dimensional space lattice are constructed by the repeated application of the basic translations that carry a unit cell into its neighbor.
Lignin. A substance related to cellulose that with cellulose forms the woody cell walls of plants and the material that cements them together. Methyl alcohol is derived from lignin in the destructive distillation of wood.
Liquidus. The temperature at which freezing begins during cooling and ends during heating under equilibrium conditions, represented by a line on a two-phase diagram.
Macroscopic. Structural details on an object large enough to be observed by the naked eye or with low magnification (about 10×).
Macrostructure. The structure of metals as revealed by macroscopic examination.
Malleability. The ability of a metal to deform permanently without rupture when loaded in compression.
Martempering. The process of quenching an austenitized ferrous alloy to a temperature just above or near the Ms point and maintaining until the temperature throughout the part is uniform. The alloy is then allowed to slowly cool in air through the range of martensite formation.
Martensite. An unstable constituent formed by heating and quenching steel. It is formed without diffusion and only below a certain temperature known as the Ms temperature. Martensite is the hardest of the transfor-

mation products of austenite, having an acicular, or needlelike, microstructure.
Metalloid. A nonmetal that exhibits some, but not all, of the properties of a metal. Examples are sulfur, silicon, carbon, phosphorus, and arsenic.
Metallurgy. The science and study of the behaviors and properties of metals and their extraction from their ores.
Methanol (methyl alcohol, wood alcohol). Produced by the destructive distillation of wood or made synthetically.
Microscopy. The use of, or investigation with, the microscope.
Microstructure. The structure of polished and etched metal specimens as seen enlarged through a microscope.
Modulus of elasticity. The ratio of the unit stress to the unit deformation (strain) of a structural material is a constant as long as the unit stress is below the yield point. Shearing modulus of elasticity is often called the modulus of rigidity.
Monomer. A single molecule or a substance consisting of single molecules. The basic unit in a polymer.
Muffled furnace. A gas or oil fired furnace in which the fuel-air ratio is controlled to maintain a carbonizing atmosphere for the purpose of limiting the decarburization of heated steels.
Nitrided steel. A process of case hardening in which a special ferrous alloy is heated in an atmosphere of ammonia or is in contact with other nitrogenous material. In this process, surface hardening is achieved by the absorption of nitrogen without quenching.
Nonferrous. Metals other than iron and iron alloys.
Normalizing. To homogenize and produce a uniform structure in alloy steels by heating above the transformation range and cooling in air.
Notch toughness. The resistance to fracture of a metal specimen having a standard notch or groove when subjected to a sudden predetermined load, usually tested on an Izod-Charpy testing machine.
Oxidation. The slow or rapid reaction of oxygen with other elements; burning. In metals, the overoxidation during heating under oxidizing conditions often results in permanent damage to metals.
Oxidation corrosion. The formation of oxide scale on metals when at high temperatures. On steel, the scale that is formed is called mill scale or black scale (Fe_3O_4).
Oxidation-reduction. A chemical reaction in which one or more electrons are transferred from one atom or molecule to another. In the smelting of iron in a blast furnace, carbon atoms in the coke combine with the oxygen atoms in the iron ore to produce carbon monoxide or carbon dioxide, thus releasing the iron from the ore.
Pearlite. The lamellar mixture of ferrite and cementite in slowly cooled iron-carbon alloys as found in steel and cast iron.
Permeability. In casting of metals, the term is used to define the porosity of foundry sands in molds and the ability of trapped gases to escape through the sand.
Phase. In general, a phase is one of two or more forms, appearances, or kinds of behavior exhibited by the same entity; in metals, one of several crystalline forms of the same material.
Pig iron. The product of a blast furnace. It is a raw iron that usually contains about 4.5 percent carbon and impurities such as phosphorus, sulfur, and silicon.
Plasticity. A material that can be deformed without breaking. Clay is a completely plastic material. Metals exhibit plasticity in varying amounts.
Polymerization. A reaction in which a complex molecule (a polymer) is formed from a number of simpler molecules that can be alike or unlike.
Powder metallurgy. A process by which metal shapes are formed from metal powders by pressure and heat.
Precipitation hardening. A process of hardening an alloy by heat treatment in which a constituent precipitates from a supersaturated solid solution while at room temperature or at some slightly elevated temperature.
Quenching. The process of rapid cooling of metal alloys for the purpose of hardening. Quenching media include air, oil, water, molten metals, and fused salts.
Recovery. The relaxation and return to the original dimension of the metal after it has been stressed.
Recrystallization. A process in which the distorted grain structure of metals subjected to mechanical deformation is replaced by a new strain-free grain structure during annealing.
Refractory. Materials that will resist change of shape, weight, or physical properties at high temperatures. These materials are usually silica, fire clay, diaspore, alumina, and kaolin. They are used for furnace linings.
Rimmed steel. A low carbon steel (insufficiently deoxidized) that during solidification releases considerable quantities of gases (mainly carbon monoxide). When the mold top is not capped, a side and bottom rim of several inches forms. The solidified ingot has scattered blow holes and porosity in the center but a relatively thick skin free from blow holes.
SAE. Abbreviation for *The Society of Automotive Engineers.*
Scale. The surface oxidation on metals caused by heating in air or in other oxidizing atmospheres.
Scrap. Materials or metals that have lost their usefulness and are collected for reprocessing.
Sealant. A sealing agent having some adhesive qualities; it is used to prevent leakage.
Sintered. To cause to become a consolidated mass by heating without melting.
Slag (dross). A fused product that occurs in the melting of metals and is composed of oxidized impurities of a metal and a fluxing substance such as limestone. The slag protects the metal from oxidation of the atmosphere since it floats on the surface of the molten metal.
Slip planes. Also called slip bands. These are lines that appear on the polished surface of a plastically deformed metal. The slip bands are the result of crystal displacement defining planes in which shear has taken place.
Smelting. The process of heating ores to a high temperature in the presence of a reducing agent such as carbon (coke) and of a fluxing agent to remove the gangue.
Soaking. A prolonged heating of a metal at a predeter-

mined temperature to create a uniform temperature throughout its mass.

Solid solution. Found in metals at temperatures below the solidus. Some of the types of solid solutions are continuous, intermediate, interstitial, substitutional, and terminal.

Solidus. Seen as a line on a two-phase diagram, it represents the temperatures at which freezing ends when cooling, or melting begins when heating under equilibrium conditions.

Solubility. The degree to which one substance will dissolve in another.

Solute. A substance that is dissolved in a solution and is present in minor amounts.

Solution heat treatment. A process in which an alloy is heated to a predetermined temperature for a length of time suitable to allow a certain constituent to enter into solid solution. The alloy is then cooled quickly to hold the constituent in solution, causing the metal to be in an unstable supersaturated condition. This condition is often followed by age hardening.

Solvent. A substance capable of dissolving another substance and is the major constituent in a solution.

Sorbite. A term that was once used to denote tempered martensite having a microstructure with a granular appearance.

Specific gravity. A numerical value representing the weight of a given substance with the weight of an equal volume of water. The specific gravity for pure water is taken as 1.000.

Spheroidizing. Consists of holding carbon steel for a period of time at just under the transformation temperature. An aggregate of globular carbide is formed from other microstructures such as pearlite.

Stainless steel. An alloy of iron containing at least 11 percent chromium and sometimes nickel that resists almost all forms of rusting and corrosion.

Steel. An alloy of iron and less than 2 percent carbon plus some impurities and small amounts of alloying elements is known as plain carbon steel. Alloy steels contain substantial amounts of alloying elements such as chromium or nickel besides carbon.

Strain. The unit deformation of a metal when stress is applied.

Strain hardening. An increase in hardness and strength of a metal that has been deformed by cold working or at temperatures lower than the recrystallization range.

Strength. The ability of a metal to resist external forces. This is called tensile, compressive, or shear strength, depending on the load. See **stress.**

Stress. The load per unit of area on a stress-strain diagram. *Tensile stress* refers to an object loaded in tension denoting the longitudinal force that causes the fibers of a material to elongate. *Compressive stress* refers to a member loaded in compression, which either gives rise to a given reduction in volume or a transverse displacement of material. *Shear stress* refers to a force that lies in a parallel plane. The force tends to cause the plane of the area involved to slide on the adjacent planes. *Torsional stress* is a shearing stress that occurs at any point in a body as the result of an applied torque or torsional load.

Stress corrosion. (Also called corrosion fatigue.) A rapid deterioration of properties resulting from the repeated cyclic stressing of a metal in a corrosive medium.

Stress raiser. Can be a notch, nick, weld undercut, sharp change in section, machining grooves, or hairline cracks that provide a concentration of stresses when the metal is under tensile stress. Stress raisers pose a particular problem and can cause early failure in members that are subjected to many cycles of stress reversals.

Stress relief. The reduction of residual stress in a metal part by heating it to a given temperature and holding it there for a suitable length of time. This treatment is used to relieve stresses caused by welding, cold working, machining, casting, and quenching.

Temper. In ferrous metals, the stress relief of steels hardened by quenching for the purpose of toughening them and reducing their brittleness. In nonferrous metals, it is a condition produced by mechanical treatment such as cold working. An alloy may be cold worked to the hard temper, fully softened to the annealed temper, or two intermediate tempers.

Thermal conductivity. The quantity of heat transmitted per unit time, per unit cross section, per unit temperature gradient through a given substance. All materials are in some measure conductors of heat.

Thermal expansion. The increase of the dimension of a material that results from the increased movement of atoms caused by increased temperature gradients.

Thermal stress. Shear stress that is induced in a material due to unequal heating or cooling rates. The difference of expansion and contraction between the interior and exterior surfaces of a metal that is being heated or cooled is an example.

Thermoplastic. Capable of softening or fusing when heated and of hardening again when cooled.

Thermosetting. Capable of becoming permanently rigid when cured by heating; will not soften by reheating.

Tool steel. A special group of steels designed for specific uses such as heat resistant steels that can be heat treated to produce certain properties, mainly hardness and wear resistance.

Toughness. Generally measured in terms of notch toughness, which is a measure of a metal to resist rupture from impact loading when a notch is present. A standard test specimen containing a prepared notch is inserted into the vise of a testing machine. This device, called the Izod-Charpy testing machine, consists of a weight on a swinging arm. The arm or pendulum is released, strikes the specimen, and continues to swing forward. The amount of energy absorbed by the breaking of the specimen is measured by how far the pendulum continues to swing.

Transducer. A device by means of which energy can flow from one or more transmission systems to other transmission systems such as a mechanical force being converted into electrical energy (or the opposite effect) by means of a piezoelectrical crystal.

Transformation temperature. The temperature(s) at which ferrite or alpha iron transforms into austenite or gamma iron.

Transgranular cracking. Cracking that takes place along

crystallographic planes through the grains rather than along the grain boundaries.

Trootsite. A very fine aggregate of ore mixture of ferrite and cementite; a term formerly used to denote a very fine pearlite that had been cooled at a relatively slow rate.

Twinning. When metals are subjected to shock loads, a change in crystal orientation is often found in the crystal lattice. These differently oriented crystal structures within the grain appear as bands under the microscope.

Ultimate strength. (Also tensile strength.) The highest strength that a metal exhibits after it begins to deform plastically under load. Rupture of the material occurs either at the peak of its ultimate strength or at a point of further elongation and at a drop in stress load.

Valence. The capacity of an atom to combine with other atoms to form a molecule. The inert gases have zero valence; valence is determined by considering the positive and negative atoms as determined by their gaining or losing of valence electrons.

Viscosity. The property in fluids, either liquid or gaseous, that may be described as a resistance to flow; also the capability of continuous yielding under stress.

Void. A cavity or hole in a substance.

Weld metal. The molten area of a weld, either introduced by a filler rod or produced by the fusion of the base metal.

Widmanstatten structure. A duplex microstructure often observed in steels and certain other alloys in which one phase forms sets of parallel plates embedded in the matrix. In medium carbon steels, the structure is seen as ferrite needles growing outward from the grain boundaries.

Work hardening. Also called strain hardening in which the grains become distorted and elongated in the direction of working (rolling). This process hardens and strengthens metals but reduces their ductility. Excessive work hardening can cause ultimate brittle failure of the part. Stress relief is often used between periods of working of metals to restore their ductility.

Wrought iron. Contains 1 or 2 percent slag, which is distributed through the iron as threads and fibers imparting a tough fibrous structure. Usually contains less than 0.1 percent carbon. It is tough, malleable, and relatively soft.

Yield point. The stress at which a marked increase in deformation occurs without an increase in load stress as seen in mild or medium carbon steel. This phenomenon is not seen in nonferrous metals and other alloy steels.

Yield strength. Observed at the proportional limit of metals and is the stress at which a material deviates from that proportionality of stress to strain at a specified amount. The offset for some metals, for example, is 0.2 percent.

Index